D0858039

Food Processing Handbook

Edited by
James G. Brennan

Further of Interest

W. Pietsch

Agglomeration in Industry
Occurrence and Applications

2004
ISBN 3-527-30582-3

K. J. Heller (Ed.)

Genetically Engineered Food
Methods and Detection

2003
ISBN 3-527-30309-X

E. Ziegler, H. Ziegler (Eds.)

Handbook of Flavourings
Production, Composition, Applications, Regulations
Second, Completely Revised Edition

2006
ISBN 3-527-31406-7

J. N. Wintgens (Ed.)

Coffee: Growing, Processing, Sustainable Production
A Guidebook for Growers, Processors, Traders and Researchers

2005
ISBN 3-527-30731-1

G.-W. Oetjen

Freeze-Drying
Second, Completely Revised Edition

2004
ISBN 3-527-30620-X

O.-G. Piringer, A. L. Baner (Eds.)

Plastic Packaging Materials for Food and Pharmaceuticals

2007
ISBN 3-527-31455-5

K. Bauer, D. Garbe, H. Surburg

Common Fragrance and Flavor Materials
Preparation, Properties and Uses
Fourth, Completely Revised Edition

2001
ISBN 3-527-30364-2

F. Müller (Ed.)

Agrochemicals
Composition, Production, Toxicology, Applications

2000
ISBN 3-527-29852-5

Food Processing Handbook

Edited by
James G. Brennan

WILEY-
VCH

WILEY-VCH Verlag GmbH & Co. KGaA

Editor

James G. Brennan
16 Benning Way
Wokingham
Berks
RG40 1 XX
UK

Library of Congress Card No.: applied for

British Library Cataloguing-in-Publication Data:
A catalogue record for this book is available from the British Library.

Bibliographic information published by Die Deutsche Bibliothek
Die Deutsche Bibliothek lists this publication in the Deutsche Nationalbibliografie; detailed bibliographic data is available in the Internet at <http://dnb.ddb.de>

Typesetting K+V Fotosatz GmbH, Beerfelden
Printing and Bookbinding Markono Print Media Pte Ltd, Singapore

Printed in Singapore

Printed on acid-free paper

ISBN-13: 978-3-527-30719-7
ISBN-10: 3-527-30719-2

Contents

Food Processing Handbook. Edited by James G. Brennan

ISBN: 3-527-30719-2

Preface

There are many excellent texts available which cover the fundamentals of food engineering, equipment design, modelling of food processing operations etc. There are also several very good works in food science and technology dealing with the chemical composition, physical properties, nutritional and microbiological status of fresh and processed foods. This work is an attempt to cover the middle ground between these two extremes. The objective is to discuss the technology behind the main methods of food preservation used in today's food industry in terms of the principles involved, the equipment used and the changes in physical, chemical, microbiological and organoleptic properties that occur during processing. In addition to the conventional preservation techniques, new and emerging technologies, such as high pressure processing and the use of pulsed electric field and power ultrasound are discussed. The materials and methods used in the packaging of food, including the relatively new field of active packaging, are covered. Concerns about the safety of processed foods and the impact of processing on the environment are addressed. Process control methods employed in food processing are outlined. Treatments applied to water to be used in food processing and the disposal of wastes from processing operations are described.

Chapter 1 covers the postharvest handling and transport of fresh foods and preparatory operations, such as cleaning, sorting, grading and blanching, applied prior to processing. Chapters 2, 3 and 4 contain up-to-date accounts of heat processing, evaporation, dehydration and freezing techniques used for food preservation. In Chapter 5, the potentially useful, but so far little used process of irradiation is discussed. The relatively new technology of high pressure processing is covered in Chapter 6, while Chapter 7 explains the current status of pulsed electric field, power ultrasound, and other new technologies. Recent developments in baking, extrusion cooking and frying are outlined in Chapter 8. Chapter 9 deals with the materials and methods used for food packaging and active packaging technology, including the use of oxygen, carbon dioxide and ethylene scavengers, preservative releasers and moisture absorbers. In Chapter 10, safety in food processing is discussed and the development, implementation and maintenance of HACCP systems outlined. Chapter 11 covers the various types of control systems applied in food processing. Chapter 12 deals with envi-

Food Processing Handbook. Edited by James G. Brennan

ISBN: 3-527-30719-2

ronmental issues including the impact of packaging wastes and the disposal of refrigerants. In Chapter 13, the various treatments applied to water to be used in food processing are described and the physical, chemical and biological treatments applied to food processing wastes are outlined. To complete the picture, the various separation techniques used in food processing are discussed in Chapter 14 and Chapter 15 covers the conversion operations of mixing, emulsification and size reduction of solids.

The editor wishes to acknowledge the considerable advice and help he received from former colleagues in the School of Food Biosciences, The University of Reading, when working on this project. He also wishes to thank his wife, Anne, for her support and patience.

Reading, August 2005 *James G. Brennan*

List of Contributors

Dr. Araya Ahromrit
Assistant Professor
Department of Food Technology
Khon Kaen University
Khon Kaen 40002
Thailand

Professor Paul Ainsworth
Department of Food and Consumer Technology
Manchester Metropolitan University
Old Hall Lane
Manchester, M14 6HR
UK

Professor Dr. Ing. Ali Abd El-Aal Bakr
Food Science and Technology Department
Faculty of Agriculture
Minufiya University
Shibin El-Kom
A. R. Egypt

Dr. Pedro Bouchon
Departamento de Ingeniera Quimica y Bioprocesos
Pontificia Universidad Católica de Chile
Vicuña Mackenna 4860
Macul
Santiago
Chile

Mr. James G. Brennan (Editor)
16 Benning Way
Wokingham
Berkshire, RG40 1XX
UK

Dr. Brian P. F. Day
Program Leader –
Minimal Processing & Packaging
Food Science Australia
671 Sneydes Road (Private Bag 16)
Werribee
Victoria 3030
Australia

Dr. Bogdan J. Dobraszczyk
School of Food Biosciences
The University of Reading
P.O. Box 226
Whiteknights
Reading, RG6 6AP
UK

Dr. Alistair S. Grandison
School of Food Biosciences
The University of Reading
P.O. Box 226
Whiteknights
Reading, RG6 6AP
UK

Food Processing Handbook. Edited by James G. Brennan

ISBN: 3-527-30719-2

Dr. Senol Ibanoglu
Department of Food Engineering
Gaziantep University
Kilis Road
27310 Gaziantep
Turkey

Dr. Ashok Khare
School of Food Biosciences
The University of Reading
P.O. Box 226
Whiteknights
Reading, RG6 6AP
UK

Mr. Craig E. Leadley
Campden & Chorleywood
Food Research Association
Food Manufacturing Technologies
Chipping Campden
Gloucestershire, GL55 6LD
UK

Professor Dave A. Ledward
School of Food Biosciences
The University of Reading
Whiteknights
Reading, RG6 6AP
UK

Dr. Michael J. Lewis
School of Food Biosciences
The University of Reading
P.O. Box 226
Whiteknights
Reading, RG6 6AP
UK

Mrs. Niharika Mishra
School of Food Biosciences
The University of Reading
P.O. Box 226
Whiteknights
Reading, RG6 6AP
UK

Professor Keshavan Niranjan
School of Food Biosciences
The University of Reading
P.O. Box 226
Whiteknights
Reading, RG6 6AP
UK

Dr. Jose Mauricio Pardo
Director
Ingenieria de Produccion
Agroindustrial
Universidad de la Sabana
A. A. 140013
Chia
Columbia

Dr. Margaret F. Patterson
Queen's University, Belfast
Department of Agriculture and Rural
Development
Agriculture and Food Science Center
Newforge Lane
Belfast, BT9 5PX
Northern Ireland
UK

Mr. Nigel Rogers
Avure Technologies AB
Quintusvägen 2
Vasteras, SE 72166
Sweden

Mrs. Carol Anne Wallace
Principal Lecturer
Food Safety Management
Lancashire School of Health
& Postgraduate Medicine
University of Central Lancashire
Preston, PR1 2HE
UK

Mr. R. Andrew Wilbey
School of Food Biosciences
The University of Reading
P.O. Box 226
Whiteknights
Reading, RG6 6AP
UK

Dr. Alan Williams
Senior Technologist & HACCP
Specialist
Department of Food Manufacturing
Technologies
Campden & Chorleywood Food
Research Association Group
Chipping Campden
Gloucestershire, GL55 6LD
UK

1
Postharvest Handling and Preparation of Foods for Processing

Alistair S. Grandison

1.1
Introduction

Food processing is seasonal in nature, both in terms of demand for products and availability of raw materials. Most crops have well established harvest times – for example the sugar beet season lasts for only a few months of the year in the UK, so beet sugar production is confined to the autumn and winter, yet demand for sugar is continuous throughout the year. Even in the case of raw materials which are available throughout the year, such as milk, there are established peaks and troughs in volume of production, as well as variation in chemical composition. Availability may also be determined by less predictable factors, such as weather conditions, which may affect yields, or limit harvesting. In other cases demand is seasonal, for example ice cream or salads are in greater demand in the summer, whereas other foods are traditionally eaten in the winter months, or even at more specific times, such as Christmas or Easter.

In an ideal world, food processors would like a continuous supply of raw materials, whose composition and quality are constant, and whose prices are predictable. Of course this is usually impossible to achieve. In practice, processors contract ahead with growers to synchronise their needs with raw material production. The aim of this chapter is to consider the properties of raw materials in relation to food processing, and to summarise important aspects of handling, transport, storage and preparation of raw materials prior to the range of processing operations described in the remainder of this book. The bulk of the chapter will deal with solid agricultural products including fruits, vegetables, cereals and legumes; although many considerations can also be applied to animal-based materials such as meat, eggs and milk.

Food Processing Handbook. Edited by James G. Brennan

ISBN: 3-527-30719-2

1.2
Properties of Raw Food Materials and Their Susceptibility to Deterioration and Damage

The selection of raw materials is a vital consideration to the quality of processed products. The quality of raw materials can rarely be improved during processing and, while sorting and grading operations can aid by removing oversize, undersize or poor quality units, it is vital to procure materials whose properties most closely match the requirements of the process. Quality is a wide-ranging concept and is determined by many factors. It is a composite of those physical and chemical properties of the material which govern its acceptability to the 'user'. The latter may be the final consumer, or more likely in this case, the food processor. Geometric properties, colour, flavour, texture, nutritive value and freedom from defects are the major properties likely to determine quality.

An initial consideration is selection of the most suitable cultivars in the case of plant foods (or breeds in the case of animal products). Other preharvest factors (such as soil conditions, climate and agricultural practices), harvesting methods and postharvest conditions, maturity, storage and postharvest handling also determine quality. These considerations, including seed supply and many aspects of crop production, are frequently controlled by the processor or even the retailer.

The timing and method of harvesting are determinants of product quality. Manual labour is expensive, therefore mechanised harvesting is introduced where possible. Cultivars most suitable for mechanised harvesting should mature evenly producing units of nearly equal size that are resistant to mechanical damage. In some instances, the growth habits of plants, e.g. pea vines, fruit trees, have been developed to meet the needs of mechanical harvesting equipment. Uniform maturity is desirable as the presence of over-mature units is associated with high waste, product damage, and high microbial loads, while under-maturity is associated with poor yield, hard texture and a lack of flavour and colour. For economic reasons, harvesting is almost always a 'once over' exercise, hence it is important that all units reach maturity at the same time. The prediction of maturity is necessary to coordinate harvesting with processors' needs as well as to extend the harvest season. It can be achieved primarily from knowledge of the growth properties of the crop combined with records and experience of local climatic conditions. The 'heat unit system', first described by Seaton [1] for peas and beans, can be applied to give a more accurate estimate of harvest date from sowing date in any year. This system is based on the premise that growth temperature is the overriding determinant of crop growth. A base temperature, below which no growth occurs, is assumed and the mean temperature of each day through the growing period is recorded. By summing the daily mean temperatures minus base temperatures on days where mean temperature exceeds base temperature, the number of 'accumulated heat units' can be calculated. By comparing this with the known growth data for the particular cultivar, an accurate prediction of harvest date can be computed. In addition, by allowing fixed numbers of accumulated heat units between sowings, the harvest season can be spread, so that individual fields may be harvested at peak maturity. Sowing plans and harvest date are determined by negotiation between the growers and the processors; and the latter may even provide the equipment and labour for harvesting and transport to the factory.

An important consideration for processed foods is that it is the quality of the processed product, rather than the raw material, that is important. For minimally processed foods, such as those subjected to modified atmospheres, low-dose irradiation, mild heat treatment or some chemical preservatives, the characteristics of the raw material are a good guide to the quality of the product. For more severe processing, including heat preservation, drying or freezing, the quality characteristics may change markedly during processing. Hence, those raw materials which are preferred for fresh consumption may not be most appropriate for processing. For example, succulent peaches with delicate flavour may be less suitable for canning than harder, less flavoursome cultivars, which can withstand rigorous processing conditions. Similarly, ripe, healthy, well coloured fruit may be perfect for fresh sale, but may not be suitable for freezing due to excessive drip loss while thawing. For example, Maestrelli [2] reported that different strawberry cultivars with similar excellent characteristics for fresh consumption exhibited a wide range of drip loss (between 8% and 38%), and hence would be of widely different value for the frozen food industry.

1.2.1 Raw Material Properties

The main raw material properties of importance to the processor are geometry, colour, texture, functional properties and flavour.

1.2.1.1 Geometric Properties

Food units of regular geometry are much easier to handle and are better suited to high speed mechanised operations. In addition, the more uniform the geometry of raw materials, the less rejection and waste will be produced during preparation operations such as peeling, trimming and slicing. For example, potatoes of smooth shape with few and shallow eyes are much easier to peel and wash mechanically than irregular units. Smooth-skinned fruits and vegetables are much easier to clean and are less likely to harbour insects or fungi than ribbed or irregular units.

Agricultural products do not come in regular shapes and exact sizes. Size and shape are inseparable, but are very difficult to define mathematically in solid food materials. Geometry is, however, vital to packaging and controlling fill-in weights. It may, for example, be important to determine how much mass or how many units may be filled into a square box or cylindrical can. This would require a vast number of measurements to perform exactly and thus approximations must be made. Size and shape are also important to heat processing and freezing, as they will determine the rate and extent of heat transfer within food units. Mohsenin [3] describes numerous approaches by which the size and shape of irregular food units may be defined. These include the development of statistical techniques based on a limited number of measurements and more subjective approaches involving visual comparison of units to charted standards. Uniformity of size and shape is also important to most operations and processes. Process control to give accurately and uniformly treated products is always simpler with more uniform materials. For example, it is essential that wheat kernel size is uniform for flour milling.

Specific surface (area/mass) may be an important expression of geometry, especially when considering surface phenomena such as the economics of fruit peeling, or surface processes such as smoking and brining.

The presence of geometric defects, such as projections and depressions, complicate any attempt to quantify the geometry of raw materials, as well as presenting processors with cleaning and handling problems and yield loss. Selection of cultivars with the minimum defect level is advisable.

There are two approaches to securing the optimum geometric characteristics: firstly the selection of appropriate varieties, and secondly sorting and grading operations.

1.2.1.2 Colour

Colour and colour uniformity are vital components of visual quality of fresh foods and play a major role in consumer choice. However, it may be less important in raw materials for processing. For low temperature processes such as chilling, freezing or freeze-drying, the colour changes little during processing, and thus the colour of the raw material is a good guide to suitability for processing. For more severe processing, the colour may change markedly during the process. Green vegetables, such as peas, spinach or green beans, on heating change colour from bright green to a dull olive green. This is due to the conversion of chlorophyll to pheophytin. It is possible to protect against this by addition of sodium bicarbonate to the cooking water, which raises the pH. However, this may cause softening of texture and the use of added colourants may be a more practical solution. Some fruits may lose their colour during canning, while pears develop a pink tinge. Potatoes are subject to browning during heat processing due to the Maillard reaction. Therefore, different varieties are more suitable for fried products where browning is desirable, than canned products in which browning would be a major problem.

Again there are two approaches: i.e. procuring raw materials of the appropriate variety and stage of maturity, and sorting by colour to remove unwanted units.

1.2.1.3 **Texture**

The texture of raw materials is frequently changed during processing. Textural changes are caused by a wide variety of effects, including water loss, protein denaturation which may result in loss of water-holding capacity or coagulation, hydrolysis and solubilisation of proteins. In plant tissues, cell disruption leads to loss of turgor pressure and softening of the tissue, while gelatinisation of starch, hydrolysis of pectin and solubilisation of hemicelluloses also cause softening of the tissues.

The raw material must be robust enough to withstand the mechanical stresses during preparation, for example abrasion during cleaning of fruit and vegetables. Peas and beans must be able to withstand mechanical podding. Raw materials must be chosen so that the texture of the processed product is correct, such as canned fruits and vegetables in which raw materials must be able to withstand heat processing without being too hard or coarse for consumption.

Texture is dependent on the variety as well as the maturity of the raw material and may be assessed by sensory panels or commercial instruments. One widely recognised instrument is the tenderometer used to assess the firmness of peas. The crop would be tested daily and harvested at the optimum tenderometer reading. In common with other raw materials, peas at different maturities can be used for different purposes, so that peas for freezing would be harvested at a lower tenderometer reading than peas for canning.

1.2.1.4 **Flavour**

Flavour is a rather subjective property which is difficult to quantify. Again, flavours are altered during processing and, following severe processing, the main flavours may be derived from additives. Hence, the lack of strong flavours may be the most important requirement. In fact, raw material flavour is often not a major determinant as long as the material imparts only those flavours which are characteristic of the food. Other properties may predominate. Flavour is normally assessed by human tasters, although sometimes flavour can be linked to some analytical test, such as sugar/acid levels in fruits.

1.2.1.5 **Functional Properties**

The functionality of a raw material is the combination of properties which determine product quality and process effectiveness. These properties differ greatly for different raw materials and processes, and may be measured by chemical analysis or process testing.

For example, a number of possible parameters may be monitored in wheat. Wheat for different purposes may be selected according to protein content. Hard wheat with 11.5–14.0% protein is desirable for white bread and some wholewheat breads require even higher protein levels, 14–16% [4]. In contrast, soft or weak flours with lower protein contents are suited to chemically leavened products with a lighter or more tender structure. Hence protein levels of 8–11%

are adequate for biscuits, cakes, pastry, noodles and similar products. Varieties of wheat for processing are selected on this basis; and measurement of protein content would be a good guide to process suitability. Furthermore, physical testing of dough using a variety of rheological testing instruments may be useful in predicting the breadmaking performance of individual batches of wheat flours [5]. A further test is the Hagberg Falling Number which measures the amount of α-amylase in flour or wheat [6]. This enzyme assists in the breakdown of starch to sugars and high levels give rise to a weak bread structure. Hence, the test is a key indicator of wheat baking quality and is routinely used for bread wheat; and it often determines the price paid to the farmer.

Similar considerations apply to other raw materials. Chemical analysis of fat and protein in milk may be carried out to determine its suitability for manufacturing cheese, yoghurt or cream.

1.2.2
Raw Material Specifications

In practice, processors define their requirements in terms of raw material specifications for any process on arrival at the factory gate. Acceptance of, or price paid for the raw material depends on the results of specific tests. Milk deliveries would be routinely tested for hygienic quality, somatic cells, antibiotic residues, extraneous water, as well as possibly fat and protein content. A random core sample is taken from all sugar beet deliveries and payment is dependent on the sugar content. For fruits, vegetables and cereals, processors may issue specifications and tolerances to cover the size of units, the presence of extraneous vegetable matter, foreign bodies, levels of specific defects, e.g. surface blemishes, insect damage etc., as well as specific functional tests. Guidelines for sampling and testing many raw materials for processing in the UK are available from the Campden and Chorleywood Food Research Association (www.campden.co.uk).

Increasingly, food processors and retailers may impose demands on raw material production which go beyond the properties described above. These may include 'environmentally friendly' crop management schemes in which only specified fertilisers and insecticides are permitted, or humanitarian concerns, especially for food produced in Third World countries. Similarly animal welfare issues may be specified in the production of meat or eggs. Another important issue is the growth of demand for organic foods in the UK and Western Europe, which obviously introduces further demands on production methods, but are beyond the scope of this chapter.

1.2.3
Deterioration of Raw Materials

All raw materials deteriorate following harvest, by some of the following mechanisms:
- Endogenous enzymes: e.g. post-harvest senescence and spoilage of fruit and vegetables occurs through a number of enzymic mechanisms, including oxidation of phenolic substances in plant tissues by phenolase (leading to browning), sugar-starch conversion by amylases, postharvest demethylation of pectic substances in fruits and vegetables leading to softening tissues during ripening and firming of plant tissues during processing.
- Chemical changes: deterioration in sensory quality by lipid oxidation, non-enzymic browning, breakdown of pigments such as chlorophyll, anthocyanins, carotenoids.
- Nutritional changes: especially ascorbic acid breakdown.
- Physical changes: dehydration, moisture absorption.
- Biological changes: germination of seeds, sprouting.
- Microbiological contamination: both the organisms themselves and toxic products lead to deterioration of quality, as well as posing safety problems.

1.2.4
Damage to Raw Materials

Damage may occur at any point from growing through to the final point of sale. It may arise through external or internal forces.

External forces result in mechanical injury to fruits and vegetables, cereal grains, eggs and even bones in poultry. They occur due to severe handling as a result of careless manipulation, poor equipment design, incorrect containerisation and unsuitable mechanical handling equipment. The damage typically results from impact and abrasion between food units, or between food units and machinery surfaces and projections, excessive vibration or pressure from overlying material. Increased mechanisation in food handling must be carefully designed to minimise this.

Internal forces arise from physical changes, such as variation in temperature and moisture content, and may result in skin cracks in fruits and vegetables, or stress cracks in cereals.

Either form of damage leaves the material open to further biological or chemical damage, including enzymic browning of bruised tissue, or infestation of punctured surfaces by moulds and rots.

1.2.5
Improving Processing Characteristics Through Selective Breeding and Genetic Engineering

Selective breeding for yield and quality has been carried out for centuries in both plant and animal products. Until the 20th century, improvements were made on the basis of selecting the most desirable looking individuals, while increasingly systematic techniques have been developed more recently, based on a greater understanding of genetics. The targets have been to increase yield as well as aiding factors of crop or animal husbandry such as resistance to pests and diseases, suitability for harvesting, or development of climate-tolerant varieties (e.g. cold-tolerant maize, or drought-resistant plants) [7]. Raw material quality, especially in relation to processing, has become increasingly important. There are many examples of successful improvements in processing quality of raw materials through selective plant breeding, including:
- improved oil percentage and fatty acid composition in oilseed rape;
- improved milling and malting quality of cereals;
- high sugar content and juice quality in sugar beets;
- development of specific varieties of potatoes for the processing industry, based on levels of enzymes and sugars, producing appropriate flavour, texture and colour in products, or storage characteristics;
- brussels sprouts which can be successfully frozen.

Similarly traditional breeding methods have been used to improve yields of animal products such as milk and eggs, as well as improving quality, e.g. fat/lean content of meat. Again the quality of raw materials in relation to processing may be improved by selective breeding. This is particularly applicable to milk, where breeding programmes have been used at different times to maximise butterfat and protein content, and would thus be related to the yield and quality of fat- or protein-based dairy products. Furthermore, particular protein genetic variants in milk have been shown to be linked with processing characteristics, such as curd strength during manufacture of cheese [8]. Hence, selective breeding could be used to tailor milk supplies to the manufacture of specific dairy products.

Traditional breeding programmes will undoubtedly continue to produce improvements in raw materials for processing, but the potential is limited by the gene pool available to any species. Genetic engineering extends this potential by allowing the introduction of foreign genes into an organism, with huge potential benefits. Again many of the developments have been aimed at agricultural improvements, such as increased yield, or introducing herbicide, pest or drought resistance, but there is enormous potential in genetically engineered raw materials for processing [9]. The following are some examples which have been demonstrated:
- tomatoes which do not produce pectinase and hence remain firm while colour and flavour develop, producing improved soup, paste or ketchup;
- potatoes with higher starch content, which take up less oil and require less energy during frying;

- canola (rape seed) oil tailored to contain: (a) high levels of lauric acid to improve emulsification properties for use in confectionery, coatings or low fat dairy products, (b) high levels of stearate as an alternative to hydrogenation in manufacture of margarine, (c) high levels of polyunsaturated fatty acids for health benefits;
- wheat with increased levels of high molecular weight glutenins for improved breadmaking performance;
- fruits and vegetables containing peptide sweeteners such as thaumatin or monellin;
- 'naturally decaffeinated' coffee.

There is, however, considerable opposition to the development of genetically modified foods in the UK and elsewhere, due to fears of human health risks and ecological damage, discussion of which is beyond the scope of this book. It therefore remains to be seen if, and to what extent, genetically modified raw materials will be used in food processing.

1.3 Storage and Transportation of Raw Materials

1.3.1 Storage

Storage of food is necessary at all points of the food chain from raw materials, through manufacture, distribution, retailers and final purchasers. Today's consumers expect a much greater variety of products, including nonlocal materials, to be available throughout the year. Effective transportation and storage systems for raw materials are essential to meet this need.

Storage of materials whose supply or demand fluctuate in a predictable manner, especially seasonal produce, is necessary to increase availability. It is essential that processors maintain stocks of raw materials, therefore storage is necessary to buffer demand. However, storage of raw materials is expensive for two reasons: firstly, stored goods have been paid for and may therefore tie up quantities of company money and, secondly, warehousing and storage space are expensive. All raw materials deteriorate during storage. The quantities of raw materials held in store and the times of storage vary widely for different cases, depending on the above considerations. The 'just in time' approaches used in other industries are less common in food processing.

The primary objective is to maintain the best possible quality during storage, and hence avoid spoilage during the storage period. Spoilage arises through three mechanisms:

- living organisms such as vermin, insects, fungi and bacteria: these may feed on the food and contaminate it;

- biochemical activity within the food leading to quality reduction, such as: respiration in fruits and vegetables, staling of baked products, enzymic browning reactions, rancidity development in fatty food;
- physical processes, including damage due to pressure or poor handling, physical changes such as dehydration or crystallisation.

The main factors which govern the quality of stored foods are temperature, moisture/humidity and atmospheric composition. Different raw materials provide very different challenges.

Fruits and vegetables remain as living tissues until they are processed and the main aim is to reduce respiration rate without tissue damage. Storage times vary widely between types. Young tissues such as shoots, green peas and immature fruits have high respiration rates and shorter storage periods, while mature fruits and roots and storage organs such as bulbs and tubers, e.g. onions, potatoes, sugar beets, respire much more slowly and hence have longer storage periods. Some examples of conditions and storage periods of fruits and vegetables are given in Table 1.1. Many fruits (including bananas, apples, tomatoes and mangoes) display a sharp increase in respiration rate during ripening, just before the point of optimum ripening, known as the 'climacteric'. The onset of the climacteric is associated with the production of high levels of ethylene, which is believed to stimulate the ripening process. Climacteric fruit can be harvested unripe and ripened artificially at a later time. It is vital to maintain careful temperature control during storage or the fruit will rapidly over-ripen. Non-climacteric fruits, e.g. citrus fruit, pineapples, strawberries, and vegetables do not display this behaviour and generally do not ripen after harvest. Quality is therefore optimal at harvest, and the task is to preserve quality during storage.

With meat storage the overriding problem is growth of spoilage bacteria, while avoiding oxidative rancidity. Cereals must be dried before storage to avoid

Table 1.1 Storage periods of some fruits and vegetables under typical storage conditions (data from [25]).

Commodity	Temperature (°C)	Humidity (%)	Storage period
Garlic	0	70	6–8 months
Mushrooms	0	90–95	5–7 days
Green bananas	13–15	85–90	10–30 days
Immature potatoes	4–5	90–95	3–8 weeks
Mature potatoes	4–5	90–95	4–9 months
Onions	–1 to 0	70–80	6–8 months
Oranges	2–7	90	1–4 months
Mangoes	5.5–14	90	2–7 weeks
Apples	–1 to 4	90–95	1–8 months
French beans	7– 8	95–100	1–2 week

germination and mould growth and subsequently must be stored under conditions which prevent infestation with rodents, birds, insects or moulds.

Hence, very different storage conditions may be employed for different raw materials. The main methods employed in raw material storage are the control of temperature, humidity and composition of atmosphere.

1.3.1.1 Temperature

The rate of biochemical reactions is related to temperature, such that lower storage temperatures lead to slower degradation of foods by biochemical spoilage, as well as reduced growth of bacteria and fungi. There may also be limited bacteriocidal effects at very low temperatures. Typical Q_{10} values for spoilage reactions are approximately 2, implying that spoilage rates would double for each 10 °C rise, or conversely that shelflife would double for each 10 °C reduction. This is an oversimplification, as Q_{10} may change with temperature. Most insect activity is inhibited below 4 °C, although insects and their eggs can survive long exposure to these temperatures. In fact, grain and flour mites can remain active and even breed at 0 °C.

The use of refrigerated storage is limited by the sensitivity of materials to low temperatures. The freezing point is a limiting factor for many raw materials, as the tissues will become disrupted on thawing. Other foods may be subject to problems at temperatures above freezing. Fruits and vegetables may display physiological problems that limit their storage temperatures, probably as a result of metabolic imbalance leading to a build up of undesirable chemical species in the tissues. Some types of apples are subject to internal browning below 3 °C, while bananas become brown when stored below 13 °C and many other tropical fruits display chill sensitivity. Less obvious biochemical problems may occur even where no visible damage occurs. For example, storage temperature affects the starch/sugar balance in potatoes: in particular below 10 °C a build up of sugar occurs, which is most undesirable for fried products. Examples of storage periods and conditions are given in Table 1.1, illustrating the wide ranges seen with different fruits and vegetables. It should be noted that predicted storage lives can be confounded if the produce is physically damaged, or by the presence of pathogens.

Temperature of storage is also limited by cost. Refrigerated storage is expensive, especially in hot countries. In practice, a balance must be struck incorporating cost, shelflife and risk of cold injury. Slower growing produce such as onions, garlic and potatoes can be successfully stored at ambient temperature and ventilated conditions in temperate climates.

It is desirable to monitor temperature throughout raw material storage and distribution.

Precooling to remove the 'field heat' is an effective strategy to reduce the period of high initial respiration rate in rapidly respiring produce prior to transportation and storage. For example, peas for freezing are harvested in the cool early morning and rushed to cold storage rooms within 2–3 h. Other produce, such as leafy vegetables (lettuce, celery, cabbage) or sweetcorn, may be cooled using water sprays or drench streams. Hydrocooling obviously reduces water loss.

1.3.1.2 **Humidity**

If the humidity of the storage environment exceeds the equilibrium relative humidity (ERH) of the food, the food will gain moisture during storage, and *vice versa*. Uptake of water during storage is associated with susceptibility to growth of microorganisms, whilst water loss results in economic loss and more specific problems, such as cracking of seed coats of cereals, or skins of fruits and vegetables. Ideally, the humidity of the store would equal the ERH of the food so that moisture is neither gained nor lost, but in practice a compromise may be necessary. The water activity (a_w) of most fresh foods (e.g. fruit, vegetables, meat, fish, milk) is in the range 0.98–1.00, but they are frequently stored at a lower humidity. Some wilting of fruits or vegetable may be acceptable in preference to mould growth, while some surface drying of meat is preferable to bacterial slime. Packaging may be used to protect against water loss of raw materials during storage and transport, see Chapter 9.

1.3.1.3 **Composition of Atmosphere**

Controlling the atmospheric composition during storage of many raw materials is beneficial. The use of packaging to allow the development or maintenance of particular atmospheric compositions during storage is discussed in greater detail in Chapter 9.

With some materials, the major aim is to maintain an oxygen-free atmosphere to prevent oxidation, e.g. coffee, baked goods, while in other cases adequate ventilation may be necessary to prevent anaerobic fermentation leading to off flavours.

In living produce, atmosphere control allows the possibility of slowing down metabolic processes, hence retarding respiration, ripening, senescence and the development of disorders. The aim is to introduce N_2 and remove O_2, allowing a build up of CO_2. Controlled atmosphere storage of many commodities is discussed by Thompson [10]. The technique allows year-round distribution of apples and pears, where controlled atmospheres in combination with refrigeration can give shelflives up to 10 months, much greater than by chilling alone. The particular atmospheres are cultivar specific, but in the range 1–10% CO_2, 2–13% O_2 at 3 °C for apples and 0 °C for pears. Controlled atmospheres are also used during storage and transport of chill-sensitive crops, such as for transport of bananas, where an atmosphere of 3% O_2 and 5% CO_2 is effective in preventing premature ripening and the development of crown rot disease. Ethene (ethylene) removal is also vital during storage of climacteric fruit.

With fresh meat, controlling the gaseous environment is useful in combination with chilling. The aim is to maintain the red colour by storage in high O_2 concentrations, which shifts the equilibrium in favour of high concentrations of the bright red oxymyoglobin pigment. At the same time, high levels of CO_2 are required to suppress the growth of aerobic bacteria.

1.3.1.4 Other Considerations

Odours and taints can cause problems, especially in fatty foods such as meat and dairy products, as well as less obvious commodities such as citrus fruits, which have oil in the skins. Odours and taints may be derived from fuels or adhesives and printing materials, as well as other foods, e.g. spiced or smoked products. Packaging and other systems during storage and transport must protect against contamination.

Light can lead to oxidation of fats in some raw materials, e.g. dairy products. In addition, light gives rise to solanine production and the development of green pigmentation in potatoes. Hence, storage and transport under dark conditions is essential.

1.3.2 Transportation

Food transportation is an essential link in the food chain and is discussed in detail by Heap [11]. Raw materials, food ingredients, fresh produce and processed products are all transported on a local and global level, by land, sea and air. In the modern world, where consumers expect year-round supplies and nonlocal products, long distance transport of many foods has become commonplace and air transport may be necessary for perishable materials. Transportation of food is really an extension of storage: a refrigerated lorry is basically a cold store on wheels. However, transport also subjects the material to physical and mechanical stresses, and possibly rapid changes in temperature and humidity, which are not encountered during static storage. It is necessary to consider both the stresses imposed during the transport and those encountered during loading and unloading. In many situations, transport is multimodal. Air or sea transport would commonly involve at least one road trip before and one road trip after the main journey. There would also be time spent on the ground at the port or airport where the material could be exposed to wideranging temperatures and humidities, or bright sunlight, and unscheduled delays are always a possibility. During loading and unloading, the cargo may be broken into smaller units where more rapid heat penetration may occur.

The major challenges during transportation are to maintain the quality of the food during transport, and to apply good logistics – in other words, to move the goods to the right place at the right time and in good condition.

1.4 Raw Material Cleaning

All food raw materials are cleaned before processing. The purpose is obviously to remove contaminants, which range from innocuous to dangerous. It is important to note that removal of contaminants is essential for protection of process equipment as well as the final consumer. For example, it is essential to remove sand, stones or metallic particles from wheat prior to milling to avoid damaging the machinery. The main contaminants are:
- unwanted parts of the plant, such as leaves, twigs, husks;
- soil, sand, stones and metallic particles from the growing area;
- insects and their eggs;
- animal excreta, hairs etc.;
- pesticides and fertilisers;
- mineral oil;
- microorganisms and their toxins.

Increased mechanisation in harvesting and subsequent handling has generally led to increased contamination with mineral, plant and animal contaminants, while there has been a general increase in the use of sprays, leading to increased chemical contamination. Microorganisms may be introduced preharvest from irrigation water, manure, fertiliser or contamination from feral or domestic animals, or postharvest from improperly cleaned equipment, wash waters or cross-contamination from other raw materials.

Cleaning is essentially separation in which some difference in physical properties of the contaminants and the food units is exploited. There are a number of cleaning methods available, classified into dry and wet methods, but a combination would usually be employed for any specific material. Selection of the appropriate cleaning regime depends on the material being cleaned, the level and type of contamination and the degree of decontamination required. In practice a balance must be struck between cleaning cost and product quality, and an 'acceptable standard' should be specified for the particular end use. Avoidance of product damage is an important contributing factor, especially for delicate materials such as soft fruit.

1.4.1 Dry Cleaning Methods

The main dry cleaning methods are based on screens, aspiration or magnetic separations. Dry methods are generally less expensive than wet methods and the effluent is cheaper to dispose of, but they tend to be less effective in terms of cleaning efficiency. A major problem is recontamination of the material with dust. Precautions may be necessary to avoid the risk of dust explosions and fires.

Screens are essentially size separators based on perforated beds or wire mesh. Larger contaminants are removed from smaller food items: e.g. straw from cereal grains, or pods and twigs from peas. This is termed 'scalping', see Fig. 1.1 a. Alternatively, 'dedusting' is the removal of smaller particles, e.g. sand or dust, from larger food units, see Fig. 1.1 b.

The main geometries are rotary drums (also known as reels or trommels), and flatbed designs. Some examples are shown in Fig. 1.2.

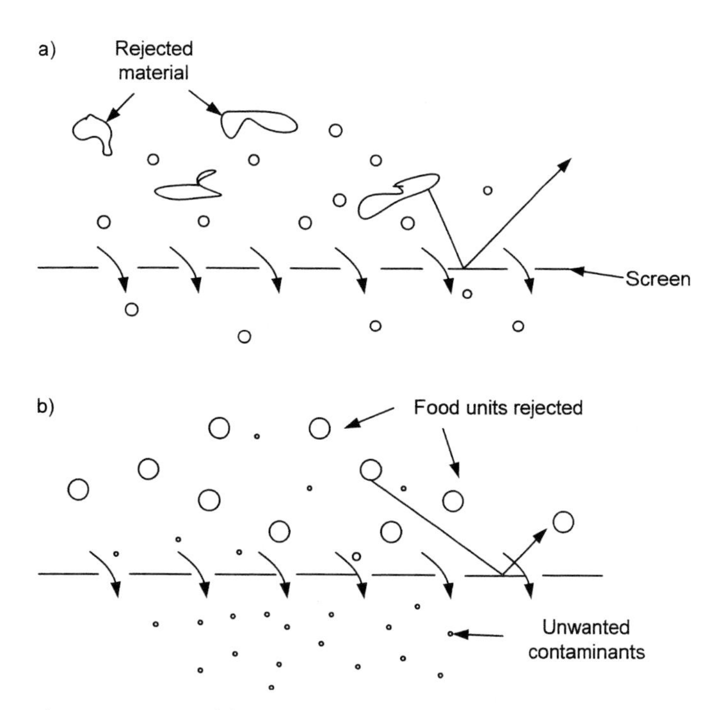

Fig. 1.1 Screening of dry particulate materials: (a) scalping, (b) dedusting.

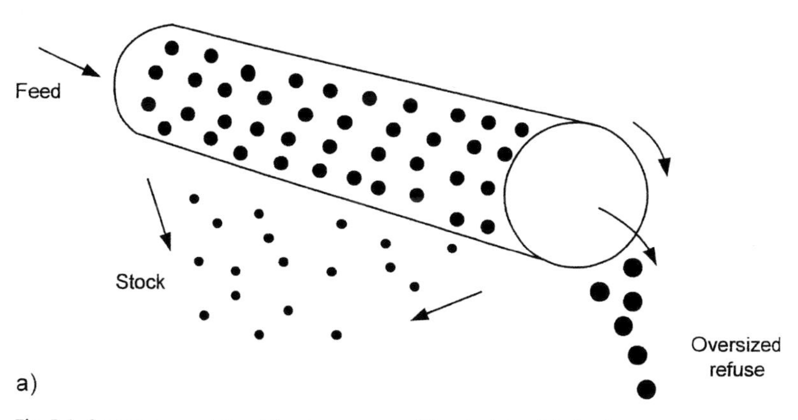

Fig. 1.2 Screen geometries: (a) rotary screen, (b) principle of flatbed screen.

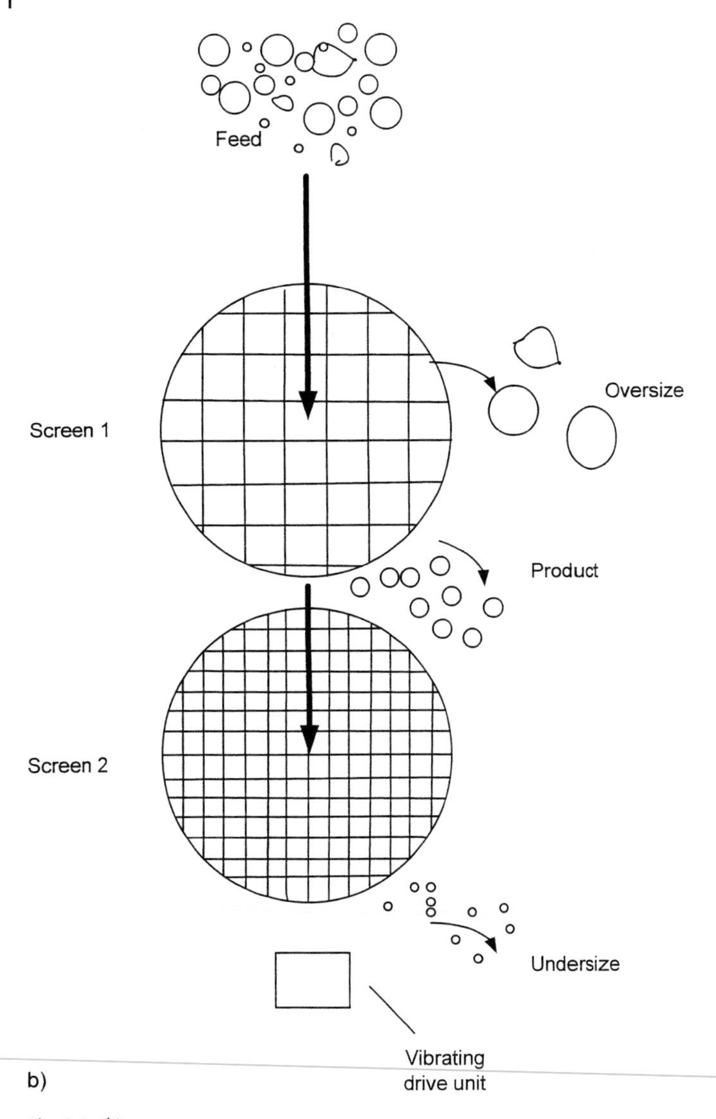

Fig. 1.2 (b)

Abrasion, either by impact during the operation of the machinery, or aided by abrasive discs or brushes, can improve the efficiency of dry screens. Screening gives incomplete separations and is usually a preliminary cleaning stage.

Aspiration exploits the differences in aerodynamic properties of the food and the contaminants. It is widely used in the cleaning of cereals, but is also incorporated into equipment for cleaning peas and beans. The principle is to feed the raw material into a carefully controlled upward air stream. Denser material will fall, while lighter material will be blown away depending on the terminal velocity. Terminal velocity in this case can be defined as the velocity of upward air stream in which a particle remains stationary; and this depends on the density and projected area of the particles (as described by Stokes' equation). By using different air velocities, it is possible to separate say wheat from lighter chaff (see Fig. 1.3) or denser small stones. Very accurate separations are possible, but large amounts of energy are required to generate the air streams. Obviously the system is limited by the size of raw material units, but is particularly suitable for cleaning legumes and cereals.

Air streams may also be used simply to blow loose contaminants from larger items such as eggs or fruit.

Magnetic cleaning is the removal of ferrous metal using permanent or electromagnets. Metal particles, derived from the growing field or picked up during transport or preliminary operations, constitute a hazard both to the consumer and to processing machinery, for example cereal mills. The geometry of magnetic cleaning systems can be quite variable: particulate foods may be passed over magnetised drums or magnetised conveyor belts, or powerful magnets may be located above conveyors. Electromagnets are easy to clean by turning off the power. Metal detectors are frequently employed prior to sensitive processing equipment as well as to protect consumers at the end of processing lines.

Electrostatic cleaning can be used in a limited number of cases where the surface charge on raw materials differs from contaminating particles. The principle can be used to distinguish grains from other seeds of similar geometry but different surface charge; and it has also been described for cleaning tea. The feed

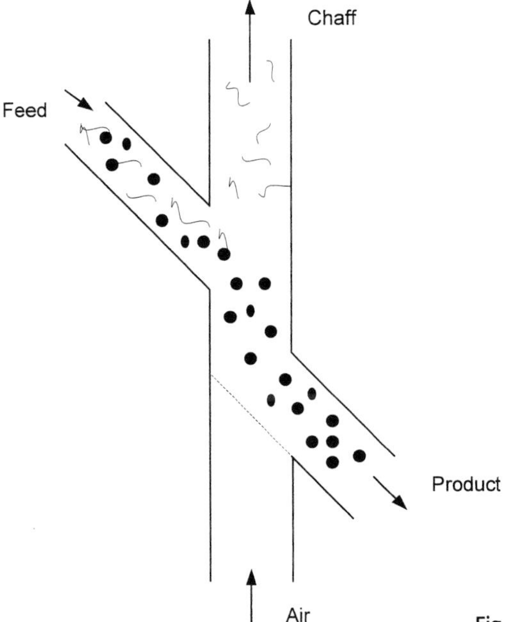

Fig. 1.3 Principle of aspiration cleaning.

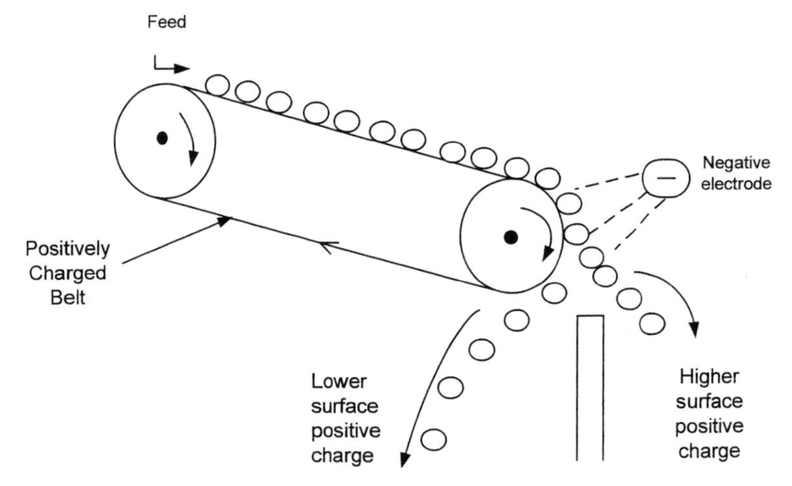

Fig. 1.4 Principle of electrostatic cleaning.

is conveyed on a charged belt and charged particles are attracted to an oppositely charged electrode (see Fig. 1.4) according to their surface charge.

1.4.2 Wet Cleaning Methods

Wet methods are necessary if large quantities of soil are to be removed; and they are essential if detergents are used. However, they are expensive, as large quantities of high purity water are required and the same quantity of dirty effluent is produced. Treatment and reuse of water can reduce costs. Employing the countercurrent principle can reduce water requirements and effluent volumes if accurately controlled. Sanitising chemicals such as chlorine, citric acid and ozone are commonly used in wash waters, especially in association with peeling and size reduction, where reducing enzymic browning may also be an aim [12]. Levels of 100–200 mg l^{-1} chlorine or citric acid may be used, although their effectiveness for decontamination has been questioned and they are not permitted in some countries.

Soaking is a preliminary stage in cleaning heavily contaminated materials, such as root crops, permitting softening of the soil and partial removal of stones and other contaminants. Metallic or concrete tanks or drums are employed; and these may be fitted with devices for agitating the water, including stirrers, paddles or mechanisms for rotating the entire drum. For delicate produce such as strawberries or asparagus, or products which trap dirt internally, e.g. celery, sparging air through the system may be helpful. The use of warm water or including detergents improves cleaning efficiency, especially where mineral oil is a possible contaminant, but adds to the expense and may damage the texture.

Spray washing is very widely used for many types of food raw material. Efficiency depends on the volume and temperature of the water and time of exposure. As a general rule, small volumes of high pressure water give the most efficient dirt removal, but this is limited by product damage, especially to more delicate produce. With larger food pieces, it may be necessary to rotate the unit so that the whole surface is presented to the spray (see Fig. 1.5 a). The two most common designs are drum washers and belt washers (see Figs. 1.5 a, b). Abrasion may contribute to the cleaning effect, but again must be limited in delicate units. Other designs include flexible rubber discs which gently brush the surface clean.

Flotation washing employs buoyancy differences between food units and contaminants. For instance sound fruit generally floats, while contaminating soil, stones or rotten fruits sink in water. Hence fluming fruit in water over a series of weirs gives very effective cleaning of fruit, peas and beans (see Fig. 1.6). A disadvantage is high water use, thus recirculation of water should be incorporated.

Froth flotation is carried out to separate peas from contaminating weed seeds and exploits surfactant effects. The peas are dipped in an oil/detergent emulsion and air is blown through the bed. This forms a foam which washes away the contaminating material and the cleaned peas can be spray washed.

Following wet cleaning, it is necessary to remove the washing water. Centrifugation is very effective, but may lead to tissue damage, hence dewatering screens or reels are more common.

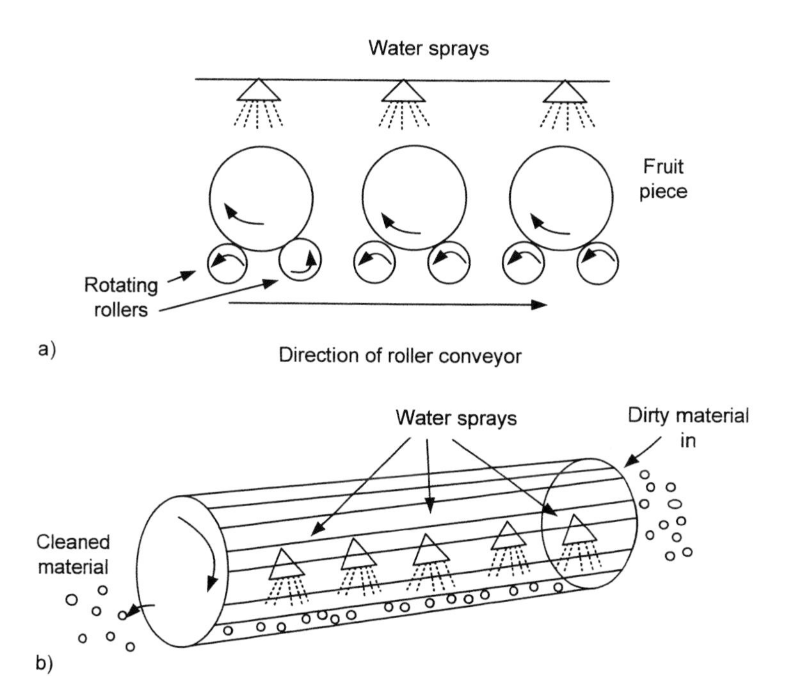

Fig. 1.5 Water spray cleaning: (a) spray belt washer, (b) drum washer.

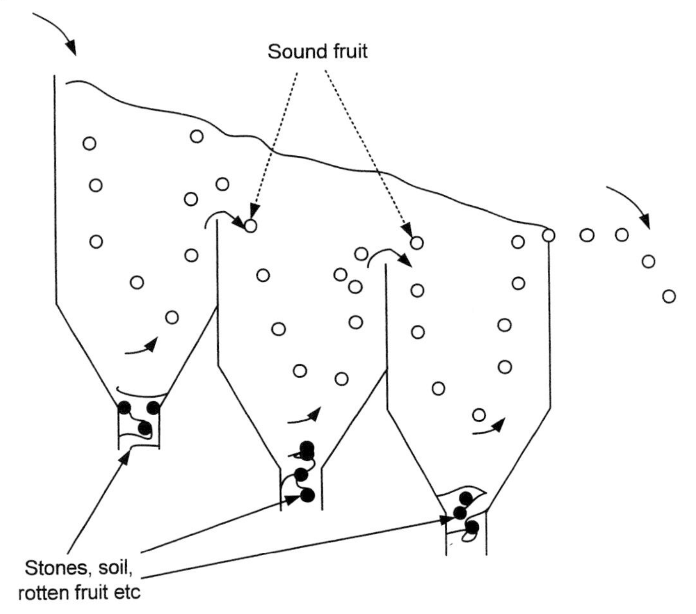

Fig. 1.6 Principle of flotation washing.

Prestorage hot water dipping has been used as an alternative to chemical treatments for preserving the quality of horticultural products. One recent development is the simultaneous cleaning and disinfection of fresh produce by a short hot water rinse and brushing (HWRB) treatment [13]. This involves placing the crops on rotating brushes and rinsing with hot water for 10–30 s. The effect is through a combination of direct cleaning action plus the lethal action of heat on surface pathogens. Fungicides may also be added to the hot water.

1.4.3
Peeling

Peeling of fruits and vegetables is frequently carried out in association with cleaning. Mechanical peeling methods require loosening of the skin using one of the following principles, depending on the structure of the food and the level of peeling required [14]:

- Steam is particularly suited to root crops. The units are exposed to high pressure steam for a fixed time and then the pressure is released causing steam to form under the surface of the skin, hence loosening it such that it can be removed with a water spray.
- Lye (1–2% alkali) solution can be used to soften the skin which can again be removed by water sprays. There is, however, a danger of damage to the product.

- Brine solutions can give a peeling effect but are probably less effective than the above methods.
- Abrasion peeling employs carborundum rollers or rotating the product in a carborundum-lined bowl, followed by washing away the loosened skin. It is effective but here is a danger of high product loss by this method.
- Mechanical knives are suitable for peeling citrus fruits.
- Flame peeling is useful for onions, in which the outer layers are burnt off and charred skin is removed by high pressure hot water.

1.5 Sorting and Grading

Sorting and grading are terms which are frequently used interchangeably in the food processing industry, but strictly speaking they are distinct operations. Sorting is a separation based on a single measurable property of raw material units, while grading is "the assessment of the overall quality of a food using a number of attributes" [14]. Grading of fresh produce may also be defined as 'sorting according to quality', as sorting usually upgrades the product.

Virtually all food products undergo some kind of sorting operation. There are a number of benefits, including the need for sorted units in weight-filling operations and the aesthetic and marketing advantages in providing units of uniform size or colour. In addition, it is much easier to control processes such as sterilisation, dehydration or freezing in sorted food units; and they are also better suited to mechanised operations such as size reduction, pitting or peeling.

1.5.1 Criteria and Methods of Sorting

Sorting is carried out on the basis of individual physical properties. Details of principles and equipment are given by Saravacos and Kostaropoulos [15], Brennan et al. [16] and Peleg [17]. No sorting system is absolutely precise and a balance is often struck between precision and flow rate.

Weight is usually the most precise method of sorting, as it is not dependent on the geometry of the products. Eggs, fruit or vegetables may be separated into weight categories using spring-loaded, strain gauge or electronic weighing devices incorporated into conveying systems. Using a series of tipping or compressed air blowing mechanisms set to trigger at progressively lesser weights, the heavier items are removed first, followed by the next weight category and so on. These systems are computer controlled and can additionally provide data on quantities and size distributions from different growers. An alternative system is to use the 'catapult' principle where units are thrown into different collecting chutes, depending on their weight, by spring-loaded catapult arms. A disadvantage of weight sorting is the relatively long time required per unit; and other methods are more appropriate with smaller items such as legumes or cereals, or if faster throughput is required.

Size sorting is less precise than weight sorting, but is considerably cheaper. As discussed in Section 1.2, the size and shape of food units are difficult to define precisely. Size categories could involve a number of physical parameters, including diameter, length or projected area. Diameter of spheroidal units such as tomatoes or citrus fruits is conventionally considered to be orthogonal to the fruit stem, while length is coaxial. Therefore rotating the units on a conveyor can make size sorting more precise. Sorting into size categories requires some sort of screen, many designs of which are discussed in detail by Slade [18], Brennan et al. [16] and Fellows [14]. The main categories of screens are fixed aperture and variable aperture designs. Flatbed and rotary screens are the main geometries of the fixed bed screen and a number of screens may be used in series or in parallel to sort units into several size categories simultaneously. The problem with fixed screens is usually contacting the feed material with the screen, which may become blocked or overloaded. Fixed screens are often used with smaller particulate foods such as nuts or peas. Variable aperture screens have either a continuous diverging or stepwise diverging apertures. These are much more gentle and are commonly used with larger, more delicate items such as fruit. The principles of some sorting screens are illustrated in Fig. 1.7.

Shape sorting is useful in cases where the food units are contaminated with particles of similar size and weight. This is particularly applicable to grain which may contain other seeds. The principle is that discs or cylinders with accurately shaped indentations will pick up seeds of the correct shape when rotated through the stock, while other shapes will remain in the feed (see Fig. 1.8).

Density can be a marker of suitability for certain processes. The density of peas correlates well with tenderness and sweetness, while the solids content of potatoes, which determines suitability for manufacture of crisps and dried products, relates to density. Sorting on the basis of density can be achieved using flotation in brine at different concentrations.

Photometric properties may be used as a basis for sorting. In practice this usually means colour. Colour is often a measure of maturity, presence of defects or the degree of processing. Manual colour sorting is carried out widely on conveyor belts or sorting tables, but is expensive. The process can be automated using highly accurate photocells which compare reflectance of food units to preset standards and can eject defective or wrongly coloured, e.g. blackened, units, usually by a blast of compressed air. This system is used for small particulate foods such as navy beans or maize kernels for canning, or nuts, rice and small fruit (see Fig. 1.9). Extremely high throughputs have been reported, e.g. 16 t h^{-1} [14]. By using more than one photocell positioned at different angles, blemishes on large units such as potatoes can be detected.

Colour sorting can also be used to separate materials which are to be processed separately, such as red and green tomatoes. It is feasible to use transmittance as a basis for sorting although, as most foods are completely opaque, very

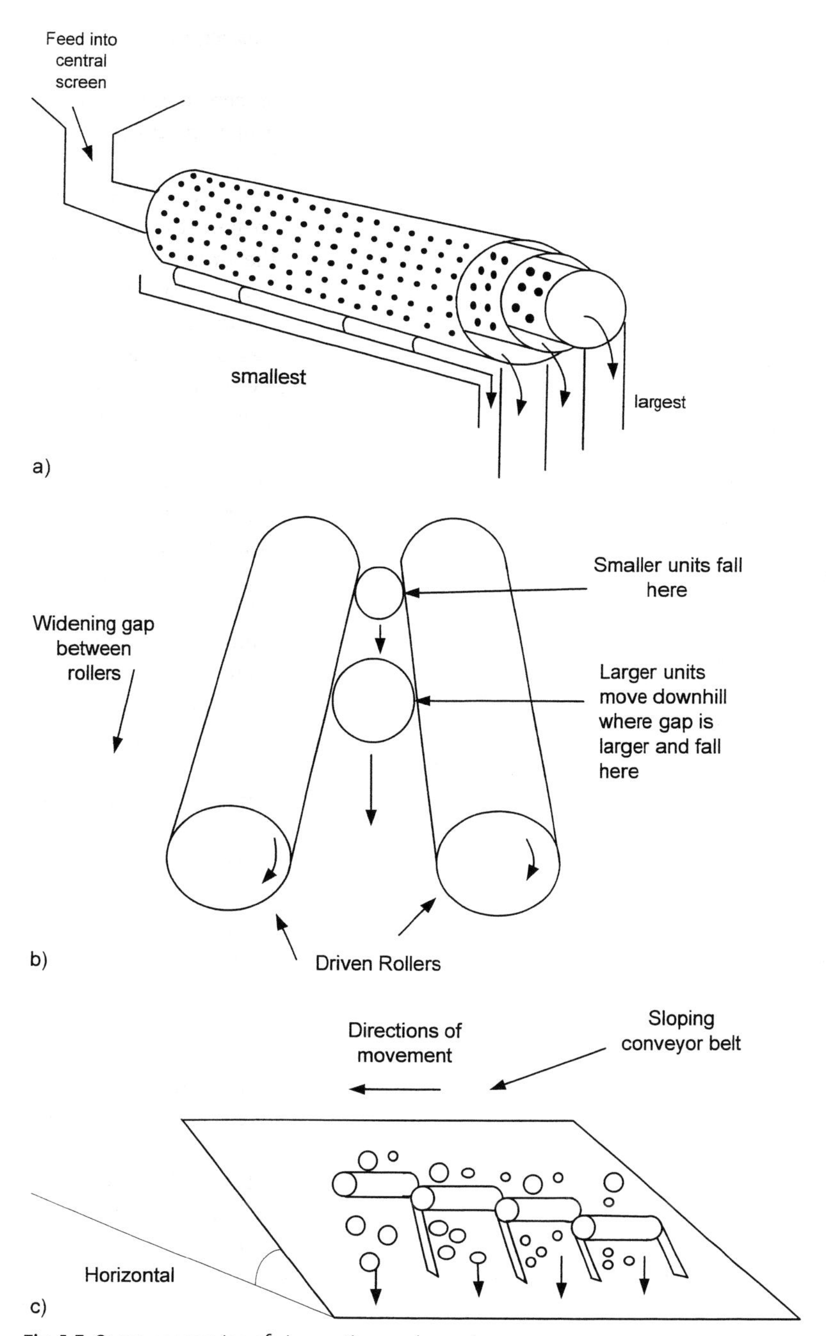

Fig. 1.7 Some geometries of size sorting equipment: (a) concentric drum screen, (b) roller size sorter, (c) belt and roller sorter.

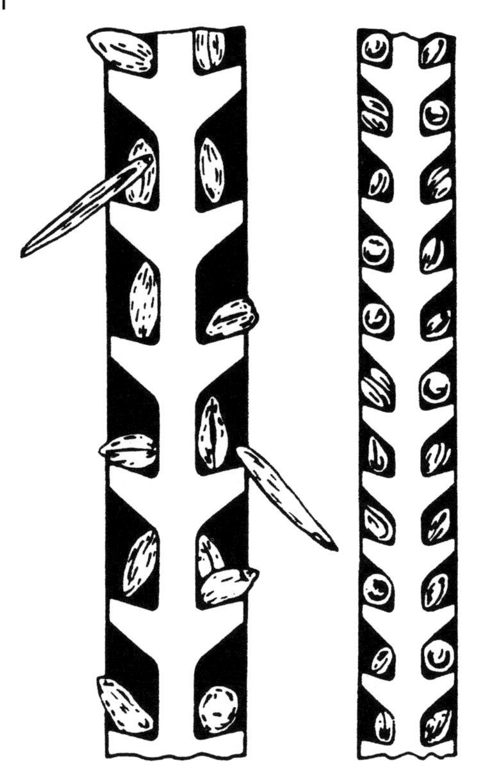

Fig. 1.8 Cross-section of disc separators for cleaning cereals.

few opportunities are available. The principle has been used for sorting cherries with and without stones and for the internal examination, or 'candling', of eggs.

1.5.2 Grading

Grading is classification on the basis of quality (incorporating commercial value, end use and official standards [15]), and hence requires that some judgement on the acceptability of the food is made, based on simultaneous assessment of several properties, followed by separation into quality categories. Appropriate inspection belts or conveyors are designed to present the whole surface to the operator. Trained manual operators are frequently used to judge the quality, and may use comparison to charted standards, or even plastic models. For example, a fruit grader could simultaneously judge shape, colour, evenness of colour and degree of russeting in apples. Egg candling involves inspection of eggs spun in front of a light so that many factors, including shell cracks, diseases, blood spots or fertilisation, can be detected. Apparently, experienced candlers can grade thousands of eggs per hour. Machine grading is only feasible where quali-

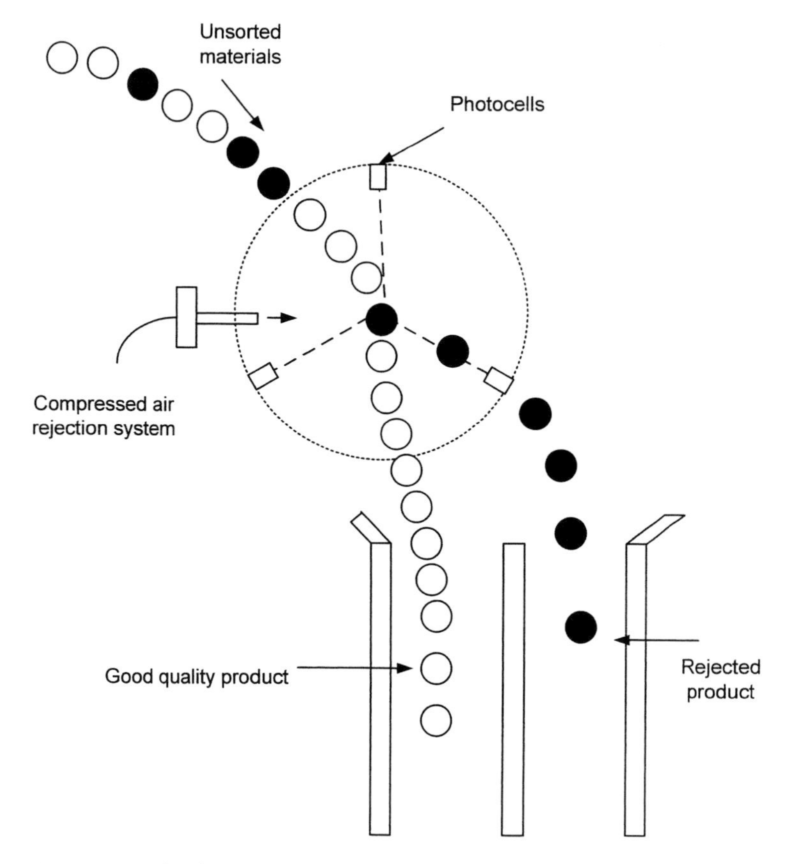

Fig. 1.9 Principle of colour sorter.

ty of a food is linked to a single physical property and hence a sorting operation leads to different grades of material. Size of peas, for example, is related to tenderness and sweetness, therefore size sorting results in different quality grades.

Grading of foods is also the determination of the quality of a batch. This can be done by human graders who assess the quality of random samples of foods such as cheese or butter, or meat inspectors who examine the quality of individual carcasses for a number of criteria. Alternatively, batches of some foods may be graded on the basis of laboratory analysis.

There is much interest in the development of rapid, nondestructive methods of assessing the quality of foods, which could be applied to the grading and sorting of foods. Cubeddu et al. [19] describe the potential application of advanced optical techniques to give information on both surface and internal properties of fruits, including textural and chemical properties. This could permit classification of fruit in terms of maturity, firmness or the presence of defects, or even more specifically, the noninvasive detection of chlorophyll, sugar and acid levels. Another promising approach is the use of sonic techniques to measure the texture of fruits and vegetables [20]. Similar applications of X-rays, lasers, infrared rays and microwaves have also been studied [15].

Numerous other miscellaneous mechanical techniques are available which effectively upgrade the material such as equipment for skinning and dehairing fish and meat, removing mussel shells, destemming and pitting fruit etc. [15].

1.6 Blanching

Most vegetables and some fruits are blanched prior to further processing operations, such as canning, freezing or dehydration. Blanching is a mild heat treatment, but is not a method of preservation *per se*. It is a pretreatment usually performed between preparation and subsequent processing. Blanching consists of heating the food rapidly to a predetermined temperature, holding for a specified time, then either cooling rapidly or passing immediately to the next processing stage.

1.6.1 Mechanisms and Purposes of Blanching

Plant cells are discrete membrane-bound structures contained within semirigid cell walls. The outer or cytoplasmic membrane acts as a skin, maintaining turgor pressure within the cell. Loss of turgor pressure leads to softening of the tissue. Within the cell are a number of organelles, including the nucleus, vacuole, chloroplasts, chromoplasts and mitochondria. This compartmentalisation is essential to the various biochemical and physical functions. Blanching causes cell death and physical and metabolic chaos within the cells. The heating effect leads to enzyme destruction as well as damage to the cytoplasmic and other membranes, which become permeable to water and solutes. An immediate effect is the loss of turgor pressure. Water and solutes may pass into and out of the cells, a major consequence being nutrient loss from the tissue. Also cell constituents, which had previously been compartmentalised in subcellular organelles, become free to move and interact within the cell.

The major purpose of blanching is frequently to inactivate enzymes, which would otherwise lead to quality reduction in the processed product. For example, with frozen foods, deterioration could take place during any delay prior to processing, during freezing, during frozen storage or during subsequent thawing. Similar considerations apply to the processing, storage and rehydration of dehydrated foods. Enzyme inactivation prior to heat sterilisation is less important as the severe processing will destroy any enzyme activity, but there may be an appreciable time before the food is heated to sufficient temperature, so quality may be better maintained if enzymes are destroyed prior to heat sterilisation processes such as canning.

It is important to inactivate quality-changing enzymes, that is enzymes which will give rise to loss of colour or texture, production of off odours and flavours or breakdown of nutrients. Many such enzymes have been studied, including a range of peroxidases, catalases and lipoxygenases. Peroxidase and to a lesser extent catalase are frequently used as indicator enzymes to determine the effectiveness of blanching. Although other enzymes may be more important in terms of their quality-changing effect, peroxidase is chosen because it is extremely easy to measure and it is the most heat resistant of the enzymes in question. More recent work indicates that complete inactivation of peroxidase may not be necessary and retention of a small percentage of the enzyme following blanching of some vegetables may be acceptable [21].

Blanching causes the removal of gases from plant tissues, especially intercellular gas. This is especially useful prior to canning where blanching helps achieve vacua in the containers, preventing expansion of air during processing and hence reducing strain on the containers and the risk of misshapen cans and/or faulty seams. In addition, removing oxygen is useful in avoiding oxidation of the product and corrosion of the can. Removal of gases, along with the removal of surface dust, has a further effect in brightening the colour of some products, especially green vegetables.

Shrinking and softening of the tissue is a further consequence of blanching. This is of benefit in terms of achieving filled weight into containers, so for example it may be possible to reduce the tinplate requirement in canning. It may also facilitate the filling of containers. It is important to control the time/temperature conditions to avoid overprocessing, leading to excessive loss of texture in some processed products. Calcium chloride addition to blanching water helps to maintain the texture of plant tissue through the formation of calcium pectate complexes. Some weight loss from the tissue is inevitable as both water and solutes are lost from the cells.

A further benefit is that blanching acts as a final cleaning and decontamination process. Selman [21] described the effectiveness of blanching in removing pesticide residues or radionuclides from the surface of vegetables, while toxic constituents naturally present (such as nitrites, nitrates and oxalate) are reduced by leaching. Very significant reductions in microorganism content can be achieved, which is useful in frozen or dried foods where surviving organisms can multiply on thawing or rehydration. It is also useful before heat sterilisation if large numbers of microorganisms are present before processing.

1.6.2
Processing Conditions

It is essential to control the processing conditions accurately to avoid loss of texture (see Section 1.6.1), weight, colour and nutrients. All water-soluble materials, including minerals, sugars, proteins and vitamins, can leach out of the tissue, leading to nutrient loss. In addition, some nutrient loss (especially ascorbic acid) occurs through thermal lability and, to a lesser extent, oxidation. Ascorbic acid is the most commonly measured nutrient with respect to blanching [21], as it covers all eventualities, being water soluble and hence prone to leaching from cells, thermally labile, as well as being subject to enzymic breakdown by ascorbic acid oxidase during storage. Wide ranges of vitamin C breakdown are observed, depending on the raw material and the method and precise conditions of processing. The aim is to minimise leaching and thermal breakdown while completely eliminating ascorbic acid oxidase activity, such that vitamin C losses in the product are restricted to a few percent. Generally steam blanching systems (see Section 1.6.3) give rise to lower losses of nutrients than immersion systems, presumably because leaching effects are less important.

Blanching is an example of unsteady state heat transfer involving convective heat transfer from the blanching medium and conduction within the food piece. Mass transfer of material into and out of the tissue is also important. The precise blanching conditions (time and temperature) must be evaluated for the raw material and usually represent a balance between retaining the quality characteristics of the raw material and avoiding over-processing. The following factors must be considered:

- fruit or vegetable properties, especially thermal conductivity, which will be determined by type, cultivar, degree of maturity etc.;
- overall blanching effect required for the processed product, which could be expressed in many ways including: achieving a specified central temperature, achieving a specified level of peroxidase inactivation, retaining a specified proportion of vitamin C;
- size and shape of food pieces;
- method of heating and temperature of blanching medium.

Time/temperature combinations vary very widely for different foods and different processes and must be determined specifically for any situation. Holding times of 1–15 min at 70–100 °C are normal.

1.6.3
Blanching Equipment

Blanching equipment is described by Fellows [14]. The two main approaches in commercial practice are to convey the food through saturated steam or hot water. Cooling may be with water or air. Water cooling may increase leaching losses but the product may also absorb water, leading to a net weight gain. Air cooling leads to weight loss by evaporation but may be better in terms of nutrient retention.

Conventional steam blanching consists of conveying the material through an atmosphere of steam in a tunnel on a mesh belt. Uniformity of heating is often poor where food is unevenly distributed; and the cleaning effect on the food is limited. However, the volumes of wastewater are much lower than for water blanching. Fluidised bed designs and 'individual quick blanching' (a three-stage process in which vegetable pieces are heated rapidly in thin layers by steam, held in a deep bed to allow temperature equilibration, followed by cooling in chilled air) may overcome the problems of nonuniform heating and lead to more efficient systems.

The two main conventional designs of hot water blancher are *reel* and *pipe* designs. In reel blanchers, the food enters a slowly rotating mesh drum which is partly submerged in hot water. The heating time is determined by the speed of rotation. In pipe blanchers, the food is in contact with hot water recirculating through a pipe. The residence time is determined by the length of the pipe and the velocity of the water.

There is much scope for improving energy efficiency and recycling water in either steam or hot water systems. Blanching may be combined with peeling and cleaning operations to reduce costs.

Microwave blanching has been demonstrated on an experimental scale but is too costly at present for commercial use.

1.7 Sulphiting of Fruits and Vegetables

Sulphur dioxide (SO_2) or inorganic sulphites (SO_3^{2-}) may be added to foods to control enzymic and nonenzymic browning, to control microbial growth, or as bleaching or reducing agents or antioxidants. The main applications are preserving or preventing discolouration of fruit and vegetables. The following sulphiting agents are permitted by European law: sulphur dioxide, sodium sulphite, sodium hydrogen sulphite, sodium metabisulphite, potassium metabisulphite, calcium sulphite, calcium hydrogen sulphite and potassium hydrogen sulphite. However, sulphites have some disadvantages, notably dangerous side effects for asthmatics; and their use has been partly restricted by the US Food and Drug Administration.

Sulphur dioxide dissolves readily in water to form sulphurous acid (H_2SO_3); and the chemistry of sulphiting agents can be summarised as follows:

$$H_2SO_3 \leftrightarrow H^+ + HSO_3^- \ (pK_1 \approx 2)$$

$$HSO_3^- \leftrightarrow H^+ + SO_3^{2-} \ (pK_1 \approx 7)$$

Most foods are in the pH range 4–7 and therefore the predominant form is HSO_3^-. Sulphites react with many food components, including aldehydes, ketones, reducing sugars, proteins and amino acids, to form a range of organic sulphites [22]. It is not clear exactly which reactions contribute to the beneficial applications of sulphites in the food industry. It should be noted that some of the reactions lead to undesirable consequences, in particular leading to vitamin breakdown. For example, Bender [23] reported losses of thiamin in meat products and fried potatoes when sulphiting agents were used during manufacture. In contrast, the inhibitory effect of sulphiting agents on oxidative enzymes, e.g.

ascorbic acid oxidase, may aid the retention of other vitamins including ascorbic acid and carotene.

Sulphites may be used to inhibit and control microorganisms in fresh fruit and fruits used in the manufacture of jam, juice or wine. In general, the antimicrobial action follows the order [22]: Gram negative bacteria > Gram positive bacteria > moulds > yeasts.

The mechanism(s) of action are not well understood, although it is believed that undissociated H_2SO_3 is the active form, thus the treatment is more effective at low pH (≥4).

A more widespread application is the inhibition of both enzymic and nonenzymic browning. Sulphites form stable hydroxysulphonates with carbonyl compounds and hence prevent browning reactions by binding carbonyl intermediates such as quinones. In addition, sulphites bind reducing sugars, which are necessary for nonenzymic browning, and inhibit oxidative enzymes including polyphenoloxidase, which are responsible for enzymic browning. Therefore, sulphite treatments can be used to preserve the colour of dehydrated fruits and vegetables. For example, sundried apricots may be treated with gaseous SO_2 to retain their natural colour [24], the product containing 2500–3000 ppm of SO_2. Sulphiting has commonly been used to prevent enzymic browning of many fruits and vegetables, including peeled or sliced apple and potato, mushrooms for processing, grapes and salad vegetables.

References

1 Seaton, H.L. **1955**, Scheduling plantings and predicting harvest maturities for processing vegetables, *Food Technology* 9, 202–209.

2 Maestrelli, A. **2000**, Fruit and vegetables: the quality of raw material in relation to freezing, in *Managing Frozen Foods*, ed. C. J. Kennedy, Woodhead Publishing, Cambridge, pp. 27–55.

3 Mohsenin, N.N. **1989**, *Physical Properties of Food and Agricultural Materials*, Gordon and Breach Science Publishers, New York.

4 Chung, O.K., Pomeranz, Y. **2000**, Cereal Processing, in *Food Proteins: Processing Applications*, ed. S. Nakai, H.W. Modler, Wiley-VCH, Chichester, pp. 243–307.

5 Nakai, S., Wing, P.L. **2000**, Breadmaking, in *Food Proteins: Processing Applications*, ed. S. Nakai, H.W. Modler, Wiley-VCH, Chichester, pp. 209–242.

6 Dobraszczyk, B.J. **2001**, Wheat and Flour, in *Cereals and Cereal Products: Chemistry and Technology*, ed. D.A.V. Dendy, B.J. Dobraszczyk, Aspen, Gaithersburg, pp. 100–139.

7 Finch, H.J.S., Samuel, A.M., Lane, G.P.F. **2002**, *Lockhart and Wiseman's Crop Husbandry*, 8th edn, Woodhead Publishing, Cambridge.

8 Ng-Kwai-Hang, K.F., Grosclaude, F. **2003**, Genetic polymorphism of milk proteins, in *Advanced Dairy Chemistry, vol. 1 – Proteins – part B*, ed. P.F. Fox, P.L.H. McSweeney, Kluwer Academic/ Plenum, New York, pp. 739– 816.

9 Nottingham, S. **1999**, *Eat Your Genes*, Zed Books, London.

10 Thompson, A.K. **1998**, *Controlled Atmosphere Storage of Fruits and Vegetables*, CAB International, Wallingford.

11 Heap, R., Kierstan, M., Ford, G. **1998**, *Food Transportation*, Blackie, London.

12 Ahvenian, R. **2000**, Ready-to-use fruit and vegetable, *Fair-Flow Europe Technical*

Manual F-FE 376A/00, Fair-Flow, London.

13 Orea, J. M., Gonzalez Urena, A. **2002**, Measuring and improving the natural resistance of fruit, in *Fruit and Vegetable Processing: Improving Quality*, ed. W. Jongen, Woodhead Publishing, Cambridge, pp. 233–266.

14 Fellows, P. J. **2000**, *Food Processing Technology: principles and practice*, 2nd edn, Woodhead Publishing, Cambridge.

15 Saravacos, G. D., Kostaropoulos, A. E. **2002**, *Handbook of Food Processing Equipment*, Kluwer Academic, London.

16 Brennan J. G., Butters, J. R., Cowell, N. D, Lilly, A. E. V. **1990**, *Food Engineering Operations*, 3rd edn, Elsevier Applied Science, London.

17 Peleg, K. **1985**, *Produce Handling, Packaging and Distribution*, AVI, Westport.

18 Slade, F. H. **1967**, *Food Processing Plant, vol.1*, Leonard Hill, London.

19 Cubeddu, R., Pifferi, Taroni, P., Torricelli, A. **2002**, Measuring Fruit and Vegetable Quality: Advanced Optical Methods, in *Fruit and Vegetable Processing: Improving Quality*, ed. W. Jongen, Woodhead Publishing, Cambridge, pp. 150–169.

20 Abbott, J. A., Affeldt, H. A., Liljedahl, L. A. **2002**, Firmness measurement in stored "Delicious" apples by sensory methods, Magness-Taylor, and sonic transmission, *Journal of the American Society of Horticultural Science*, 117, 590–595.

21 Selman, J. D. **1987**, The blanching process, in *Developments in Food Processing – 4*, ed. S. Thorne, Elsevier Applied Science, London, pp. 205–249.

22 Chang, P. Y. **2000**, Sulfites and food, in *Encyclopaedia of Food Science and Technology, vol. 4*, ed. F. J. Francis, Wiley Interscience, Chichester, pp. 2218–2220.

23 Bender, A. F. **1987**, Nutritional changes in food processing, in *Developments in Food Processing – 4*; ed. S. Thorne, Elsevier Applied Science, London, pp. 1–34.

24 Ghorpade, V. M., Hanna, M. A., Kadam, S. S. **1995**, Apricot, in *Handbook of Fruit Science and Technology: Production, Composition, Storage and Processing*, ed. D. K. Salunke, S. S. Kadam, Marcel Dekker, New York, pp. 335–361.

25 Aked, J. **2002**, Maintaining the post-harvest quality of fruits and vegetables, in *Fruit and Vegetable Processing: Improving Quality*, ed. W. Jongen, Woodhead Publishing, Cambridge, pp. 119–149.

2
Thermal Processing

Michael J. Lewis

2.1
Introduction

Thermal processing involves heating food, either in a sealed container or by passing it through a heat exchanger, followed by packaging. It is important to ensure that the food is adequately heat treated and to reduce postprocessing contamination (ppc). The food should then be cooled quickly and it may require refrigerated storage or be stable at ambient temperature. The heating process can be either batch or continuous. In all thermal processes, the aim should be to heat and cool the product as quickly as possible. This has economic implications and may also lead to an improvement in quality. Heat or energy (J) is transferred from a high to a low temperature, the rate of heat transfer being proportional to the temperature difference. Therefore, high temperature driving forces will promote heat transfer. SI units for rate of heat transfer ($J\ s^{-1}$ or W) are mainly used but Imperial units ($BTU\ h^{-1}$) may also be encountered [1]. The heating medium is usually saturated steam or hot water. For temperatures above 100 °C, steam and hot water are above atmospheric pressure. Cooling is achieved using either mains water, chilled water, brine or glycol solution. Regeneration is used in continuous processes to further reduce energy utilisation (see Section 2.2.3).

2.1.1
Reasons for Heating Foods

Foods are heated for a number of reasons, the main one being to inactivate pathogenic or spoilage microorganisms. It may also be important to inactivate enzymes, to avoid the browning of fruit by polyphenol oxidases and minimise flavour changes resulting from lipase and proteolytic activity. The process of heating a food also induces physical changes and chemical reactions, such as starch gelatinisation, protein denaturation or browning, which in turn affect the sensory characteristics, such as colour, flavour and texture, either advanta-

Food Processing Handbook. Edited by James G. Brennan

ISBN: 3-527-30719-2

geously or adversely. For example, heating pretreatments are used in the production of evaporated milk to prevent gelation and age-thickening and for yoghurt manufacture to achieve the required final texture in the product. Heating processes may also change the nutritional value of the food.

Thermal processes vary considerably in their intensity, ranging from mild processes such as thermisation and pasteurisation through to more severe processes such as in-container sterilisation. The severity of the process affects both the shelf life and other quality characteristics.

Foods which are heat-treated can be either solid or liquid, so the mechanisms of conduction and convection may be involved. Solid foods are poor conductors of heat, having a low thermal conductivity, and convection is inherently a much quicker process than conduction. Fluids range from those having a low viscosity (1–10 mPa s), through to highly viscous fluids; and the presence of particles (up to 25 mm in diameter) further complicates the process, as it becomes necessary to ensure that both the liquid and solid phases are at least adequately and if possible equally heated. The presence of dissolved air in either of the phases is a problem as it becomes less soluble as temperature increases and can come out of solution. Air is a poor heat transfer fluid and hot air is rarely used as a heating medium. Attention should be paid to removing air from steam e.g. venting of steam retorts and removing air from sealed containers (exhausting).

Heating is also involved in many other operations, which will not be covered in such detail in this chapter, such as evaporation and drying (see Chapter 3). It is also used for solids; in processing powders and other particulate foods: for example extrusion, baking (see Chapter 8) and spice sterilisation.

2.1.2
Safety and Quality Issues

The two most important issues connected with thermal processing are *food safety* and *food quality*. The major safety issue involves inactivating pathogenic microorganisms which are of public health concern. The World Health Organisation estimates that there are over 100 million cases of food poisoning each year and that one million of these result in death. These pathogens show considerable variation in their heat resistance: some are heat-labile, such as *Campylobacter, Salmonella, Lysteria* and of more recent concern *Escherichia coli* 0157, which are inactivated by pasteurisation, while of greater heat resistance is *Bacillus cereus*, which may survive pasteurisation and also grow at low temperatures. The most heat-resistant pathogenic bacterial spore is *Clostridium botulinum*. As well as these major foodborne pathogens, it is important to inactivate those microorganisms which cause food spoilage, such as yeasts, moulds and gas-producing and souring bacteria. Again there is considerable variation in their heat resistance, the most heat-resistant being the spores of *Bacillus stearothermophilus*. The heat resistance of any microorganism changes as the environment changes, for example pH, water activity or chemical composition changes; and foods themselves provide such a complex and variable environment. New microorganisms may also be encountered, such as *Bacillus sporothermodurans* (see Section 2.5.1). Therefore it is important to be aware of the type of microbial flora associated with all raw materials which are to be heat-treated.

After processing, it is very important to avoid reinfection of the product, generally known as ppc, which can cause problems in both pasteurisation and sterilisation. Therefore, raw materials and finished products should not be allowed in close proximity to each other. Other safety issues are concerned with natural toxins, pesticides, herbicides, antibiotics and growth hormones and environmental contaminants. Again, it is important that steps are taken to ensure that these do not finish up in the final product. Recently, there have been some serious cases of strong allergic reactions, with some deaths, shown by some individuals to foods such as peanuts and shellfish. These are all issues which also need to be considered for heat-treated foods.

Quality issues revolve around minimising chemical reactions and loss of nutrients and ensuring that sensory characteristics (appearance, colour, flavour and texture) are acceptable to the consumer. Quality changes, which may result from enzyme activity, must also be considered. There may also be conflicts between safety and quality issues. For example, microbial inactivation and food safety is increased by more severe heating conditions, but product quality in general deteriorates. To summarise, it is important to understand reaction kinetics and how they relate to:

- microbial inactivation
- chemical damage
- enzyme inactivation
- physical changes.

2.1.3 Product Range

The products covered in this book include those which can be filled into containers and subsequently sealed and heat-treated and those which can be processed by passing them through a continuous heat exchanger. This latter category includes milks and milk-based drinks, fruit and vegetable juices, purees, soups and sauces (both sweet and savoury) and a range of products containing particulate matter, up to about 25 mm diameter. There are two distinct market sectors. The first involves those products which are given a mild heat treatment and then kept refrigerated throughout storage: these are covered in Section 2.4. The second involves those which are sterilised and stored at ambient temperature: these are covered in Section 2.5. The relative importance of these two sectors varies from country to country for different products and from country to country for the same product. For example, in England and Wales, pasteurised milk, which is stored chilled, accounts for about 87% of liquid milk consumption, whilst UHT and sterilised milk accounts for about 10%. However, UHT milk accounts for a much greater proportion of milk consumed in other countries, for example France (78%) and Germany (58%), so it is important to note that there are regional differences and preferences [2].

In general, heat processing eliminates the need to use further additives to extend the shelf life, although additives may help improve the sensory characteristics or make processes less susceptible to fouling. In addition to reactions taking place during the heat treatment, chemical, enzymatic and physical changes will continue to take place during storage. Microorganisms which survive the heat treatment may also grow if conditions are favourable. Pasteurised products are normally kept refrigerated during storage to retard microbial growth; and low temperatures must be maintained throughout the Cold Chain. In contrast, sterilised products are not normally refrigerated and are stored at ambient temperature. This may vary considerably throughout the world, ranging from below 0 °C to above 50 °C. All the reactions mentioned above are temperature-dependent and considerable changes may take place during the storage period. One example is browning of milk and milk products, which is very significant after 4 months storage at 40 °C. Changes during storage are discussed in more detail for pasteurised products in Section 2.4.3 and for sterilised products in Section 2.7.1.

2.2 Reaction Kinetics

2.2.1 Microbial Inactivation

All thermal processes involve three distinct periods: a heating period, holding period and cooling period. In theory, all three periods may contribute to the reactions taking place, although in situations where heating and cooling are rapid, the holding period is the most significant. However, procedures are needed to evaluate each of these periods individually, to determine the overall effect. By far the easiest to deal with is the holding period, as this takes place at constant temperature. Then it needs to be established how reaction rates are affected by changes in temperature during heating and cooling. To simplify the analysis, microbial inactivation is first measured at constant temperature. This is usually followed by observing how microbial inactivation changes with temperature.

2.2.2 Heat Resistance at Constant Temperature

When heat inactivation studies are carried out at constant temperature, it is often observed that microbial inactivation follows first order reaction kinetics i.e. the rate of inactivation is directly proportional to the population. This can be illustrated by plotting the log of the population against time (see Fig. 2.1) and finding that there is a straightline relationship.

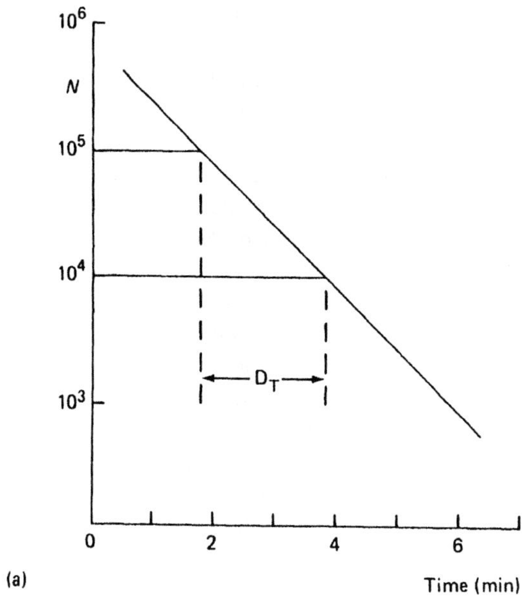

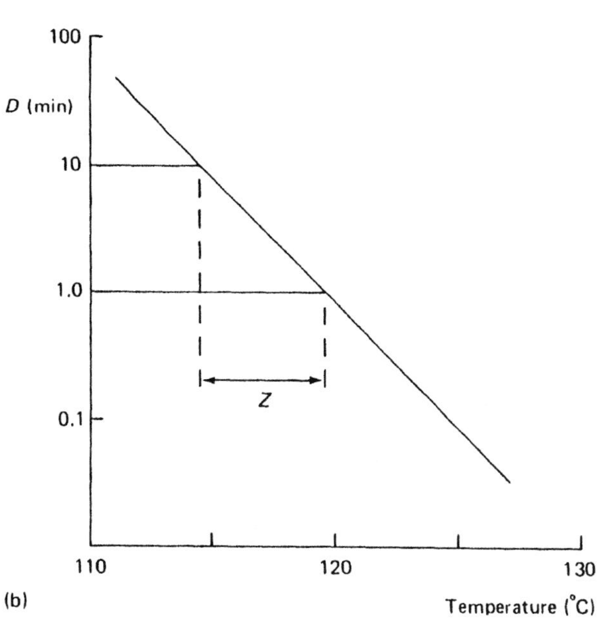

Fig. 2.1 (a) Relationship between the population of microorganisms and time at a constant heating temperature, (b) relationship between the decimal reduction time and temperature, to determine the z value; from [1] with permission.

The heat resistance of an organism is characterised by its decimal reduction time (D_T), which is defined as the time required to reduce the population by 90% or by one order of magnitude or one log cycle, i.e. from 10^4 to 10^3, at a constant temperature, *T*.

Every microorganism has its own characteristic heat resistance and the higher its *D* value, the greater is its heat resistance. Heat resistance is also affected by a wide range of other environmental factors, such as pH, water activity and the presence of other solutes, such as sugars and salts.

The extent of microbial inactivation is measured by the number of decimal reductions which can be achieved. This is given by $\log(N_0/N)$, where N_0 is the initial population and *N* is the final population; and it is determined from the following equation:

$$\log\left(\frac{N_0}{N}\right) = \frac{\text{Heating time}}{D_T} \tag{2.1}$$

There are two important aspects associated with first order reaction kinetics: it is not theoretically possible to achieve 100% reduction and, for a specific heat treatment, the final population increases as the initial population increases.

Example: for an organism with a D_{70} value of 10 s, heating for 10 s at 70 °C will achieve a 90% reduction in the population, 20 s heating will achieve 2*D* (99%), 30 s will achieve 3*D* (99.9%) and 60 s will achieve 6*D* (99.9999%) reduction.

Although it is not theoretically possible to achieve 100% reduction, in practical terms this may appear to be the case. For sterilisation processes, the term *commercial sterility* is used, rather than absolute sterility, to indicate that there will always be a small chance that one or more microorganisms will survive the heat treatment. Increasing the severity of the heating process, e.g. by prolonging the time period or by using higher temperatures, will reduce the chance of finding survivors of any particular bacterium.

Thus, if the initial population had been 10^6 ml^{-1}, after 80 s heating the final population would be 10^{-2} ml^{-1}. It may be difficult to imagine a fraction of an organism, but another way of expressing this is one organism per 100 ml. Thus, if the sample had been packaged in 1-ml portions, it should be possible to find one surviving microorganism in every 100 samples analysed; i.e. 99 would be free of microorganisms. The same heat treatment given to a raw material with a lower count would give one surviving microorganism in every 10 000 ml. Thus if 10 000 (1-ml) samples were analysed, 9999 would be free of viable microorganisms. Note that just one surviving bacterium in a product or package can give rise to a spoiled product or package. In practice, deviations from first order reaction kinetics are often encountered and the reasons for this are discussed in more detail by Gould [4].

For pasteurisation processes the temperature range of interest is 60–90 °C. Sterilisation is a more severe process and temperatures in excess of 100 °C are needed to inactivate heat-resistant spores. Sterilisation processes can be either in a sealed container, 110–125 °C for 10–120 min, or by continuous flow techniques, using temperatures in the range 135–142 °C for several seconds. In both cases, the amount of chemical reaction is increased and the flavour is different to pasteurised milk. The product has a shelf life of up to 6 months at ambient storage conditions.

2.3 Temperature Dependence

As mentioned, most processes do not take place at constant temperature, but involve heating and cooling periods. Therefore, although it is easy to evaluate the effect of the holding period on the heat resistance and the lethality, i.e. the number of decimal reductions of a process, see Eq. (2.1), it is important to appreciate that the heating and cooling may also contribute to the overall lethality. It can easily be demonstrated that reaction rates increase as temperature increases and microbial inactivation is no exception. Food scientists use a parameter known as the z value, to describe temperature dependence. This is based on the observation that, over a limited temperature range, there is a linear relationship between the log of the decimal reduction time and the temperature (see Fig. 2.1).

This is used to define the z value for inactivation of that particular microorganism as follows: the z value is the temperature change which results in a tenfold change in the decimal reduction time. The z value for most heat-resistant spores is about 10 °C, whereas the z value for vegetative bacteria is considerably lower, usually between 4 °C and 8 °C. A low z value implies that the reaction in question is very temperature-sensitive. In general, microbial inactivation is very temperature-sensitive, with inactivation of vegetative bacteria being more temperature-sensitive than heat-resistant spores.

In contrast to microbial inactivation, chemical reaction rates are much less temperature-sensitive than microbial inactivation, having higher z values (20–40 °C; see Table 2.1). This is also the case for many heat-resistant enzymes, although heat-labile enzymes such as alkaline phosphatase or lactoperoxidase are exceptions to this rule. This difference for chemical reactions and microbial inactivation has some important implications for quality improvement when using higher temperatures for shorter times; and this is discussed in more detail in Section 2.5.3.

The relationship between D values at two different temperatures and the z value is given by:

$$\log\left(\frac{D_1}{D_2}\right) = \left[\frac{(\theta_2 - \theta_1)}{z}\right] \tag{2.2}$$

An alternative way of using z values is for comparing processes. For example, if it is known that a temperature of 68 °C is effective for 10 min and the z value is 6 °C; then equally effective processes would also be 62 °C for 100 min or 74 °C for 1 min, i.e. 1 min at θ is equivalent to 0.1 min at $(\theta + z)$ or 10 min at $(\theta - z)$.

Table 2.1 Values of D and z for microbial inactivation, enzyme inactivation and some chemical reactions; from [6] with permission.

Microbe	D_{121} (°C)	z (°C)
Bacillus stearothermophilus NCDO 1096, milk	181.0	9.43
B. stearothermophilus FS 1518, conc. milk	117.0	9.35
B. stearothermophilus FS 1518, milk	324.0	6.7
B. stearothermophilus NCDO 1096, milk	372.0	9.3
B. subtilis 786, milk	20.0	6.66
B. coagulans 604, milk	60.0	5.98
B. cereus, milk	3.8	35.9
Clostridium sporogenes PA 3679, conc. milk	43.0	11.3
C. botulinum NCTC 7272	3.2	36.1
C. botulinum (canning data)	13.0	10.0
Proteases inactivation	0.5–27.0 min at 150 °C	32.5–28.5
Lipases inactivation	0.5–1.7 min at 150 °C	42.0–25.0
Browning	–	28.2; 21.3
Total whey protein denaturation, 130–150 °C	–	30.0
Available lysine	–	30. 1
Thiamin (B_1) loss	–	31.4–29.4
Lactulose formation	–	27.7–21.0

Thus Eq. (2.2) can be rewritten, replacing decimal reduction time (D) by the processing time (t), as:

$$\log\left(\frac{t_1}{t_2}\right) = \left[\frac{(\theta_2 - \theta_1)}{z}\right] \tag{2.3}$$

In the context of milk pasteurisation, if the holder process (63 °C for 30 min) is regarded as being equivalent to the high temperature short time (HTST) process of 72 °C for 15 s, the z value would be about 4.3 °C.

Another approach is to use this concept of equivalence, together with a reference temperature. For example, 72 °C is used for pasteurisation and 121.1 °C and 135 °C for sterilisation processes (see later). Perhaps best known are the standard lethality tables, used in the sterilisation of low-acid foods.

Thus, the lethality at any experimental temperature (T) can be compared to that at the reference temperature (θ), using the following equation:

$$\log L = \frac{(T - \theta)}{z} \tag{2.4}$$

Thus for a standard reference temperature of 121.1 °C and an experimental temperature of 118 °C, using z=10 °C, then L=0.49. Thus, 1 min at 118 °C would be equivalent to 0.49 min at 121.1 °C. Note that a temperature drop of 3 °C will halve the lethality.

Q_{10} is another parameter used to measure temperature dependence. It is defined as the ratio of reaction rate at $T+10$ to that at T, i.e. the increase in reaction rate caused by an increase in temperature of 10 °C. Some D and z values for heat-resistant spores and chemical reactions are given in Table 2.1.

It is interesting that, in spite of the many deviations reported, this log linear relationship still forms the basis for thermal process calculations in the food industry. Gould [4] has surmised that there is a strong view that this relationship remains at least a very close approximation of the true thermal inactivation kinetics of spores. Certainly, the lack of major problems when sterilisation procedures are properly carried out according to the above principles has provided evidence over many years that the basic rationale, however derived, is sound even though it may be cautious.

2.3.1 Batch and Continuous Processing

A process such as pasteurisation can be done batchwise or continuously. Batch processing involves filling the vessel, heating, holding, cooling, emptying the vessel, filling into containers and cleaning the vessel. Holding times may be up to 30 min. An excellent account of batch pasteurisation is provided by Cronshaw [5]. Predicting the heating and cooling times involves unsteady state heat transfer and illustrates the exponential nature of the heat transfer process. The heating time is determined by equating the rate of heat transfer from the heating medium to the rate at which the fluid absorbs energy. Thus:

$$UA(\theta - \theta) = mc\frac{d\theta}{dt} \tag{2.5}$$

which on integration becomes:

$$t = \frac{mc}{UA}\ln\left[\frac{(\theta_h - \theta_l)}{(\theta_h - \theta_f)}\right] \tag{2.6}$$

where m is mass (kg), c is specific heat (J kg^{-1} K^{-1}), A is surface area (m^2), U is OHTC (W m^{-2} K^{-1}) and t is the heating time (s) required to raise the temperature from θ_l (initial temperature) to θ_f (final temperature) using a heating medium temperature, θ_h.

The dimensionless temperature ratio represents the ratio of the initial temperature driving force to the approach temperature. The concept of approach temperature, i.e. how close the product approaches the heating or cooling medium temperature, is widely used in continuous heat exchangers. Heating and cooling times can be long.

Batch processes were easy to operate, flexible, are able to deal with different size batches and different products and this is still the case; also, if well mixed, no distribution of residence times, which is a problem with continuous processes.

Heating and cooling rates are slower and the operation is more labour intensive; it will involve filling, heating, holding, cooling, emptying, cleaning and disinfecting, which may take up to 2 h. Postprocessing contamination (ppc) should be avoided in the subsequent packaging operations.

The alternative process for both pasteurisation and sterilisation involves continuous processes. Some advantages of continuous processes are as follows:

- Foods can be heated and cooled more rapidly compared to in-container processes. This improves the economics of the process and the quality of the product.
- There are none of the pressure constraints which apply to heating products in sealed containers. This allows the use of higher temperatures and shorter times, which results in less damage to the nutrients and improved sensory characteristics, these being: appearance, colour, flavour and texture.
- Continuous processes provide scope for energy savings, whereby the hot fluid is used to heat the incoming fluid. This is known as regeneration and saves both heating and cooling costs (see Section 2.2.3).

Heating processes can be classified as direct or indirect. The most widely used is indirect heating, where the heat transfer fluid and the liquid food are separated by a barrier. For in-container sterilisation, this will be the wall of the bottle and, for continuous processes, the heat exchanger plate or tube wall. In direct processes, steam is the heating medium and the steam comes into direct contact with the product (see Section 2.2.3).

The mechanisms of heat transfer are by conduction in solids and convection in liquids. Thermal conductivity ($W\,m^{-1}\,K^{-1}$) is the property which measures the rate of heat transfer due to conduction. Metals are good conductors of heat, although stainless steel has a much lower value ($\sim 20\,W\,m^{-1}\,K^{-1}$) than both copper ($\sim 400\,W\,m^{-1}\,K^{-1}$) and aluminium ($\sim 220\,W\,m^{-1}\,K^{-1}$). However, it is much higher than glass ($\sim 0.5\,W\,m^{-1}\,K^{-1}$). In general, foods are poor conductors of heat ($\cong 0.5\,W\,m^{-1}\,K^{-1}$) and this can be a problem when heating particulate systems (see Section 2.5.3).

The efficiency of heat transfer by convection is measured by the heat film coefficient. Condensing steam has a much higher heat film coefficient (as well as a high latent heat of vaporisation) than hot water, which is turn is higher than hot air. Inherently, heat transfer by convection is faster than heat transfer by conduction.

In the indirect process, there are three resistances to the transfer of heat from the bulk of a hot fluid to the bulk of a cold fluid, two due to convection and one due to conduction. The overall heat transfer coefficient (U) provides a measure of the efficiency of the heat transfer process and takes into account all three resistances. It can be calculated from:

$$\frac{1}{U} = \frac{1}{h_1} + \frac{1}{h_2} + \frac{L}{k} \tag{2.7}$$

where h_1 and h_2 are the heat film coefficients (W m^{-2} K^{-1}) for the hot fluid and the cold fluid respectively, L is heat exchanger wall thickness (m) and k is the thermal conductivity of the plate or tube wall (W m^{-1} K^{-1}).

The higher the value of U, the more efficient is the heat exchange system. Each one of the terms in Eq. (2.7) represents a resistance. The highest of the individual terms is known as the limiting resistance. This is the one that controls the overall rate of heat transfer. Thus, to improve the performance of a heat exchanger, it is best to focus on the limiting resistance.

The basic design equation for a heat exchanger is as follows:

$$Q = UA\Delta\theta_m \tag{2.8}$$

where Q is the duty or rate of heat transfer (J s^{-1}), $\Delta\theta_m$ (°C or K) is the log mean temperature difference and A is the surface area (m^2).

The duty (Q) is obtained form the following expression:

$$Q = m'c\Delta\theta \tag{2.9}$$

where m' is mass flow rate (kg s^{-1}), c is specific heat capacity (J kg^{-1} K^{-1}) and $\Delta\theta$ is the change in product temperature (K or °C).

In a continuous heat exchanger, the two fluids can either flow in the same direction (co-current) or in opposite directions (counter-current). Counter-current is the preferred direction, as it results in a higher mean temperature driving force and a closer approach temperature.

One of the main practical problems with indirect heating is fouling. This is the formation of deposits on the wall of the heat exchanger. These can introduce one or two additional resistances to heat transfer and lead to a reduction in U. Fouling may be the result of deposits from the food or deposits from the service fluids in the form of sediment from steam, hardness from water or microbial films from cooling water. Fouling may result in a decrease in product temperature and eventually in the product being underprocessed. A further problem is a reduction in the cross-sectional area of the flow passage, which leads to a higher pressure drop. Fouled deposits also need to be removed at the end of the process as they may serve as a breeding ground for bacteria, particularly thermoduric bacteria. For example, for milk, fouling becomes more of a problem as the processing temperature increases and the acidity of the milk increases. Fouling is discussed in more detail by Lewis and Heppell [6]. Heat exchangers also need to be cleaned and disinfected after their use (see Section 2.5.3).

2.3.2 Continuous Heat Exchangers

The viscosity of the product is one major factor which affects the choice of the most appropriate heat exchanger and the selection of pumps. The main types of indirect heat exchanger for fluids such as milk, creams and liquid egg are the

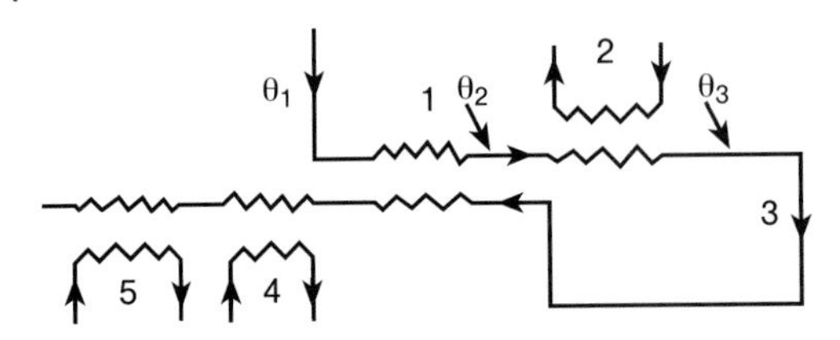

Fig. 2.2 Heat exchanger sections for a high temperature short time (HTST) pasteuriser: 1, regeneration; 2, hot water section; 3, holding tube; 4, mains water cooling; 5, chilled water cooling; from [3] with permission.

plate heat exchanger and the tubular heat exchanger. A high product viscosity gives rise to a high pressure drop, which can cause a problem in the cooling section, especially when phase transitions such as gelation or crystallisation take place. For more viscous products, such as starch-based desserts, a scraped surface heat exchanger may be used (see Section 2.4.3.3).

One of the main advantages of continuous systems over batch systems is that energy can be recovered in terms of regeneration. The layout for a typical regeneration section is shown in Fig. 2.2. The hot fluid (pasteurised or sterilised) can be used to heat the incoming fluid, thereby saving on heating and cooling costs. The regeneration efficiency (RE) is defined as follows:

$$\text{RE} = 100 \times (\text{amount of heat supplied by regeneration/amount of heat required assuming no regeneration}) \quad (2.10)$$

Regeneration efficiencies up to 95% can be obtained, which means that a pasteurised product would be heated up to almost 68 °C by regeneration. Although high regeneration efficiencies result in considerable savings in energy, it necessitates the use of larger surface areas, resulting from the lower temperature driving force and a slightly higher capital cost for the heat exchanger. This also means that the heating and cooling rates are also slower, and the transit times longer, which may affect the quality, especially in UHT processing.

Plate heat exchangers (PHEs; see Fig. 2.3) are widely used both for pasteurisation and sterilisation processes. They have a high OHTC and are generally more compact than tubular heat exchangers. Their main limitation is pressure, with an upper limit of about 2 MPa. The normal gap width between the plates is between 2.5 mm and 5.0 mm, but wider gaps are available for viscous liquids. The narrower the gap, the more pressure required and wide gap plates are not in regular use for UHT treatment of low-acid foods. In general, a PHE is the cheapest option and the one most widely used for low viscosity fluids. However, maintenance costs may be high, as gaskets may need replacing at regular intervals and the integrity of the plates need to be checked at regular intervals, especially those in the regeneration section, where a cracked or leaking plate may allow raw product, e.g. raw milk, to contaminate already pasteurised milk or sterilised milk.

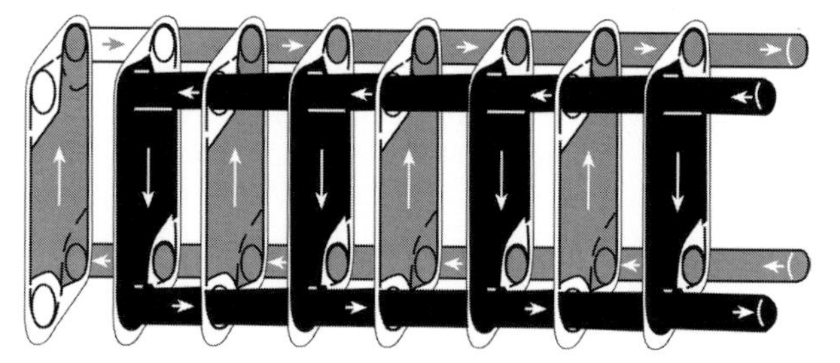

Fig. 2.3 Flow through a plate heat exchanger, by courtesy of APV.

Tubular heat exchangers (see Fig. 2.4) have a lower OHTC than plates and generally occupy a larger space. They have slower heating and cooling rates with a longer transit time through the heat exchanger. In general they have fewer seals and provide a smoother flow passage for the fluid. A variety of tube designs is available to suit different product characteristics. These designs include single tubes with an outer jacket, double or multiple concentric tubes or shell and tube types. Most UHT plants use a multitube design. They can withstand higher pressures than PHEs. Although, they are still susceptible to fouling, high pumping pressures can be used to overcome the flow restrictions. Thus, tubular heat exchangers give longer processing times with viscous materials and with products which are more susceptible to fouling.

For products containing fat, such as milk and cream, homogenisation (see Chapter 15) must be incorporated to prevent fat separation. This may be upstream or downstream. For UHT processes, downstream homogenisation requires the process to be achieved under aseptic conditions and provides an additional risk of recontaminating the product. Upstream homogenisation removes the need to operate aseptically, but is thought to produce a less stable emulsion.

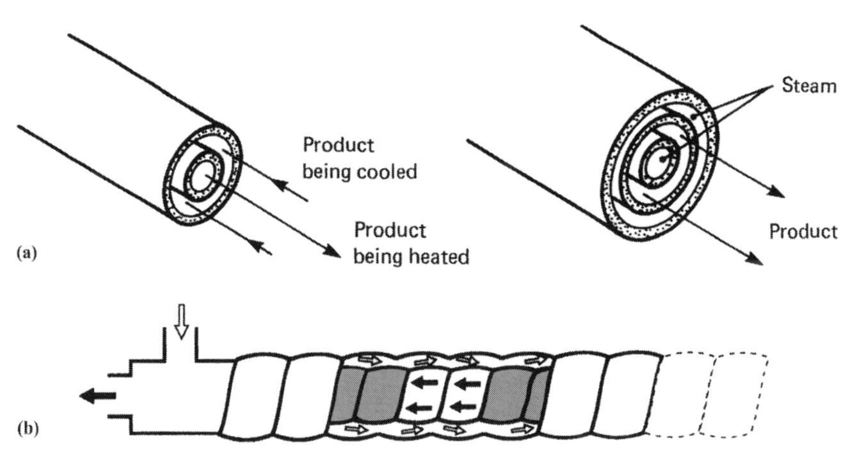

Fig. 2.4 Types of concentric tube heat exchangers: (a) plain wall, (b) corrugated spiral wound; from [7] with permission.

In direct processes, the product is preheated up to a temperature of 75 °C, often by regeneration, before being exposed to culinary steam to achieve a temperature of 140–145 °C. The steam should be free of chemical contaminants and saturated to avoid excessive dilution of the product (which is between about 10% and 15%). This heating process is very rapid. The product is held for a few seconds in a holding tube. Added water is removed by flash cooling, which involves a sudden reduction in pressure to bring the temperature of the product down to between 75 °C and 80 °C. This sudden fall in temperature is accompanied by the removal of some water vapour. There is a direct relationship between the fall in temperature and the amount of water removed. The final temperature and hence the amount of water removed is controlled by the pressure (vacuum) in the flash cooling chamber. This cooling process, like the heating, is very rapid. As well as removing the added water (as vapour), flash cooling removes other volatile components, which, in the case of UHT milk, gives rise to an improvement in the flavour. Direct processes employ a short sharp heating profile and result in less chemical damage, compared to an equivalent indirect process of similar holding time and temperature. They are also less susceptible to fouling and will give long processing times, but their regeneration efficiencies are usually below 50%. Direct processes are usually employed for UHT rather than pasteurisation processes.

There are two principle methods of contacting the steam and the food liquid. Steam can either be injected into the liquid (injection processes) or liquid can be injected into the steam (infusion). There is a school of thought that claims that infusion is less severe than injection since the product has less contact with hot surfaces. However, direct experimental evidence is scant. There is no doubt that direct processes (both injection and infusion) produce a less intense cooked flavour than any indirect process, although claims that direct UHT milk is indistinguishable from pasteurised milk are not always borne out. Successful operation depends upon maintaining a steady vacuum, as the flash cooling vessel operates at the boiling point of the liquid. If the pressure fluctuates, the boiling point also fluctuates; and this leads to boiling over if the pressure suddenly drops, for whatever reason. Thus, maintaining a steady vacuum is a major control point in the operation of these units. Note that some indirect UHT plants may incorporate a de-aeration unit, which operates under the same principle. The effects of heating and cooling profiles will be compared for UHT products in Section 2.4.3.2.

In continuous processes, there is a distribution of residence times. It is important to know whether the flow is streamline or turbulent, as this influences heat transfer rates and the distribution of residence times within the holding tube and also the rest of the plant. This can be established by evaluating the Reynolds number (Re), where:

$$\mathrm{Re} = \frac{vD\rho}{\mu} = \frac{4Q\rho}{\Pi\mu D} \tag{2.11}$$

where v is average fluid velocity (m s^{-1}), ρ is fluid density (kg m^{-3}), D is pipe diameter (m), μ is fluid viscosity (Pas) and Q is volumetric flow rate (m^3 s^{-1}).

Note that the average residence time (based upon the average velocity) can be determined from t_{av}=volume of tube/volumetric flow rate.

For viscous fluids, the flow in the holding tube is likely to be streamline, i.e. its Reynolds Number (Re) is less than 2000 and there is a wide distribution of residence times. For Newtonian fluids, the minimum residence time is half the average residence time. Turbulent flow (Re>4100) will result in a narrower distribution of residence times, with a minimum residence time of 0.83 times the average residence time. Fig. 2.5 illustrates residence time distributions for three situations, namely plug flow, streamline flow and turbulent flow. Plug flow is

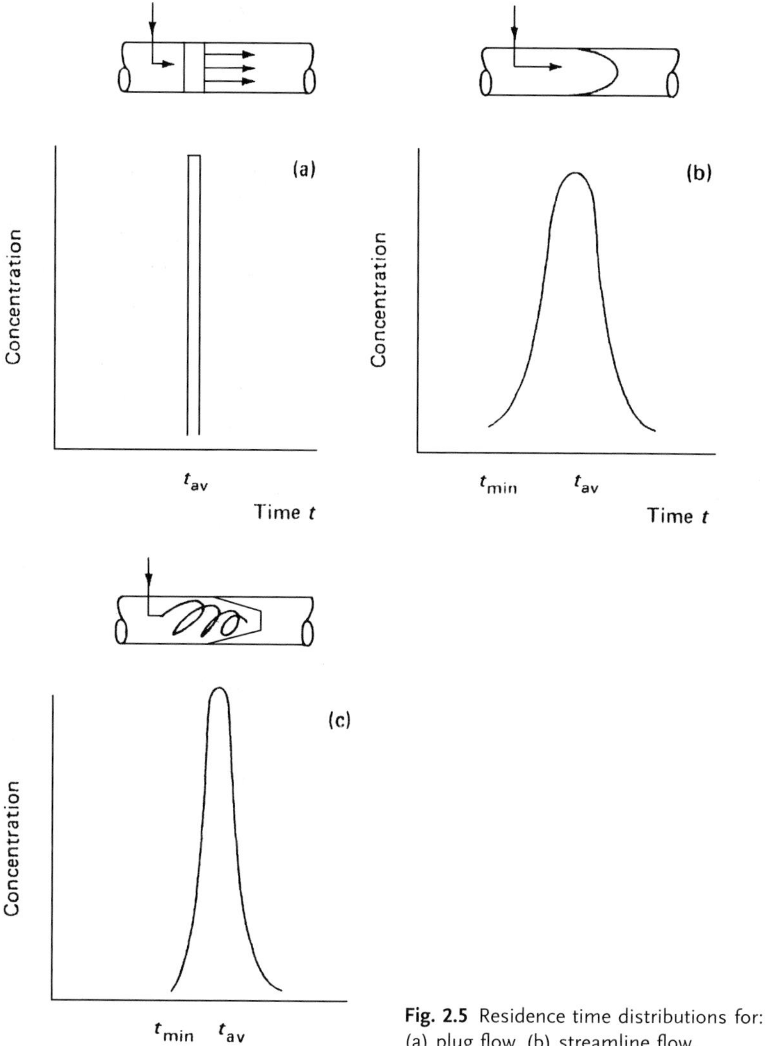

Fig. 2.5 Residence time distributions for: (a) plug flow, (b) streamline flow, (c) turbulent flow; from [1] with permission.

the ideal situation with no spread of residence times, but for both streamline and turbulent flow, the minimum residence time should be greater than the stipulated residence time, to avoid under-processing. Residence time distributions and their implications for UHT processing are discussed in more detail by Lewis and Heppell [6] and Burton [7].

2.4 Heat Processing Methods

The main types of heat treatment will now be covered, namely pasteurisation and sterilisation. A third process is known as thermisation.

2.4.1 Thermisation

Thermisation is a mild process which is designed to increase the keeping quality of raw milk. It is used mainly when it is known that it may not be possible to use raw milk immediately for conversion to other products, such as cheese or milk powder. The aim is to reduce psychrotropic bacteria, which can release heat resistant protease and lipase enzymes into the milk. These enzymes are not inactivated during pasteurisation and may give rise to off flavours if the milk is used for cheese or milk powders. Temperatures used are 58–68 °C for 15 s. Raw milk thus treated can be stored at a maximum of 8 °C for up to 3 days [8]. It is usually followed later by a more severe heat treatment. Thermised milk should show the presence of alkaline phosphatase, to distinguish it from pasteurised milk. To the author's knowledge, there is no equivalent process for other types of foods.

2.4.2 Pasteurisation

Pasteurisation is a mild heat treatment, which is used on a wide range of different types of food products. The two primary aims of pasteurisation are to remove pathogenic bacteria from foods, thereby preventing disease and to remove spoilage (souring) bacteria to improve its keeping quality. It largely stems from the discovery of Pasteur in 1857 that souring in milk could be delayed by heating milk to 122–142 °F (50.0–61.0 °C), although it was not firmly established that the causative agents of spoilage and disease were microorganisms until later in that century. However, even earlier than this, foods were being preserved in sealed containers by a process of sterilisation, so for some considerable time foods were being supplied for public consumption without an understanding of the mechanism of preservation involved. In fact, the first stage in the history of pasteurisation between 1857 and the end of the 19th century might well be called the medical stage, as the main history in heat-treating milk came chiefly from the medical profession interested in infant feeding. By 1895, it was recognised that a thoroughly satisfactory product can only be secured where a definite quantity of milk is heated for a definite length of time at a definite temperature. By 1927, North and Park [9] had established a wide range of time/temperature conditions for inactivating tubercle bacilli, ranging from 130 °F (54 °C) for 60 min up to 212 °F (100 °C) for 10 s. The effectiveness of heat treatments was determined by taking samples of milk, which had been heavily infected with tubercle organisms and then subjected to different time/temperature combinations, and inoculating the samples into guinea pigs and noting those conditions which did not kill the animals. The use of alkaline phosphatase as an indicator was first investigated in 1933 and is now still standard practice. Pasteurisation is now accepted as the simplest method to counter milk-borne pathogens and has now become commonplace, although there are still some devotees of raw milk. The IDF [10] definition of pasteurisation is as follows: "pasteurisation is a process applied to a product with the objective of minimising possible health hazards arising from pathogenic microorganisms associated with the product (milk) which is consistent with minimal chemical, physical and organoleptic changes in the product". This definition is also applicable to products other than milk, including, creams, icecream mix, eggs, fruit juices, fermented products, soups and other beverages.

Even in the early days of pasteurisation, milk only had a short shelf life, as domestic refrigeration was not widespread. This occurred in the 1940s and had an almost immediate impact on keeping quality. Initially there was also considerable resistance to the introduction of milk pasteurisation, not dissimilar to that now being encountered by irradiation [11]. Pasteurisation does not inactivate all microorganisms: those which survive pasteurisation are termed thermodurics and those which survive a harsher treatment (80–100 °C for 30 min) are termed spore formers. Traditionally it was a batch process – the Holder process – at 63 °C for 30 min, but this was followed by the introduction and acceptance of continuous HTST processes.

2.4.2.1 **HTST Pasteurisation**

HTST processes were investigated in the late 1920s. It was approved for milk in the USA in 1933 and approval for the process was granted in the UK in 1941. Continuous operations offer a number of advantages, such as faster heating and cooling rates, shorter holding times and regeneration, which saves both heating and cooling costs and contributes to the low processing costs incurred in thermal processing operations, compared to many novel techniques. Scales of operations on continuous heat exchangers range between 500 $l\,h^{-1}$ and 50 000 $l\,h^{-1}$, with experimental models down to 50 $l\,h^{-1}$. Continuous processing introduces some additional complications which have been well resolved, including flow control, flow diversion and distribution of residence times.

Schematics for the flow of fluid through the heat exchanger and the heat exchange sections are shown in Fig. 2.2 and Fig. 2.6. The fluid first enters the re-

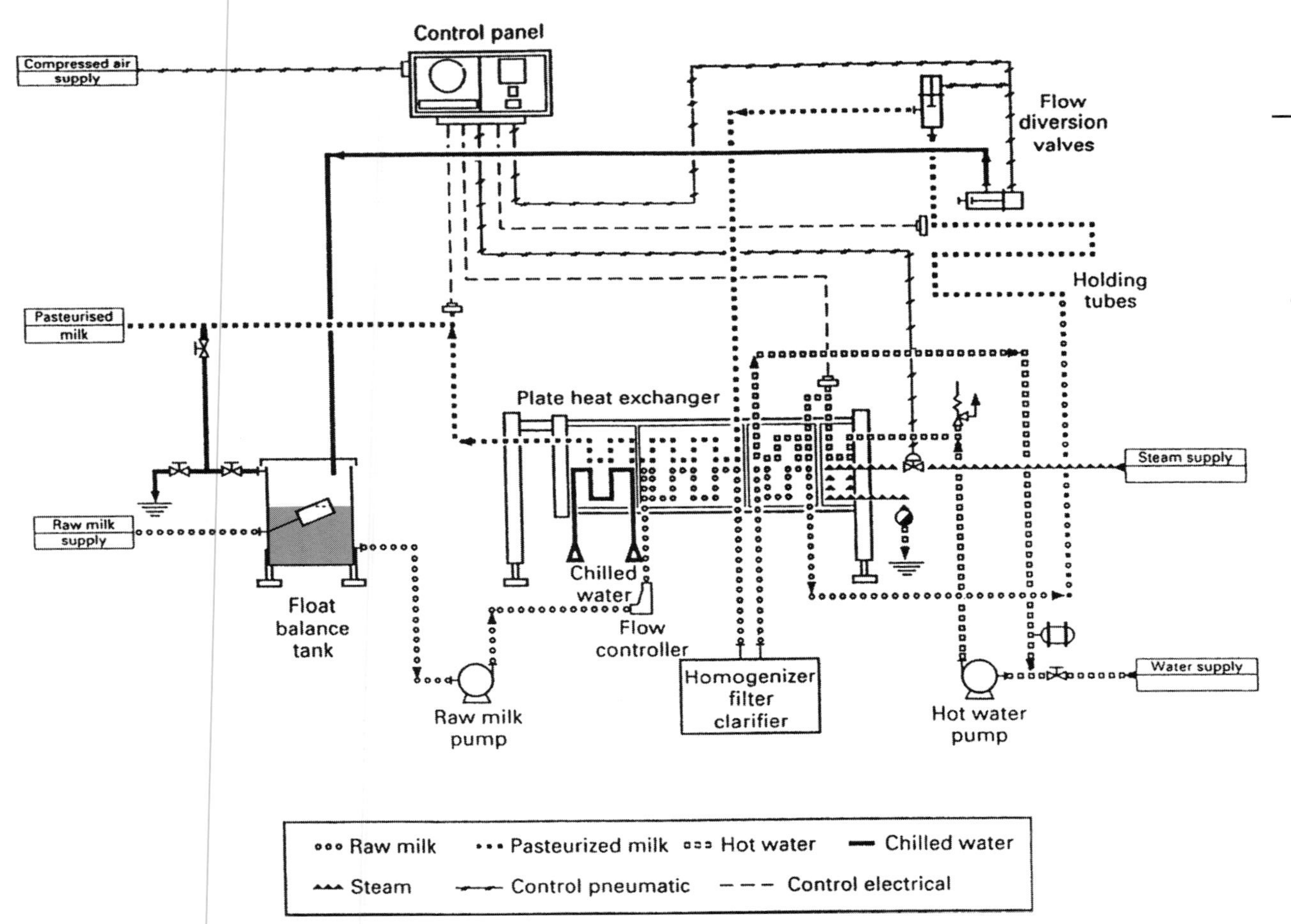

Fig. 2.6 Typical milk pasteurisation system; from [17] with permission.

generation section (A), where it is heated from θ_1 to θ_2 by the fluid leaving the holding tube. It then enters the main heating section, where it is heated to the pasteurisation temperature, θ_3. It then passes through the holding tube. The tube is constructed such that the minimum residence time exceeds the stipulated residence time and this can be determined experimentally at regular intervals. It then passes back into the regeneration section, where it is cooled to θ_4. This is followed by further cooling sections, employing mains water and chilled water. The mains water section is usually dispensed with where it may heat the product rather than cool it, for example at high RE or high mains water temperature. As RE increases, the size and capital cost of the heat exchanger increases; heating rate decreases, which may affect quality, but this is more noticeable in UHT sterilisation [6]. Heating profiles tend to become more linear at high RE efficiencies. Other features are a float-controlled balance tank, to ensure a constant head to the feed pump and a range of screens and filters to remove any suspended debris from the material.

In most pasteurisers, one pump is used. It is crucial that the flow rate remains constant, despite any disturbances in feed composition temperature, or changes in the system characteristics. The two most common options are a centrifugal pump with a flow controller or a positive displacement pump. If the product is to be homogenised, the homogeniser itself is a positive pump and is sized to control the flow rate. In the majority of pasteurisers, the final heating process is provided by a hot water set. Steam is used to maintain the temperature of the hot water at a constant value, somewhere between 2 °C and 10 °C higher than the required pasteurisation temperature. Electrical heating can be used, typically in locations where it would be costly or difficult to install a steam generator (boiler). The holding system is usually a straightforward holding tube, with a temperature probe at the beginning and a flow diversion valve at the end. The position of the temperature probe in the holding tube is one aspect for consideration. When positioned at the beginning, there is more time for the control system to respond to underprocessed fluid, i.e. the time it takes to pass through the tube, but it will not measure the minimum temperature obtained, as there will be a reduction in temperature due to heat loss as the fluid flows through the tube. This could be reduced by insulating the holding tube, but it is not generally considered to be a major problem in commercial pasteurisers.

Ideally temperature control should be within ±0.5 °C. Note that a temperature error of 1 °C, will lead to a reduction of about 25% in the process lethality (calculated for z=8 °C). From the holding tube, in normal operation, it goes back into regeneration, followed by final cooling. The pasteurised product may be packaged directly or stored in bulk tanks.

In principle, a safe product should be produced at the end of the holding tube. However, there may be an additional contribution to the total lethality of the process from the initial part of the cooling cycle. Thereafter, it is important to prevent recontamination, both from dirty pipes and from any recontamination with raw feed.

Should this occur, it is known as postpasteurisation contamination; and this can be a major determinant of keeping quality. Failures of pasteurisation processes have resulted from both causes. The most serious incidents have caused food poisoning outbreaks and have arisen where pasteurised milk has been recontaminated with raw milk, which must have unfortunately contained pathogens. Such contamination may arise for a number of reasons, all of which involve a small fraction of raw milk not going through the holding tube (see Fig. 2.7). One explanation lies with pinhole leaks or cracks in the plates, which may appear with time, due to corrosion. With plate heat exchangers, the integrity of the plates needs testing. This is most critical in the regeneration section; where there is a possibility of contamination from raw to treated, i.e. from high to low pressure.

An additional safeguard is the incorporation of an extra pump, to ensure that the pressure on the pasteurised side is higher than that on the raw side, but this further complicates the plant. In some countries, this requirement may be incorporated into the heat treatment regulations. Another approach is the use of double-walled plates, which also increases the heat transfer area by about 15–20% due to the air gap. These and other safety aspects have been discussed by Sorensen [13]. Pinholes in the heating and cooling sections could lead to product dilution or product contamination in the hot water or chilled water sections; and this may result in an unexpected microbiological hazards. It is important that plates are regularly pressure-tested and the product tested, by measuring the depression of its freezing point, to ensure that it has not been diluted. Similar problems may arise from leaking valves, either in recycle or detergent lines.

It is now easier to detect whether pasteurised milk has been recontaminated with raw milk, since the introduction of more sensitive instrumentation for detecting phosphatase activity. It is claimed that raw milk contamination as low as 0.01% can be detected. Miller Jones [14] documented a major pasteurisation failure in America, where over 16000 people were infected with salmonella and ten died. The cause was believed to be due to ppc, caused by a section of the plant which was not easy to drain and clean, hence leading to recontamination of already pasteurised milk.

Some further considerations of the engineering aspects are provided by Kessler [10], SDT [15] and Hasting [16]. Fouling is not considered to be such a problem compared to that found in UHT processing. However, with longer proces-

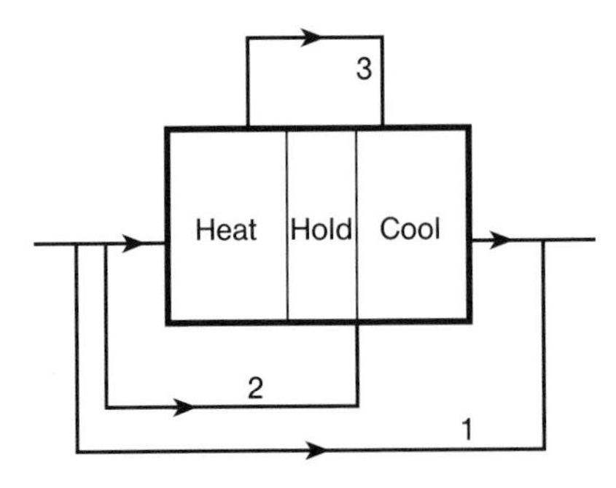

Fig. 2.7 Bypass routes in a commercial pasteuriser: 1, via cleaning routes; 2, via flow diversion route; 3, via regeneration section (e.g. pinhole leak in plates); from [12] with permission.

sing times and poorer quality raw materials, it may have to be accounted for and some products such as eggs may be more prone to fouling. One important aspect is a reduction in residence time due to fouling in the holding tube [16].

The trend is for HTST pasteurisation plants to run for much longer periods (16–20 h), before cleaning and shutdown. Again, monitoring phosphatase activity at regular intervals throughout is useful to ensure uniform pasteurisation throughout and for detecting more subtle changes in plant performance, which may lead to a better estimation of when cleaning is required. However, it has also been suggested that there is an increase in thermophilic bacteria due to an accumulation of such bacteria in the regenerative cooling section arising toward the end of such long processing runs.

Also very important are issues of fouling, cleaning and disinfecting, which are all paramount to the economics of the process.

2.4.2.2 Tunnel (Spray) Pasteurisers

Tunnel or spray pasteurisers are widely used in the beverage industry, for continuous heating and cooling of products in sealed containers. They are ideal for high volume throughput. Examples of such products are soft and carbonated drinks, juices, beers and sauces. Using this procedure, ppc should be very much reduced, the major cause being due to defective seams in the containers. There are three main stages in the tunnel, heating, holding and cooling; and in each stage water at the appropriate temperature is sprayed onto the container. Since heating rates are not as high as for plate or tubular heat exchangers, these processes are more suited to longer time/lower temperature processes. The total transit time may be about 1 h, with holding temperatures between 60 °C and 70 °C for about 20 min. Pearse [17] cites 60 °C for 20 min as a proven time/temperature profile. There is some scope for regeneration in these units.

2.4.3 Sterilisation

2.4.3.1 In-Container Processing

Sterilisation of foods by the application of heat can either be in sealed containers or by continuous flow techniques. Traditionally it is an in-container process, although there have been many developments in container technology since the process was first commercialised at the beginning of the 19th century. Whatever the process, the main concerns are with food safety and quality. The most heat-resistant pathogenic bacterium is *Clostridium botulinum*, which does not grow below pH 4.5. On this basis, the simplest classification is to categorise foods as either as acid foods (pH $<$ 4.5) or low-acid foods (pH $>$ 4.5). Note that a broader classification has been used for canning: low-acid (pH $>$ 5.0), medium-acid (pH 4.5–5.0), acid (pH 4.5–3.7), high-acid (pH $<$ 3.7). However, as mentioned earlier, the main concern is with foods at pH $>$ 4.5. For such foods, the minimum recommended process is to achieve 12*D* reductions for *C. botulinum*. This is known as the 'minimum botulinum cook'. This requires heating at 121 °C for 3 min, measured at the slowest heating point. The evidence for this producing a safe process for sterilised foods is provided by the millions of units of heat-preserved foods consumed worldwide each year, without any botulinum-related problems.

The temperature of 121.1 °C (250 °F) is taken as a reference temperature for sterilisation processes. This is used in conjunction with the z value for *C. botulinum*, which is taken as 10 °C, to construct standard lethality tables (see Table 2.2). Since lethalities are additive, it is possible to sum the lethalities for a process and determine the total integrated lethal effect, which is known as the *Fo* value.

In mathematical terms, F_0 is defined as the total integrated lethal effect, i.e.:

$$F_0 = \int L dt \tag{2.12}$$

It is expressed in minutes at a reference temperature of 121.1 °C, using the standard lethality tables derived for a z value of 10 °C.

For canned products, it is determined by placing a thermocouple at the slowest heating point in the can and measuring the temperature throughout the sterilisation process. This is known as the general method and is widely used to evaluate the microbiological severity of an in-container sterilisation process. Other methods are available based on knowing the heating and cooling rates of the products.

The F_0 values recommended for a wide range of foods are given in Table 2.3. It can be seen from these values that some foods need well in excess of the minimum botulinum cook, i.e. an *Fo* value of 3 (with values ranging between 4 and 18) to achieve commercial sterility. This is because there are some other bacterial spores which are more heat resistant than *C. botulinum*. The most heat-resistant of these is the thermophile *Bacillus stearothermophilus*, which has a decimal reduction time of about 4 min at 121 °C. Of recent interest is the mesophilic spore-forming bacterium *B. sporothermodurans* [18]. Some heat resistance values for other important spores are summarised by Lewis and Heppell [6] and more detailed compilations are given by Burton [7], Holdsworth [19] and Walstra et al. [20]. Such heat-resistant spores may cause food spoilage, either through the production of acid (souring) or the production of gas. Again, such spores will not grow below pH 4.5 and many of them are inhibited at higher pH values than this: e.g. *B. stearothermophilus*, which causes flat/sour spoilage, will not continue to grow below about pH 5.2.

The severity of the process (*Fo* value) selected for any food depends upon the nature of the spoilage flora associated with the food, the numbers likely to be present in that food and to a limited extent on the size of the container, since more organisms will go into a larger container. Such products are termed commercially sterile, the target spoilage rate being less than 1:10000. It should be remembered that canning and bottling operations are high-speed operations,

Table 2.2 Lethality values, using a reference temperature of 121.1 °C, $z = 10\,°C$. Lethality values are derived from $\log L = (T - \theta)/z)$, where L = number of minutes at reference temperature equivalent to 1 min at experimental temperature, T = experimental temperature, θ = reference temperature (121.1 °C); from [6] with permission.

Processing temperature		Lethality
(°C)	(°F)	(*L*)
110	230.0	0.078
112	237.2	0.195
114	237.2	0.195
116	240.8	0.309
118	244.4	0.490
120	248.0	0.776
121	249.8	0.977
121.1	250.0	1.000
122	251.6	1.230
123	253.4	1.549
124	255.2	1.950
125	257.0	2.455
126	258.8	3.090
127	260.6	3.890
128	262.4	4.898
129	264.2	6.166
130	266.0	7.762
131	267.8	9.772
132	269.6	12.30
133	271.4	15.49
134	273.2	19.50
135	275.0	24.55
136	276.8	30.90
137	278.6	38.90
138	280.4	48.98
139	282.2	61.66
140	284.0	77.62
141	285.8	97.72
142	287.6	123.0
143	289.4	154.9
144	291.2	195.0
145	293.0	245.5
146	294.8	309.2
147	296.6	389.0
148	298.4	489.8
149	300.2	616.6
150	302.0	776.24

Table 2.3 F_o values which have been successfully used for products on the UK market, from [23] with permission.

Product	Can size(s)	F_o values
Baby foods	Baby food	3–5
Beans in tomato sauce	All	4–6
Peas in brine	Up to A2	6
	A2 to A10	6–8
Carrots	All	3–4
Green beans in brine	Up to A2	4–6
	A2 to A10	6–8
Celery	A2	3–4
Mushrooms in brine	A 1	8–10
Mushrooms in butter	Up to A1	6–8
Meats in gravy	All	12–15
Sliced meat in gravy	Ovals	10
Meat pies	Tapered, flat	10
Sausages in fat	Up to 1 lb	4–6
Frankfurters in brine	Up to 16Z	3–4
Curries, meats and vegetables	Up to 16Z	8–12
Poultry and game, whole in brine	A2.5 to A10	15–18
Chicken fillets in jelly	Up to 16 oz	6–10
'Sterile' ham	1 lb, 2 lb	3–4
Herrings in tomato	Ovals	6–8
Meat soups	Up to 16Z	10
Tomato soup, not cream of	All	3
Cream soups	A1 to 16Z	4–5
	Up to A10	6–10
Milk puddings	Up to 16Z	4–10
Cream	4 oz to 6 oz	3–4

with the production of up to 50 000 containers every hour from one single product line. It is essential that each of those containers is treated exactly the same and that products treated on every subsequent day are also subjected to the same conditions.

The philosophy for ensuring safety and quality in thermal processing is to identify the operations where hazards may occur (critical control points) and devise procedures for controlling these operations to minimise the hazards (see Chapter 10). Of crucial importance is the control of all those factors which affect heat penetration into the product and minimise the number of heat-resistant spores entering the can prior to sealing. It is also important to ensure that the closure (seal) is airtight, thereby eliminating ppc.

Since it is not practicable to measure the temperatures in every can, the philosophy for quality assurance involves verifying that the conditions used throughout the canning process lead to the production of a product which is commercially sterile and ensuring that these conditions are reproduced on a daily basis.

Processing conditions such as temperature and time are critical control points. Others are raw material quality (especially counts of heat-resistant spores), and controlling all factors affecting heat penetration. These include filling temperature, size of headspace, ratio of solids to liquid, liquid viscosity, venting procedures and reducing ppc by seal integrity, cooling water chlorination and avoiding the handling of wet cans (after processing) and drying them quickly. It is also important to avoid large pressure differentials between the inside and outside of a container. Drying cans quickly after cooling and reducing manual han-

Fig. 2.8 The canning process; from [21] with permission.

dling of cans are very important for minimising ppc. In-container sterilisation involves the integration of a number of operations, all of which will contribute to the overall effectiveness of the process. These are summarised both in Fig. 2.8 and by Jackson and Shinn [21] and will be further discussed below.

Types of Container For in-container processes, there is now a wide range of containers available. The most common container is the can and its lid. There have been many developments since its inception as mild steel coated with tin plate: these include soldered or welded cans, two- and three-piece cans, tinfree steel and aluminium and there are many different lacquers to prevent chemical interactions between the metal and the foods (see Chapter 9). These modifications have been made to effect cost reductions and provide greater convenience. Cans are able to tolerate reasonable pressure differentials. Glass bottles and jars are common. Sterilised milk was traditionally produced in glass bottles but products in glass need to be heated and cooled more slowly to avoid breakage of the containers. Other containers used for in-container sterilisation include flexible pouches, plastic trays and bottles. All these materials have different wall thicknesses, different thermal conductivities and different surface area to volume ratios, all of which can influence heat transfer rates and thus the quality of the final heat-treated product.

Supply of Raw Materials It is essential that there is a supply of raw materials of the right quality and quantity: the requirements for contracts between the supplier and the food processor which protect both their interests. For fruit and vegetables, appropriate varieties should be selected for canning, as they must be able to withstand the heat treatment without undue softening or disintegration. Also important are their sensory characteristics, spore counts and chemical contaminants. Food to be processed should be transported quickly to the processing factory.

Preliminary Operations Preliminary operations depend upon the type of food and could include inspection, preparation, cleaning, peeling, destoning and size reduction. Where it is used, water quality is important and there will be considerable waste for disposal. Note that some preprepared vegetables may have been sulphited (see Chapter 1) and for sterilisation in metal cans these should be avoided, as sulphite may strip tinplate (see Chapter 9).

Blanching is an important operation, using hot water or steam (see Chapter 1). Different products have different time/temperature combinations. Blanching inactivates enzymes and removes intracellular air, thereby helping to minimise the internal pressure generated on heating. It also increases the density of food and softens cell tissue, which facilitates filling and it further cleans the product as well as removing vegetative organisms. It may lead to some thermal degradation of nutrients and some leaching losses for hot water blanching. An excellent detailed review on blanching is given by Selman [22].

Filling is an important operation, both for the product and any brine syrup or sauce that may accompany it. It is important to achieve the correct filled and drained weights and headspace. When using hot filling, the filling temperature must be controlled, as variations will lead to variations in the severity of the overall sterilisation process.

Sauces, brines and syrups may be used; and their composition may be covered by Codes of Practice. One of the main reasons for their use is to improve heat transfer.

Exhausting is another important process. It involves the removal of air prior to sealing, helping to prevent excessive pressure development in the container during heat treatment, which would increase the likelihood of damaging the seal. Four methods are available for exhausting: mechanical vacuum, thermal exhausting, hot filling and steam flow closing.

The next process is **sealing**, i.e. producing an airtight (hermetic) seal. For cans, this is achieved by a double rolling process (see Chapter 9) and the integrity of the seal is checked by visual inspection and by tearing down the seal and looking at the overlap and tightness. Can seamers can handle from 50 to 2000 cans per minute.

Containers are **sterilised** in retorts, which are large pressure vessels. Batch and continuous retorts are available and the heating medium is either steam, pressurised hot water, or steam/air mixtures. Some examples are shown in Figs. 2.9, 2.10. For steam, there is a fixed relationship between its pressure and temperature, given by steam tables, Lewis [1] and Holdsworth [19].

There should be an accurate system for recording temperatures and an indicating thermometer. A steam pressure gauge should be incorporated, as this will act indirectly as a second device for monitoring temperature. Discrepancies between temperature and pressure readings could suggest some air in the steam or that the instruments are incorrect [6]. Venting involves the removal of air from the retort and venting conditions need to be established for each individual retort. Every product will have its own unique processing time and temperature; and these would have been established to ensure that the appropriate F_0 value is achieved for that product. Ensuring $12D$ reductions for *C. botulinum* (safety) does ensure that a food is safe but more stringent conditions may be required for commercial sterility. The processing time starts when the temperature in the retort reaches the required processing temperature.

Cooling is a very important operation and containers should be cooled as quickly as possible down to a final temperature of 35–40 °C. As the product cools, the pressure inside the can falls and it is important to ensure that the pressure in the retort falls at about the same rate. This is achieved by using a combination of cooling water and compressed air to avoid a sudden fall in pressure caused by steam condensation. Water quality is important and it should be free of pathogenic bacteria. This can be assured by chlorination, but an excessive amount should be avoided as this may cause container corrosion. It is also important to avoid too much manual handling of wet cans to reduce the levels of ppc.

The containers are then labelled and stored. A small proportion may be incubated at elevated temperatures to observe for blown containers.

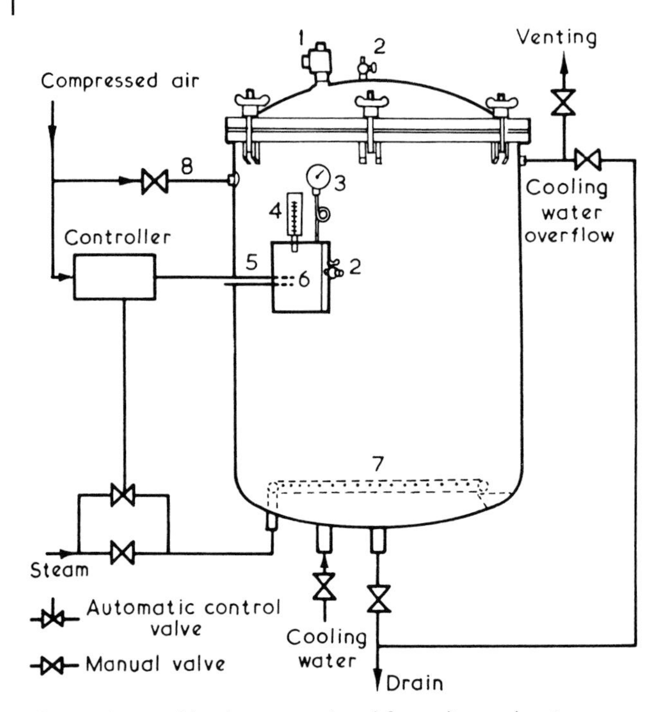

Fig. 2.9 A vertical batch retort equipped for cooling under air pressure: 1, safety valve; 2, valve to maintain a steam bleed from retort during processing; 3, pressure gauge; 4, thermometer; 5, sensing element for contoller; 6, thermometer box; 7, steam spreader; 8, air inlet for pressure cooling; from [23] with permission of the authors.

Quality Assurance Strict monitoring of all these processes is necessary to ensure that in-container sterilisation provides food which is commercially sterile. A summary of the requirements are as follows.

The target spoilage rate is <1 in 10^4 containers. There should be strict control of raw material quality, control of all factors affecting heat penetration and final product assessment (filled and drained weights, sensory characteristics and regular seal evaluation). Use should be made of hazard analysis critical control points (HACCP) and advice given in 'Food and drink – good manufacturing practice' [24].

One of the main problems with in-container processing is that there is considerable heat damage to the nutrients and changes in the sensory characteristics, which can be assessed by the cooking value [19], where a reference temperature of 100 °C is used with a z value of 20–40 °C (typically about 33 °C). Also summarised [19] are the z values for the sensory characteristics, which range between 25 °C and 47 °C. Further information on canned food technology is provided by the following excellent reference works: Stumbo [25], Hersom and Hul-

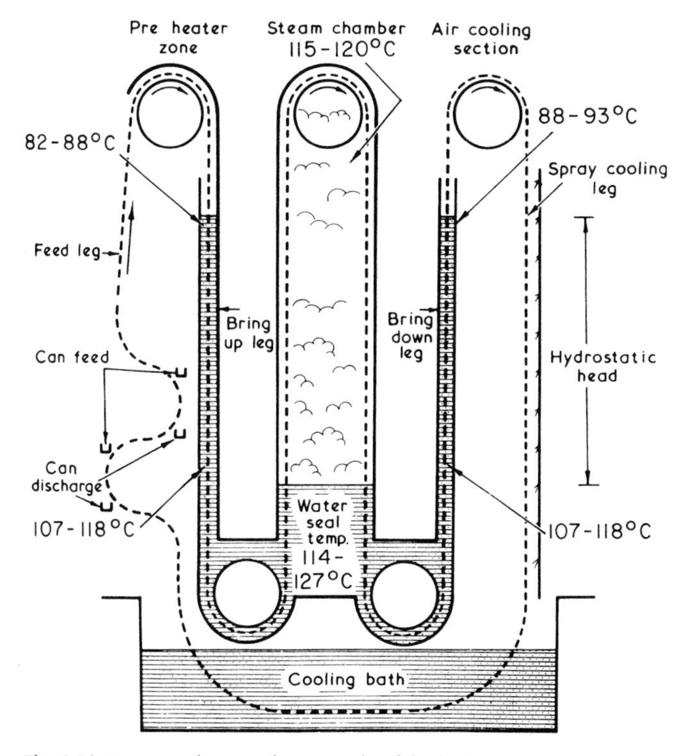

Fig. 2.10 Diagram showing the principle of the hydrostatic steriliser; from [23] with permission of the authors.

land [26], Jackson and Shinn [21], Rees and Bettison [27], Footitt and Lewis [28] and Holdsworth [19].

It is noteworthy that most fruit and other acidic products, e.g. pickles and fermented products, need a less harsh heat treatment; and processing conditions of 100 °C for 10–20 min are usually applied.

2.4.3.2 **UHT Processing**

More recently, continuous sterilisation processes have been introduced. Ultra-high temperature (UHT) or aseptic processing involves the production of a commercially sterile product by pumping the product through a heat exchanger. To ensure a long shelf life, the sterile product is packed into presterilised containers in a sterile environment (see Chapter 9). An airtight seal is formed, which prevents reinfection, in order to provide a shelf life of at least 3 months at ambient temperature. It has also been known for a long time that the use of higher temperatures for shorter times results in less chemical damage to important nutrients and functional ingredients within foods, thereby leading to an improvement in product quality [7].

Sterilisation of the product is achieved by rapid heating to temperatures about 140 °C and holding for several seconds, followed by rapid cooling. Ideally, heating and cooling should be as quick as possible. Indirect and direct heating methods are available (see Section 2.3). UHT products are in a good position to be able to improve the quality image of heat processed, ambient stable foods.

Safety and Spoilage Considerations From a safety standpoint, the primary objective is the production of commercially sterile products, with an extended shelf life. The main concern is inactivation of the most heat resistant pathogenic spore, namely *C. botulinum*. The safety criteria used for UHT processing should be based upon those well established for canned and bottled products. The minimum *Fo* value for any low acid food should be 3. Fruit juices and other acidic products require a less stringent process or may be heat-treated in their containers, in either tunnels or oventype equipment.

The time/temperature conditions required to achieve the minimum botulinum cook can be estimated at UHT temperatures. At a temperature of 141 °C, a time of 1.8 s would be required. There is experimental evidence to show that the data for botulinum can be extended up to 140 °C [29]. For UHT products, an approximate value of *Fo* can be obtained from the holding temperature (T, °C) and minimum residence time (t, s): Eq. (2.13). In practice, the real value is higher than this estimated value because of the lethality contributions from the end of the heating period and the beginning of the cooling period as well as some additional lethality from the distribution of residence times.

$$Fo = 10^{\frac{(T-121.1)}{10}} \cdot \frac{t}{60} \tag{2.13}$$

Therefore, the botulinum cook should be a minimum requirement for all low acid foods, even those where botulinum has not been a problem, e.g. for most dairy products. In the UK, there are statutory heat treatment regulations for some UHT products:

- milk, 135 °C for 1 s
- cream, 140 °C for 2 s
- milkbased products, 140 °C for 2 s
- icecream mix, 148.9 °C for 2 s.

In some cases, lower temperatures and longer times can be used, provided it can be demonstrated that the process renders the product free from viable microorganisms and their spores. If no guidelines are given, recommended *Fo* values for similar canned products would be an appropriate starting point (typically 6–10 for dairy products). Sufficient pressure must also be applied in order to achieve the required temperature. A working pressure in the holding tube in excess of 100 kPa over the saturated vapour pressure, corresponding to the UHT temperature, has been suggested.

UHT processes, like canned products, are also susceptible to ppc. This does not usually give rise to a public health problem, although contamination with pathogens cannot be ruled out. However, high levels of spoiled product do not improve its quality image, particularly if not detected before it is released for sale. Contamination may arise from the product being reinfected in the cooling section of the plant, or in the pipelines leading to the aseptic holding/buffer tank or the aseptic fillers. This is avoided by heating all points downstream of the holding tube at 130 °C for 30 min. The packaging material may have defects, or the seals may not be airtight, or the packaging may be damaged during subsequent handling All these could result in an increase in spoilage rate (see Section 2.5).

Process Characterisation: Safety and Quality Aspects As in other thermal processes, the requirements for safety and quality are in conflict, as a certain amount of chemical change will occur during adequate sterilisation of the food. Therefore, it is important to ask what is meant by quality and what is the scope for improving the quality. One aspect of quality, which has already been discussed, is reducing microbial spoilage. A second important aspect is minimising chemical damage and reducing nutrient loss. In this aspect, UHT processing offers some distinct advantages over in-container sterilisation. Chemical reactions are less temperature-sensitive, so the use of higher temperatures, combined with more rapid heating and cooling help to reduce the amount of chemical reaction. This has been well documented by Kessler [31, 32] and more recently by Browning et al. [33]. For example, reactions such as colour changes, hydroxymethyl furfural formation, thiamine loss, whey protein denaturation and lactulose formation will all be higher for in-container sterilisation compared to UHT processing.

There is also a choice of indirect heat exchangers available, such as plates, tubular and scraped surface, as well as direct steam injection or infusion plants, all of which will heat products at different rates and shear conditions. To better understand the quality of products produced from a UHT process, knowledge of the temperature/time profile for the product is required. Some examples of such profiles are shown for a number of different UHT process plants are shown in Fig. 2.11. There are considerable differences in the heating and cooling rates for indirect processes and between the direct and indirect processes due to steam injection and flash cooling. Because of these differences similar products processed on different plants may well be different in quality. A more detailed discussion is given by Burton [7].

Two other parameters introduced for UHT processing of dairy products, but which could be more widely used for other UHT products, are B^* and C^* values [31]. The reference temperature used (135 °C) is much closer to UHT processing temperatures, than that used for *Fo* (121 °C) or cooking value (100 °C) estimations.

B^* is a microbial parameter used to measure the total integrated lethal effect of a process. A process giving $B^*=1$ would be sufficient to produce nine decimal reductions of mesophilic spores and would be equivalent to 10.1 s at 135 °C.

C^* is a parameter to measure the amount of chemical damage taking place during the process. A process giving $C^*=1$ would cause 3% destruction of thiamine and would be equivalent to 30.5 s at 135 °C.

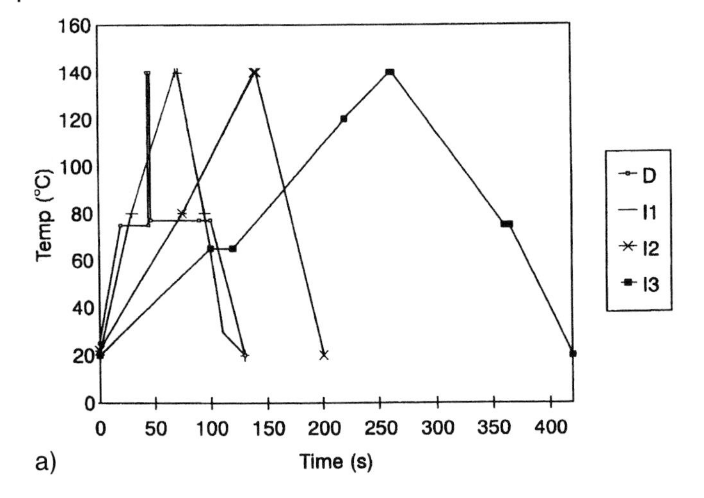

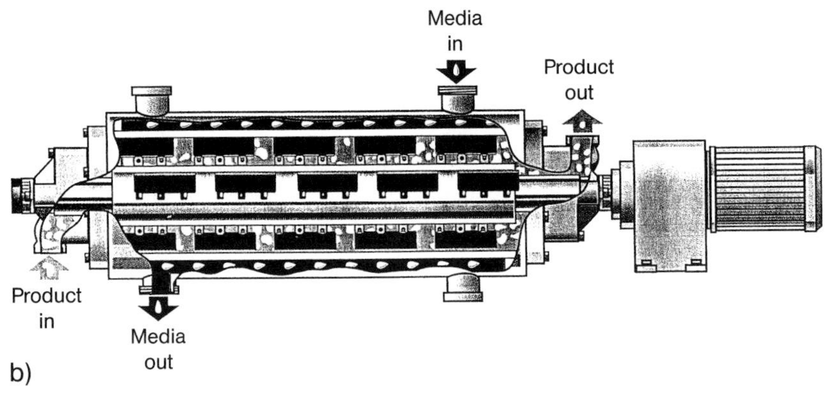

Fig. 2.11 (a) Temperature/time profiles for different UHT plants; from [6] with permission. (b) Cutaway view of a horizontal scraped surface heat exchanger; by courtesy of APV.

Again the criterion in most cases is to obtain a high B* and a low C* value.

Calculations of B^* and C^* based on the minimum holding time (t, s) and temperature (T, °C) are straightforward:

$$B^* = 10^{\frac{(T-135.0)}{10.5}} \cdot \frac{t}{10.1} \tag{2.14}$$

$$\text{and } C^* = 10^{\frac{(T-135.0)}{31.4}} \cdot \frac{t}{30.5} \tag{2.15}$$

Browning et al. [33] evaluated a standard temperature/holding time combination of 140 °C for 2 s for heating at cooling times from 1 s to 120 s.

It is noticeable that the amount of chemical change increased significantly as the heating and cooling times increased and that the longer heating and cooling times gave rise to quite severe microbiological processing conditions, i.e. high B^* and Fo values. At a heating period of about 8 s, the amount of chemical damage done during heating and cooling exceeds that in the holding tube. It is this considerable increase in chemical damage which is more noticeable in terms of decreasing the quality of the product. However this may be beneficial in those circumstances where a greater extent of chemical damage may be required; i.e. for inactivating enzymes or for heat-inactivation of natural toxic components, e.g. trypsin inhibitor in beans or softening of vegetable tissue (cooking). Differences in temperature/time processes arise due to use of different heat exchangers and the extent of energy savings by regeneration.

Chemical damage could be further reduced by using temperatures in excess of 145 °C. The best solution would be the direct process, with its accompanying rapid heating and cooling. Steam is mixed with the product, preheated to about 75 °C, by injection or infusion. The steam condenses and becomes an ingredient in the product. Steam utilisation is between about 10–15% (mass/mass). There are special requirements for the removal of impurities from the steam, such as water droplets, oil and rust. Heating is almost instantaneous. The condensed steam is removed by flash cooling, if required. It will also remove heat-induced volatile components, e.g. hydrogen sulphide and other low molecular weight compounds containing sulphur which are thought to be responsible for the initial cooked flavour in milk. There is also a reduction in the level of dissolved oxygen, which may improve the storage stability of the product. Advantages of this process are reduced chemical damage and a less intense cooked flavour for many products. One problem would be the very short holding times required and the control of such short holding times. In theory, it should be possible to obtain products with very high B^* and low C^* values, at holding times of about 1 s. For indirect processes, the use of higher temperatures may be limited by fouling considerations and it is important to ensure that the heat stability of the formulation is optimised. It may be worthwhile developing simple tests to assess heat stability. The alcohol stability test is one such test which is useful for milk products. Generally, direct systems give longer processing runs than indirect processes.

Raw Material Quality and Other Processing Conditions In terms of controlling the process, the following aspects will also merit some attention. Aspects of raw material quality relate to an understanding of the physical properties of the food, through to spore loadings and chemical composition. Of particular concern would be high levels of heat-resistant spores and enzymes in the raw materials, as these could lead to increased spoilage and stability problems during storage; dried products such as milk other dairy powders, cocoa, other functional powders, and spices are examples to be particularly careful with. Quality assurance programmes must ensure that such poor quality raw materials are avoided. The product formulation is also important; the nature of the principle ingredients; the levels of sugar, starch, salt as well as the pH of the mixture, particularly if there is appreciable amounts of protein. Some thought should be given to water quality, particularly the mineral content. Reproducibility in metering and weighing ingredients is also important, as is ensuring that powdered materials are properly dissolved or dispersed and that there are no clumps, which may protect heat resistant spores. Homogenisation conditions may be important. Is it necessary to homogenise and if so at what pressures? Should the homogeniser be positioned upstream or downstream of the holding tube? Will two-stage homogenisation offer any advantages? Homogenisation upstream offers the advantage of breaking down any particulate matter to facilitate heat transfer, as well as avoiding the need to keep the homogeniser sterile during processing. All of these aspects will influence both the safety and the quality of the products.

It is important to ensure that sterilisation and cleaning procedures have been properly accomplished. The plant should be sterilised downstream of the holding tube at 130 °C for 30 min. Cleaning should be adequate (detergent concentrations and temperatures) to remove accumulated deposits and the extent of fouling should be monitored, if possible. Steam barriers should be incorporated if some parts of the equipment are to be maintained sterile, whilst other parts are being cleaned.

All the important experimental parameters should be recorded. This will help ensure that any peculiarities can be properly investigated. Regular inspection and maintenance of equipment are important, particularly eliminating leaks. All staff involved with the process should be educated in order to understand the principles and be encouraged to be diligent and observant. With experience, further hazards will become apparent and methods for controlling them introduced. The overall aim should be to further reduce spoilage rates and to improve the quality of the product.

It is recognised that UHT processing is more complex than conventional thermal processing [24]. The philosophy of UHT processing should be based upon preventing and reducing microbial spoilage by understanding and controlling the process. One way of achieving this by using the principles of Hazard analysis critical control points (HACCP [34], also see Chapter 10). The hazards of the process are identified and procedures adapted to control them. An acceptable initial target spoilage rate of less than one in 10^4 should be aimed for. Such low spoilage rates require very large numbers of samples to be taken to verify that the process is being performed and controlled at the desired level. Initially a new process should be verified by 100% sampling. Once it is established that the process is under control, sampling frequency can be reduced and sampling plans can be designed to detect any spasmodic failures. More success will result from targeting high risk occurrences, such as start up, shut down and product changeovers. Holding time and temperature are perhaps the two most critical parameters. Recording thermometers should be checked and calibrated regularly, and accurate flow control is crucial (as for pasteurisation).

2.4.3.3 Special Problems with Viscous and Particulate Products

Continuous heat processing of viscous and particulate products provides some special problems. For viscous products it may be possible to use a tubular heat exchanger, but it is more probable that a scraped surface heat exchanger will be required. Fig. 2.11 b shows a typical scraped surface heat exchanger. This incorporates a scraper blade which continually sweeps product away from the heat transfer surface. These heat exchangers are mechanically more complex, with seals at the inlet and outlet ends of the scraper blade shaft. Overall heat transfer coefficients are low, the flow is streamline with a consequent increase in the spread of residence times and the fluid being heated may also be nonNewtonian. Heating and cooling rates are fairly slow and the process is generally more expensive to run, because of the higher capital and maintenance costs and less scope for regeneration. They may be used for pasteurisation or sterilisation processes. Two specialist uses of this equipment are for freezing ice-cream and for margarine and lowfat spread manufacture. In most cases, increasing the agitation speed improves heat transfer efficiency by increasing the overall heat transfer coefficient. However, when cooling products, some may become very viscous, e.g. due to product crystallisation; and higher agitation speeds may create additional frictional heat, making the product warmer rather than cooling. This is known as viscous dissipation. Scraped surface heat exchangers can also handle particulate systems, up to 25 mm diameter.

Problems arise with particulate systems because the solid phase conducts heat slowly so it will take longer to sterilise the particles compared to the liquid. Also, determination of the heat film coefficient is difficult due to the uncertainty in the relative velocity of the solid with respect to the liquid. There can also be problems determining the residence time distribution of the solid particles.

One special system is Ohmic heating, whereby particulate material is pumped through a nonconducting tube in which electrodes are placed. An electric current is passed through the material and the heating effect is caused by resistive heating and is determined by the electrical conductivity of the food as well as the applied voltage. If the solid particles are the same conductivity as the liquid, there should be no difference in heating rates between them. Unfortunately there is no rapid cooling process to accompany it and little scope for energy regeneration.

A second approach is to have a selective holding tube system, whereby larger particles are held up in the holding tube for a longer period of time. A third approach involves heating the solid and liquid phases separately and recombining them. This is the feature of the Jupiter heating system. These systems are discussed in more detail by Lewis and Heppell [6].

Dielectric heating, particularly microwaves, may also be used for pasteurisation and sterilisation of foods in continuous processes. For foods containing solid particles, there are some advantages in terms of being able to generate heat within the particles. Factors affecting the rate of heating include the field strength and frequency of the microwave energy and the dielectric loss factor of the food. However, dielectric heating processes are much more complex compared to well established HTST and UHT processes and the benefits that might result have to be weighed against the increased costs. One major drawback is that of non-uniform heating. This is a critical aspect in pasteurisation and sterilisation processes, where it is a requirement that all elements of the food reach a minimum temperature for a minimum time; for example 70 °C for 2 min to inactivate *E. coli* 0157. If part of the food only reaches 65 °C, even though other parts may well be above 70 °C, serious underprocessing will occur. Identifying with greater certainty where the slower heating points are will help to improve matters. Brody [35] provides further reading on microwave food pasteurisation.

2.5 Filling Procedures

For pasteurisation and extended shelf-life products, clean filling systems are used and for UHT products aseptic systems are required. There are a number of aseptic packaging systems available. They all involve putting a sterile product into a sterile container in an aseptic environment. Pack sizes range from individual portions (14 ml) and retail packs (125 ml to 1 l), through to bag in the box systems up to 1000 l. The sterilising agent is usually hydrogen peroxide (35% at about 75–80 °C), the contact time is short and the residual hydrogen peroxide is decomposed using hot air. The aim is to achieve a 4*D* process for spores. Superheated steam has been used for sterilisation of cans in the Dole process. Irradiation may be used for plastic bags (see Chapter 9).

Since aseptic packaging systems are complex, there is considerable scope for packaging faults to occur, which leads to spoiled products. Where faults occur, the spoilage microorganisms are random and could include those microorganisms which would be expected to be inactivated by UHT processing; and these often result in blown packages.

Packages should be inspected regularly to ensure that they are airtight, again focusing upon those more critical parts of the process, e.g. startup, shutdown, product changeovers and, for carton systems, reel splices and paper splices. Sterilisation procedures should be verified. The seal integrity of the package should be monitored as well as the overall microbial quality of packaging material itself. Care should be taken to minimise damage during subsequent handling. All these could result in an increase in spoilage rate.

2.6 Storage

UHT products are commonly stored at room (ambient) temperature and good quality products should be microbiologically stable. Nevertheless, chemical reactions and physical changes can take place which will change the quality of the product. Particularly relevant are oxidation reactions and Maillard browning, both of which can lead to a deterioration in the sensory characteristics of the product. These are discussed in more detail by Burton [7] and Lewis and Heppell [6].

References

1 Lewis, M. J. **1990**, *Physical Properties of Foods and Food Processing Systems*, Woodhead Publishing, Cambridge.

2 *EEC Facts and Figures* **1994**, Residary Milk Marketing Board, Thames Ditton.

3 Lewis, M. J. **1994**, Physical Properties of Dairy Products, in *Modern Dairy Technology, vol. 1*, ed. R. K. Robinson, Elsevier Applied Science, London, pp. 1–60.

4 Gould, G. W. **1989**, Heat Induced Injury and Inactivation, in *Mechanisms of Action of Food Preservation Procedures*, ed. G. W. Gould, Elsevier Applied Science, London, pp. 11–42.

5 Cronshaw, H. B. **1947**, *Dairy Information*, Dairy Industries, London.

6 Lewis, M. J., Heppell, N. **2000**, *Continuous Thermal Processing of Foods: Pasteurization and UHT Sterilization*, Aspen Publishers, Gaithersburg.

7 Burton, H. **1988**, *UHT Processing of Milk and Milk Products*, Elsevier Applied Science, London.

8 International Dairy Federation **1984**, The Thermization of Milk, *International Dairy Federation Bulletin* 182.

9 North and Park, **1927**, cited in Cronshaw [5], p. 616.

10 International Dairy Federation **1986**, Monograph on Pasteurised Milk, *International Dairy Federation Bulletin* 200.

11 Satin, M. **1996**, *Food Irradiation – A Guidebook*, Technomic Publishing Co., Lancaster.

12 Lewis, M. J. **1999**, Microbiological Issues Associated with Heat-Treated Milks, *International Journal of Dairy Technology*, 52, 121–125.

13 Sorensen, K. R. **1996**, APV Heat Exchanger; AS Introduces a New Security System for Pasteuriser Installations, in *IDF Heat Treatments and Alternative Methods*, IDF/FILS No. 9602, pp. 179–183.

14 Miller Jones, J. **1992**, *Food Safety*, Egan Press, St Paul.

15 Society of Dairy Technology **1983**, *Pasteurizing Plant Manual*, Society of Dairy Technology, Hungtingdon.

16 Hasting, A. P. M. **1992**, Practical Considerations in the Design, Operation and Control of Food Pasteurisation Processes, *Food Control*, 3, 27–32.

17 Pearse, M. A. **1993**, Pasteurization of Liquid Products, in *Encyclopeadia of Food Science, Food Technology and Nutrition*, Academic Press, London, pp. 3441–3450.

18 Hammer, P., Lembke, F., Suhren, G., Heeschen, W. **1996**, Characterisation of heat resistant mesophilic *Bacillus* species affecting the quality of UHT milk, in *IDF Heat Treatments and Alternative Methods*, IDF/FIL No 9602, pp. 9–25.

19 Holdsworth, S. D. **1997**, *Thermal Processing of Packaged Foods*, Blackie Academic and Professional, London.

20 Walstra, P., Guerts, T. J., Noomen, A., Jellema, A., van Boekel, M. A. J. S. **2001**, *Dairy Technology: Principles of Milk Properties and Processes*, Marcel Dekker, New York.

21 Jackson, J. M., Shinn, B. M., (eds.) **1979**, *Fundamentals of Food Canning Technology*, AVI, Westport.

22 Selman, J. D. **1987**, The blanching process, in *Developments in Food Preservation – 4*, ed. S. Thorne, Elsevier Applied Science, London, pp. 205–249.

23 Brennan, J. G., Butters, J. R., Cowell, N. D., Lilly, A. E. V. **1990**, *Food Engineering Operations*, 3rd edn, Applied Science Publishers, London.

24 Institute of Food Science and Technology **1991**, *Food and Drink – Good Manufacturing Practice: a Guide to its Responsible Management*, 3rd edn, Institute of Food Science and Technology, London.

25 Stumbo, C. R. **1965**, *Thermobacteriology in Food Processing*, Academic Press, New York.

26 Hersom, A. C., Hulland, E. D. **1980**, *Canned Foods*, Churchill Livingstone, Edinburgh.

27 Rees, J. A. G., Bettison, J. **1991**, *Processing and Packaging of Heat Preserved Foods*, Blackie, Glasgow.

28 Footitt, R. J., Lewis, A. S. (eds.) **1995**, *The Canning of Fish and Meat*, Blackie Academic and Professional, Glasgow.

29 Gaze, J. E., Brown, K. L. **1988**, The Heat Resistance of Spores of *Clostridium botulinum* 213B over the Temperature Range 120–140°C, *International Journal of Food Science and Technology*, 23, 373–378.

30 Lewis M. J. **1999**, Ultra-High Temperature Treatments, in *Encyclopedia of Food Microbiology*, ed. R. K. Robinson, C. A. Batt, P. D. Patel, Academic Press, New York, pp. 1023–1030.

31 Kessler, H. G. **1981**, *Food Engineering and Dairy Technology*, Verlag A. Kessler, Freising.

32 Kessler, H. G. **1989**, Effect of Thermal Processing of Milk, in *Developments in Food Preservation – 5*, ed. S. Thorne, Elsevier Applied Science, London, pp. 91–130.

33 Browning, E., Lewis, M. J., MacDougall, D. **2001**, Predicting Safety and Quality Parameters for UHT-Processed Milks, *International Journal of Dairy Technology*, 54, 111–120.

34 International Commission on Microbiological Specifications for Foods **1988**, *Application of the Hazard Analysis Critical Control Point (HACCP) System to Ensure Microbiological Safety (Micro-Organisms in Foods vol. 4)*, Blackwell Scientific Publications, Oxford.

35 Brody, A. L. **1992**, Microwave Food Pasteurisation, *Food Technology International Europe*, 67–72.

3
Evaporation and Dehydration

James G. Brennan

3.1
Evaporation (Concentration, Condensing)

3.1.1
General Principles

Most food liquids have relatively low solids contents. For example, whole milk contains approximately 12.5% total solids, fruit juice 12%, sugar solution after extraction from sugar beet 15%, solution of coffee solutes after extraction from ground roasted beans 25%. For various reasons, some of which are discussed below in Section 3.1.5, it may be desirable to increase the solids content of such liquids. The most common method used to achieve this is to "boil off" or evaporate some of the water by the application of heat. Other methods used to concentrate food liquids are freeze concentration and membrane separation. If evaporation is carried out in open pans at atmospheric pressure, the initial temperature at which the solution boils will be some degrees above 100 °C, depending on the solids content of the liquid. As the solution becomes more concentrated, the evaporation temperature will rise. It could take from several minutes to a few hours to attain the solids content required. Exposure of the food liquid to these high temperatures for these lengths of time is likely to cause changes in the colour and flavour of the liquid. In some cases such changes may be acceptable, or even desirable, for example when concentrating sugar solutions for toffee manufacture or when reducing gravies. However, in the case of heat-sensitive liquids such as milk or fruit juice, such changes are undesirable. To reduce such heat damage, the pressure above the liquid in the evaporator may be reduced below atmospheric by means of condensers, vacuum pumps or steam ejectors (see Section 3.1.2.7). Since a liquid boils when the vapour pressure it exerts equals the external pressure above it, reducing the pressure in the evaporator lowers the temperature at which the liquid will evaporate. Typically, the pressure in the evaporator will be in the range 7.5–85.0 kPa absolute, corresponding to evaporation temperatures in the range 40–95 °C. The use of lower pressures is usually uneconomic. This is known as *vacuum*

Food Processing Handbook. Edited by James G. Brennan

ISBN: 3-527-30719-2

evaporation. The relatively low evaporation temperatures which prevail in vacuum evaporation mean that reasonable temperature differences can be maintained between the heating medium, saturated steam, and the boiling liquid, while using relatively low steam pressures. This limits undesirable changes in the colour and flavour of the product. For aqueous liquids, the relationship between pressure and evaporation temperature may be obtained from thermodynamic tables and psychrometric charts. Relationships are available for estimating the evaporating temperatures of nonaqueous liquids at different pressures [1, 2].

Another factor which affects the evaporation temperature, is known as the boiling point rise (BPR) or boiling point elevation (BPE). The boiling point of a solution is higher than that of the pure solvent at the same pressure. The higher the soluble solids content of the solution, the higher its boiling point. Thus, the initial evaporation temperature will be some degrees above that corresponding to pressure in the evaporator, depending on the soluble solids content of the feed. However, as evaporation proceeds and the concentration of the soluble solids increases, the evaporation temperature rises. This is likely to result in an increase in changes in the colour and flavour of the product. If the temperature of the steam used to heat the liquid is kept constant, the temperature difference between it and the evaporating liquid decreases. This reduces the rate of heat transfer and hence the rate of evaporation. To maintain a constant rate of evaporation, the steam pressure may be increased. However, this is likely to result in a further decrease in the quality of the product. Data on the BPR in simple solutions and some more complex foods is available in the literature in the form of plots and tables. Relationships for estimating the BPR with increase in solids concentrations have also been proposed [1, 3]. BPR may range from $<1\,^{\circ}\mathrm{C}$ to $10\,^{\circ}\mathrm{C}$ in food liquids. For example, the BPR of a sugar solution containing 50% solids is about $7\,^{\circ}\mathrm{C}$.

In some long tube evaporators (see Section 3.1.2.3), the evaporation temperature increases with increase in the depth of the liquid in the tubes in the evaporator, due to hydrostatic pressure. This can lead to overheating of the liquid and heat damage. This factor has to be taken into account in the design of evaporators and in selecting the operating conditions, in particular, the pressure of the steam in the heating jacket.

The viscosity of most liquids increases as the solids content increases during evaporation. This can lead to a reduction in the circulation rates and hence the rates of heat transfer in the heating section of the evaporator. This can influence the selection of the type of evaporator for a particular liquid food. Falling film evaporators (see Section 3.1.2.3) are often used for moderately viscous liquids. For very viscous liquids, agitated thin film evaporators are used (see Section 3.1.2.5). Thixotropic (or time-dependent) liquids, such as concentrated tomato juice, can pose special problems during evaporation. The increase in viscosity can also limit the maximum concentration attainable in a given liquid.

Fouling of the heat transfer surfaces may occur in evaporators. This can result in a decrease in the rate of heat transfer and hence the rate of evaporation. It can also necessitate expensive cleaning procedures. Fouling must be taken into account in the design of evaporators and in the selection of the type of evaporator for a given duty. Evaporators that feature forced circulation of the liquid or agitated thin films are used for liquids that are susceptible to fouling.

Some liquids are prone to foaming when vigorously boiling in an evaporator. Liquids which contain surface active foaming agents, such as the proteins in skimmed milk, are liable to foam. This can reduce rates of heat transfer and hence rates of evaporation. It may also result in excessive loss of product by entrainment in the vapour leaving the heating section. This in turn can cause contamination of the cooling water to spray condensers and lead to problems in the disposal of that effluent. In some cases, antifoaming agents may be added to the feed to reduce foaming. Care must be taken not to infringe any regulations by the addition of such aids.

Volatile aroma and flavour compounds may be lost during vacuum evaporation, resulting in a reduction in the organoleptic quality of products such as fruit juices or coffee extract. In the case of fruit juices, this loss may be partly offset by adding some of the original juice, known as "cut back juice", to the concentrate. Alternatively, the volatiles may be stripped from the vapour, concentrated and added to the concentrated liquid (see Section 3.1.4.3) [1, 5].

3.1.2 Equipment Used in Vacuum Evaporation

A single-effect vacuum evaporator has the following components.

A heat exchanger, known as a *calandria,* by means of which the necessary sensible and latent heat is supplied to the feed to bring about the evaporation of some of the liquid. Saturated steam is the usual heating medium but hot water and other thermal fluids are sometimes used. Tubular and plate exchangers of various designs are widely used. Other, more sophisticated designs are available, including agitated thin film models, expanding flow chambers and centrifugal exchangers.

A device to separate the vapour from the concentrated liquid phase. In vacuum evaporators, mechanical devices such as chambers fitted with baffles or meshes and cyclone separators are used to reduce entrainment losses.

A condenser to convert the vapour back to a liquid and a pump, steam ejector or barometric leg to remove the condensate, thus creating and maintaining the partial vacuum in the system.

Most evaporators are constructed in stainless steel except where there are extreme corrosion problems.

The following types of evaporators are used in the food industry.

3.1.2.1 Vacuum Pans

A hemispherical pan equipped with a steam jacket and sealed lid, connected to a vacuum system, is the simplest type of vacuum evaporator in use in industry. The heat transfer area per unit volume is small and so the time required to reach the desired solids content can run into hours. Heating occurs by natural convection. However, an impeller stirrer may be introduced to increase circulation and reduce fouling. Small pans have a more favourable heat transfer area to volume ratio. They are useful for frequent changes of product and for low or variable throughputs. They are used in jam manufacture, the preparation of sauces, soups and gravies and in tomato pulp concentration.

3.1.2.2 Short Tube Vacuum Evaporators

This type of evaporator consists of a calandria made up of a bundle of short vertical tubes surrounded by a steam jacket, located near the bottom of a large vessel (see Fig. 3.1). The tubes are typically 25–75 mm in diameter and 0.5–2.0 m long. The liquid being concentrated normally covers the calandria. Steam condensing on the outside of the tubes heats the liquid causing it to rise by natural convection. Some of the water evaporates and flows to the condenser. The liquid circulates down through the larger, cooler tube in the centre of the bundle, known as the *downcomer*. This type of evaporator is suitable for low to moderate viscosity liquids, which are not very heat-sensitive. With viscous liquids heat transfer rates are low, hence residence times are relatively long and there is a high risk of fouling. Sugar solutions, glucose and malt extract are examples of products concentrated in this type of evaporator. It can also be used for crystallisation operations. For this application an impeller may be located in the downcomer to keep the crystals in suspension.

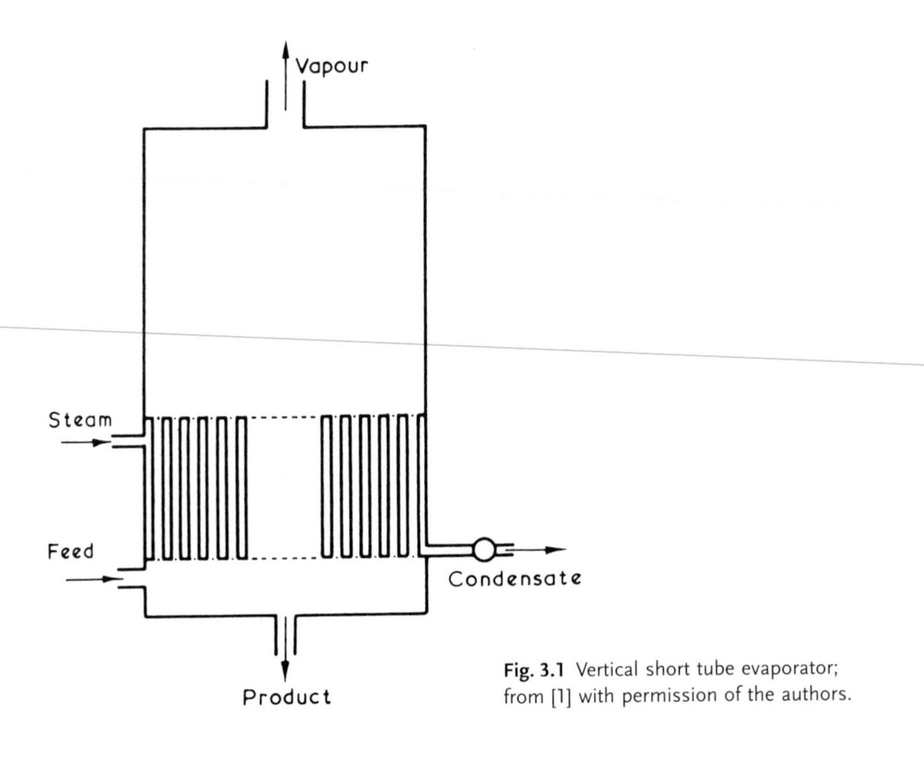

Fig. 3.1 Vertical short tube evaporator; from [1] with permission of the authors.

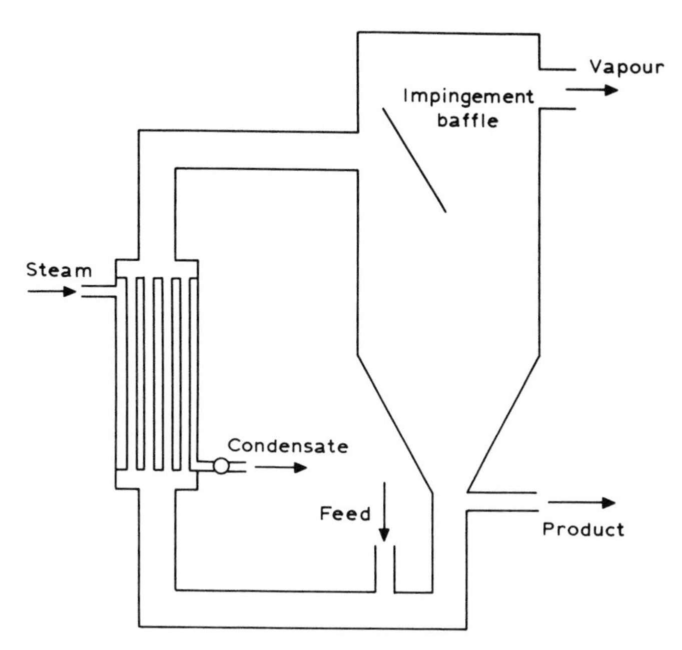

Fig. 3.2 Natural circulation evaporator; from [1] with permission of the authors.

In another design of a short tube evaporator the calandria is external to separator chamber and may be at an angle to the vertical (see Fig. 3.2). The liquid circulates by natural convection within the heat exchanger and also through the separation chamber. The liquid enters the separation chamber tangentially. A swirling flow pattern develops in the chamber, generating centrifugal force, which assists in separating the vapour from the liquid. The vigorous circulation of the liquid results in relatively high rates of heat transfer. It also helps to break up any foam which forms. The tubes are easily accessible for cleaning. Such evaporators are also used for concentrating sugar solutions, glucose and malt and more heat-sensitive liquids such as milk, fruit juices and meat extracts.

A pump may be introduced to assist in circulating more viscous liquids. This is known as forced circulation. The choice of pump will depend on the viscosity of the liquid. Centrifugal pumps are used for moderately viscous materials, while positive displacement pumps are used for very viscous liquids.

3.1.2.3 **Long Tube Evaporators**

These consist of bundles of long tubes, 3–15 m long and 25–50 mm in diameter, contained within a vertical shell into which steam is introduced. The steam condensing on the outside of the tubes provides the heat of evaporation. There are three patterns of flow of the liquid through such evaporators.

In the *climbing film evaporator* the preheated feed is introduced into the bottom of the tubes. Evaporation commences near the base of the tubes. As the vapour expands, ideally, it causes a thin film of liquid to rise rapidly up the inner walls of the tubes around a central core of vapour. In practice, slugs of liquid and vapour bubbles also rise up the tubes. The liquid becomes more concentrated as it rises. At the top, the liquid-vapour mixture enters a cyclone separator. The vapour is drawn off to a condenser and pump, or into the heating jacket of another calandria, in a multiple-effect system (see Section 3.1.3). The concentrated liquid may be removed as product, recycled through the calandria or fed to another calandria, in a multiple-effect system. The residence time of the liquid in the tubes is relatively short. High rates of heat transfer are attainable in this type of evaporator, provided there are relatively large temperature differences between the heating medium and the liquid being concentrated. However, when these temperature differences are low, the heat transfer rates are also low. This type of evaporator is suitable for low viscosity, heat-sensitive liquids such as milk and fruit juices.

In the *falling film evaporator* the preheated feed is introduced at the top of the tube bundle and distributed to the tubes so that a thin film of the liquid flows down the inner surface of each tube, evaporating as it descends. Uniform distribution of the liquid so that the inner surfaces of the tubes are uniformly wetted is vital to the successful operation of this type of evaporator [6]. From the bottom of the tubes, the liquid-vapour mixture passes into a centrifugal separator and from there the liquid and vapour streams are directed in the same manner as in the climbing film evaporator. High rates of liquid flow down the tubes are attained by a combination of gravity and the expansion of the vapour, resulting in short residence times. These evaporators are capable of operating with small temperature differences between the heating medium and the liquid and can cope with viscous materials. Consequently, they are suitable for concentrating heat-sensitive foods and are very widely used in the dairy and fruit juice processing sections of the food industry today.

A *climbing-falling film evaporator* is also available. The feed is first partially concentrated in a climbing film section and then finished off in a falling film section. High rates of evaporation are attainable in this type of plant.

3.1.2.4 **Plate Evaporators**

In these evaporators the calandria is a plate heat exchanger, similar to that used in pasteurising and sterilising liquids (see Chapter 2). The liquid is pumped through the heat exchanger, passing on one side of an assembly of plates, while steam passes on the other side. The spacing between plates is greater than that in pasteurisers to accommodate the vapour produced during evaporation. The liquid usually follows a climbing-falling film flow pattern. However, designs featuring only a falling film flow pattern are also available. The mixture of liquid and vapour leaving the calandria passes into a cyclone separator. The vapour from the separator goes to a condenser or into the heating jacket of the next stage, in a multiple-effect system. The concentrate is collected as product or goes to another stage. The advantages of plate evaporators include: high liquid velocities leading to high rates of heat transfer, short residence times and resistance to fouling. They are compact and easily dismantled for inspection and maintenance. However, they have relatively high capital costs and low throughputs. They can be used for moderately viscous, heat-sensitive liquids such as milk, fruit juices, yeast and meat extracts.

3.1.2.5 **Agitated Thin Film Evaporators**

For very viscous materials and/or materials which tend to foul, heat transfer may be increased by continually wiping the boundary layer at the heat transfer surface. An agitated thin film evaporator consists of a steam jacketed shell equipped with a centrally located, rotating shaft carrying blades which wipe the inner surface of the shell. The shell may be cylindrical and mounted either vertically or horizontally. Horizontal shells may be cone-shaped, narrowing in the direction of flow of the liquid. There may be a fixed clearance of 0.5–2.0 mm between the edge of the blades and the inner surface of the shell. Alternatively, the blades may float and swing out towards the heat transfer surface as the shaft rotates, creating a film of liquid with a thickness as little as 0.25 mm. Most of the evaporation takes place in the film that forms behind the rotating blades. Relatively high rates of heat transfer are attained and fouling and foaming are inhibited. However, these evaporators have relatively high capital costs and low throughputs. They are used as single-effect units, with relatively large temperature differences between the steam and the liquid being evaporated. They are often used as "finishers" when high solids concentrations are required. Applications include tomato paste, gelatin solutions, milk products, coffee extract and sugar products.

3.1.2.6 **Centrifugal Evaporators**

In this type of evaporator a rotating stack of cones is housed in a stationary shell. The cones have steam on alternate sides to supply the heat. The liquid is fed to the undersides of the cones. It forms a thin film, which moves quickly across the surface of the cones, under the influence of centrifugal force, and rapid evaporation occurs. Very high rates of heat transfer, and so very short residence times, are attained. The vapour and concentrate are separated in the shell surrounding the cones. This type of evaporator is suitable for heat-sensitive and viscous materials. High capital costs and low throughputs are the main limitations of conical evaporators. Applications include fruit and vegetable juices and purees and extracts of coffee and tea.

3.1.2.7 Ancillary Equipment

Vapour-Concentrate Separators The mixture of vapour and liquid concentrate leaving the calandria needs to be separated and entrainment of droplets of the liquid in the vapour minimised. Entrained droplets represent a loss of product. They can also reduce the energy value of the vapour which would make it less effective in multiple-effect systems or when vapour recompression is being used (see Sections 3.1.3., 3.1.4). Separation may be brought about by gravity. If sufficient headspace is provided above the calandria, as in Fig. 3.1, the droplets may fall back into the liquid. Alternatively, this may occur in a second vessel, known as a flash chamber. Baffles or wire meshes may be located near the vapour outlet. Droplets of liquid impinge on these, coalesce into larger droplets and drain back into the liquid under gravity. The mixture of vapour and liquid may be directed tangentially, at high velocity, either by natural or forced circulation, into a cyclone separator. Centrifugal force is developed and the more dense liquid droplets impinge on the inner wall of the chamber, lose their kinetic energy and drain down into the liquid. This type of separator is used in most long tube, plate and agitated thin film evaporators.

Condensers and Pumps The water vapour leaving the calandria contains some noncondensable gases, which were in the feed or leaked into the system. The water vapour is converted back to a liquid in a condenser. Condensers may be of the indirect type, in which the cooling water does not mix with the water vapour. These are usually tubular heat exchangers. They are relatively expensive. Indirect condensers are used when volatiles are being recovered from the vapour or to facilitate effluent disposal. Direct condensers, known as jet or spray condensers, are more widely used. In these a spray of water is mixed with the water vapour, to condense it. A condensate pump or barometric leg is used to remove the condensed vapour. The noncondensable gases are removed by positive displacement vacuum pumps or steam jet ejectors [1, 3].

3.1.3 Multiple-Effect Evaporation (MEE)

In a single-effect evaporator it takes 1.1–1.3 kg of steam to evaporate 1.0 kg of water. This is known as the *specific steam consumption* of an evaporator. The specific steam consumption of driers used for liquid foods is much higher than this. A spray drier has a specific steam consumption of 3.0–3.5 kg of steam to evaporate 1.0 kg of water. That of a drum drier is slightly lower. Thus, it is common practice to concentrate liquid foods by vacuum evaporation before drying them in such equipment. It is also common practice to preheat the liquid to its evaporation temperature before feeding it to the evaporator. This improves the specific steam consumption. The vapour leaving a single-effect evaporator contains useful heat. This vapour may be put to other uses, for example, to heat water for cleaning. However, the most widely used method of recovering heat

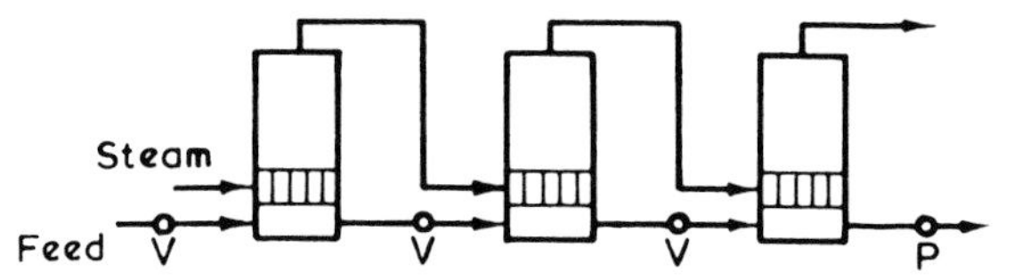

Fig. 3.3 The principle of multiple-effect evaporation, with forward feeding; adapted from [1] with permission of the authors.

from the vapour leaving an evaporator is multiple-effect evaporation, the principle of which is shown in Fig. 3.3.

The vapour leaving the separator of the first evaporator, effect 1, enters the steam jacket of effect 2, where it heats the liquid in that effect, causing further evaporation. The vapour from effect 2 is used to heat the liquid in effect 3 and so on. The vapour from the last effect goes to a condenser. The liquid also travels from one effect to the next, becoming more concentrated as it does so. This arrangement is only possible if the evaporation temperature of the liquid in effect 2 is lower than the temperature of the vapour leaving effect 1. This is achieved by operating effect 2 at a lower pressure than effect 1. This principle can be extended to a number of effects. The pressure in effect 1 may be atmospheric or slightly below. The pressure in the following effects decreases with the number of effects. The specific steam consumption in a double effect evaporator is in the range 0.55–0.70 kg of steam per 1.0 kg of water evaporated, while in a triple effect it is 0.37–0.45 kg. However, the capital cost of a multiple-effect system increases with the number of effects. The size of each effect is normally equivalent to that of a single-effect evaporator with the same capacity, working under similar operating conditions. Thus a three-effect unit will cost approximately three times that of a single effect. It must be noted that MEE does not increase the throughput above that of a single effect. Its purpose is to reduce the steam consumption. Three to five effects are most commonly used. Up to seven effects are in use in some large milk and fruit juice processing plants.

The flow pattern in Fig. 3.3 is known as forward feeding and is the most commonly used arrangement. Other flow patterns, including backward and mixed feeding, are also used. Each has its own advantages and limitations [1, 5].

3.1.4 Vapour Recompression

This procedure involves compressing some or all of the vapour from the separator of an evaporator to a pressure that enables it to be used as a heating medium. The compressed vapour is returned to the jacket of the calandria. This reduces the amount of fresh steam required and improves the specific steam consumption of the evaporator. The vapour may be compressed thermally or mechanically.

In *thermal vapour recompression* (TVR) the vapour from the separator is divided into two streams. One stream goes to the condenser or to the next stage of a multiple-effect system. The other enters a steam jet compressor fed with fresh high pressure steam. As it passes through the jet of the compressor the pressure of the fresh steam falls and it mixes with the vapour from the evaporator. The vapour mix then passes through a second converging-diverging nozzle where the pressure increases. This high pressure mixture then enters the jacket of the calandria.

In *mechanical vapour recompression* (MVR) all the vapour from the separator is compressed in a mechanical compressor which can be driven by electricity, a gas turbine or a steam turbine. The compressed vapour is then returned to the jacket of the calandria. Both methods are used in industry. MVR is best suited to large capacity duties. Both TVR and MVR are used in conjunction multiple-effect evaporators. Vapour recompression may be applied to one or more of the effects. A specific steam consumption of less than 0.10 kg steam per 1.0 kg of water evaporated is possible. A seven-effect falling film system, with TVR applied to the second effect, the compressed vapour being returned to the first effect, was reported by Pisecky [7]. See also [1, 3–10].

3.1.5
Applications for Evaporation

The purposes for which evaporation is used in the food industry include: to produce concentrated liquid products (for sale to the consumer or as ingredients to be used in the manufacture of other consumer products), to preconcentrate liquids for further processing and to reduce the cost of transport, storage and in some cases packaging, by reducing the mass and volume of the liquid.

3.1.5.1 Concentrated Liquid Products

Evaporated (Unsweetened Condensed) Milk
The raw material for this product is whole milk which should of good microbiological quality. The first step is standardisation of the composition of the raw milk so as to produce a finished product with the correct composition. The composition of evaporated milk is normally 8% fat and 18% solids not fat (snf) but this may vary from country to country. Skim milk and/or cream is added to the whole milk to achieve the correct fat:snf ratio. To prevent coagulation during heat processing and to minimise age thickening during storage, the milk is stabilised by the addition of salts, including phosphates, citrates and bicarbonates, to maintain pH 6.6–6.7. The milk is then heat treated. This is done to reduce the microbiological load and to improve its resistance to coagulation during subsequent sterilisation. The protein is denatured and some calcium salts are precipitated during this heat treatment. This results in stabilisation of the milk. The usual heat treatment is at 120–122 °C for several minutes, in tubular or plate heat exchangers. This heat treatment also influences the viscosity of the final product, which is an important quality attribute. The milk is then con-

centrated by vacuum evaporation. The evaporation temperature is in the range 50–60 °C and is usually carried out in multiple-effect, falling film, evaporator systems. Plate and centrifugal evaporators may also be used. Two or three effects are common, but up to seven effects have been used. The density of the milk is monitored until it reaches a value that corresponds to the desired final composition of the evaporated milk. The concentrated milk is then homogenised in two stages in a pressure homogeniser, operating at 12.5–25.0 MPa. It is then cooled to 14 °C. The concentrated milk is then tested for stability by heating samples to sterilisation temperature. If necessary, more stabilising salts, such as disodium or trisodium phosphates, are added to improve stability. The concentrated milk is then filled into containers, usually tinplate cans, and sealed by double seaming. The filled cans are then sterilised in a retort at 110–120 °C for 15–20 min. If a batch retort is used, it should have a facility for continually agitating the cans during heating, to ensure that any protein precipitate formed is uniformly distributed throughout the cans. The cans are then cooled and stored at not more than 15 °C.

As and alternative to in-package heat processing, the concentrated milk may be UHT treated at 140 °C for about 3 s (see Chapter 2) and aseptically filled into cans, cartons or 'bag in box' containers.

Evaporated milk has been used as a substitute for breast milk, with the addition of vitamin D, in cooking and as a coffee whitener. A lowfat product may be manufactured using skimmed or semiskimmed milk as the raw material. Skimmed milk concentrates are used in the manufacture of ice cream and yoghurt. Concentrated whey and buttermilk are used in the manufacture of margarine and spreads.

Sweetened Condensed Milk

This product consists of evaporated milk to which sugar has been added. It normally contains about 8% fat, 20% snf and 45% sugar. Because of the addition of the sugar, the water activity of this sweetened concentrate is low enough to inhibit the growth of spoilage and pathogenic microorganisms. Consequently, it is shelf stable, without the need for sterilisation. In the manufacture of sweetened condensed milk, whole milk is standardised and heat treated in a similar manner to evaporated milk. At this point granulated sugar may be added, or sugar syrup may be added at some stage during evaporation. Usually, evaporation is carried out in a two- or three-stage, multiple-effect, falling film, evaporator at 50–60 °C. Plate and centrifugal evaporators may also be used. If sugar syrup is used, it is usually drawn into the second effect evaporator. The density of the concentrate is monitored during the evaporation. Alternatively, the soluble solids content is monitored, using a refractometer, until the desired composition of the product is attained. The concentrate is cooled to about 30 °C. Finely ground lactose crystals are added, while it is vigorously mixed. After about 60 min of mixing, the concentrate is quickly cooled to about 15 °C. The purpose of this seeding procedure is to ensure that, when the supersaturated solution of lactose crystallises out, the crystals formed will be small and not cause grittiness in the product. The cooled product is then filled into cans, cartons or tubes and sealed in an appropriate manner.

Sweetened condensed milk is used in the manufacture of other products such as ice cream and chocolate. If it is to be used for these purposes, it is packaged in larger containers such as 'bag in box' systems, drums or barrels [10, 11].

Concentrated fruit and vegetable juices are also produced for sale to the consumer (see Section 3.1.5.3).

3.1.5.2 Evaporation as a Preparatory Step to Further Processing

One important application for evaporation is to preconcentrate liquids which are to be dehydrated by spray drying, drum drying or freeze drying. As stated in Section 3.1.3 above, the specific steam consumption of such dryers is greater than single-effect evaporators and much greater than multiple-effect systems, particularly, if MVR or TVR is incorporated into one or more of the effects. Whole milk, skimmed milk and whey are examples of liquids which are preconcentrated prior to drying. Equipment similar to that used to produce evaporated milk (see Section 3.1.5.1) is used and the solids content of the concentrate is in the range 40–55%.

Instant Coffee Beverages, such as coffee and tea, are also available in powder form, so called instant drinks. In the production of instant coffee, green coffee beans are cleaned, blended and roasted. During roasting, the colour and flavour develop. Roasting is usually carried out continuously. Different types of roasted beans, light, medium and dark, are produced by varying the roasting time. The roasted beans are then ground in a mill to a particle size to suit the extraction equipment, usually in the range 1000–2000 μm. The coffee solubles are extracted from the particles, using hot water as the solvent. Countercurrent, static bed or continuous extractors are used (see Chapter 15). The solution leaving the extractor usually contains 15–28% solids. After extraction, the solution is cooled and filtered. This extract may be directly dried by spray drying or freeze drying. However, it is more usual to concentrate the solution to about 60% solids by vacuum evaporation. Multiple-effect falling film systems are commonly used. The volatile flavour compounds are stripped from the solution, before or during the evaporation, in a similar manner to that used when concentrating fruit juice (see Section 3.1.5.3) and added back to the concentrate before drying. Coffee powder is produced using a combination of spray drying and fluidised bed drying (see Section 3.2.12.1). Alternatively, the concentrated extract may be frozen in slabs, the slabs broken up and freeze dried in a batch or continuous freeze drier (see Section 3.2.6) [12–14].

Granulated Sugar Vacuum evaporation is used in the production of granulated sugar from sugar cane and sugar beet. Sugar juice is expressed from sugar cane in roller mills. In the case of sugar beet, the sugar is extracted from sliced beet, using heated water at 55–85 °C, in a multistage, countercurrent, static bed or moving bed extractor (see Chapter 15). The crude sugar juice, from either source, goes through a series of purification operations. These include screening and carbonation. Lime is added and carbon dioxide gas is bubbled through the juice. Calcium carbonate crystals are formed. As they settle, they carry with them a lot of the insoluble impurities in the juice. The supernatant is taken off and filtered. Carbonation may be applied in two or more stages during the purification of the juice. The juice may be treated with sulphur dioxide to limit nonenzymic browning. This process is known as sulphitation. The treated juice is again filtered. Various type of filters are used in processing sugar juice including plate and frame, shell and tube and rotary drum filters (see Chapter 15). The purified juice is concentrated up to 50–65% solids by vacuum evaporation. Multiple-effect systems are employed, usually with five effects. Vertical short tube, long tube and plate evaporators are used. The product from the evaporators is concentrated further in vacuum pans or single-effect short tube evaporators, sometimes fitted with an impeller in the downcomer. This is known as *sugar boiling*. Boiling continues until the solution becomes supersaturated. It is then *shocked* by the addition of a small amount of seeding material, finely ground sugar crystals, to initiate crystallisation. Alternatively, a slurry of finely ground sugar crystals in isopropyl alcohol may be added at a lower degree of supersaturation. The crystals are allowed to grow, under carefully controlled conditions, until they reach the desired size and number. The slurry of sugar syrup and crystals is discharged from the evaporator into a temperature-controlled tank, fitted with a slow-moving mixing element. From there it is fed to filtering centrifugals, or basket centrifuges (see Chapter 15) where the crystals are separated from the syrup and washed. The crystals are dried in heated air in rotary driers (see Section 3.2.3.7) and cooled. The dry crystals are conveyed to silos or packing rooms. The sugar syrup from the centrifuges is subjected to further concentration, seeding and separation processes, known as second and third boilings, to recover more sugar in crystal form [1, 15–18].

3.1.5.3 The Use of Evaporation to Reduce Transport, Storage and Packaging Costs

Concentrated Fruit and Vegetable Juices Many fruit juices are extracted, concentrated by vacuum evaporation, and the concentrate frozen on one site, near the growing area. The frozen concentrate is then shipped to several other sites where it is diluted, packaged and sold as chilled fruit juice. Orange juice is the main fruit juice processed in this way [20]. The fruit is graded, washed and the juice extracted using specialist equipment described in [21, 22]. The juice contains about 12.0% solids at this stage. The extracted juice is then *finished*. This involves removing bits of peel, pips, pulp and rag from the juice by screening and/or centrifugation. The juice is then concentrated by vacuum evaporation. Many types of evaporator have been used, including high vacuum, low temperature systems. These had relatively long residence times, up to 1 h in some cases. To inactivate enzymes, the juice was pasteurised in plate heat exchangers before being fed to the evaporator. One type of evaporator, in which evaporation took place at temperatures as low as 20 °C, found use for juice concentration. This operated on a heat pump principle. A refrigerant gas condensed in the heating jacket of the calandria, releasing heat, which caused the liquid to evaporate. The liquid refrigerant evaporated in the jacket of the condenser, taking heat from the water vapour, causing it to condense [1].

Modern evaporators for fruit juice concentration, work on a high temperature, short time (HTST) principle. They are multiple-effect systems, comprised of up to seven falling film or plate evaporators. The temperatures reached are high enough to inactivate enzymes, but the short residence times limit undesirable changes in the product. They are operated with forward flow or mixed flow feeding patterns. Some designs of this type of evaporator are known as thermally accelerated short time or TASTE evaporators.

Volatile compounds, which contribute to the odour and flavour of fruit juices, are lost during vacuum evaporation, resulting in a concentrate with poor organoleptic qualities. It is common practice to recover these volatiles and add them back to the concentrate. Volatiles may be stripped from the juice, prior to evaporation, by distillation under vacuum. However, the most widely used method is to recover these volatiles after partial evaporation of the water. The vapours from the first effect of a multiple-effect evaporation system consist of water vapour and volatiles. When these vapours enter the heating jacket of the second effect, the water vapour condenses first and the volatiles are taken from the jacket and passed through a distillation column, where the remaining water is separated off and the volatiles are concentrated [20–23].

Orange juice is usually concentrated up to 65% solids, filled into drums or 'bag in box' containers and frozen in blast freezers. Unfrozen concentrated juice may also be transported in bulk in refrigerated tankers or ships' holds. Sulphur dioxide may be used as a preservative. Concentrated orange juice may be UHT treated and aseptically filled into 'bag in box' containers or drums. These concentrates may be diluted back to 12% solids, packed into cartons or bottles and sold as chilled orange juice. They may also be used in the production of squashes, other soft and alcoholic drinks, jellies and many other such products.

A frozen concentrated orange juice may also be marketed as a consumer product. This usually has 42–45% solids and is made by adding fresh juice, *cut-back juice,* to the 65% solids concentrate, together with some recovered volatiles. This is filled into cans and frozen. Concentrate containing 65% solids, with or without added sweetener, may be pasteurised and hot-filled into cans, which are rapidly cooled. This is sold as a chilled product. UHT treated concentrate may also be aseptically filled into cans or cartons for sale to the consumer from the ambient shelf.

Juices from other fruits, including other citrus fruits, pineapples, apples, grapes, blackcurrants and cranberries, may be concentrated. The procedures are similar to those used in the production of concentrated orange juice. When clear concentrates are being produced, enzymes are used to precipitate pectins, which are then separated from the juice.

Vacuum evaporation is also applied to the production of concentrated tomato products. Tomatoes are chopped and/or crushed and subjected to heat treatment, which may be the hot break or cold break process [24, 25]. Skin and seeds are removed and the juice extracted in a cyclone separator. The juice is then concentrated by vacuum evaporation. For small-scale production, vacuum pans may be employed. For larger throughputs, two- or three-effect tubular or plate evaporators may be used. If highly concentrated pastes are being produced, agitated thin film evaporators may be used as finishers.

Glucose syrup, skimmed milk and whey are among other food liquids that may be concentrated, to reduce weight and bulk, and so reduce transport and storage costs.

3.2 Dehydration (Drying)

3.2.1 General Principles

Dehydration is the oldest method of food preservation practised by man. For thousands of years he has dried and/or smoked meat, fish, fruits and vegetables, to sustain him during out of season periods in the year. Today the dehydration section of the food industry is large and extends to all countries of the globe. Drying facilities range from simple sun or hot air driers to high capacity, sophisticated spray drying or freeze drying installations. A very large range of dehydrated foods is available and makes a significant contribution to the convenience food market.

In this chapter the terms dehydration and drying are used interchangeably to describe the removal of most of the water, normally present in a foodstuff, by evaporation or sublimation, as a result of the application of heat. The main reason for drying a food is to extend its shelf life beyond that of the fresh material, without the need for refrigerated transport and storage. This goal is achieved by reducing the available moisture, or water activity (a_w; see Section 3.2.16) to a level which inhibits the growth and development of spoilage and pathogenic microorganisms, reducing the activity of enzymes and the rate at which undesirable chemical changes occur. Appropriate packaging is necessary to maintain the low a_w during storage and distribution.

Drying also reduces the weight of the food product. Shrinkage, which occurs often during drying, reduces the volume of the product. These changes in weight and volume can lead to substantial savings in transport and storage costs and, in some cases, the costs of packaging. However, dehydration is an energy intensive process and the cost of supplying this energy can be relatively high, compared to other methods of preservation.

Changes detrimental to the quality of the food may also occur during drying. In the case of solid food pieces, shrinkage can alter the size and shape of the pieces. Changes in colour may also occur. When the food pieces are rehydrated, their colour and texture may be significantly inferior to those of the fresh material. Dry powders may be slow to rehydrate. Changes in flavour may occur during drying solid or liquid foods, as a result of losing volatile flavour compounds and/or the development of cooked flavours. A reduction in the nutritional value of foods can result from dehydration. In particular, loss of vitamins C and A may be greater during drying than in canning or freezing.

Dehydration is usually described as a simultaneous heat and mass transfer operation. Sensible and latent heat must be transferred to the food to cause the water to evaporate. Placing the food in a current of heated air is the most widely used method of supplying heat. The heat is transferred by convection from the air to the surface of the food and by conduction within the food. Alternatively, the food may be placed in contact with a heated surface. The heat is transferred by conduction to the surface of the food in contact with the heated surface and within the food. There is limited use of radiant, microwave and radio frequency energy in dehydration. Freeze drying involves freezing the food and removal of the ice by sublimation. This is usually achieved by applying heat, by conduction or radiation, in a very low pressure environment. In osmotic drying food pieces are immersed in a hypertonic solution. Water moves from the food into the solution, under the influence of osmotic pressure.

3.2.2 Drying Solid Foods in Heated Air

When a wet material is placed in a current of heated air, heat is transferred to its surface, mainly by convection. The water vapour formed is carried away from the drying surface in the air stream. Consider a model system in which a wet material, consisting of an inert solid wetted with pure water, in the form of a thin slab, is placed in a current of heated air, flowing parallel to one of its large faces. The temperature, humidity and velocity of the air are maintained constant. It is assumed that all the heat is transferred to the solid from the air by convection and that drying takes place from one large face only. If the rate of change of moisture content is plotted against time, as in Fig. 3.4, the drying curve may be seen to consist of a number of stages or periods.

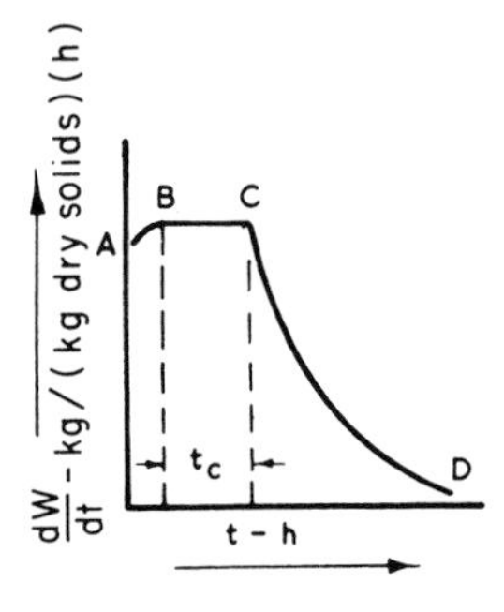

Fig. 3.4 Drying curve for a wet solid in heated air at constant temperature, humidity and velocity; adapted from [1] with permission of the authors.

Period A-B is a settling down or *equilibration period*. The surface of the wet solid comes into equilibrium with the air. This period is usually short compared to the total drying time. Period B-C is known as the *constant rate period*. Throughout this period the surface of the solid is saturated with water. As water evaporates from the surface, water from within the solid moves to the surface, keeping it in a saturated state. The rate of drying during this period remains constant. So also does the surface temperature, at a value corresponding to the wet bulb temperature of the air. By considering heat and mass transfer across the drying surface, a model for the prediction of the duration of the constant rate period of drying can be developed as in Eq. (3.1):

$$t_c = \frac{(W_0 - W_c)\rho_s L_s l}{h_c(\theta_a - \theta_s)} \tag{3.1}$$

where t_c is the duration of the constant rate period, W_0 is the initial moisture content of the wet solid (dry weight basis, dwb), W_c is the moisture content at the end of the constant rate period (dwb), ρ_s is the density of the material, L_s is the latent heat of evaporation at θ_s, l is the thickness of the slab, h_c is heat transfer coefficient for convection heating, θ_a is the dry bulb temperature of the air and θ_s is the wet bulb temperature of the air.

From Eq. (3.1) it can be seen that the main factors that influence the rate of drying during the constant rate period are the air temperature and humidity and the area of the drying surface. The air velocity also has an influence, as the higher this is the greater the value of h_c. As long as this state of equilibrium exists, high rates of evaporation may be maintained, without the danger of overheating the solid. This is an important consideration when drying heat-sensitive foods. Some foods exhibit a constant rate period of drying. However, it is usually short compared to the total drying time. Many foods show no measurable constant rate period.

As drying proceeds, at some point (represented by C in Fig. 3.4) the movement of water to the surface is not enough to maintain the surface in a saturated condition. The state of equilibrium at the surface no longer holds and the rate of drying begins to decline. Point C is known as the critical point and the period C-D is the *falling rate period*. From point C on, the temperature at the surface of the solid rises and approaches the dry bulb temperature of the air as drying nears completion. Hence, it is towards the end of the drying cycle that any heat damage to the product is likely to occur. Many research workers claim to have identified two or more falling rate periods where points of inflexion in the curve have occurred. There is no generally accepted explanation for this phenomenon. During the falling rate period, the rate of drying is governed by factors which affect the movement of moisture within the solid. The influence of external factors, such as the velocity of the air, is reduced compared to the constant rate period. In the dehydration of solid food materials, most of the drying takes place under falling rate conditions.

Numerous mathematical models have been proposed to represent drying in the falling rate period. Some are empirical and were developed by fitting relationships to data obtained experimentally. Others are based on the assumption that a particular mechanism of moisture movement within the solid prevails. The best known of these is based on the assumption that moisture migrates within the solid by diffusion as a result of the concentration difference between the surface and the centre of the solid. It is assumed that Fick's second law of diffusion applies to this movement. A well known solution to this law is represented by Eq. (3.2):

$$\frac{W - W_e}{W_c - W_e} = \frac{8}{\pi^2}\left[\exp\left\{-Dt\left(\frac{\pi}{2l}\right)^2\right\}\right] \tag{3.2}$$

where W is the average moisture content at time t (dwb), W_e is the equilibrium moisture content (dwb), W_c is the moisture content at the start of the falling rate period (dwb), D is the liquid diffusivity and l is the thickness of the slab.

In Eq. (3.2) it is assumed that the value of D is constant throughout the falling rate period. However, many authors have reported that D decreases as the moisture content decreases. Some authors who reported the existence of two or more falling rate periods successfully applied Eq. (3.2) to the individual periods but used a different value of D for each period. Many other factors may change the drying pattern of foods. Shrinkage alters the dimensions of food pieces. The presence of cell walls can affect the movement of water within the solids. The density and porosity of the food material may change during drying. The thermal properties of the food material, such as specific heat and thermal conductivity, may change with a change in moisture content. As water moves to the surface, it carries with it any soluble material, such as sugars and salts. When the water evaporates at the surface, the soluble substances accumulate at the drying surface. This can contribute to the formation of an impervious dry layer at the surface, which impedes drying. This phenomenon is known as case hardening. The diffusion theory does not take these factors into account and so has had only limited success in modelling falling rate drying. Many more complex models have been proposed, which attempt to take some of these changes into account, see [32] as an example [1, 4, 5, 26–32].

3.2.3
Equipment Used in Hot Air Drying of Solid Food Pieces

3.2.3.1 Cabinet (Tray) Drier

This is a multipurpose, batch-operated hot air drier. It consists of an insulated cabinet, equipped with a fan, an air heater and a space occupied by trays of food. It can vary in size from a bench-scale unit holding one or two small trays of food to a large unit taking stacks of large trays. The air may be directed by baffles to flow the across surface of the trays of food or through perforated trays and the layers of food, or both ways. The moist air is partly exhausted from the cabinet and partly recycled by means of dampers. Small cabinet driers are used in laboratories, while larger units are used as industrial driers, mainly for drying sliced or diced fruits and vegetables. A number of large cabinets may be used in parallel, with a staggered loading sequence, to process relatively large quantities of food, up to 20 000 t day^{-1} of raw material [1, 26, 29, 30, 34].

3.2.3.2 **Tunnel Drier**

This type of drier consists of a long insulated tunnel. Tray loads of the wet material are assembled on trolleys which enter the tunnel at one end. The trolleys travel the length of the tunnel and exit at the other end. Heated air also flows through the tunnel, passing between the trays of food and/or through perforated trays and the layers of food. The air may flow parallel to and in the same direction as the trolleys, as shown in Fig. 3.5. This is known as a concurrent tunnel. Other designs featuring countercurrent, concurrent-countercurrent and crossflow of air are available. Each pattern of airflow has its advantages and limitations. The trolleys may move continuously through the tunnel. Alternatively, the movement may be semicontinuous. As a trolley full of fresh material is introduced into one end of the tunnel, a trolley full of dried product exits at the other end. Tunnels may be up to 25 m in length and about 2 m×2 m in cross-section. Tunnel driers are mainly used for drying sliced or diced fruits and vegetables [1, 26, 30, 33, 34].

3.2.3.3 **Conveyor (Belt) Drier**

In this type of drier the food material is conveyed through the drying tunnel on a perforated conveyor, made of hinged, perforated metal plates or wire or plastic mesh. The heated air usually flows through the belt and the layer of food, upward in the early stages of drying and downward in the later stages. The feed is applied to the belt in a layer 75–150 mm deep. The feed must consist of particles that form a porous bed allowing the air to flow through it. Conveyors are typically 2–3 m wide and up to 50 m long. The capacity of a conveyor drier is much less than that of a tunnel drier, occupying the same floor space. As shrinkage occurs during drying, the thickness of the layer of food on the conveyor becomes less. Thus the belt is being used less efficiently as drying proceeds. The use of multistage conveyor drying is common. The product from the first conveyor is redistributed, in a thicker layer, on the second conveyor. This

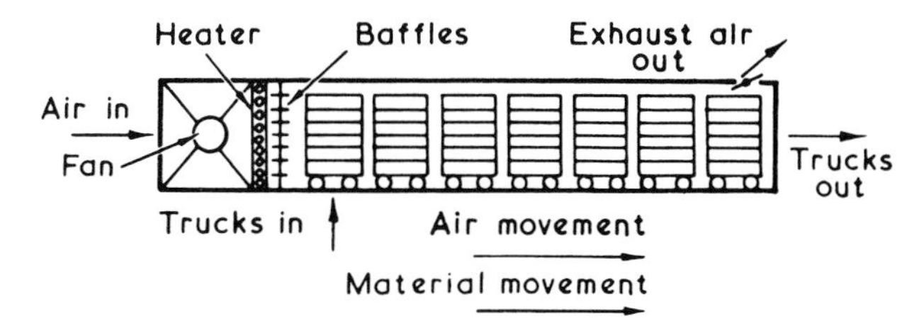

Fig. 3.5 Principle of concurrent tunnel drier; from [1] with permission of the authors.

may be extended to three stages. In this way, the conveyor is used more efficiently, compared to a single-stage unit. On transfer of the particles from one stage to the next, new surfaces are exposed to the heated air, improving the uniformity of drying. The air temperature and velocity may be set to different levels in each stage. Thus, good control may be exercised over the drying, minimising heat damage to the product. However, even when using two or more stages, drying in this type of drier is relatively expensive. Consequently, they are often used to remove moisture rapidly, in the early stages of drying, and the partly dried product is dried to completion in another type of drier. Diced vegetables, peas, sliced beans and grains are examples of foods dried in conveyor driers [1, 26, 30, 33–35].

3.2.3.4 Bin Drier

This is a throughflow drier, mainly used to complete the drying of particulate material partly dried in a tunnel or conveyor drier. It takes the form of a vessel fitted with a perforated base. The partly dried product is loaded into the vessel to up to 2 m deep. Dry, but relatively cool air, percolates up through the bed slowly, completing the drying of the product over an extended period, up to 36 h. Some migration of moisture between the particles occurs in the bin. This improves the uniformity of moisture content in the product [1, 26, 33, 34].

3.2.3.5 Fluidised Bed Drier

This is another throughflow, hot air drier which operates at higher air velocities than the conveyor or bin drier. In this type of drier heated air is blown up through a perforated plate which supports a bed of solid particles. As the air passes through the bed of particles, a pressure drop develops across the bed. As the velocity of the air increases the pressure drop increases. At a particular air velocity, known as the incipient velocity, the frictional drag on the particles exceeds the weight of the particles. The bed then expands, the particles are suspended in the air and the bed starts to behave like a liquid, with particles circulating within the bed. This is what is meant by the term fluidised bed. As the air velocity increases further, the movement of the particles becomes more vigorous. At some particular velocity, particles may detach themselves from the surface of the bed temporarily and fall back onto it. At some higher velocity, known as the entrainment velocity, particles are carried away from the bed in the exhaust air stream. Fluidised bed driers are operated at air velocities between the incipient and entrainment values. The larger and more dense the particles are, the higher the air velocity required to fluidise them. Particles in the size range 20 μm to 10 mm in diameter can usually be fluidised. The particles must not be sticky or prone to mechanical damage. Air velocities in the range 0.2–5.0 m s^{-1} are used. There is very close contact between the heated air and the particles, which results in high rates of heat transfer and relatively short drying times.

Fluidised beds may be operated on a batch or continuous basis. Batch units are used for small-scale operations. Because of the mixing which occurs in such beds, uniform moisture contents are attainable. Continuous fluidised bed driers used in the food industry are of the plug flow type shown in Fig. 3.6. The feed enters the bed at one end and the product exits over a weir at the other. As the particles dry, they become lighter and rise to the surface of the bed and are discharged over the weir. However, if the size, density and moisture content of the feed particles are not uniform, the moisture content of the product may also not be uniform.

The use of multistage fluidised bed driers is common. The partly dried product from the first stage is discharged over the weir onto the second stage and so on. Up to six stages have been used. The temperature of the air may be controlled at a different level in each stage. Such systems can result in savings in energy and better control over the quality of the product, as compared with a single-stage unit. Fluidised beds may be mechanically vibrated. This enables them to handle particles with a wider size range than a standard bed. They can also accommodate sticky products and agglomerated particles better than a standard bed. The air velocities needed to maintain the particles in a fluidised state are less than in a stationary bed. Such fluidised beds are also known as *vibro-fluidisers*. Another design of the fluidised bed is known as the spouted bed drier (see Fig. 3.7). Part of the heated air is introduced into the bottom of the bed in the form of a high velocity jet. A spout of fast-moving particles is formed in the centre of the bed. On reaching the top of the bed, the particles return slowly to the bottom of the bed in an annular channel surrounding the spout. Some of

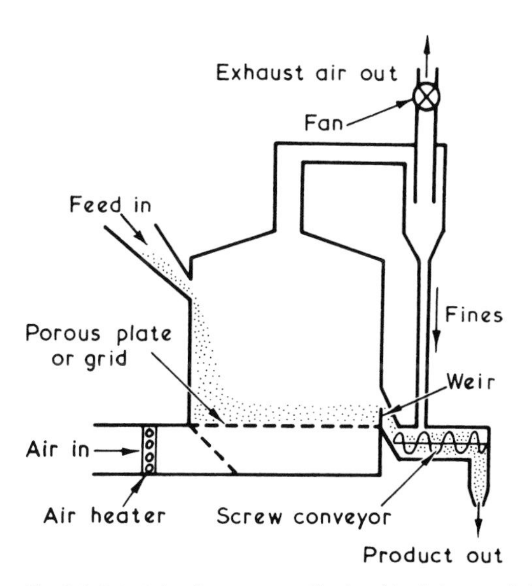

Fig. 3.6 Principle of continuous fluidised bed drier, with fines recovery system; from [1] with permission of the authors.

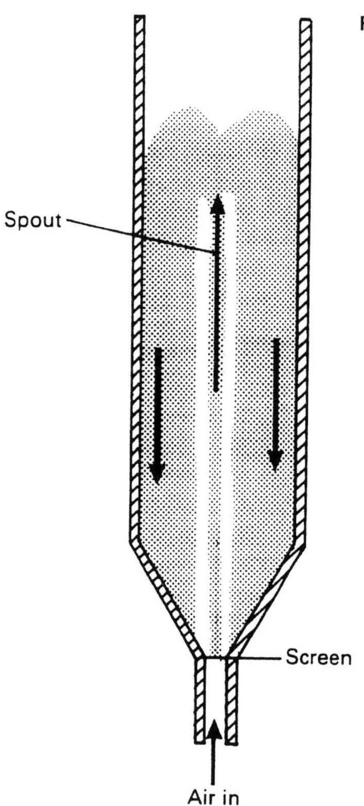

Fig. 3.7 Principle of spouted bed drier.

the heated air flows upward through the slow-moving channel, countercurrent to the movement of the particles. High rates of evaporation are attained in the spout, while evaporative cooling keeps the particle temperature relatively low. Conditions in the spout are close to constant rate drying (see Section 3.2.2). Drying of the particles is completed in the annular channel. This type of drier can handle larger particles than the conventional fluidised bed. In some spouted bed driers the air is introduced tangentially into the base of the bed and a screw conveyor is located at its centre to control the upward movement of the particles. Such driers are suitable for drying relatively small particles. In the toroidal bed drier the heated air enters the drying chamber through blades or louvres, creating a fast-moving, rotating bed of particles (see Fig. 3.8). High rates of heat and mass transfer in the bed enable rapid drying of relatively small particles. Peas, sliced beans, carrots and onions, grains and flours are some foods that have been dried in fluidised bed driers [1, 3, 5, 26, 30, 36–39].

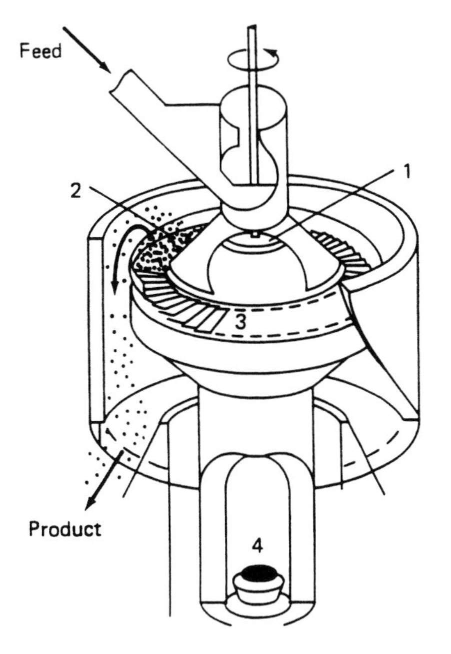

Fig. 3.8 Principle of toroidal bed drier.

3.2.3.6 **Pneumatic (Flash) Drier**

In this type of drier the food particles are conveyed in a heated air stream through ducting of sufficient length to give the required drying time. It is suitable for relatively small particles and high air velocities are used, in the range 10–40 m s^{-1}. The dried particles are separated from the air stream by cyclone separators or filters. The ducting may be arranged vertically or horizontally. Single vertical driers are used mainly for removing surface moisture. Drying times are short, in the range 0.5–3.5 s and they are also known as flash driers. When internal moisture is to be removed, longer drying times are required. Horizontal pneumatic driers, or vertical driers, consisting of a number of vertical columns in series, may be used for this purpose. The ducting may be in the form of a closed loop. The particles travel a number of times around the loop until they reach the desired moisture content. The dried particles are removed from the air stream by means of a cyclone. Fresh air is introduced continuously into the loop through a heater and humid air is continuously expelled from it. This type of drier is known as a pneumatic ring drier. Grains and flours are the main products dried in pneumatic driers [1, 5, 26, 40].

3.2.3.7 **Rotary Drier**

The most common design of rotary drier used for food application is known as the direct rotary drier. This consists of a cylindrical shell, set at an angle to the horizontal. The shell rotates at 4–5 rpm. Wet material is fed continuously into the shell at its raised end and dry product exits over a weir at the lower end. Baffles or flights are fitted to the inner surface of the shell. These lift the material up as the shell rotates and allow it to fall down through a stream of heated air, that may flow concurrent or countercurrent to the direction of movement of the material. The feed material must consist of relatively small particles, which are free-flowing and reasonably resistant to mechanical damage. Air velocities in the range 1.5–2.5 m s^{-1} are used and drying times are from 5 min to 60 min.

One particular design of rotary drier is known as the *louvred drier*. The wall of the shell is made up of overlapping louvre plates, through which the heated air is introduced. As the shell rotates, the bed of particles is gently rolled within it, causing mixing to take place and facilitating uniform drying. The movement is less vigorous than in the conventional rotary drier and causes less mechanical damage to the particles. Grains, flours, cocoa beans, sugar and salt crystals are among the food materials dried in rotary driers [26, 41, 42].

3.2.4 Drying of Solid Foods by Direct Contact With a Heated Surface

When a wet food material is placed in contact with a heated surface, usually metal, the sensible and latent heat is transferred to the food and within the food mainly by conduction. Most of the evaporation takes place from the surface, which is not in contact with the hot surface. The drying pattern is similar to that which prevails during drying in heated air (see Section 3.2.2). After an equilibration period, there a constant rate period, during which water evaporates at a temperature close to its boiling point at the prevailing pressure. The rate of drying will be higher than that in heated air at the same temperature. When the surface begins to dry out, the rate of drying will decrease and the temperature at the drying surface will rise and approach that of the heated metal surface, as drying nears completion. If drying is taking place at atmospheric pressure, the temperature at the drying surface in the early stages will be in excess of 100 °C. There is little or no evaporative cooling as is the case in hot air drying. The temperature of the heated metal surface will need to be well above 100 °C in order to achieve reasonable rates of drying and low product moisture content. Exposure of a heat sensitive food to such high temperatures for prolonged periods, up to several hours, is likely to cause serious heat damage. If the pressure above the evaporating surface is reduced below atmospheric, the evaporation temperature will be reduced and hence the temperature of the heated metal surface may be reduced. This is the principle behind *vacuum drying*.

If it is assumed that drying takes place from one large face only and that there is no shrinkage, the rate of drying, at any time t, may be represented by the following equation:

$$\frac{dw}{dt} = \frac{(W_0 - W_f)M}{t} = \frac{K_c A(\theta_w - \theta_e)}{L_e} \tag{3.3}$$

where $\frac{dw}{dt}$ = rate of drying (rate of change of weight) at time t, W_0 is the initial moisture content of the material (dwb), W_f is the final moisture content of the material (dwb), M is the mass of dry solids on the surface, t is the drying time, K_c is the overall heat transfer coefficient for the complete drying cycle, A is the drying area, θ_w is the temperature of the heated metal surface (wall temperature), θ_e is the evaporating temperature and L_e is the latent heat of evaporation at θ_e [1, 5, 26, 43, 44].

3.2.5 Equipment Used in Drying Solid Foods by Contact With a Heated Surface

3.2.5.1 Vacuum Cabinet (Tray or Shelf) Drier

This drier consists of a vacuum chamber connected to a condenser and vacuum pump. The chamber is usually cylindrical and has one or two access doors. It is usually mounted in a horizontal position. The chamber is equipped with a number of hollow plates or shelves, arranged horizontally. These shelves are heated internally by steam, hot water or some other thermal fluid, which is circulated through them. A typical drying chamber may contain up to 24 shelves, each measuring 2.0 × 1.5 m. The food material is spread in relatively thin layers on metal trays. These trays are placed on the shelves, the chamber is sealed and the pressure reduced by means of the condenser and vacuum pump. Absolute pressures in the range 5–30 kPa are created, corresponding to evaporation temperatures of 35–80 °C. Drying times can range from 4 h to 20 h, depending on the size and shape of the food pieces and the drying conditions. The quality of vacuum dried fruits or vegetables is usually better than air dried products. However, the capital cost of vacuum shelf driers is relatively high and the throughput low, compared to most types of hot air drier [1, 5, 26, 34, 43, 44].

3.2.5.2 Double Cone Vacuum Drier

This type of drier consists of a hollow vessel in the shape of a double cone. It rotates about a horizontal axis and is connected to a condenser and vacuum pump. Heat is applied by circulating steam or a heated fluid through a jacket fitted to the vessel. The pressures and temperatures used are similar to those used in a vacuum shelf drier (see Section 3.2.5.1). It is suitable for drying particulate materials, which are tumbled within the rotating chamber as heat is applied through the jacket. Some mechanical damage may occur to friable materials due to the tumbling action. If the particles are sticky at the temperatures being used, they may form lumps or balls or even stick to the wall of the vessel. Operating at low speeds or intermittently may reduce this problem. This type of drier has found only limited application in the food industry [26, 43, 44].

3.2.6
Freeze Drying (Sublimation Drying, Lyophilisation) of Solid Foods

This method of drying foods was first used in industry in the 1950s. The process involves three stages: (a) freezing the food material, (b) subliming the ice (primary drying) and (c) removal of the small amount of water bound to the solids (secondary drying or desorption). Freezing may be carried out by any of the conventional methods including blast, immersion, plate or liquid gas freezing (see Chapter 4). Blast freezing in refrigerated air is most often used. It is important to freeze as much of the water as possible. This can be difficult in the case of material with a high soluble solids content, such as concentrated fruit juice. As the water freezes in such a material, the soluble solids content of remaining liquid increases, and so its freezing point is lowered. At least 95% of the water present in the food should be converted to ice, to attain successful freeze drying.

Ice will sublime when the water vapour pressure in the immediate surroundings is less than the vapour pressure of ice at the prevailing temperature. This condition could be attained by blowing dry, refrigerated air across the frozen material. However, this method has proved to be uneconomic on a large scale. On an industrial scale, the vapour pressure gradient is achieved by reducing the total pressure surrounding the frozen food to a value lower than the ice vapour pressure. The vapour pressure of ice at $-20\,^{\circ}$C is about 135 Pa, absolute. Industrial freeze drying is carried out in vacuum chambers operated at pressures in the range 13.5–270.0 Pa, absolute. The main components of a batch freeze drier are a well sealed vacuum chamber fitted with heated shelves, a refrigerated condenser and a vacuum pump or pumps (see Fig. 3.9). The refrigerated condenser removes the water vapour formed by sublimation. The water vapour freezes on to the condenser, thus maintaining the low water vapour pressure in the chamber. The vacuum pump(s) remove the noncondensable gases.

The heated shelves supply the heat of sublimation. Heat may be applied from above the frozen food by radiation or from below by conduction or from both directions.

Once sublimation starts, a dry layer will form on the top surface of the food pieces. Together, the rate at which water vapour moves through this dry layer

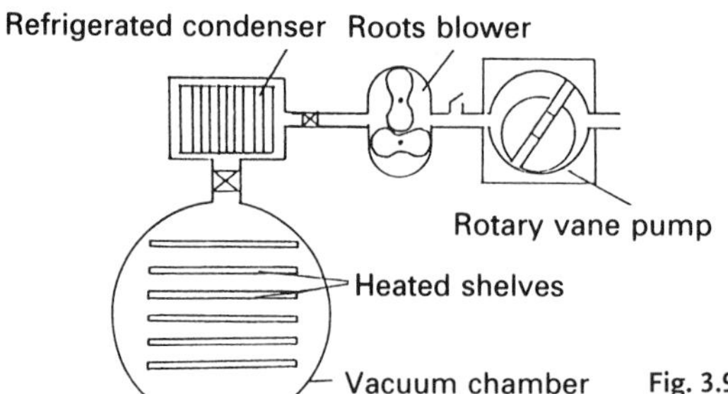

Fig. 3.9 Main components of a batch freeze drier; from [26] with permission.

and the rate at which heat travels through the dry layer and/or the frozen layer determine the rate of drying. If a slab-shaped solid is being freeze dried from its upper surface only and if the heat is supplied only from above, through the dry layer, a state of equilibrium will be attained between the heat and mass transfer. Under these conditions, the rate of drying (rate of change of weight, $\frac{dw}{dt}$), may be represented by the following model:

$$\frac{dw}{dt} = \frac{Ak_d(\theta_d - \theta_i)}{L_s l} = \frac{Ab(p_i - p_d)}{l} = A\rho_s(W_0 - W_f)\frac{dl}{dt} \qquad (3.4)$$

where A is the drying area normal to the direction of flow of the vapour, k_d is the thermal conductivity of the dry layer, θ_d is the temperature at the top surface of the dry layer, θ_i is the temperature at the ice front, L_s is the heat of sublimation, l is the thickness of the dry layer, b is the permeability of the dry layer to water vapour, p_I is the water vapour pressure at the ice front; p_d is the water vapour pressure at the top surface of the dry layer, ρ_s is the density of the dry layer, W_0 is the initial moisture content of the material (dwb), W_f is the final moisture content of the material (dwb) and $\frac{dl}{dt}$=rate of change of thickness of the dry layer. When heat is supplied from below, or from both above and below, the equilibrium between the heat and mass transfer no longer exists. More complex models representing these situations are available in the literature [30, 45, 46].

The advantages of freeze drying as compared with other methods lie in the quality of the dried product. There is no movement of liquid within the solid during freeze drying. Thus, shrinkage does not occur and solutes do not migrate to the surface. The dried product has a light porous structure, which facilitates rehydration. The temperature to which the product is exposed is lower than in most other methods of drying, so heat damage is relatively low. There is good retention of volatile flavour compounds during freeze drying. However, some damage to the structure may occur during freezing which can result in some structural collapse on rehydration and a poor texture in the rehydrated product. Some denaturation of proteins may occur due to pH changes and concentration of solutes during freezing. Freeze dried food materials are usually hygroscopic, prone to oxidation and fragile. Relatively expensive packaging may be necessary, compared to other types of dried foods. The capital cost of freeze drying equipment is relatively high and so are the energy costs. It is the most expensive method of drying food materials [1, 5, 26, 30, 45–47].

3.2.7 Equipment Used in Freeze Drying Solid Foods

3.2.7.1 Cabinet (Batch) Freeze Drier

This is a batch operated drier. The vacuum chamber is cylindrical, mounted horizontally and fitted with doors front and back. The vacuum is created and maintained by a refrigerated condenser backed up by vacuum pumps (see Fig. 3.9).

Once the frozen food is sealed into the chamber, the pressure must be reduced rapidly to avoid melting of the ice. Once the vacuum is established, the low pressure must be maintained throughout the drying cycle, which may range from 4 h to 12 h. The vacuum system must cope with the water vapour produced by sublimation and noncondensable gases, which come from the food or through leaks in the system. The refrigerated condenser, which may be a plate or coil, is located inside the drying chamber, or in a smaller chamber connected to the main chamber by a duct. The water vapour freezes onto the surface of the plate or coil. The temperature of the refrigerant must be below the saturation temperature, corresponding to the pressure in the chamber. This temperature is usually in the range –10 °C to –50 °C. As drying proceeds, ice builds up on the surface of the condenser. This reduces its effectiveness. If it is not to be defrosted during the drying cycle, then a condenser with a large surface area is required. Defrosting may be carried out during the cycle by having two condensers. These are located in separate chambers, each connected to the main drying chamber via a valve. The condensers are used alternately. While one is isolated from the drying chamber and is defrosting, the other is connected to the main chamber and is condensing the water vapour. The roles are reversed periodically during the drying cycle.

Usually two vacuum pumps are used in series to cope with the noncondensable gases. The first may be a Roots blower or an oil-sealed rotary pump. The second is a gas-ballasted, oil-sealed rotary pump. In the early days of industrial freeze drying, multistage steam ejectors were used to evacuate freeze drying chambers. These could handle both the water vapour and noncondensable gases. However, most of these have been replaced by the system described above, because of their low energy efficiency.

Heat is normally supplied by means of heated shelves. The trays containing the frozen food are placed between fixed, hollow shelves, which are heated internally by steam, heated water or other thermal fluids. Heat is supplied by conduction from below the trays and by radiation from the shelf above. Plates and trays are designed to ensure good thermal contact. Ribbed or finned trays are used to increase the area of the heated surface in contact with granular solids, without impeding with the escape of the water vapour. The use of microwave heating in freeze drying is discussed in Section 3.2.9 [1, 5, 26, 45–48].

3.2.7.2 **Tunnel (SemiContinuous) Freeze Drier**

This is one type of freeze drier capable of coping with a wide range of piece sizes, from meat and fish steaks down to small particles, while operating on a semicontinuous basis. It is comprised of a cylindrical tunnel, 1.5–2.5 m in diameter. It is made up of sections. The number of sections depends on the throughput required. Entry and exit locks are located at each end of the tunnel. Gate valves enable these locks to be isolated from the main tunnel. To introduce a trolleyload of frozen material into the tunnel, the gate valve is closed, air is let into the entry lock, the door is opened and the trolley pushed into the lock. The door is then closed and the lock is evacuated by a dedicated vacuum system. When the pressure is reduced to that in the main tunnel, the gate valve is opened and the trolley enters the main tunnel. A similar procedure is in operation at the dry end of the tunnel, to facilitate the removal of a trolleyload of dried product. Fixed heater places are located in the tunnel. The food material, in trays or ribbed dishes, passes between these heater plates as the trolley proceeds through the tunnel. Since both the vacuum and heat requirements decrease as drying proceeds, the vacuum and heating systems are designed to match these requirements. The main body of the tunnel is divided into zones by means of vapour restriction plates. Each zone is serviced by its own vacuum and heating system. Each section of the tunnel can process 3–4 t of frozen material per 24 h [26, 45, 49].

3.2.7.3 **Continuous Freeze Driers**

There are a number of designs of continuous freeze driers suitable for processing granular materials. In one design, a stack of circular heater plates is located inside a vertical, cylindrical vacuum chamber. The frozen granules enter the top of the chamber alternately through each of two entrance locks, and fall onto the top plate. A rotating central vertical shaft carries arms which sweep the top surface of the plates. The arm rotating on the top plate pushes the granules outward and over the edge of the plate, onto the plate below, which has a larger diameter than the top plate. The arm on the second plate pushes the granules inward towards a hole in the centre of the plate, through which they fall onto the third plate, which has the same diameter as the top plate. In this way, the granules travel down to the bottom plate in the stack. From that plate, they fall through each of two exit locks alternately and are discharged from the chamber.

In another design, the frozen granules enter at one end of a horizontal, cylindrical vacuum chamber via an entrance lock, onto a vibrating deck. This carries them to the other end of the chamber. They then fall onto a second deck, which transports them back to the front end of the chamber, where they fall onto another vibrating deck. In this way, the granules move back and forth in the chamber until they are dry. They are then discharged from the chamber through a vacuum lock. Heat is supplied by radiation from heated platens located above the vibrating decks. Another design is similar to the above but conveyor belts, rather than vibrating decks, move the granules back and forth within the chamber [30, 49].

3.2.7.4 **Vacuum Spray Freeze Drier**

A prototype of this freeze drier, developed for instant coffee and tea is described by Mellor [45]. The preconcentrated extract is sprayed into a tall cylindrical vacuum chamber, which is surrounded by a refrigerated coil. The droplets freeze by evaporative cooling, losing about 15% of their moisture. The partially dried particles fall onto a moving belt conveyor in the bottom of the chamber and are carried through a vacuum tunnel where drying is completed, by the application of radiant heat. The dry particles are removed from the tunnel by means of a vacuum lock.

The production of instant coffee is the main industrial application of freeze drying. Freeze dried fruits, vegetables, meat and fish are also produced, for inclusion in ready meals and soups. However, they are relatively expensive compared to similar dried foods produced by hot air or vacuum drying.

3.2.8 Drying by the Application of Radiant (Infrared) Heat

When thermal radiation is directed at a body, it may be absorbed and its energy converted into heat, or reflected from the surface of the body, or transmitted through the material. It is the absorbed energy that can provide heat for the purposes of drying. Generally, in solid materials all the radiant energy is absorbed in a very shallow layer beneath the surface. Thus radiant drying is best suited to drying thin layers or sheets of material or coatings. Applications include textiles, paper, paints and enamels. In the case of food materials, complex relationships exist between their physical, thermal and optical properties. These in turn influence the extent to which radiant energy is absorbed by foods. The protein, fat and carbohydrate components of foods have their own absorption patterns. Water in liquid, vapour or solid form also has characteristic absorption patterns that influence the overall the absorption of radiant energy. It is very difficult to achieve uniform heating of foods by radiant heat. Control of the heating rate is also a problem. Infrared heating is not normally used, in the food industry, for the removal of water in bulk from wet food materials. It has been used to remove surface moisture from sugar or salt crystals and small amounts of water from low moisture particles such as breadcrumbs and spices. These are conveyed in thin layers beneath infrared heaters. Shortwave lamps are used for very heat-sensitive materials. Longwave bar heaters are used for more heat-resistant materials. Radiant heating is used in vacuum driers and freeze driers, usually in combination with heat transferred by conduction from heated shelves [26, 50].

3.2.9 Drying by the Application of Dielectric Energy

There is some confusion in the literature regarding terminology when describing dielectric and microwave energy. In this section, the term dielectric is used to represent both the radio frequency (RF) and microwave (MW) bands of the electromagnetic spectrum. RF energy is in the frequency range 1–200 MHz and MW from 300 MHz to 300 GHz. By international agreement, specific frequencies have been allocated for industrial use. These are: RF 27.12 MHz and 13.56 MHz, MW 2450 MHz and a band within the range 896–915 MHz. These are known as ISM (industrial, scientific and medical) bands. Dielectric heating is used for cooking, thawing, melting and drying. The advantages of this form of heating over more conventional methods are that heat generation is rapid and occurs throughout the body of the food material. This is known as volumetric heating. Water is heated more rapidly than the other components in the food. This is an added advantage when it is used for drying foods. The depth to which electromagnetic waves penetrate into a material depends on their frequency and the characteristics of each material. The lower the frequency and so the longer the wavelength, the deeper the penetration. The energy absorbed by a wet material exposed to electromagnetic waves in the dielectric frequency range depends on a characteristic of the material known as its dielectric loss factor, which depends on the distribution of dipoles in the material. The higher the loss factor, the more energy is absorbed by the material. The loss factor of a material is dependent on its moisture content, temperature and, to some extent, its structure. The loss factor of free water is greater than bound water. The loss factors of both free and bound water are greater than that of the dry matter. Heating both by RF and MW methods is mainly due to energy absorbed by water molecules. However, the mechanism whereby this energy is absorbed is different for the two frequency ranges. In RF heating, heat is generated by the passage of an electric current through the water. This is due to the presence of ions in the water, which give it a degree of electrical conductivity. In MW heating, dipolar molecules in the water are stressed by the alternating magnetic field and this results in the generation of heat. At frequencies corresponding to RF heating, the conductivity and hence the energy absorbed increases with increasing temperature. However, in the case of MW frequencies, the loss factor and so the energy absorbed, decreases with increasing temperature. At both RW and MW frequencies, the rapid generation of heat within the material leads to the rapid evaporation of water. This gives rise to a total pressure gradient which causes a rapid movement of liquid water and water vapour to the surface of the solid. This mechanism results in shorter drying times and lower material temperatures, compared to hot air or contact drying. Drying is uniform, as thermal and concentration gradients are comparatively small. There is less movement of solutes within the material and overheating of the surface is less likely than is the case when convected or conducted heat is applied. There is efficient use of energy as the water absorbs most of the heat. However, too high a heating rate can cause scorching or burning of the material. If water becomes entrapped within the material, rupture of solid pieces may occur, due to the development of high pressure within them.

Equipment for the generation of RF and MW is described in the literature [51–53]. Equipment for applying RF and MW differ from each other. A basic RF (platen) applicator consists of two metal plates, between which the food is placed or conveyed. The plates are at different electrical voltages. This is also known as a throughfield applicator and is used mainly for relatively thick objects. In a stray- or fringefield applicator, a thin layer of material passes over electrodes, in the form of bars, rods or plates, of alternating polarity. In a staggered, throughfield applicator, bars are located above and below the product to form staggered throughfield arrays. This type is suitable for intermediate thickness products. In a basic batch MW applicator, microwaves are directed from the generator into a metal chamber via a waveguide or coaxial cable. The food is placed in the chamber. To improve the uniformity of heating, the beam of microwaves may be disturbed by a mode stirrer, which resembles a slow turning fan. This causes reflective scattering of the waves.

Alternatively, the food may be placed on a rotating table in the chamber. In one continuous applicator, known as a leaky waveguide applicator, microwaves are allowed to leak in a controlled manner from slots or holes cut in the side of a waveguide. A thin layer of product passes over the top of the slots. In the slotted waveguide applicator, the product is drawn through a slot running down the centre of the waveguide. This is also suitable for thin layers of material.

Dielectric heating is seldom used as the main source of heat for drying wet food materials. It is mainly used in conjunction with heated air.

Dielectric heating may be used to preheat the feed to a hot air drier. This quickly raises the temperature of the food and causes moisture to rise to the surface. The overall drying time can be reduced in this way. It may also be applied in the early stages of the falling rate period of drying, or towards the end of the drying cycle, to reduce the drying time. It is more usual to use it near the end of drying. RF heating is used in the later stage of the baking of biscuits (postbaking) to reduce the moisture content to the desired level. This significantly shortens the baking time compared to completing the baking in a conventional oven. A similar procedure has been used for breakfast cereals. MW heating is used, in combination with heated air of high humidity, to dry pasta. Cracking of the product is avoided and the drying time is shortened from 8 h to 1 h, compared to drying to completion in heated air. MW heating has been used after frying of potato chips and crisps to attain the desired moisture content, without darkening the product, if the sugar content of the potatoes is high. It has also been used as a source of heat when vacuum drying pasta. The use of MW heating in freeze drying has been the subject of much research. Ice absorbs energy more rapidly than dry matter, which is an advantage. However, ionisation of gases can occur in the very low pressure conditions and this can lead to plasma discharge and heat damage to the food. Using a frequency of 2450 MHz can prevent this. If some of the ice melts, the liquid water will absorb energy so rapidly that it may cause solid food particles to explode. Very good control of the MW heating is necessary to avoid this happening. MW heating in freeze drying has not yet been used on an industrial scale, mainly because of the high cost involved [1, 5, 26, 52–54].

3.2.10 Osmotic Dehydration

When pieces of fresh fruits or vegetables are immersed in a sugar or salt solution, which has a higher osmotic pressure than the food, water passes from the food into the solution under the influence of the osmotic pressure gradient; and the water activity of the food is lowered. This method of removing moisture from food is known as osmotic dehydration (drying). This term is misleading as the end product is seldom stable and further processing is necessary to extend its shelf life. Osmotic concentration would be a more accurate description of this process. During osmosis the cell walls act as semipermeable membranes, releasing the water and retaining solids. However, these membranes are not entirely selective and some soluble substances, such as sugars, salts, organic acids and vitamins, may be lost from the cells, while solutes from the solution may penetrate into the food. Damage that occurs to the cells during the preparation of the pieces, by slicing or dicing, will increase the movement of soluble solids. The solutes, which enter the food from the solution, can assist in the reduction of the water activity of the food. However, they may have an adverse effect on the taste of the end product.

In the case of fruits, sugars with or without the addition of salt, are used to make up the osmotic solution, also known as the hypertonic solution. Sucrose is commonly used, but fructose, glucose, glucose/fructose and glucose/polysaccharide mixtures and lactose have been used experimentally, with varying degrees of success. The inclusion of 0.5–2.0% of salt in the sugar solution can increase the rate of osmosis. Some other low molecular weight compounds such as malic acid and lactic acid have been shown to have a similar effect. Sugar solutions with initial concentrations in the range 40–70% are used. In general, the higher the solute concentration the greater the rate and extent of drying. The higher the sugar concentration the more sugar will enter the food. This may result in the product being unacceptably sweet. The rate of water loss is high initially, but after 1–2 h it reduces significantly. It can take days before equilibrium is reached. A typical processing time to reduce the food to 50% of its fresh weight is 4–6 h.

In the case of vegetables, sodium chloride solutions in the range 5–20% are generally used. At high salt concentrations, the taste of the end product may be adversely affected. Glycerol and starch syrup have been used experimentally, as solutes for osmotic drying of vegetables.

In general, the higher the temperature of the hypertonic solution the higher the rate of water removal. Temperatures in the range 20–70 °C have been used. At the higher temperatures, there is a danger that the cell walls may be damaged. This can result in excessive loss of soluble material, such as vitamins, from the food. Discoloration of the food may occur at high temperature. The food may be blanched, in water or in the osmotic solution, to prevent browning. This may affect the process in different ways. It can speed up water removal in some large fruit pieces, due to relaxation of the structural bonds in the fruit. In the case of some small fruit pieces, blanching may reduce water loss and increase the amount of solute entering into the fruit from the solution. In general, the lower the weight ratio of food to solution the greater the water loss and solids gain. Ratios of 1:4 to 1:5 are usually employed. The smaller the food pieces the faster the process, due to the larger surface area. However, the smaller the pieces the more cell damage is likely to occur when cutting them and, hence, the greater the amount of soluble solids lost from the food. Promoting movement of the solution relative to the food pieces should result in faster osmosis. However, vigorous mixing is likely to lead to cell damage. Delicate food pieces may remain motionless in a tank of solution. Some improvement in the rate of drying may be obtained by recirculating the solution through the tank by means of a pump. In large-scale installations, the food pieces may be contained within a basket, which is immersed in the tank of solution. The basket is vibrated by means of an eccentric drive. Alternatively, the food pieces may be packed into a tall vessel and the solution pumped through the porous bed of solids.

Reuse of the hypertonic solution is desirable to make osmotic drying an economic process. Insoluble solids may be removed by filtration and the solution concentrated back to its original soluble solids content by vacuum evaporation. Discoloration may limit the number of times the solution can be reused. A mild heat treatment may be necessary to inactivate microorganisms, mainly yeasts, that may build up in the solution.

As stated above, the products from osmotic drying are usually not stable. In the case of fruits and vegetables, the osmosed products have water activities in the range 0.90–0.95. Consequently, further processing is necessary. Drying in heated air, vacuum drying or freeze drying may be employed to stabilise such products. Alternatively, they may be frozen [26, 30, 55].

3.2.11
Sun and Solar Drying

In this section the term 'sun drying' is used to describe the process whereby some or all of the energy for drying of foods is supplied by direct radiation from the sun. The term 'solar drying' is used to describe the process whereby solar collectors are used to heat air, which then supplies heat to the food by convection.

For centuries, fruit, vegetables, meat and fish have been dried by direct exposure to the sun. The fruit or vegetable pieces were spread on the ground on leaves or mats while strips of meat and fish were hung on racks. While drying in this way, the foods were exposed to the vagaries of the weather and to contamination by insects, birds and animals. Drying times were long and spoilage of the food could occur before a stable moisture content was attained. Covering the food with glass or a transparent plastic material can reduce these problems. A higher temperature can be attained in such an enclosure compared to those reached by direct exposure to the sun. Most of the incident radiation from the sun will pass through such transparent materials. However, most radiation from hot surfaces within the enclosure will be of longer wavelength and so will not readily pass outwards through the transparent cover. This is known as the 'greenhouse effect' and it can result in shorter drying times as compared with those attained in uncovered food exposed to sunlight. A transparent plastic tent placed over the food, which is spread on a perforated shelf raised above the ground, is the simplest form of covered sun drier. Warm air moves by natural convection through the layer of food and contributes to the drying. A simple sun drier of sturdier construction is shown in Fig. 3.10.

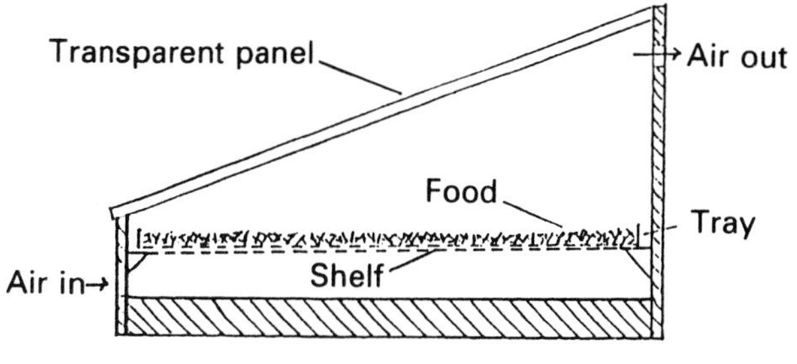

Fig. 3.10 Simple sun drier.

The capacity of such a drier may be increased by incorporating a solar collector. The warm air from the collector passes up through a number of perforated shelves supporting layers of food and is exhausted near the top of the chamber. A chimney may be fitted to the air outlet to increase the rate of flow of the air. The taller the chimney, the faster the air will flow. If a power supply is available, a fan may be incorporated to improve the airflow still further. Heating by gas or oil flames may be used in conjunction with solar drying. This enables heating to continue when sunlight is not available. A facility for storing heat may also be incorporated into solar driers. Tanks of water and beds of pebbles or rocks may be heated via a solar collector. The stored heat may then be used to heat the air entering the drying chamber. Drying can proceed when sunlight is not available. Heat storing salt solutions or adsorbents may be used instead or water or stones. Quite sophisticated solar drying systems, incorporating heat pumps, are also available [26, 30, 56–58].

3.2.12
Drying Food Liquids and Slurries in Heated Air

3.2.12.1 Spray Drying

This is the method most commonly used to dry liquid foods and slurries. The feed is converted into a fine mist or spray. This is known as atomisation and the spray forming device as an atomiser. The droplet size is usually in the range 10–200 μm, although for some applications larger droplets are produced. The spray is brought into contact with heated air in a large drying chamber. Because of the relatively small size of the droplets, a very large surface area is available for evaporation of the moisture. Also, the distance that moisture has to migrate to the drying surface is relatively short. Hence, the drying time is relatively short, usually in the range 1–20 s. Evaporative cooling at the drying surface maintains the temperature of the droplets close to the wet bulb temperature of the drying air, i.e. most of the drying takes place under constant rate conditions (see Section 3.2.2). If the particles are removed quickly from the drying chamber once they are dried, heat damage is limited. Hence, spray drying can be used to dry relatively heat sensitive materials.

The main components of a single stage spray drier are shown in Fig. 3.11.

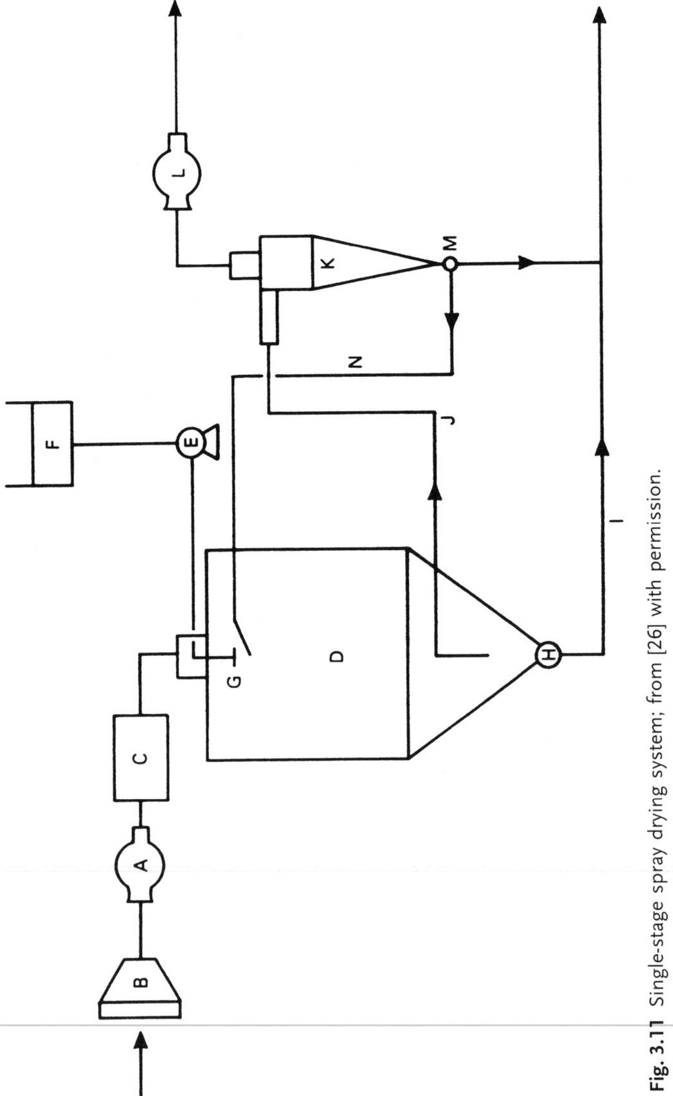

Fig. 3.11 Single-stage spray drying system; from [26] with permission.

Inlet fan A draws air in through filter B and then through heater C into drying chamber D. Pump E delivers the feed from tank F to the atomiser G. This converts the feed into a spray which then contacts the heated air in drying chamber D, where drying takes place. Most of the dry powder is removed from the chamber through valve H and pneumatically conveyed through duct I to a storage bin. The air leaves the chamber through duct J and passes through one or more air/powder separators K to recover the fine powder carried in the air. This powder may be added to the main product stream through valve M or returned to the wet zone of the drying chamber through duct N.

The air may be heated indirectly through a heat exchanger using oil or gas fuel or steam. In recent years, natural gas has been used to heat the air directly. There is still some concern about possible contamination of the food with nitrate and nitrite compounds, in particular *N*-nitrosodimethylamine, which has been shown to be harmful. The use of low NO_x (nitrogen oxide) burners reduces this problem. However, the quality of the air should be monitored when direct heating is used.

It is important that the droplets produced by the atomiser are within a specified size range. If the droplets vary too much in size, drying may not be uniform. The drying conditions must be set so that the larger droplets reach the desired moisture content. This may result in smaller droplets being overexposed to the heated air. Droplet size can affect some important properties of the dry powder, such as its rehydration behaviour and flow properties. There are three types of atomiser: centrifugal atomiser, pressure nozzle and two-fluid nozzle. A centrifugal atomiser consists of a disc, bowl or wheel on the end of a rotating shaft. The liquid is fed onto the disc near its centre of rotation. Under the influence of centrifugal force, it moves out to the edge of the disc and is spun off, initially in the form of threads, which then break up into droplets. There are many designs of disc in use. An example is shown in Fig. 3.12. Disc diameters range over 50–300 mm and they rotate at speeds in the range 50 000 rpm to 10 000 rpm, respectively. They are capable of producing uniform droplets. They can handle viscous feeds and are not subject to blocking or abrasion by insoluble solid particles in the feed.

A pressure nozzle features a small orifice, with a diameter in the range 0.4–4.0 mm, through which the feed is pumped at high pressure, in the range 5.0–50.0 MPa. A grooved core insert, sited before the orifice, imparts a spinning motion to the liquid, producing a hollow cone of spray. Pressure nozzles are capable of producing droplets of uniform size, if the pumping pressure is maintained steady. However, they are subject to abrasion and/or blocking by insoluble solid particles in the feed. They are best suited to handling homogeneous liquids, of relatively low viscosity.

A two-fluid nozzle, also known as a pneumatic nozzle, features an annular opening through which a gas, usually air, exits at high velocity. The feed exits through an orifice concentric with the air outlet. A venturi effect is created and the liquid is converted into a spray. The feed pumping pressure is lower than that required in a pressure nozzle. Such nozzles are also subject to abrasion

Fig. 3.12 Centrifugal atomiser; by courtesy of Niro.

and blocking if the feed contains insoluble solid particles. The droplets produced by two-fluid nozzles are generally not as uniform in size as those from the other two types of atomiser, especially when handling high viscosity liquids. They are also best suited to handling homogeneous liquids.

There are numerous designs of drying chamber used in industry. Three types are shown in Fig. 3.13. Figure 3.13 a depicts a tall cylindrical tower with a conical base. Both heated air and feed are introduced at the top of the chamber and flow concurrently down through the tower. This design is best suited to drying relatively large droplets of heat sensitive liquids, especially if the powder particles are sticky and tend to adhere to the wall of the chamber. The wall of the conical section of the chamber may be cooled to facilitate removal of the powder. Figure 3.13 b depicts another concurrent chamber, but featuring a shorter cylindrical body. The air enters tangentially at the top of the chamber and follows a downward, spiral flow path. The feed is introduced into the top of the chamber, through a centrifugal atomiser. Because of the spiral flow pattern followed by the air, particles tend to be thrown against he wall of the chamber. Consequently, this type of chamber is best suited to drying foods, which are not very heat sensitive or sticky. The chamber depicted in Fig. 3.13 c features a mixed flow pattern. The heated air is directed upward initially and contacts the spray of liquid from the centrifugal atomiser. The air, containing the droplets, then travels down to the bottom of the chamber. The risk of heat damaging the product is greater than in concurrent chambers. This design is not widely used for food dehydration. Many other designs of spray drying chambers are used in industry including flat-bottomed cylindrical and horizontal boxlike versions.

In industrial spray driers, most of the dry powder is removed from the bottom of the chamber through a rotary valve or vibrating device. Some powders, containing high amounts of sugar or fat, may tend to adhere to the wall of the chamber. Various devices are used to loosen such deposits. Pneumatically operated hammers, which strike the outside wall of the chamber, may be used. Brushes, chains or air brooms may sweep the inner surface of the chamber. The temperature of the wall of the chamber may be reduced by removing some of the insulation or by drawing cool air through a jacket covering some or all of the chamber wall. This can also reduce product build up on the wall.

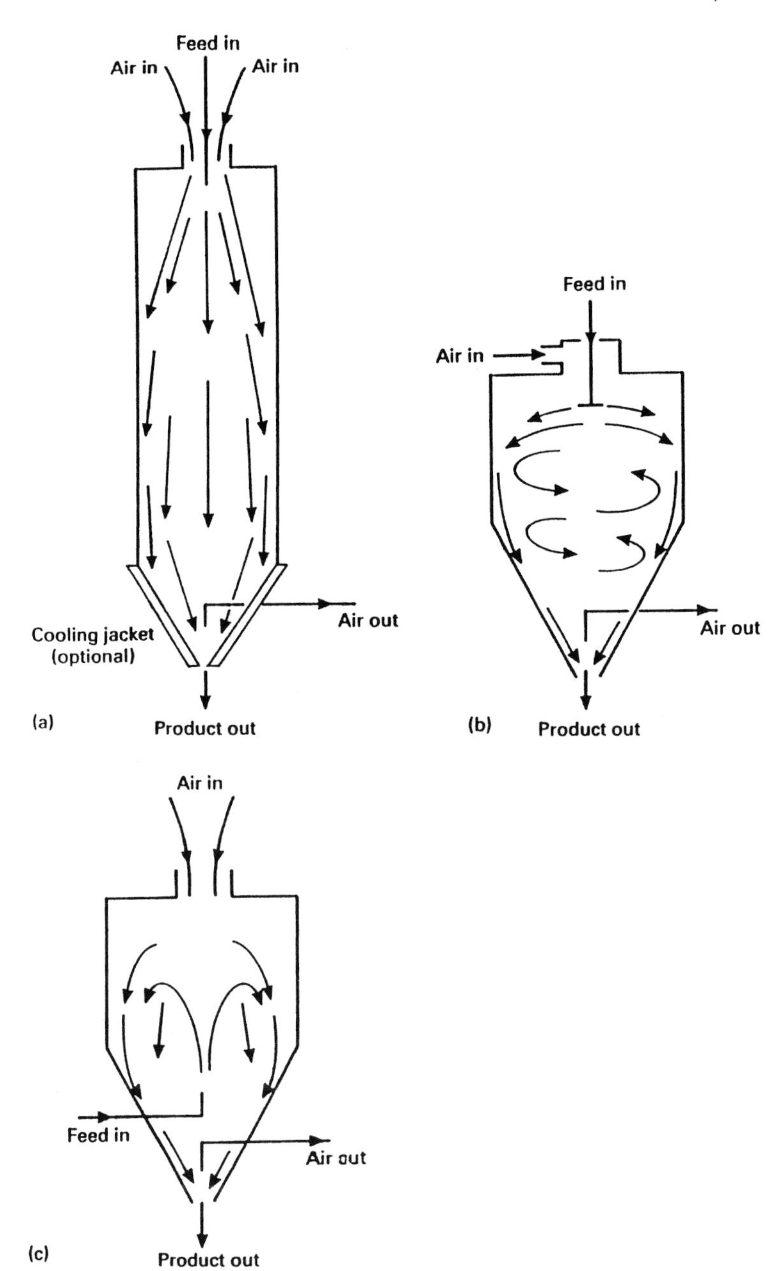

Fig. 3.13 Different designs of spray drying chambers: (a) concurrent with straight-line flow path, (b) concurrent with spiral flow path, (c) mixed flow; from [60] with permission.

The exhaust air from the chamber carries with it some of the fine particles of product. These need to be separated from the air, as they represent valuable product and/or may contaminate the environment close to the plant. Large dry cyclone separators are often used, singly or in pairs in series, for this purpose. Fabric filters are also employed for this duty. Powder particles may be washed out of the exhaust air with water or some of the liquid feed and recycled to the drying chamber. Such devices are known as wet scrubbers. An advantage of this method is that heat may also be recovered from the air and used to preheat or preconcentrate the feed. Electrostatic precipitators could also be used for recovering fines from the air leaving the drying chamber. However, they are not widely employed in the food industry. Combinations of the above separation methods may be used. The air, after passing through the cyclones, may then pass through a filter or scrubber to remove very fine particles and avoid contaminating the outside environment. Fines recovered by cyclones or filters may be added to the main stream of product. However, powders containing large amounts of small particles, less than 50 μm in diameter, are difficult to handle and may have poor rehydration characteristics. When added to a hot or cold liquid, fine particles tend to form clumps which float on the surface of the liquid and are difficult to disperse. It is now common practice to recycle the fines back into the wet zone of the drier where they collide with droplets of the feed to form small agglomerates. Such agglomerates disperse more readily than individual small particles when added to a liquid.

It is now common practice to remove the powder from the spray drying chamber before it reaches its final moisture content and to complete the drying in another type of drier. Vibrating fluidised bed driers are most often used as secondary driers (see Section 3.2.3.5). Some agglomeration of the powder particles occurs in this second stage, which can improve its rehydration characteristics. A second fluidised bed drier may be used to cool the agglomerated particles. Spray driers are available which permit multistage drying in one unit. In the integrated fluidised bed drier, a fluidised bed in the shape of a ring is located at the base of the drying chamber. Most of the drying takes place in the main drying chamber. Drying is completed in the inbuilt fluidised bed [1, 5, 7, 12, 26, 30, 59–61].

3.2.13 Drying Liquids and Slurries by Direct Contact With a Heated Surface

3.2.13.1 Drum (Roller, Film) Drier

The principle of this type of drier is that the feed is applied in a thin film to the surface of a rotating hollow cylinder, heated internally, usually with steam under pressure. As the drum rotates, heat is transferred to the food film, mainly by conduction, causing moisture to evaporate. The dried product is removed by means of a knife which scrapes the surface of the drum. The pattern of drying is similar to that of solids drying in heated air drying i.e. there are constant and falling rate periods (see Section 3.2.2). However, the temperature

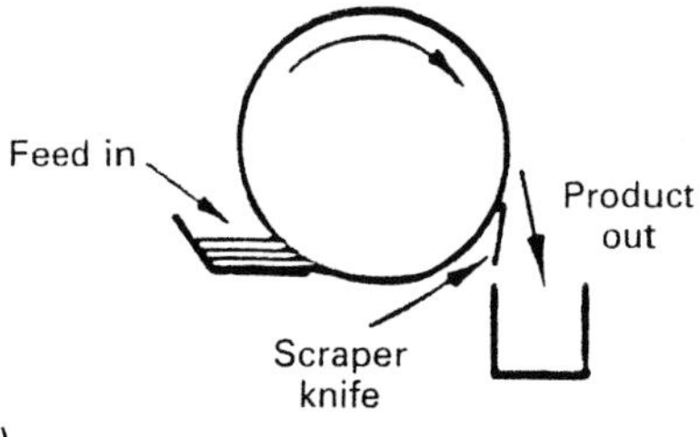

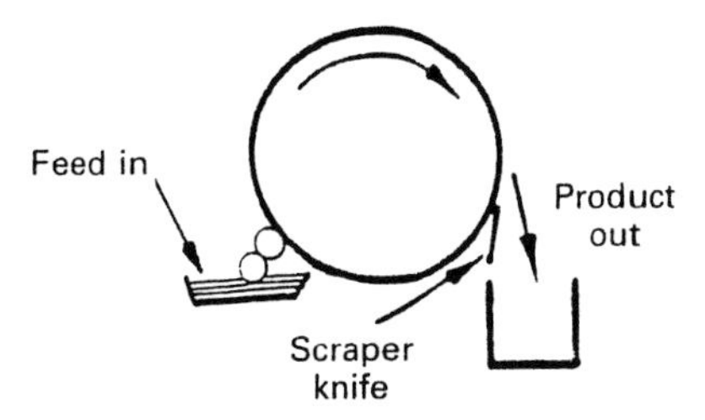

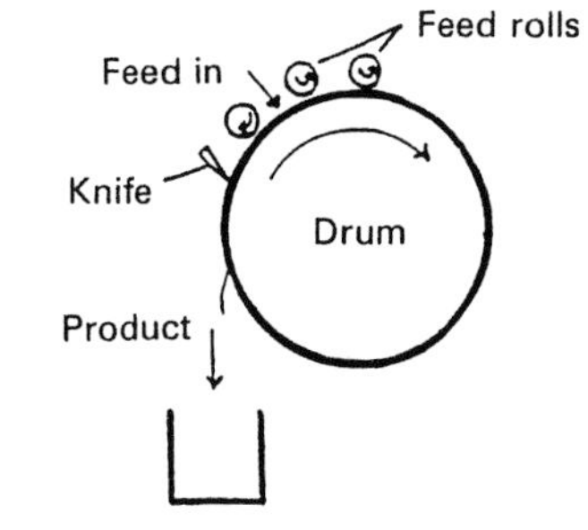

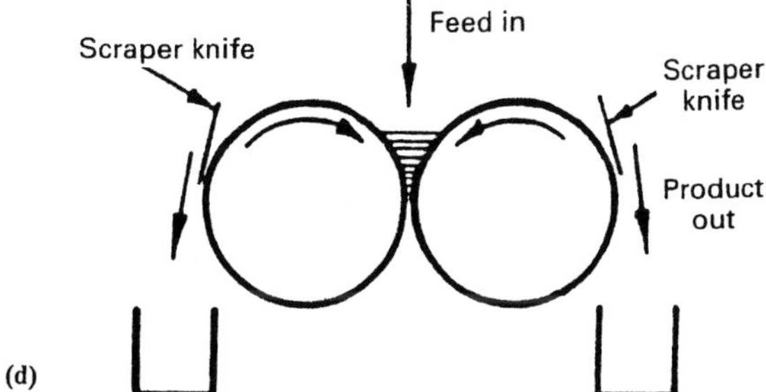

Fig. 3.14 Different designs of drum driers: (a) single drum with dip feed, (b) single drum with unheated roller feed, (c) single drum with multiroller feed (d) double drum; adapted from [26] with permission.

to which the food is exposed is usually higher than in heated air drying, as there is not the same degree of evaporative cooling. Drying times are relatively short, usually in the range 2–30 s. In a single drum drier, the feed may be applied to the drum by different means. The drum may dip into a trough containing the feed (Fig. 3.14a). This feed system is most suited to low viscosity liquids. For more viscous liquids and slurries, unheated rollers may apply the feed to the drum (Fig. 3.14b). A multiple roller system is used for high viscosity liquids and pastes (Fig. 3.14c). The layer of feed becomes progressively thicker as it passes beneath successive feed rollers. The feed may also be sprayed or splashed onto the drum surface. A double drum drier (Fig. 3.14d) consists of two drums rotating towards each other at the top. The clearance between the drums is adjustable. The feed is introduced into the trough formed between the drums and a film is applied to the surface of both drums as they rotate. This type of drum drier is suitable for relatively low viscosity liquids, that are not very heat-sensitive.

Drums are 0.15–1.50 m in diameter and 0.2–3.0 m long. They rotate at speeds in the range 3–20 rpm. They may be made from a variety of materials. For food applications, chromium-plated cast iron or stainless steel drums are mostly used. The drum surface temperature is usually in the range 110–165 °C. As the food approaches these temperatures towards the end of drying, it is likely to suffer more heat damage than in spray drying. For very heat-sensitive materials, one or two drums may be located inside a vacuum chamber operated at an absolute pressure from just below atmospheric down to 5 kPa. The drum is heated by vacuum steam or heated water, with a surface temperature in the range 100 °C to 35 °C. Provision must be made for introducing the feed into and taking the product out of the vacuum chamber and for adjusting knives, feed rollers, etc. from outside the chamber. Such vacuum drum driers are expensive to purchase and maintain and are not widely used in the food industry [1, 5, 26, 43, 62].

3.2.13.2 Vacuum Band (Belt) Drier

In this type of vacuum drier a metal belt or band, moving in a clockwise direction, passes over a heated and cooled drum inside a vacuum chamber (see Fig. 3.15).

The band may be continuous and made of stainless steel or fine stainless steel wire mesh. Alternatively, it may be made of hinged, stainless steel plates. The feed, in the form of a viscous liquid or paste, is applied to the band by means of a roller, after entering the vacuum chamber via a valve. The cooled product is scraped off the band and removed from the chamber through a rotary valve, a sealed screw or into two vacuum receivers, working alternately. Radiant heaters or heated platens, in contact with the belt, supplement the heat supplied by the heated drum. The chamber pressure and band temperature are in similar ranges to those used in vacuum drum driers (see Section 3.2.13.1). A number of bands, positioned one above the other, may be located in one vacuum chamber. Vacuum band driers are expensive to purchase and maintain and

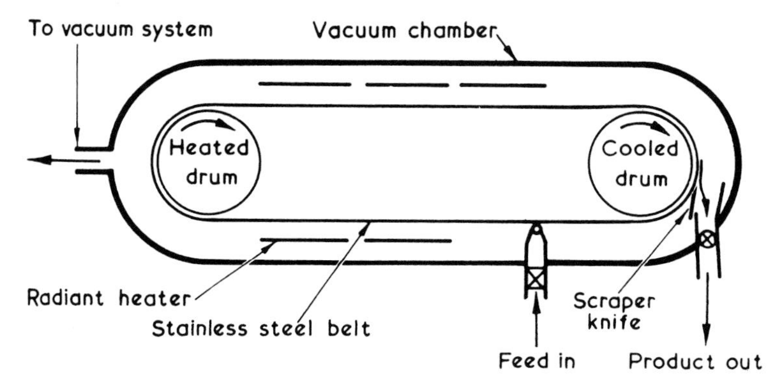

Fig. 3.15 Principle of vacuum band (belt) drier; from [1] with permission of the authors.

are only used for very heat-sensitive materials, which can bear the high costs involved [1, 5, 26, 30, 43].

3.2.14
Other Methods Used for Drying Liquids and Slurries

Concentrated liquids have been dried by heated air on *conveyor driers*, using a technique known as *foam mat drying*. The liquid is made into a foam, by the addition of a small amount, 1% or less, of a foaming agent, such as soya protein, albumin, fatty acid esters of sucrose and glycerol monostearate, and the incorporation of air or other gases by injection or mixing. The foam is spread in thin layers or strips on a wire mesh belt and conveyed through the drier. Relatively rapid drying can be achieved, of the order of 1 h in air at 100 °C, yielding a porous dry product with good rehydration properties [26, 63].

Liquids have been dried in *spouted beds* of inert particles. Metal and glass spheres and various plastic particles have been used as inert material. These have been fluidised with heated air to form spouted beds (see Section 3.2.3.5). The liquid is sprayed onto the particles and is heated by convection from the heated air and conduction from the hot particles, causing moisture to evaporate. When dry, the film is released from the particles by abrasion and impact between particles and separated from the exhaust air by a cyclone separator [57].

Liquids, usually in concentrated form, have been dried in *vacuum cabinet driers* (see Section 3.2.5.1). A technique known as *vacuum puff drying* has been used in this type of drier and in vacuum band driers. The concentrated liquid is spread in a thin layer on trays and placed in the vacuum chamber. When the vacuum is drawn, bubbles of water vapour and entrapped air form within the liquid and expand and the liquid froths up to form a foam. By careful control of pressure, temperature and viscosity of the liquid, it may be made to expand to occupy a space up to 20 times that of the original material. When heat is applied, the foam dries rapidly to form a porous dry product, with good rehydration properties [26, 63].

Concentrated liquids are *freeze dried.* The concentrate is frozen into slabs, which are broken up into pieces and dried in batch or continuous freeze driers (see Section 3.2.6).

3.2.15
Applications of Dehydration

The following are examples of the many foods which are preserved in dehydrated form.

3.2.15.1 Dehydrated Vegetable Products

Many vegetables are available in dehydrated form. A typical process involves:

← wet and dry cleaning
← peeling, if necessary, by mechanical, steam or chemical methods
← slicing, dicing or shredding
← blanching
← sulphiting, if necessary
← dewatering
← drying
← conditioning, if used
← milling or kibbling, if used
← screening
← packaging

Most vegetables are *blanched* prior to drying. See Chapter 1 for a discussion of the purposes and methods of blanching. Many vegetables are *sulphured* or *sulphited* prior to drying. Again see Chapter 1 for a discussion of this step in the process. Many different types of driers are used for drying vegetables including: cabinets, single or two stage tunnels, conveyor and fluidised bed driers (see Section 3.2.3). Air inlet temperatures vary from product to product but are usually in the range 50–110 °C. Tunnel, conveyor and fluidised bed driers may be divided into a number of drying zones, each zone being controlled at a different temperature, to optimise the process. If the drying is not completed in the main drier, the product may be *finish dried* or *conditioned* in a bin drier, supplied with dry air at 40–60 °C. Among the vegetables dried as outlined above are green beans, bell peppers, cabbage, carrot, celery, leeks, spinach and swedes. Some vegetables such as garlic, mushrooms, green peas and onions are not sulphited. Herbs such as parsley, sage and thyme may be dried without blanching or sulphiting. Vegetables may be dried in vacuum cabinet driers and in freeze driers, to yield products of superior quality to those produced by air drying. However, such products will be more expensive.

Vegetable purees may be dried. Cooked and pureed carrot and green peas maybe drum dried to produce a flaked product. Very finely divided, cooked carrot or green peas may be spray dried to a fine powder [26, 34, 65, 66].

A number of dried potato products are available. Dehydrated, diced potato is produced by a process similar to that outlined above. After blanching the potato pieces should be washed with a water spray to remove gelatinised starch from their surfaces. In addition to sulphite, calcium salts may be added to increase the firmness of the rehydrated dice. Cabinet, tunnel or conveyor driers may be used to dry the potato. Conveyor driers are most widely used. In recent years, fluidised bed driers have been applied to this duty. Finish drying in bins is often practiced. Potato flakes are produced by drum drying cooked, mashed potatoes. Two-stage cooking is followed by mashing or ricing. Sufficient sulphite is mixed with the mash to give 150–200 ppm in the dried product. An emulsion is made up containing, typically, monoglyceride emulsifier, sodium acid pyrophosphate, citric acid and an antioxidant and mixed into the mash. In some cases milk powder may also be added. The mash is then dried on single drum driers equipped with a feed roll and up to four applicator rolls. Steam at 520–560 kPa, absolute, is used to heat the drums. After drying, the dried sheet is broken up into flakes. Potato flour is made from poor quality raw potatoes which are cooked, mashed, drum dried and milled. Potato granules may also be produced from cooked, mashed potatoes. After cooking, the potato slices are carefully mashed. Some dry granules may be 'added back' to the mash, which then has sulphite added to give 300–600 ppm in the dried product. A second granulation stage then follows. It is important that as little as possible rupture of cells occurs at this stage. The granules are cooled to 15.5–26.5 °C and held at that temperature for about 1 h. During this 'conditioning' some retrogradaton of the starch occurs. A further gentle granulation then takes place and the granules are dried in pneumatic and/or fluidised bed driers to a moisture content of 6–7% (wwb [26, 34, 67]).

Sliced tomatoes may be sun dried. The slices are exposed to the fumes of burning sulphur in a chamber or dipped in, or sprayed with sulphite solution before sun drying. In recent years, there has been an increase in demand for sun dried tomatoes which are regarded as being of superior quality to those dried by other means. Tomato slices are also air dried in cabinet or tunnel driers to a moisture content of 4% (wwb). The dried slices tend to be hygroscopic and sticky. They are usually kibbled or milled into flakes for inclusion in dried soup mixes or dried meals.

Tomato juice may be spray dried. The juice is prepared by the 'hot or cold break process' (see Section 3.1.5.3) and concentrated by vacuum evaporation up to 26–48% total solids content, depending on the preparation procedure, before it is spray dried. The powder is hygroscopic and sticky when hot and tends to adhere to the wall of the drying chamber. A tall drying chamber downward, concurrent flow may be used. Alternatively, a chamber with a shorter body featuring a downward, concurrent, spiral flow path may be used. The wall of the chamber is fitted with a jacket through which cool air is circulated, to reduce wall deposition. An air inlet temperature in the range 140–150 °C is used and the product has a moisture content of 3.5% (wwb [12, 26, 34]).

3.2.15.2 Dehydrated Fruit Products

Many fruits are sun dried, including pears, peaches and apricots. However, hot air drying is also widely applied to fruits such as apple slices, apricot halves, pineapple slices and pears in halves or quarters. A typical process for such fruits involves:

← Washing
← grading
← peeling/coring, if required
← trimming, if required
← sulphiting
← cutting
← resulphiting
← drying
← conditioning, if required
← packing

Fruits are not usually blanched. However, a procedure known as dry-blanch-dry is sometimes used for some fruits such as apricots, peaches and pears. After an initial drying stage, in which the fruit is reduced to about half its initial weight, it is steam blanched for 4–5 min. It is then further dried and conditioned down to its final moisture content. Fruits are usually sulphured by exposure to the fumes of burning sulphur.

Cabinet or tunnel driers are most commonly used at the drying stage(s). Air inlet temperatures in the range 50–75 °C are used. Tunnel driers may be divided into two or three separate drying zones, each operating at a different air temperature.

Grapes are sun dried on a large scale to produce raisins or sultanas. They are also dried in heated air in cabinet or tunnel driers. The grapes are dipped in a hot 0.25% NaOH bath, washed and heavily sulphured. prior to drying down to a moisture content of 10–15% (wwb). They are usually conditioned in bins or sweat boxes to attain a uniform moisture content. Fumigation may be necessary during conditioning to kill any infestation. The 'golden bleached' appearance of the raisins or sultanas is due to the high SO_2 content which is in the range 1500–2000 ppm.

Whole plums are dried in cabinet or tunnel driers to produce prunes. The fruits are not sulphured. They are dried down to a moisture content in the range 16–19% (wwb) but may be 'stabilised' by rehydration in steam to a moisture content of 20–22% (wwb).

Fruit purees may drum dried to produce a flake or powder product. Bananas may treated with SO_2, pulped, pasteurised, further treated with SO_2, homogenised and dried on a single drum drier to a moisture content of 4% (wwb). Other fruit purees may be drum dried, including apricot, mango and peach. Some of these products are hygroscopic and sticky. Additives, such as glucose syrup, may have to be mixed into the puree to facilitate removal of the product from the drum and its subsequent handling.

Fruit juices may be spray dried. Concentrated orange juice is an example. To avoid the powder sticking to the wall of the drying chamber, additives are used. The most common one is liquid glucose with a dextrose equivalent in the range 15–30. It may be added in amounts up to 75% of the concentrate, calculated on a solids basis. Other additives such as skim milk powder and carboxymethyl celluloses have been used. However, these limit the uses to which the dry powder can be put. It is important that the concentrated juice is well homogenised, if it contains insoluble solid particles. Both centrifugal and nozzle atomisers are used. Tall chambers, featuring downward, concurrent flow patterns are favoured. The dried powder may be cooled on a fluidised bed to facilitate handling. Volatiles, separated from the juice before or after concentration, may be added to the powder to enhance its flavour. Other fruit juices which may be spray dried include lemon, mango, peach and strawberry [26, 34, 64, 68, 69].

3.2.15.3 **Dehydrated Dairy Products**

Skim or separated milk is mainly produced by spray drying. The raw whole milk is centrifuged to yield skim milk with 0.05% fat. The skim milk is then heat treated. The degree of heat treatment at this stage determines whether the powder produced is classed as low-heat, medium-heat or high-heat powder. The more severe the heat treatment, the lower the amount of soluble whey proteins (albumin and globulin) that remain in the powder. Low-heat powder is used in recombined milk products such as cheese and baby foods. Medium-heat powder is used in the production of recombined concentrated milk products. High-heat powder is mainly used in the bakery and chocolate industries. The milk is then concentrated to a total solids content in the range 40–55%, by multiple-effect evaporation. Falling film evaporators are most widely used in recent years. Various designs of spray drying chamber and atomiser can be used for drying skim milk, as it not a difficult material to dry. Air inlet and out temperatures used are in the ranges 180–230 °C and 80–100 °C, respectively. The moisture content of the powder is in the range 3.5–4.0% (wwb). It is common practice to recycle the fine powder from the separators into the wet zone of the drier. The powder may be removed from the drying chamber at a moisture content of 5–7% (wwb) and the drying completed in a vibrated fluidised bed drier. This promotes some agglomeration of the powder particles and improves its rehydration characteristics. Instant milk powder may be produced by rewetting the powder particles, usually with steam, mixing them to promote agglomeration and redrying them down to a stable moisture content. This can be carried out in a fluidised bed. The powder may be cooled in a second fluidised bed to facilitate handling.

Whole milk powder is also produced by spray drying in a similar manner to skim milk. Air inlet and outlet temperatures are usually in the ranges 175–200 °C and 75–95 °C, respectively. Whole milk powder is rather sticky when hot and can form a deposit on the chamber wall. Hammers, which are located outside the chamber and tap the wall of the chamber at intervals, may be used to assist in the removal of the powder. Whole milk powder may be agglomerated. However, this may not be as effective as in the case of skim milk powder. Fat may migrate to the surface of the particles. This reduces their wettability. The addition of small quantities of lecithin, a surface active agent, to whole milk powder, can improve its rehydration characteristics, particularly in cold liquids.

Other dairy products which are spray dried are buttermilk and whey. Whey powder is hygroscopic and difficult to handle, as the lactose is in a amorphous state. A crystallisation process before and/or after drying can alleviate this problem.

Relatively small amounts of milk and whey are drum dried. Double drum driers are usually used. The dry products are mainly used for animal feed. However, because of its good water binding properties, drum dried milk is used in some precooked foods [7, 9–12, 26, 60].

3.2.15.4 **Instant Coffee and Tea**

These products are produced by spray drying or freeze drying. The extract from ground roasted beans (see Section 3.1.5.2) is preconcentrated by vacuum evaporation before drying. Tall spray drying chambers, featuring a downward, concurrent flow pattern and a nozzle atomiser, are favoured as the powder is sticky. Air inlet and outlet temperatures in the ranges 250–300 °C and 105–115 °C, respectively, are used. Fine powder from the separators is recycled into the wet zone of the drier. Agglomeration may be attained in a fluidised bed. As an alternative to spray drying, the preconcentrated extract may be frozen into slabs, the slabs broken into pieces and freeze dried in batch or continuous equipment.

Instant tea may also be produced by spray drying the extract from the leaves. Similar equipment to that used for instant coffee is employed but at lower air inlet temperatures (200–250 °C). The concentrated extract may be freeze dried in a similar manner to coffee extract.

3.2.15.5 **Dehydrated Meat Products**

Cooked minced meat may be hot air dried in cabinet, conveyor, fluidised bed and rotary driers, down to a moisture content of 4–6% (wwb). Chicken, beef, lamb and pork may be dehydrated in this way. Chicken meat is the most stable in dried form, while pork is the least stable. The main cause of deterioration in such dried products is oxidation of fat leading to rancidity. Cooked minced meat may also be dried in vacuum cabinet driers to give better quality products, than hot air dried meat, but at a higher cost. Both raw and cooked meat in the form of steaks, slices, dice or mince may be freeze dried down to a moisture content of 1.5–3.0% (wwb), at a still higher cost. Dehydrated meat products are mainly used as ingredients in dried soup mixes, sauces and ready meals. They are also used in rations for troops and explorers [26, 34].

3.2.15.6 **Dehydrated Fish Products**

The traditional methods of extending the shelf life of fish are salting and smoking. Salting could be regarded as a form of osmotic drying whereby salt is introduced into the flesh of fish to reduce its water activity (see Section 3.2.16). Some water may evaporate during or after salting. Smoking involves exposure of the fish to smoke from burning wood. This may be done at relatively low temperature, c. 30 °C, and is known as cold smoking. A moisture loss of 10–11% may occur during smoking. Cold smoked fish products can be chilled, which gives them a shelf life of about 7 days. Hot smoking is carried out at temperatures up to 120 °C. Hot smoked products may have a sufficiently low water activity to be stable without refrigeration. Unsalted and unsmoked fish may be hot air dried in cabinet and tunnel driers, using relatively low air temperatures, down to 30 °C. However, this is not widely practised commercially. Freeze dried fish products are also available. Because of the high cost of freeze drying, only relatively expensive fish products, such as prawns and shrimps, are dried in this way. Dried fish products are used as ingredients in soup mixes and ready meals. Many fishery byproducts are produced in dried form, including fish hydrolysate, fish meal and fish protein concentrate [26, 71, 72].

3.2.16
Stability of Dehydrated Foods

When considering the stability of dehydrated foods it is not the total moisture content that is critical but rather the amount of moisture that is available to support microbial growth, enzymic and chemical activity. It is generally accepted that a proportion of the total moisture present in a food is strongly bound to individual sites on the solid components and an additional amount is less firmly bound, but is still not readily available as a solvent for various soluble food components. In studying the availability of water in food, a fundamental property known as water activity, a_w, is measured. This property is defined by the expression:

$$a_w = \frac{p_v}{p_w} \tag{3.5}$$

where p_v is the water vapour pressure exerted by a solution or wet solid and p_w is the vapour pressure of pure water at the same temperature. This expression also describes the relative humidity of an air-water vapour mixture. A plot of moisture content as a function of water activity, at a fixed temperature, is known as a sorption isotherm (Fig. 3.16). Isotherms may be prepared either by adsorption, i.e. placing a dry material in contact with atmospheres of increasing relative humidity, or by desorption, i.e. placing a wet material in contact with atmospheres of decreasing relative humidity. Thus two different curves may be obtained from the same material. This hysteresis effect is typical of many foods.

Food isotherms are often divided into three regions, denoted by A, B and C in Fig. 3.16. In region A, water molecules are strongly bound to specific sites on

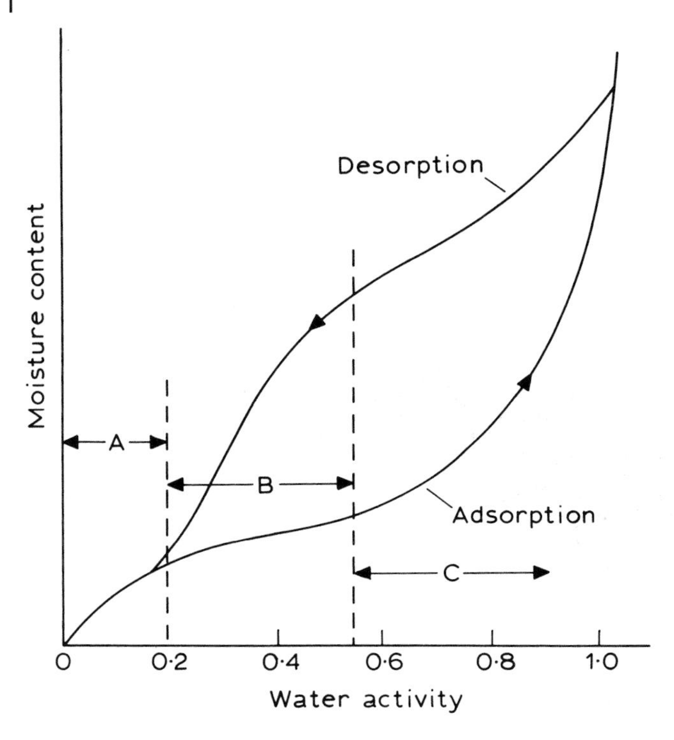

Fig. 3.16 Adsorption and desorption isotherms showing hysteresis; from [1].

the solid. Such sites may be hydroxyl groups in polysaccharides, carbonyl and amino groups in proteins and others on which water is held by hydrogen bonding, ion-dipole bonds or other strong interactions. This bound water is regarded as being unavailable as a solvent and hence does not contribute to microbial, enzymic or chemical activity. It is in the a_w range 0–0.35, and is known as the monomolecular or monolayer value. Monolayer moisture content in foods is typically in the range 0.05–0.11 (dwb). Above region A, water may still be bound to the solid but less strongly than in region A. Region B is said to consist of a multilayer region and region C is one in which structural and solution effects account for the lowering of the water vapour pressure. However, this distinction is dubious, as these effects can occur over the whole isotherm. Above region A, weak bonding, the influence of capillary forces in the solid structure and the presence of soluble solids in solution all have the effect of reducing the water vapour pressure of the wet solid. All these effects occur at a moisture content below 1.0 (dwb). Most foods exhibit a water vapour pressure close to that of pure water when the moisture content is above 1.0 (dwb). Temperature affects the sorption behaviour of foods. The amount of adsorbed water at any given value of a_w decreases with increase in temperature. Knowledge of the sorption characteristics of a food is useful in predicting its shelf life. In many cases the most stable moisture content corresponds to the monolayer value (Fig. 3.17).

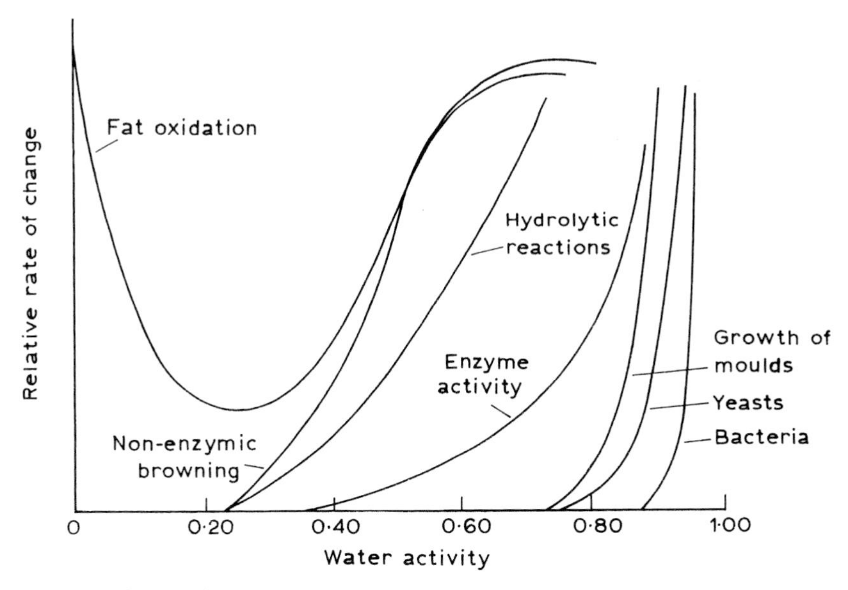

Fig. 3.17 Influence of water activity on the stability of foods; adapted from [70].

In many foods the rate of oxidation of fat is minimum in the a_w range 0.20–0.40. The rate of nonenzymic browning is highest in the a_w range 0.40–0.60. Below this range, reaction is slow due to the lack of mobility of the water. Hydrolytic reactions are also most rapid in the a_w range 0.40–0.70. The activity of enzymes starts to increase above the monolayer region and, at an a_w value above 0.80, it accelerates rapidly. Most moulds will not grow below an a_w value of 0.70, yeasts below 0.80. Most bacteria will not grow below an a_w of 0.90. Halophilic (salt loving) bacteria can grow at lower a_w. It is important to note that many other factors influence the activity of microorganisms including temperature, pH and availability of oxygen and nutrients and these may affect their behaviour at different a_w levels [1, 4, 5, 26, 70].

References

1 Brennan, J. G., Butters, J. R., Cowell, N. D., Lilly, A. E. V. **1990**, *Food Engineering Operations*, 3rd edn, Elsevier Applied Science, London.

2 Haywood, R. W. **1990**, *Thermodynamic Tables in SI (metric) Units*, 3rd edn, Cambridge University Press, Cambridge.

3 Hartel, R. A. **1992**, Evaporation and Freeze Concentration, in *Handbook of Food Engineering*, ed. D. R. Heldman, D. B. Lund, Marcel Dekker, New York.

4 Singh, R. P., Heldman, D. R. **1993**, *Introduction to Food Engineering, 2nd edn*, Academic Press, London.

5 Fellows, P. **2000**, *Food Processing Technology, 2nd edn*, Woodhead Publishing, Cambridge.

6 Burkart, A., Wiegand, B. **1987**, Quality and Economy in Evaporator Technology,

in *Food Technology International Europe*, ed. A. Turner, Sterling Publications International, London, pp. 35–39.
7 Pisecky, J. **1995**, Evaporation and Spray Drying, in *Handbook of Industrial Drying*, vol. 1, 2nd edn, ed. A. S. Mujumdar, Marcel Dekker, New York.
8 Walstra, P., Geurts, T. J., Jellema, A., van Boekel, M. A. J. S. **1999**, *Dairy Technology, Principles of Milk Properties and Processes*, Marcel Dekker, New York.
9 Knipschild M. E., Anderson G. G. **1994**, Drying of Milk and Milk Products, in *Modern Dairy Technology*, vol. 1, ed. R. K. Robinson, Chapman & Hall, London, pp. 159–254.
10 Early R. **1998**, Milk Concentrates and Milk Powders, in *The Technology of Dairy Products*, 2nd edn, ed. R. Early, Blackie Academic & Professional, Glasgow, pp. 228–300.
11 Lawley, R. **2001**, *LFRA Microbiology Handbook – Dairy Products*, 2nd edn, Leatherhead Food RA, Leatherhead.
12 Masters, K. **1991**, *Spray Drying Handbook*, 5th edn, Longman Scientific and Technical, New York.
13 Clarke, R. J. **1987**, Extraction, in *Coffee Technology*, vol. 2, ed. R. J. Clarke, R. Macrae, Elsevier Applied Science, London, pp. 109–145.
14 Clarke, R. J. **1987**, Drying, in *Coffee Technology*, vol. 2, ed. R. J. Clarke, R. Macrae, Elsevier Applied Science, London, pp. 147–199.
15 McGinnis, R. A. (ed.) **1971**, *Beet-Sugar Technology*, 2nd edn, Beet Sugar Development Foundation, Fort Collins.
16 Meade, G. P., Chen, J. C. P. **1977**, *Cane Sugar Handbook*, 10th edn, John Wiley & Sons, New York.
17 Perk, C. G. M. **1973**, *The Manufacture of Sugar from Sugarcane*, C. G. M. Perk, Durban.
19 Hugot, E. **1986**, *Handbook of Cane Sugar Engineering*, 3rd edn, Elsevier, Amsterdam.
20 Rao, M. A., Vitali, A. A. **1999**, Fruit Juice Concentration and Preservation, in *Handbook of Food Preservation*, ed. M. S. Rahman, Marcel Dekker, New York, pp. 217–258.
21 Nelson, P. E., Tressler, D. K. **1980**, *Fruit and Vegetable Juice Processing Technology*, 3rd edn, AVI Publishing Company, Westport.
22 Rebeck, H. M. **1990**, Processing of Citrus Juices, in *Production and Packaging of Non-Carbonated Fruit Juices and Fruit Beverages*, ed. P. R. Ashurst, Blackie Academic and Professional, Glasgow, pp. 221–252.
23 Kale, P. N., Adsule, P. G. **1995**, Citrus in *Handbook of Fruit Science and Technology*, ed. D. K. Salunkhe, S. S. Kadam, Marcel Dekker, New York, pp. 39–65.
24 Goose, P. G., Binsted, R. **1973**, *Tomato Paste and Other Tomato Products*, Food Trade Press, London.
25 Madhavi, D. L., Salunkhe, D. K. **1998**, Tomato in *Handbook of Vegetable Science and Technology*, Marcel Dekker, New York, pp. 171–201.
26 Brennan, J. G. **1994**, *Food Dehydration – a Dictionary and Guide*, Butterworth-Heinemann, Oxford.
27 Okos, M. R., Narsimhan, G., Singh, R. K., Weitnauer A. C. **1992**, Food Dehydration, in *Handbook of Food Engineering*, ed. D. R. Heldman, D. B. Lund, Marcel Dekker, New York, pp. 437–562.
28 Mujumdar, A. S. **1997**, Drying Fundamentals, in *Industrial Drying of Foods*, ed. C. G. J. Baker, Blackie Academic & Professional, London, pp. 7–30.
29 Sokhansanj, S., Jayes, D. S. **1995**, Drying of Foodstuffs, in *Handbook of Industrial Drying*, 2nd edn, ed. A. S. Mujumdar, Marcel Dekker, New York, pp. 589–625.
30 Barbosa-Canovas, G. V., Vega-Mercado, H. **1996**, *Dehydration of Foods*, Chapman & Hall, New York.
31 Toledo, R. T. **1991**, *Fundamentals of Food Process Engineering*, 2nd edn, Van Nostrand Reinhold, New York.
32 Wang, N., Brennan, J. G. **1994**, A Mathematical Model of Simultaneous Heat and Moisture Transfer During Drying of Potato, *Journal of Food Engineering*, 24, 47–60.
33 Brown, A. H., Van Arsdel, W. B., Lowe, E., Morgan Jr, A. I. **1973**, Air Drying and Drum Drying, in *Food Dehydration*, vol. 1, 2nd edn, ed. W. B. Van Arsdel,

M. J. Copley, A. I. Morgan Jr., AVI Publishing Company, Westport, pp. 82–160.
34 Greensmith, M. **1998**, *Practical Dehydration*, 2nd edn, Woodhead Publishing, Cambridge.
35 Sturgeon, L. F. **1995**, Conveyor Dryers, in *Handbook of Industrial Drying*, vol. 1, 2nd edn, ed. A. S. Mujumdar, Marcel Dekker, New York, pp. 525–537.
36 Hovmand, S. **1995**, Fluidised Bed Drying, in *Handbook of Industrial Drying*, vol. 1, 2nd edn, ed. A. S. Mujumdar, Marcel Dekker, New York, pp. 195–248.
37 Pallai, E., Szentmarjay, T., Mujumdar, A. S. **1995**, Spouted Bed Drying, in *Handbook of Industrial Drying*, vol. 1, 2nd edn, ed. A. S. Mujumdar, Marcel Dekker, New York, pp. 453–488.
38 Brennan, J. G. **2003**, Fluidized Bed Drying, in *Encyclopedia of Food Science and Nutrition*, 2nd edn, ed. B. Caballero, L. C. Trugo, P. M. Finglas, Academic Press, London, pp. 1922–1929.
39 Bahu, R. E. **1997**, Fluidized Bed Dryers, in *Industrial Drying of Foods*, ed. C. G. J. Baker, Blackie Academic and Professional, London, pp. 65–88.
40 Kisakurek, B. **1995**, Flash Drying, in *Handbook of Industrial Drying*, vol. 1, 2nd edn, ed. A. S. Mujumdar, Marcel Dekker, New York, pp. 503–524.
41 Kelly, J. J. **1995**, Rotary Drying, in *Handbook of Industrial Drying*, vol. 1, 2nd edn, ed. A. S. Mujumdar, Marcel Dekker, New York, pp. 161–184.
42 Barr, D. J., Baker, C. G. J. **1997**, Specialized Drying Systems, in *Industrial Drying of Foods*, ed. C. G. J. Baker, Blackie Academic and Professional, London, pp. 179– 209.
43 Oakley, D. **1997**, Contact Dryers, in *Industrial Drying of Foods*, ed. C. G. J. Baker, Blackie Academic and Professional, London, pp. 115–133.
44 Anon **1992**, Dryers: Technology and Engineering, in *Encyclopedia of Food Science and Technology*, vol. 1, ed. Y. H. Hui, John Wiley & Sons, Chichester, pp. 619–656.
45 Mellor, J. D. **1978**, *Fundamentals of Freeze-Drying*, Academic Press, New York.
46 Athanasios, I., Liapis, I., Bruttini, R. **1995**, Freeze Drying, in *Handbook of Industrial Drying*, vol. 1, 2nd edn, ed. A. S. Mujumdar, Marcel Dekker, New York, pp. 309–344.
47 Snowman, J. W. **1997**, Freeze Dryers, in *Industrial Drying of Foods*, ed. C. G. J. Baker, Blackie Academic and Professional, London, pp. 134–155.
48 Dalgleish, J. McN. **1990**, *Freeze-Drying for the Food Industries*, Elsevier Applied Science, London.
49 Lorentzen, J. **1975**, Industrial Freeze Drying Plants for Food, in *Freeze Drying and Advanced Food* Technology, ed. S. A. Goldblith, L. Rey, H. H. Rothmayr, Academic Press, London, pp. 429–443.
50 Ratti, C., Mujumdar, A. S. **1995**, Infrared Drying, in *Handbook of Industrial Drying*, vol. 1, 2nd edn, ed. A. S. Mujumdar, Marcel Dekker, New York, pp. 567–588.
51 Edgar, R. H. **2001**, Consumer, Commercial, and Industrial Microwave Ovens and Heating Systems, in *Handbook of Microwave Technology for Food Applications*, ed. A. K. Datta, R. C. Anantheswaran, Marcel Dekker, New York, pp. 215–277.
52 Schiffmann, R. F. **1995**, Microwave and dielectric drying, in *Handbook of Industrial Drying*, vol. 1, 2nd edn, ed. A. S. Mujumdar, Marcel Dekker, New York, pp. 345–372.
53 Jones, P. L., Rowley, A. T. **1997**, Dielectric Dryers, in *Industrial Drying of Foods*, ed. C. G. J. Baker, Blackie Academic & Professional, London, pp. 156–178.
54 Schiffmann, R. F. **2001**, Microwave Processes for the Food Industry, in *Handbook of Microwave Technology for Food Applications*, ed. A. K. Datta, R. C. Anantheswaran, Marcel Dekker, New York, pp. 215–299.
55 Lewicki, P. P., Das Gupta, D. K. **1995**, Osmotic Dehydration of Fruits and Vegetables, in *Handbook of Industrial Drying*, vol. 1, 2nd edn, ed. A. S. Mujumdar, Marcel Dekker, New York, pp. 691–713.
56 Bolin, H. R., Salunkhe, D. K. **1982**, Fluid Dehydration by Solar Energy, *CRC Critical Reviews in Food Science and Technology*, 16, 327–354.
57 Brennan, J. G. **1989**, Dehydration of Foods, in *Water and Food Quality*, ed. T. M. Hardman, Elsevier Applied Science, London, pp. 33–70.

58 Imrie, L. **1997**, Solar Dryers, in *Industrial Drying of Foods*, ed. C. G. J. Baker, Blackie Academic & Professional, London, pp. 210–241.

59 Masters, K **1997**, Spray Dryers, in *Industrial Drying of Foods*, ed. C. G. J. Baker, Blackie Academic & Professional, London, pp. 90–114.

60 Brennan, J. G. **2003**, Spray Drying, in *Encyclopedia of Food Science and Nutrition*, 2nd edn, ed. B. Caballero, L. C. Trugo, P. M. Finglas, Academic Press, London, pp. 1929– 1938.

61 Filkova, I., Mujumdar, S. S. **1995**, Industrial Spray Drying Systems, in *Handbook of Industrial Drying*, vol. 1, 2nd edn, ed. A. S. Mujumdar, Marcel Dekker, New York, pp. 263–308.

62 Moore, J. G. **1995**, Drum Dryers, in *Handbook of Industrial Drying*, vol. 1, 2nd edn, ed. A. S. Mujumdar, Marcel Dekker, New York, pp. 249–262.

63 Salumkhe, D. K., Bolin, H. R., Reddy, N. R. **1991**, *Storage, Processing and Nutritional Quality of Fruits and Vegetables*, vol. 2, 2nd edn, CRC Press, Boca Raton.

64 Jayaraman, K. S., Das Gupta, D. K. **1995**, Drying of Fruits and Vegetables, in *Handbook of Industrial Drying*, vol. 1, 2nd edn, ed. A. S. Mujumdar, Marcel Dekker, New York, pp. 643–690.

65 Luh, B. S., Woodruff, J. G. **1975**, *Commercial Vegetable Processing*, AVI Publishing Company, Westport.

66 Feinberg, B. **1973**, Vegetables, in *Food Dehydration*, vol. 2, 2nd edn,ed. W. B. Van Arsdel, M. J. Copley, A. I. Morgan, AVI Publishing Company, Westport, pp. 1–82.

67 Talburt, W. F., Smith, O. **1975**, *Potato Processing*, 3rd edn, AVI Publishing Company, Westport.

68 Woodroof, J. G., Luh, B. S. **1975**, *Commercial Fruit Processing*, AVI Publishing Company, Westport.

69 Nury, F. S., Brekke, J. E., Bolin, H. R. **1973**, Fruits, in *Food Dehydration*, vol. 2, 2nd edn, ed. W. B. Van Arsdel, M. J. Copley, A. I. Morgan, AVI Publishing Company, Westport, pp. 158–198.

70 Labuza, T. P. **1977**, The Properties of Water in Relationship to Water Binding in Foods. A Review, *Journal of Food Processing and Preservation*, 1, 176–190.

71 Aitken, A., Mackies, I. M., Merritt, J. H., Windsor, M. L. **1981**, *Fish Processing and Handling*, 2nd edn, Her Majesty's Stationary Office, Edinburgh.

72 Windsor, M., Barlow, S. **1981**, *Introduction to Fishery By-products*, Fishing News Books, Farnham.

4
Freezing

Jose Mauricio Pardo and Keshavan Niranjan

4.1
Introduction

To describe refrigeration process as a mere temperature decrease in the product is simplistic. Many studies in literature show the influence of this preservation process on important quality attributes of food, such as texture, colour, flavour and nutrient content [1–4]. It is clear that refrigeration affects biological materials in various ways, depending on their chemical composition, microstructure and physical properties. Additionally, processing parameters such as the cooling method used, the cooling rate and the final temperature play an important role in defining the keeping quality of food.

In this chapter, the term refrigeration covers both *chilling* and *freezing*, which can be distinguished on the basis of the final temperature to which a material is cooled and the type of heat removed. In chilling, only sensible heat is extracted, whereas freezing involves the crystallisation of water which requires the removal of latent heat and, therefore, the expenditure of more energy and time to complete the process.

This chapter is broadly divided into three sections: the first relates to the refrigeration equipment used to generate low temperatures and the mechanisms by which the heat removed from the food is transferred to a 'heat sink', the second section deals with the kinetics of this process and the third section addresses the effect of refrigeration on food quality.

4.2
Refrigeration Methods and Equipment

In earlier days, low temperatures were achieved by using ice obtained from high mountains, the polar regions of the earth, or that saved during winter. In the Roman empire, the use of natural ice was widespread; and food, water and beverages placed in isolated cabinets were cooled by the latent heat required to melt the ice (ca. 333 kJ kg^{-1}). Nowadays, industrial cooling systems can be divided into four

Food Processing Handbook. Edited by James G. Brennan

ISBN: 3-527-30719-2

main categories: (a) plate contact systems, (b) gas contact systems, (c) immersion and liquid contact systems and (d) cryogenic freezing systems.

The first three systems involve indirect cooling, i.e. the food and the refrigerant are not brought into direct contact. An inherent advantage of indirect heat removal is that the final temperature of the product can be easily controlled and, therefore, these methods can be used for both chilling and freezing. However, cryogenic methods are only used to freeze foods.

4.2.1 Plate Contact Systems

In this kind of equipment the food is placed in contact with a cold surface; and the temperature difference transfers heat from the food. It is common to find plate freezers in which the sample is positioned between two cold metallic plates. Pressure is applied to the plates in order to ensure good contact between the cold surfaces and the object to be cooled. It is clear that the presence of regular shapes in the food produce better results in this type of equipment. Moreover, since high levels of overall heat transfer coefficients can be achieved, plate refrigerators are commonly used in production lines for solid or packaged food.

Other types of contact surfaces involve conveyor belts and rotating drums. In the former, the food (usually packaged) is placed on a temperature-controlled metallic belt conveyor and is expected to reach the desired low temperature at the end of the conveyor run. The refrigeration time is therefore controlled by the speed of the belt. The drum refrigerator, in contrast, is used to cool or freeze high viscosity liquids by introducing them at the top of the rotating refrigerated drum and scraping the product off as the drum rotates through about 270 degrees.

Both drum and belt refrigerators use secondary refrigerants such as brine in order to cool down their metallic surfaces, while the plate type uses direct evaporation of a refrigerant to lower its temperature.

Gas Contact Refrigerators In this type of equipment, a cold gas (usually air) flows through the food absorbing heat from it. Depending on the air velocity and temperature, such a system can be used either for freezing or for cold storage. Even though air blast refrigeration systems are widely used for food processing [5], it does not mean that this kind of equipment is suitable for all types of materials.

A commonly used configuration is the refrigeration tunnel, where food is transported through a cold chamber on a conveyor. Cold air, forced inside the tunnel, establishes contact with the food and causes the heat transfer. Higher air velocities can achieve higher heat transfer rates, but there are limits [6]. Above a certain range of velocities, heat transfer through the food controls the overall rate; and this should be considered in order to avoid unnecessary wastage of energy. In the case of particulate or granular materials, high speed air can be used to fluidise the material, which can result in a faster temperature drop [3].

The above procedures are sometimes combined with vacuum cooling, in which the food is introduced in a chamber where the pressure is reduced. Due to the drop in pressure, a fraction of the water present in the product vaporises (usually surface water) carrying with it latent heat which reduces the temperature rapidly. The food is then transferred to an air cooled chamber to further reduce its temperature, or sent directly for storage. Although vacuum cooling is rapid, it is more expensive than other refrigeration methods. However, this technique is more appropriate for highly priced products where quality considerations dominate.

4.2.3 Immersion and Liquid Contact Refrigeration

In this method, the product is brought into direct contact with a fast flowing chilled liquid. It therefore tends to attain the temperature of the liquid rapidly. Chilled water, brines (salt and sugar solutions) and other type of liquids such as alcohol or ethylene glycol are commonly used. The food (sometimes unpacked) can be brought into contact with the flowing liquid in two different ways: by immersion or by spraying. In spray contacting, the liquid is normally sprayed over the food, which makes heat transfer more efficient. Further, less liquid is used, cutting costs and lowering environmental impact.

4.2.4 Cryogenic freezing

This type of refrigeration differs from other procedures because it does not depend on external low temperature production systems. Low temperatures are produced due to the phase change of the cryogenic liquids themselves which are brought into contact with the food in freezing cabinets. The process is very rapid, but at the same time very expensive. In some cases, cryogenic freezing is combined with air contact freezing as a two-stage process, in order to increase freezing rates whilst reducing process costs. In this process, also known as the cryomechanical process, a cold hard crust is produced on the material by cryogenic freezing, before sending it to a cold chamber to finish off the solidification process.

The most common cryogenic liquid used is nitrogen, which boils at –195.8 °C. It is odourless, colourless and chemically stable; however, high costs restrict its use to high value products.

4.3 Low Temperature Production

Low temperatures are commonly achieved by mechanical refrigeration systems and by the use of cryogenic fluids [7]. However, other ways have been studied and used. Some of these are described in Table 4.1 [8].

Table 4.1 Some nontraditional methods of producing low temperatures; from [8].

Name	Description	Applications
Peltier cooling effect	In 1834, Peltier observed the inverse thermocouple effect: if electricity is forced into a thermocouple circuit, one of the metals cools down	Home freezers, water coolers and air conditioning systems in the USSR and USA since 1949
Vortex cooling effect	In 1931, G. Ranqe discovered the vortex effect in which the injection of air into a cylinder at a tangent produces a spinning expansion of the air accompanied by the simultaneous production of cold and warm air streams.	Cooling chocolate
Acoustic cooling	A hollow cylindrical tube filled with helium and xenon is used in combination with a 300 Hz speaker and a Helmholtz resounder. Sound waves compress the gas mixture and it is heated up. During decompression the gas cools down and absorbs heat from the structure. Inside the system, the noise is high (180 db). A well insulated system is not more noisy than any mechanical system.	Expensive home freezers have been produced with this environmental ly friendly system.
Cooling with gadoline	Gadoline is an element that increases its temperature while exposed to magnetic fields; and when exposure ceases, it reduces its temperature to a value lower than the initial.	A prototype refrigerator has been built.
Cooling with hydrogen	Dr. Feldman from Thermal Electric Devices has used hydrogen to refrigerate. The 'HyFrig' device uses metal hydrides that are capable of absorbing large amounts of hydrogen (gas) and produce a significant decrease in temperature when the hydrogen is eliminated.	Vehicle air conditioning

4.3.1 Mechanical Refrigeration Cycle

This method takes advantage of the latent heat needed for a refrigerant to change phase. The refrigerant – which is a fluid evaporating at very low temperatures – is circulated in an evaporation-condensation cycle and the energy it absorbs during evaporation is used to transfer heat from the refrigeration chamber (which contains the food) to a heat sink.

Basic components of typical refrigeration system are depicted in Fig. 4.1 (a compressor, two heat exchangers, an expansion engine and a refrigerant).

The most efficient refrigeration cycle is known as Carnot cycle, which consists of the following two steps:

- frictionless and adiabatic compression and expansion (constant entropy);
- heat rejection and absorption with a refrigerant at constant temperature.

It should be pointed out that frictionless compressions and expansions are highly ideal; therefore, the Carnot cycle is only useful to describe a perfect process. It is commonly used to evaluate the extent to which a real system deviates from ideal behaviour.

From Fig. 4.1, it can be seen that the cycle follows a specific path. The four stages of the cycle can be described as follows:

- Compression (points 1–2). Initially, the refrigerant is in a gaseous state (point 1). Work is done on it by the compressor, when its pressure and temperature are elevated, resulting in a superheated gas with increased enthalpy (point 2).

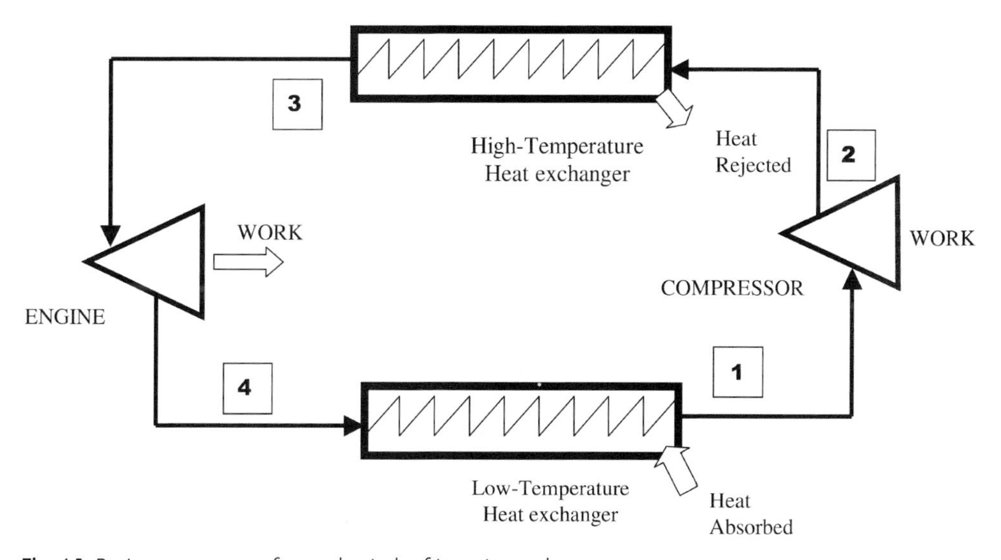

Fig. 4.1 Basic components of a mechanical refrigeration cycle.

- Condensation (points 2–3). The superheated gas enters a heat exchanger, commonly referred to as a high temperature exchanger or condenser. Using either air cooled or water cooled atmospheres, the refrigerant gives up heat to the surroundings and condenses to form a saturated liquid. By the time all the refrigerant has liquified, its temperature may fall below the condensation point and a subcooled liquid may be obtained.
- Expansion (points 3–4). The liquefied refrigerant enters the expansion engine, which separates the high and low pressure regions of the system. As the refrigerant passes through the engine, it experiences a pressure drop together with a drop in temperature. A mixture of liquid and gas leaves this process.
- Evaporation (points 4–1). Due to the energy received in the heat exchanger, also known as the low temperature exchanger or evaporator, the refrigerant evaporates. The saturated vapour may gain more energy and become superheated before entering the compressor and continuing the cycle.

The Pressure and Enthalpy Diagram The basic refrigeration cycle shown in Fig. 4.1 allows one to have a glimpse of the changes undergone by the refrigerant (energy absorption and rejection). However, other charts and diagrams are more useful while quantifying these changes. Using pressure-enthalpy charts, it is possible to estimate the changes in refrigerant properties as it goes through the process. Such charts are extremely useful in designing a refrigeration system. The most common charts depict variations in enthalpy (x axis) with pressure (y axis), and are known as p-h diagrams. However, there are other useful charts such as T-s diagrams that show variation of entropy (x axis) with temperature (y axis). Figures 4.2 and 4.3 show the Carnot refrigeration cycle on a p-h diagram and T-s diagram, respectively. The domes depicted in these figures

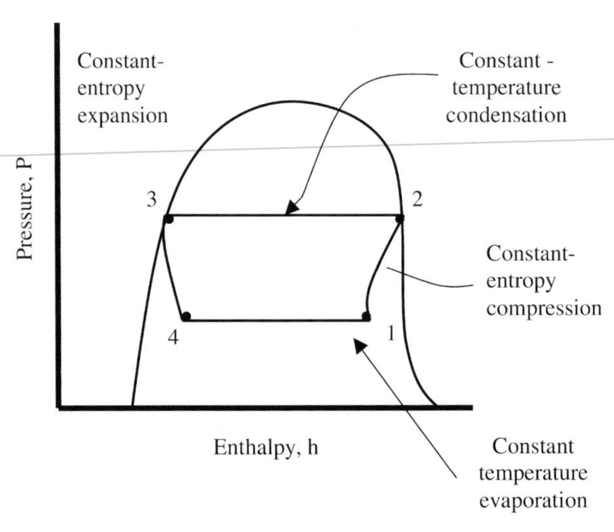

Fig. 4.2 Pressure-enthalpy diagram of the Carnot cycle.

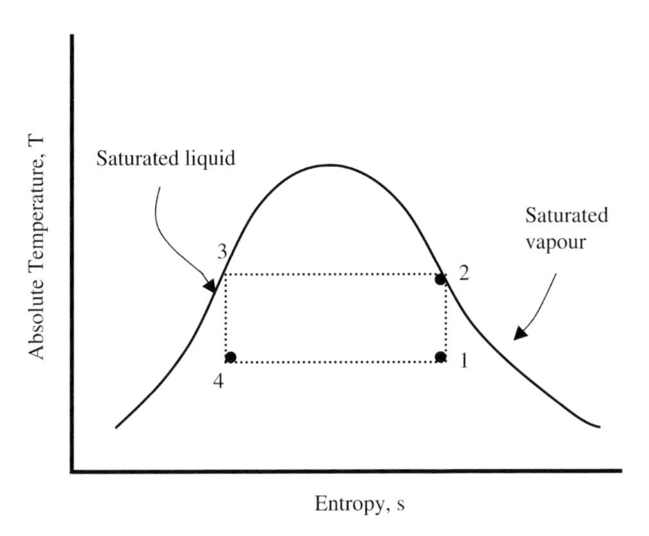

Fig. 4.3 Temperature-entropy diagram of the Carnot cycle.

represent the saturation points. To the left of the dome is the subcooled liquid region; and to the right lies the superheated vapour region. The region beneath the dome represents two phase (gas-liquid) mixtures.

From Figs. 4.2 and 4.3 it is clear that condensation (points 2–3) and evaporation (points 4-1) occur under constant temperature and pressure conditions, while expansion (points 3–4) and compression (points 1–2) occur under constant entropy conditions, as discussed earlier.

4.3.1 2 **The Real Refrigeration Cycle (Standard Vapour Compression Cycle)**

If attempts are made to operate equipment under the conditions proposed by the Carnot Cycle, severe mechanical problems are encountered, such as continuous valve breakdown, lack of good lubrication in the compression cycle, etc. Therefore a real refrigeration cycle should be adapted, considering the mechanical possibilities [7]. The main differences between the ideal Carnot cycle and that achieved in reality will be described in the following paragraphs.

Compression: Wet vs Dry The compression process observed in the ideal cycle (Fig. 4.2, points 1–2) is commonly described as wet compression because it starts with a mixture of gas and vapour (point 1), and occurs completely in the two-phase region (beneath the dome). The presence of a liquid phase can diminish lubrication effectiveness of some compressors. Moreover, droplets can also damage valves. Therefore, a real compression process should be carried out in the dry region shown in Fig. 4.4 (process 1′–2′).

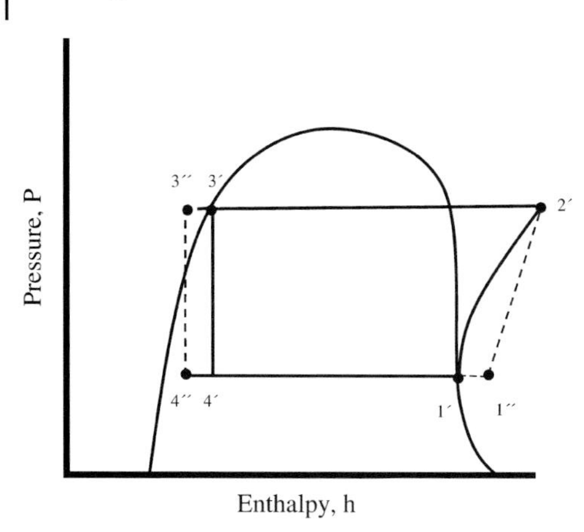

Fig. 4.4 Standard refrigeration cycle and modified standard cycle.

Expansion: Engine vs Valve Expansion process 3–4 described in the ideal cycle (see Fig. 4.2) assumes that the engine extracts energy from the refrigerant and uses it to reduce the compressor work. Practical problems include finding a suitable engine, controlling it and transferring its power to the compressor; therefore, it is uncommon to find expansion engines in refrigeration systems nowadays. Instead, the pressure drop is accomplished by a throttle valve (also called the expansion valve). Due to this change, the energy loss during expansion is negligible, thus $h'_3 = h'_4$. The changes made from cycle 1–2–3–4 to cycle 1′–2′–3′–4′ (see Fig. 4.4) results in a cycle known as the standard vapour compression cycle. Even this standard cycle has been modified further, in order to adapt it to more efficient conditions, e.g. cycle 1″–2′–3″–4″, where the vapour leaves the evaporator superheated and the liquid leaves the condenser subcooled.

4.3.2
Equipment for a Mechanical Refrigeration System

As mentioned earlier in this chapter, the basic components of a refrigeration system are the heat exchangers (evaporator and condenser), the compressor, the expansion valve and the refrigerant.

4.3.2.1 Evaporators

Inside this heat exchanger, the liquid refrigerant absorbs heat from the air in the cold room and vaporises. Depending on their design, evaporators can be classified into two types: direct expansion and indirect expansion evaporators. In the first type, the refrigerant vaporises inside the coils which are in direct contact with either the atmosphere or the object being cooled. In contrast, indirect contact evaporators use a carrier fluid (secondary refrigerant) that transfers heat from the atmosphere or the object that is being refrigerated to the evaporator coils. Although this type of evaporator costs more, it is useful when several locations are to be refrigerated with a single refrigeration system.

The most common type of industrial evaporator is that in which the air is forced through a row of fins. This air absorbs heat from food and returns to the evaporator where it rejects the heat gained. The heat transfer within such an evaporator can be described as a typical multilayer heat transfer with resistances in series. Heat must move from the air to the outer wall of the coil (convective resistance), then to the inner wall of the coil (conductive resistance) and finally to the refrigerant (convective resistance). These resistances can be defined as follows:

$$R_{cv} = \frac{1}{HA_o} \tag{4.1}$$

$$R_{cd} = \frac{\Delta x}{KA_{mean}} \tag{4.2}$$

where R_{cv} and R_{cd} are the convection and conduction resistances, respectively, H and K are heat transfer coefficient for convection and thermal conductivity and A is the related heat transfer area. It is clear from these equations that an increase in the heat transfer area will reduce the resistances, which is desirable. Additionally, it is known that the higher resistance is found in the outer layer of the coils and therefore it is common to see fins installed in the evaporator, in order to extend the contact area and decrease the resistance to heat transfer.

4.3.2.2 **Condensers**

Three main types of condensers can be found in refrigeration systems (see Fig. 4.5):

- air cooled
- water cooled
- evaporative

The purpose of each piece of equipment is to reject the heat absorbed by the refrigerant and transfer it to another medium such as air or water.

Similar to evaporators, air cooled condensers also use fins or plates to increase heat transfer efficiency. It is common to find fans attached to the condenser in order to increase air flow and therefore heat transfer rate. This type of equipment is easy and inexpensive to maintain and is therefore found in household appliances.

In water cooled condensers, the refrigerant flows inside tubes. Double pipe condensers have been used for many years in the food industry, however shell and tube condensers are more common nowadays.

Evaporative condensers operate like cooling towers. Water is pumped from a reservoir at the base of the condenser and sprayed onto the coils. Water evaporation extracts heat from the coils and a fast flowing air stream takes away the va-

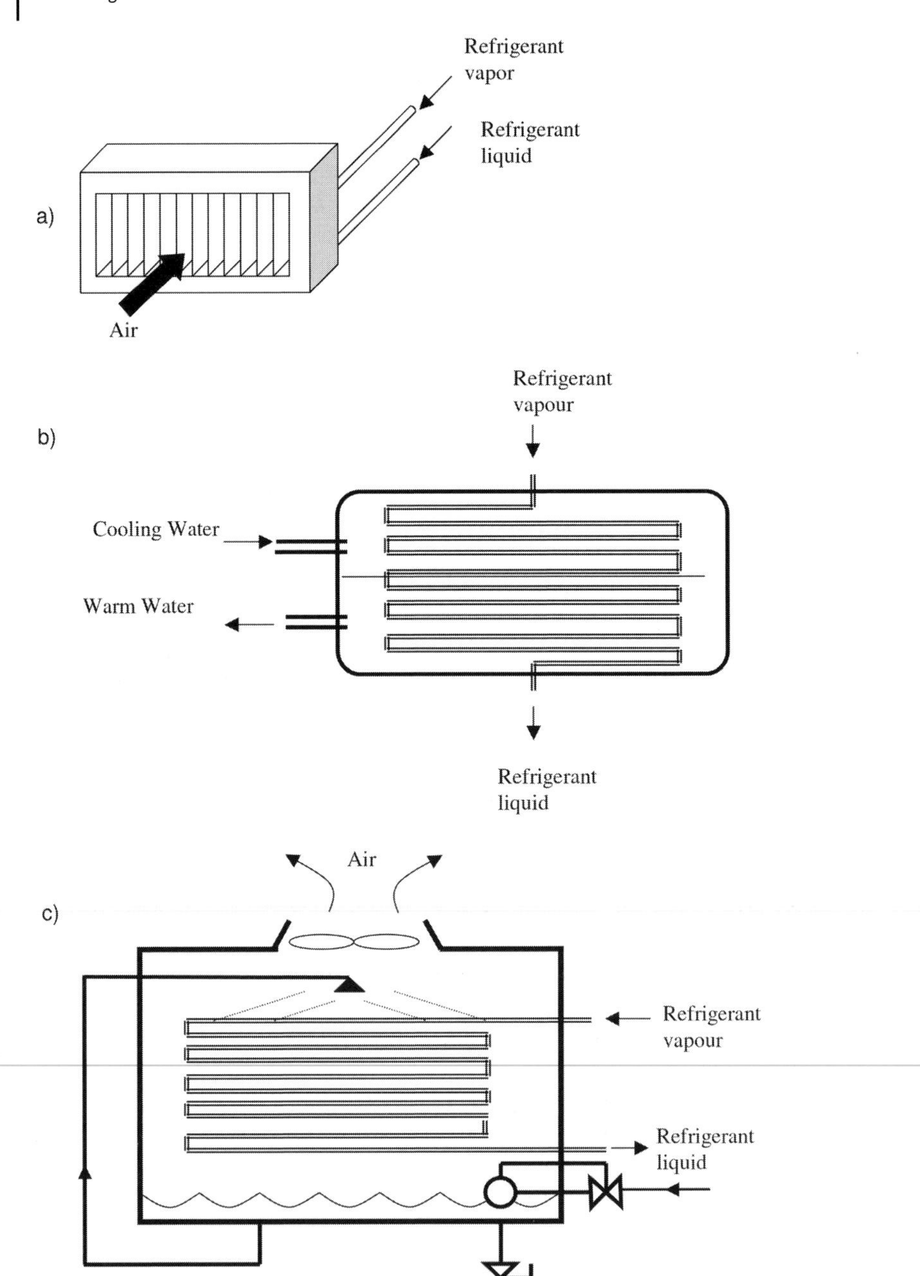

Fig. 4.5 Three different types of condensers: (a) air cooling, (b) water cooling, (c) evaporative.

pour, favouring heat and mass transfer. The water which failes to evaporate returns to the pan and is recycled with the pump. These types of systems are very efficient, although they require a lot of space.

4.3.2.3 Compressors

Refrigeration systems can be classified into three types depending on the method employed for compression:

- Mechanical compression. The most widely used system is the mechanical compression system, which is described in detail below.
- Thermal compression (absorption-desorption). This method is preferred when low cost/low pressure steam or waste heat is available. It is usually used to chill water (7–10 °C) and for capacities in the range from 300 kW to 5 MW.
- Pressure difference (or ejectors). These systems are used for similar applications as thermal compression systems and have lower initial and maintenance costs. However, they are not as common as thermal compression systems.

Mechanical Compression Compressors can be divided into two groups: positive displacement compressors (PDC) and dynamic compressors. PDC increase the pressure by reducing the volume. Typical compressors in this group are the reciprocating compressors, which dominate in the range up to 300 kW applications and tend to be the first choice due to their lower costs. Rotary compressors also belong to the PDC group. Screw type rotary compressors are found in the range 300–500 kW. Other rotary compressors such as the Vane [9] are found in applications in low capacity equipment.

In contrast, centrifugal compressors, which belong to the dynamic group, are found in equipment in the range 200–10 000 kW. These compressors have impellers turning at high speed and impart energy to the refrigerant. Some of this kinetic energy gained by the refrigerant is converted to pressure energy. These types of devices are more expensive than the reciprocating variety. However, they tend to be more economical to maintain. Further, they are generally small in size and are recommended when space is limited and long running periods are desired with minimum maintenance.

4.3.2.4 Expansion Valves

They are essentially flow controlling devices that separate the high pressure zone from the low pressure zone in a refrigeration system. Common types are:

- Manually operated. These are used to set the volume of refrigerant flowing from the high to the low pressure side.
- Automatic valves. This group can be divided into float, thermostatic and constant pressure valves. The float valve is the most cost effective of the three due to its simplicity and low maintenance costs, although it can only be used in flooded type condensers. However, thermostatic valves are the most widely used in the refrigeration industry.

4.3.2.5 **Refrigerants**

These can be classified into two groups: primary and secondary refrigerants. In the primary group belong those that vaporise and condense as they absorb and reject heat. Secondary refrigerants, in contrast, are heat transfer fluids commonly known as carriers. The choice of refrigerant is not easy, and there is a long list of commercial options. Desirable characteristics of these fluids are summarised in Table 4.2.

The refrigerants used in earlier days tended to be easily and naturally found: air, ammonia, CO_2, SO_2, ether, etc. Ammonia is the only refrigerant from this list that is still being used by the industry, however recent interests have been shown in CO_2 [10]. Ammonia shows several thermodynamically desirable properties, such as a high latent heat of vaporisation. Additionally, it has economic and environmental advantages; however, its irritating and intoxicating effects are major drawbacks. Other types of refrigerants became important in the 20th century due to efficiency and safety considerations. Halocarbons such as R12 (Freon, dichorofluormethane), R22 (cholodifluormethane), R30 (methylene chloride) and others are still being used in air conditioning (building and car) and small refrigeration systems. During the 1970s, it was postulated that chorofluorcarbons (CFCs) have damaging effects on the ozone layer. Due to their high stability in the lower atmosphere, they tend to migrate to the upper atmosphere where the chlorine portion splits and reacts with the ozone. The use of this kind of refrigerants started to decline in the 1990s due to the influence of global environmental agreements such as the Montreal treaty [11].

Hydrofluorocarbons (HFCs) are compounds containing carbon, hydrogen and fluorine. Some of these chemicals are accepted by industry and scientists as alternatives to CFCs and hydrochlorofluorocarbons. The most commonly used HFCs are HFC 134a and HFC 152a. Because HFCs contain no chlorine, they do not directly affect stratospheric ozone and are therefore classified as substances with low ozone depletion potential. Although it is believed that HFCs do not deplete ozone, these compounds have other adverse environmental effects, such as infrared absorptive capacity. Concern over these effects may make

Table 4.2 Desirable characteristics of modern refrigerants.

Safe	Nontoxic; not explosive; nonflammable
Environmentally friendly	Ozone friendly; no greenhouse potential; leaks should be easily detected; easily disposable
Low cost	Low cost per unit of mass; high latent heat/cost ratio
System compatible	Chemical stability; noncorrosive
Good thermodynamic and physical properties	High latent heat; low freezing temperature; high critical temperature (higher than ambient); low viscosity

it necessary to regulate the production and use of these compounds at some point in the future. Such restrictions have been proposed in the Kyoto protocol [12].

4.3.3 Common Terms Used in Refrigeration System Design

There are some useful expressions to define the capacity and efficiency of a refrigeration system and its components. Some of them are based on the variation of refrigerant energy and therefore show how pressure-enthalpy diagrams are useful tools for the design of refrigeration systems.

4.3.3.1 Cooling Load

This is defined as the rate of heat removal from a given space. As pointed out earlier in this chapter, ice was the main source of low temperatures when refrigeration was first commercialised; and therefore a typical unit for cooling load – known as a tonne of refrigeration – is equivalent to the latent heat of fusion of 1 t of ice in 24 h, which is equivalent to 3.52 kW. In designing a refrigeration system, several other sources of heat, in addition to the demand by the food itself, should be considered. Table 4.3 summarises some of the most relevant factors.

4.3.3.2 Coefficient of Performance (COP)

This is expressed as the ratio of the refrigeration effect obtained to the work done in order to achieve it. It is calculated by dividing the amount of heat absorbed by the refrigerant as it flows through the evaporator by the heat equivalent of the work done by the compressor. Thus in enthalpy values:

Table 4.3 Basic sources of heat to be considered in a cold room design.

Source of heat	Conside rations
Heat introduced through ceiling, floor and walls	Conduction of heat through walls and insulation materials
Heat transferred by the food material	Sensible heat; respiration heat (applied to refrigerated vegetables); latent heat (when freezing occurs)
Heat transferred by people	Working time; number of people
Heat transferred by engines working inside the cold room	Working time; number of engines; total power
Heat transferred by illumination bulbs	Working time; number of bulbs; total power
Heat introduced by air renewal	Number of times the door is opened; other programmed changes of air

$$COP = \frac{h_2 - h_1}{h_2 - h_3} \tag{4.3}$$

The COP value in a real refrigeration cycle is always less than that for the Carnot cycle. Therefore, industry is constantly striving to improve COP. It should be pointed out that methods to increase COP should be studied for every refrigerant separately.

4.3.3.3 **Refrigerant Flow Rate**

This quantity is useful for design purposes and depends on the cooling load in the following way:

$$\dot{m} = \frac{Q}{h_1 - h_4} \tag{4.4}$$

where $\dot{m}$ is the refrigerant flow rate (kg s^{-1}) and Q is the total cooling load (kW).

4.3.3.4 **Work Done by the Compressor**

The energy used for using the rise in enthalpy of the refrigerant is:

$$Wc = \dot{m}\,(h_2 - h_1) \tag{4.5}$$

An efficiency term should be introduced into the above equation when calculating the work done by the compressor.

4.3.3.5 **Heat Exchanged in the Condenser and Evaporator**

As heat exchange in both condenser and evaporator occur at constant pressure and temperature, the amount of heat rejected or gained by the refrigerant in each process can be estimated as follows:

$$Q_c = \dot{m}\,(h_2 - h_3) \tag{4.6}$$

$$Q_e = \dot{m}\,(h_1 - h_4) \tag{4.7}$$

4.4 Freezing Kinetics

As explained earlier, from a thermodynamic point of view, freezing is a more complex process than cooling due to the phase change involved:

- cooling and undercooling of the liquid sample
- nucleation
- growth
- further cooling of the frozen material.

The state diagram shown in Fig. 4.6 is useful to describe a typical water freezing process which follows the cooling path A-B-C-D-E.

Starting at room temperature (point A), the sample is taken to point B without phase change (undercooling) due to an energy barrier that must be surmounted before nucleation starts. The formation of stable nuclei is governed by the net free energy of formation, which is the summation of the surface and volume energy terms. Higher undercooling favours the formation of a higher number of nuclei, which could have an effect on the final crystal size distribution. Due to the latent heat liberated, a temperature shift would be expected (points B-C in Fig. 4.6). Once stable ice nuclei are formed, they continue to grow. Progress in crystal growth (points C-D in Fig. 4.6) depend on the supply of water molecules from the liquid phase to the nuclei (mass transfer) and the removal of latent heat (heat transfer). After all of the water is frozen, the temperature decreases again (points D-E in Fig. 4.6).

There are some differences between the phase diagrams of food and water. Due to the interaction of water with solutes present in the food, a freezing point depression occurs. Additionally, ice crystallisation causes a progressive concentration of the solution that remains unfrozen. Thus, as more ice is formed, the amorphous matrix increases its solid content and viscosity and reduces its freezing point, as seen from the path C′-E′. This concentration and ice formation continues until a temperature is reached at which no more water will freeze; this temperature is known as the glass transition temperature of the amorphous concentrated solution (T_g' [13], point E in Fig. 4.6). At this temperature, the concentrated solution vitrifies due to the high viscosity of this matrix (10^{12} Pa s); and molecular movement is negligible (diffusion rates fall to a few microns per year) and therefore in food materials a fraction of unfrozen water is unavoidable.

Both water and temperature affect molecular mobility, due to their plasticising capacity. Therefore, diffusion coefficients increase when water content or

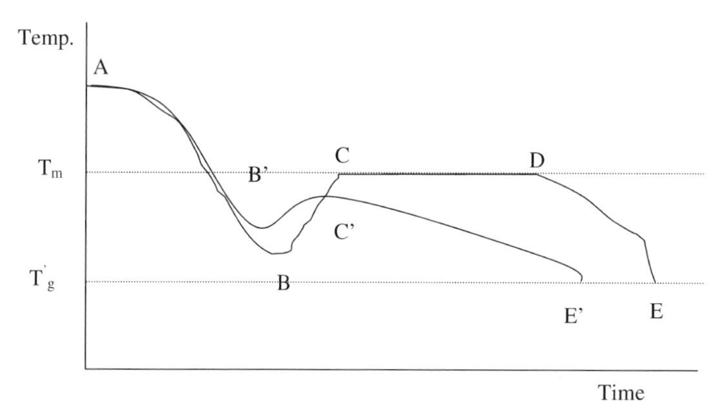

Fig. 4.6 A time-temperature curve during typical freezing. Path A-B-C-D-E corresponds to the freezing of water; and path A-B′-C′-E′ corresponds to that for a solution.

temperature is increased. William et al. [14] observed that the decrease in viscosity above T_g for various glassforming substances could be modelled with what became the well known WLF equation:

$$\log \frac{\mu}{\mu_o} = \frac{c_1(T - T_o)}{c_2 + T - T_o} \tag{4.8}$$

where μ and μ_o are the viscosities at temperatures T and T_o, and the so called universal constants c_1 and c_2 are equal to 17.44 and 51.6, respectively. The validity of this equation over a temperature range of $T_g < T < (T_g+100)$ has been tested and related to other physical changes such as collapse, crispiness, crystallisation, ice formation and deteriorative reactions [15]. However, the validity of the so called universal constants has been questioned.

4.4.1 Formation of the Microstructure During Solidification

The overall driving force for solidification is a complex balance of thermodynamic and mass transfer factors, which dominate in different parts of the process [16]. Crystal growth tends to be parallel and opposite to the direction of heat transfer; and therefore finger-like dendrites can be formed if unidirectional heat transfer is imposed [17].

Under unidirectional freezing, the most important crystal form is the dendrite [17]. Under directional cooling, cell-like structures can grow (see Fig. 4.7).

However, anisotropy in heat and mass transfer properties tends to favour dendritic cells and dendrite growth (Fig. 4.7 b, c). The formation of a dendrite begins with the breakdown of the planar solid-liquid interface. Dendritic tips reject solute in all directions and therefore spaces between dendrites accumulate the rejected solute, favouring the formation of cell-type dendrites (Fig. 4.7 a). Secondary branches form afterwards if the conditions of the media allow.

A combination of heat and mass transfer governs the movement of the freezing front. During unidirectional cooling, heat transfer tends to be faster than mass transfer due to the high thermal conductivity of ice and low mass diffusion coefficients. Therefore, solute diffusion will be the limiting factor of growth and an undercooling in the tip region will be observed. This supercooling is known as constitutional undercooling and has been used to develop models on the primary spacing between dendrites, i.e. the space between the tips of two dendrites [18–22]. All these models have, in common, the following general expression that correlates interdendritic spacing with freezing kinetics:

$$L \propto R^a G^b \tag{4.9}$$

where L is the mean interdendritic distance R is the freezing front rate, G is the temperature gradient between the ice front and the freezing plate and a, b are constants.

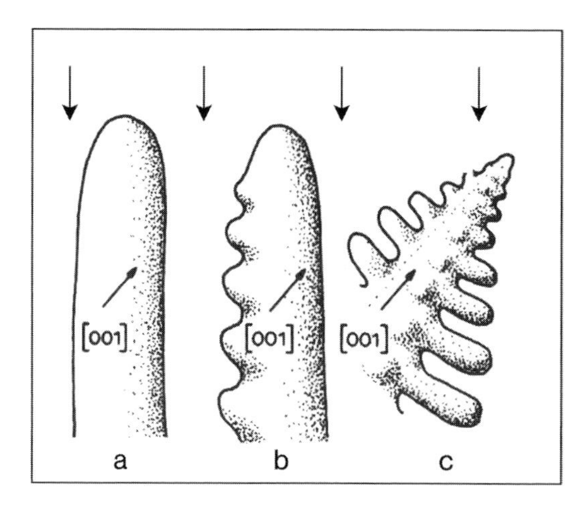

Fig. 4.7 Different ice crystal shapes: (a) represents a cell which is only possible in unidirectional freezing, (b) this shape is known as dendritic cell and (c) this shape is described as dendrite; after [17].

4.4.2 Mathematical Models for Freezing Kinetics

The kinetics of freezing has been widely studied and mathematically defined as a heat conduction process with phase change. The estimation of freezing times requires analysis of conductive heat flow through frozen and unfrozen layers, in addition to the heat transfer from the sample to the environment. The most common analyses are based on either Plank's or Neumann's models, which are exact solutions valid for unidirectional freezing under the assumption of an isothermal phase change [23, 24]. While Plank's model assumes quasi-steady state heat transfer, the Neumann's model is more generally applicable and is based on unsteady state conduction through frozen and unfrozen layers. In addition to analytical methods, approximate analytical procedures are reviewed by Singh and Mannapperuma [25].

4.4.2.1 Neumann's Model

Under conditions of unidirectional freezing, it has been established that the position of the freezing front varies with time as follows [17, 22, 26]:

$$x = c\sqrt{t} \tag{4.10}$$

where x is the position of the ice front, t is time and c is a kinetic constant. This equation is valid under the assumption of a planar freezing front.

The equations following from Neumann's model are listed in Table 4.4. They can be solved analytically to result in a relationship similar to Eq. (4.10):

$$x = 2\delta\sqrt{a_1 t} \tag{4.11}$$

where a_1 is the thermal diffusivity of the frozen zone, and δ represents the Neumman dimensionless characteristic number.

Table 4.4 Basic heat equations involved in Neumman's model.

	Frozen layer	Liquid layer
Conservation equation	$\frac{\partial T_1}{\partial t} = a_1 \frac{\partial^2 T_1}{\partial x^2} 0 < x < s$	$\frac{\partial T_2}{\partial t} = a_2 \frac{\partial^2 T_2}{\partial x^2}$ $s < x < x_t$
Initial conditions and boundary conditions	$T_1 = T_f,\ x = 0,\ t > 0$ $T_1 = T_m,\ x = s,\ t > 0$	$T_2 = T_0$ for all $x,\ t = 0$ $T_2 = T_0, x \to \infty,\ t > 0$ $T_2 = Tm, x = s,\ \ t > 0$
At the ice front	$-K_1 \frac{\partial T_1}{\partial x} + K_2 \frac{\partial T_2}{\partial x} + \rho_1 \Delta H_f \frac{ds}{dt},\ x = s$	

4.4.2.2 **Plank's Model**

Plank's analytical solution is recognised as the first equation proposed for predicting freezing times. It is based on two assumptions: (a) the sample is initially at its freezing point and (b) there is constant temperature in the unfrozen region. These assumptions imply that each layer in the unfrozen region remains at a constant temperature until the freezing front reaches it. In addition, Plank's model uses convective heat transfer as a boundary condition. Plank's model has been applied to different basic geometries and the general solution is as follows:

$$t = \frac{\rho \Delta H}{T_f - T_a} \left(\frac{PD}{h} + \frac{RD^2}{K} \right) \tag{4.12}$$

where P and R are geometric constants with values which change according to the shape of the sample. For a slab $P = \frac{1}{2},\ R = \frac{1}{8}$; for a cylinder $P = \frac{1}{4},\ R = \frac{1}{8}$; and for a sphere $P = \frac{1}{6},\ R = \frac{1}{24}$.

Pham [27] extended Plank's solution to account for sensible heat removal. He divided the freezing process into three stages: precooling, freezing and subcooling.

4.4.2.3 **Cleland's Model**

This model [28] estimates freezing times of food samples based on the following assumptions: (a) the conditions of the surrounding are constant, (b) the sample is found initially at a uniform temperature, (c) the final temperature has a fixed value and (d) Newton's cooling law describes the heat transfer at the surface.

Freezing time can be estimated using the following equation:

$$t_{slab} = \frac{R}{h} \left[\frac{\Delta H_1}{\Delta T_1} + \frac{\Delta H_2}{\Delta T_2} \right] \left(1 + \frac{N_{Bi}}{2} \right) \tag{4.13}$$

This equation is valid for Biot numbers between 0.02 and 11, Stefan numbers between 0.11 and 0.36 and Plank numbers between 0.03 and 0.61. Table 4.5 contains the equations used to estimate ΔH_1, ΔH_2, ΔT_1, ΔT_2.

Table 4.5 Equations and variables involved in Cleland's model.

$\Delta H_1 = C_u\,(T_i - T_3)$	C_u: unfrozen volumetric specific heat capacity $\frac{J}{m^3 K}$
$\Delta H_2 = H_L + C_f\,(T_3 - T_f)$	C_f: frozen volumetric heat capacity $\frac{J}{m^3 K}$
$\Delta T_1 = \frac{(T_i + T_3)}{2} - T_a$	H_L: latent heat of fusion $\frac{kJ}{kg}$
$T_3 = 1.8 + 0.263\ T_f + 0.105\ T_a$	
$\Delta T_2 = T_3 - T_a$	T_i: initial temperature
	T_a: chamber temperature

The basic model is defined for a slab, but freezing times for samples with various geometrical shapes can be estimated by approximating these shapes to an ellipsoid. A shape factor E is used to relate the estimated freezing time for the slab to that of the ellipsoid as seen from the following equations:

$$t_{ellipsoid} = \frac{t_{slab}}{E} \tag{4.14}$$

$$E = 1 + \frac{\left(1 + \frac{2}{N_{Bi}}\right)}{\left(\beta_1^2 + \frac{2\beta_1}{N_{Bi}}\right)} + \frac{\left(1 + \frac{2}{N_{Bi}}\right)}{\left(\beta_2^2 + \frac{2\beta_2}{N_{Bi}}\right)} \tag{4.15}$$

$$\beta_1 = \frac{A}{\pi R^2} \tag{4.16}$$

4.5 Effects of Refrigeration on Food Quality

Market trends show that consumers are paying more attention to flavour and nutritional attributes of food rather than to texture and colour characteristics; and therefore refrigerated products are gaining more importance in the marketplace due to the ability of this technology to maintain the two abovementioned quality factors [29]. Foods gain from refrigerated storage (frozen and chilled) due to the positive effect of low temperatures on molecular movement, microbial growth and chemical reaction rates. At temperatures marginally above zero, quality is well preserved for short periods (days or weeks). However, for longer periods, frozen storage is well suited because reactions continue at very low rates and microbial growth is virtually stopped to a point where the microbial population can be reduced [1].

During freezing, ice crystals can be formed in the space between cells and intracellular water can migrate, provided the cooling rate is slow enough [20]. This movement of water can produce irreversible changes in cell size. Additionally, it can damage membranes, causing the loss of water and enzymes that are responsible for colour and odour changes during thawing. Therefore rapid freezing, the use of cryoprotectants such as sugar and pretreatments such as blanching can improve the quality of the frozen product because cell wall damage and enzyme activity can both be reduced. Reid [30] suggests four processes that can help to explain the damages of vegetable tissue during freezing: cold (temperatures above 0 °C), solute concentration, dehydration and ice crystal injuries. These injuries also apply to animal tissues. However, their relative importance is different: while solute concentration causes more damage to the latter, crystal injuries have a bigger effect on texture and therefore are more important in vegetable tissues.

References

1 Rahman, S. **2003**, *Handbook of Food Preservation*, Marcel Dekker, New York.

2 Jeremiah, L. (ed.) **1995**, *Freezing Effects on Food Quality*, Marcel Dekker, New York.

3 Mallet, C. P. **1993**, *Frozen Food Technology*, Chapman Hall, London.

4 Jul, M. **1984**, *The Quality of Frozen Foods*, Academic Press, London.

5 Duiven, J., Binard, P. **2002**, *Bulletin of the International Institute of Refrigeration*, 2002 (2), 3–15.

6 Incropera, F., Dewitt, D. **1996**, *Fundamentals of Heat and Mass Transfer*, 4th edn, John Wiley & Sons, New York.

7 Wang, S. K. **1994**, *Handbook of Air Conditioning and Refrigeration*, McGraw-Hill, New York.

8 James, S. **2001**, Rapid Chilling of Food – a Wish or a Fact, *Bulletin of the International Institute of Refrigeration*, 3, 13.

9 Stoecker, W. F. **1988**, *Industrial Refrigeration*, Business News Publishing Company, Troy-Michingan.

10 Pearson, A. **2001**, New Developments in Industrial Refrigeration, *ASHRAE Journal*, 43, 54–60.

11 UNEP **2003**, *Handbook for the International Treaties for the Protection of the Ozone Layer*, 6th edn, available at: http://www.unep.org./ozone/Handbook2003.sht.

12 UNFCC **1977**, *Full Text of the Convention*, available at: http://unfcc.int/resource/conv/conv.html.

13 Slade, L., Levine, H. **1991**, Beyond Water Activity: Recent Advances Based on an Alternative Approach to the Assessment of Food Quality and Safety, *Critical Reviews in Food Science and Nutrition*, 30, 115–360.

14 Williams, M., Landel, R., Ferry, J. D. **1955**, The Temperature Dependence of Relaxation Mechanisms in Amorphous Polymers and Other Glass-Forming Liquids, *Journal of American Chemical Society*, 77, 3701–3707.

15 Roos, H. **1955**, Glass Transition – Relate Physicochemical Changes in Food, *Food Technology*, 49, 97– 102.

16 Sahagian, M., Douglas, G. **1995**, Fundamental Aspects of the Freezing Process, in *Freezing Effects on Food Quality*, ed. L. Jeremiah, Marcel Dekker, New York, pp. 1–50.

17 Kurz, W., Fisher, D. **1987**, *Fundamentals of Solidification*, Trans Tech Publications, Aedermansdorf.

18 Rohatgi, P., Adams, C. **1967**, Effects of Freezing Rates on Dendritic Solidification of Ice From Aqueous Solutions, *Transactions of the Metallurgical Society, AIME*, 239, 1729–1737.

19 Kurz, W., Fisher, D. J. **1981**, Dendrite Growth at the Limit of Stability: Tip Radius and Spacing, *Acta Metallurgica*, 29, 11–20.

20 Bomben, J., King, C. 1982, Heat and Mass Transport in Freezing of Apple Tissue, *Journal of Food Technology*, 17, 615–632.

21 Woinet, B., Andrieu, J., Laurant, M. **1998**, Experimental and Theoretical Study of Model Food Freezing. Part I.

Heat transfer modelling, *Journal of Food Engineering*, 35, 381–393.

22 Pardo, J., Suess, F., Niranjan, K. **2002**, An Investigation Into the Relationship Between Freezing Rate and Mean Ice Crystal Size for Coffee Extracts, *Transactions IchemE*, 80 (Part C).

23 Carslaw, H., Jaeger, J. **1959**, *Conduction of Heat in Solids*, 3rd edn, Clarendon, Oxford.

24 Ozilgen, M. **1998**, *Food Process Modeling and Control*, Gordon & Breach Science Publishers, Amsterdam.

25 Singh, R. P., Mannaperuma, J. D. **1990**, Developments in Food Freezing, in *Biotechnology and Food Process Engineering*, ed. H. Schwartzberg, A. Roa, Marcel Dekker, New York, pp. 309–358.

26 Woinet, B., Andrieu, J., Laurant, M. **1998**, Experimental and Theoretical Study of Model Food Freezing. Part II. Characterization and Modelling of the Ice Crystal Size, *Journal of Food Engineering*, 35, 395–407.

27 Pham, Q. T. **1984**, An extension to Plank's equation for predicting freezing times of foodstuffs of simple shapes, *International Journal of Refrigeration*, 7, 377–383.

28 Cleland, D. J., Cleland, A. C., Earle, R. L. **1987**, Prediction of Freezing and Thawing Times for Multi-Dimensional Shapes by Simple Formulae: I Regular Shapes, *International Journal of Refrigeration*, 10, 156–164.

29 Kadel, A. **2001**, Recent Advances and Future Research Needs in Postharvest Technology of Fruits, *Bulletin of the International Institute of Refrigeration*, 2001, 3–13.

30 Reid, D. S. **1993**, Basic Physical Phenomena of the Freezing and Thawing of Animal and Vegetable Tissue, in *Frozen Food Technology*, ed. C. P. Mallet, Blackie Academic & Professional, London, pp. 1–19.

5 Irradiation

Alistair S. Grandison

5.1 Introduction

Irradiation has probably been the subject of more controversy and adverse publicity prior to its implementation than any other method of food preservation. In many countries, including the UK, irradiation has been viewed with suspicion by the public, largely as a result of adverse and frequently misinformed reporting by the media. In the UK this has resulted in the paradoxical situation where irradiation of many foods is permitted, but not actually carried out. On the other hand, many nonfood items such as pharmaceuticals, cosmetics, medical products and plastics are routinely irradiated. Yet, the process provides a means of improving the quality and safety of certain foods, while causing minimal chemical damage.

Irradiation of food is not a new idea. Since the discovery of X-rays in the late 19th century, the possibility of controlling bacterial populations by radiation has been understood. Intensive research on food irradiation has been carried out for over 50 years, and the safety and 'wholesomeness' of irradiated food have been established to the satisfaction of most scientists.

5.2 Principles of Irradiation

Irradiation literally means exposure to radiation. In practice three types of radiation may be used for food preservation: Gamma (γ) rays, X-rays or high-energy electron beams (β particles). These are termed ionising radiations. Although the equipment and properties differ, the three radiation types are all capable of producing ionisation and excitation of the atoms in the target material, but their energy is limited so that they do not interact with the nuclei to induce radioactivity. Gamma rays and X-rays are part of the electromagnetic spectrum, and are identical in their physical properties, although they differ in origin.

Food Processing Handbook. Edited by James G. Brennan

ISBN: 3-527-30719-2

The energy of constituent particles or photons of ionising radiations is expressed in electron volts (eV), or more conveniently in MeV (1 MeV = 1.602×10^{-13} J). One eV is equal to the kinetic energy gained by an electron on being accelerated through a potential difference of 1 V. An important and sometimes confusing distinction exists between radiation energy and dose. When ionising radiations penetrate a food, energy is absorbed. This is the 'absorbed dose' and is expressed in Grays (Gy), where 1 Gy is equal to an absorbed energy of 1 J kg^{-1}. Thus, while radiation energy is a fixed property for a particular radiation type, the absorbed dose varies in relation to the intensity of radiations, exposure time and composition of the food.

Gamma rays are produced from radioisotopes and hence they have fixed energies. In practice Co 60 is the major isotope source. This isotope is specifically manufactured for irradiation and is not a nuclear waste product. An alternative radioisotope is Cs 137, which is a byproduct of nuclear fuel reprocessing, and is used very much less widely than Co 60.

Electron beams and X-rays are machine sources, which are powered by electricity. Hence they exhibit a continuous spectrum of energies depending on the type and conditions of the machinery. They hold a major advantage over isotopes in that they can be switched on and off and can in no way be linked to the nuclear industry.

The mode of action of ionising radiation can be considered in three phases:

- the primary physical action of radiation on atoms;
- the chemical consequences of these physical actions;
- the biological consequences to living cells in food or contaminating organisms.

5.2.1
Physical Effects

Although the energies of the three radiation types are comparable, and the results of ionisation are the same, there are differences in their mode of action as shown in Fig. 5.1.

High-energy electrons interact with the orbital electrons of the medium, giving up their energy. The orbital electrons are either ejected from the atom entirely, resulting in ionisation, or moved to an orbital of higher energy, resulting in excitation. Ejected (secondary) electrons of sufficient energy can go on to produce further ionisations and excitations in surrounding atoms (see Fig. 5.1 a).

X-rays and γ rays can be considered to be photons; and the most important interaction is by the Compton effect (see Fig. 5.1 b, c). The incident photons eject electrons from atoms in the target material, giving up some of their energy and changing direction. A single photon can give rise to many Compton effects and can penetrate deeply into the target material. Ejected electrons which have sufficient energy (approx. 100 eV) go on in turn to cause many further excitations and ionisations. Such electrons are known as delta rays. It has been calculated that a single Compton effect can result in 30–40 000 ionisations and 45–80 000 excitations [1].

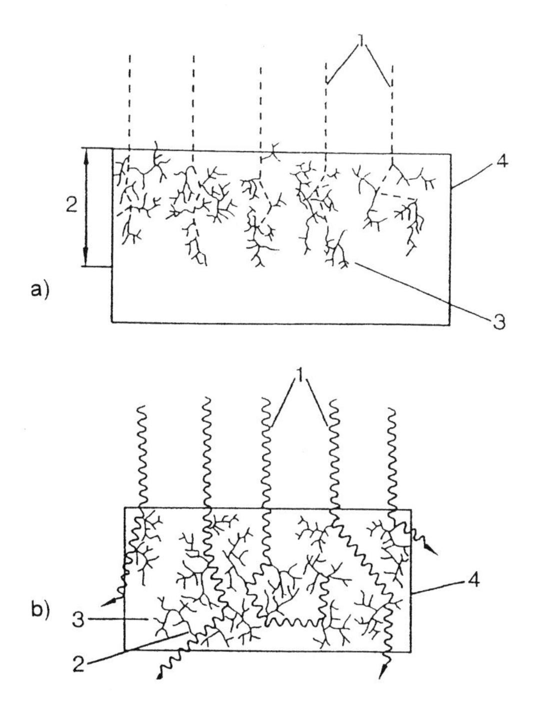

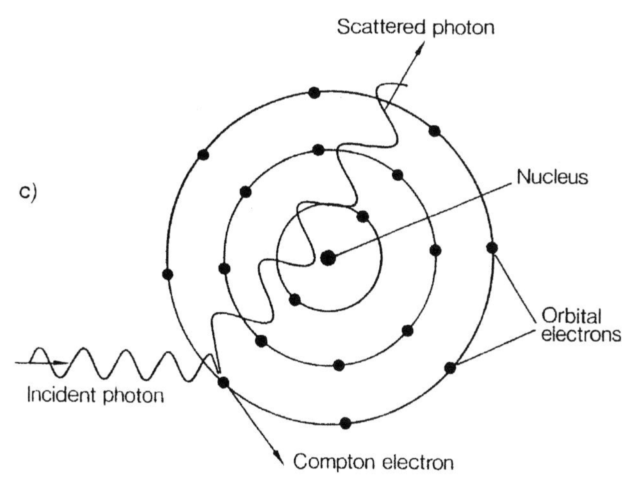

Fig. 5.1 Interaction of radiation with matter, adapted from Diehl [3]. (a) Electron radiation: 1, primary electron beam; 2, depth of penetration; 3, secondary electrons; 4, irradiated medium. (b) Gamma or X-radiation: 1, γ or X-ray photons; 2, Compton electrons; 3, secondary electrons; 4, irradiated medium. (c) Compton effect.

The radiation-induced chemical changes produced by γ or X-ray photons or by electron beams are exactly the same, because ionisations and excitations are ultimately produced by high-energy electrons in both cases. However, an important difference between photons and high-energy electrons is the depth of penetration into the food. Electrons give up their energy within a few centimetres of the food surface, depending on their energy, whereas γ or X-rays penetrate much more deeply, as illustrated in Fig. 5.1 a, b. Depth-dose distribution curves for electrons and γ rays are shown in Fig. 5.2 for irradiation of water, which acts as a reasonable model for foods.

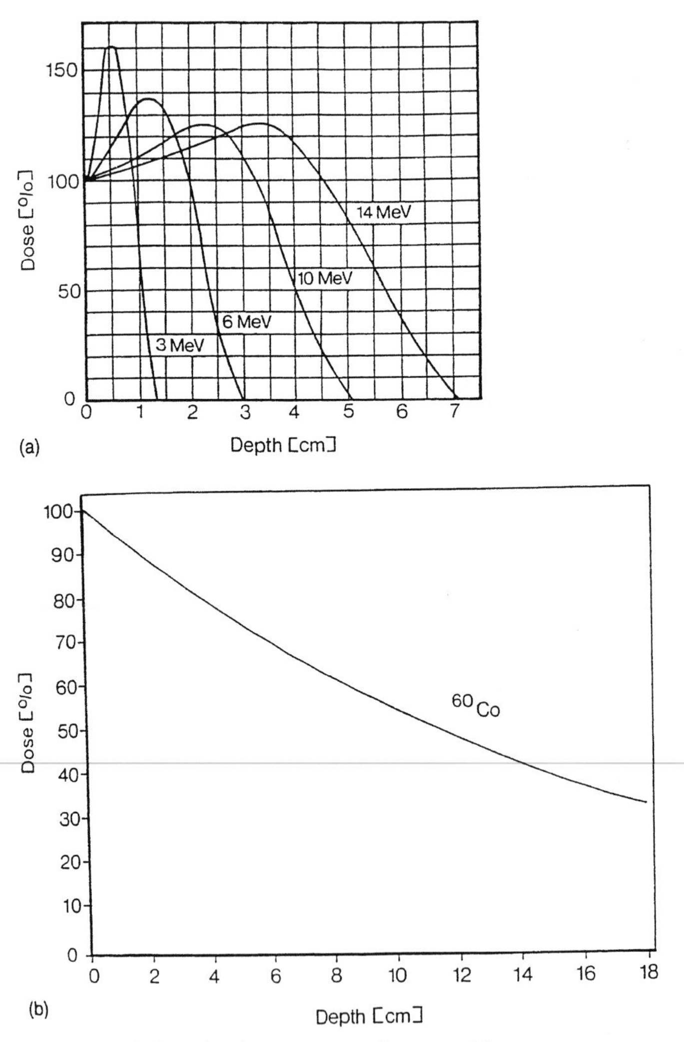

Fig. 5.2 Depth-dose distribution in water from one side: (a) electrons of different energy, (b) γ radiation; adapted from Diehl [3].

The difference in the shape of the curve between the two radiation types reflects the difference in their mechanism of action. With electrons, it can be seen that the effective dose is greatest at some distance into the food, as more secondary electrons are produced at that distance. The primary electrons lose their energy by interacting with water, so that the practical limit for electron irradiation is about 4 cm using the maximum permitted energy (10 MeV). Gamma rays, however, become depleted continuously as they penetrate the target and the effective depth is much greater. X-rays follow a similar pattern to Co 60 radiations. Two-sided irradiation permits the treatment of thicker packages of food (see Fig. 5.3) but it is still quite limited for electrons (approx. 8 cm).

The net result of these primary physical effects is a deposition of energy within the material, giving rise to excited molecules and ions. These effects are nonspecific, with no preference for any particular atoms or molecules. Hall et al. [2] estimated that the timescale of these primary effects is 10^{-14} s. It is an essential feature that radiations do not have sufficient energy to interact with the nuclei, dislodging protons or neutrons, otherwise radioactivity could be induced in the

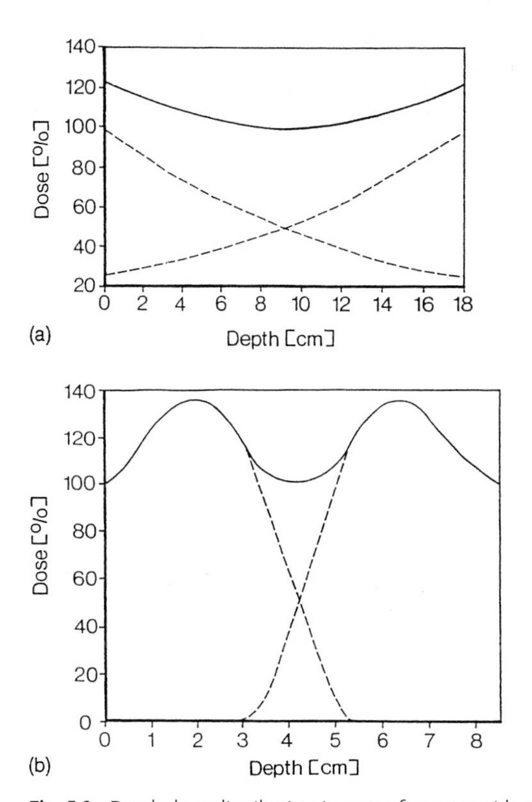

Fig. 5.3 Depth-dose distribution in water from two sides: (a) γ radiation, (b) 10 MeV electrons. Dashed lines indicate dose distribution for one-sided irradiation; adapted from Diehl [3].

target atoms. For this reason, energies are limited to 5 MeV and 10 MeV for X-rays and electrons respectively. Cobalt 60 produces γ rays of fixed energy well below that required to interfere with the nucleus.

5.2.2
Chemical Effects

The secondary chemical effects of irradiation result from breakdown of the excited molecules and ions and their reaction with neighbouring molecules, giving a cascade of reactions. The overall process by which these reactions produce stable end-products is known as 'radiolysis'. These reactions are considered in detail elsewhere [3] and only a brief summary is given here. The ions and excited molecules contain abnormal amounts of energy, which is lost through a combination of physical processes (fluorescence, conversion to heat, or transfer to neighbouring molecules) and chemical reactions. This occurs irrespective of whether they are free or molecular components. The primary reactions include isomerisation and dissociation within molecules and reactions with neighbouring species. The new products formed include free radicals, i.e. atoms or molecules with one or more unpaired electrons, which are available to form a chemical bond and are thus highly reactive. Because most foods contain substantial quantities of water and contaminating organisms contain water in their cell structure, the radiolytic products of water are particularly important. These include hydrogen, hydroxy radicals ($H^\bullet$, $^\bullet OH$), e_{aq}^-, H_2, H_2O_2 and H_3O^+. These species are highly chemically reactive and react with many substances, although not water molecules. Hall et al. [2] estimated the timescale of the secondary effects to be within 10^{-2} s. In most foods, free radicals have a short lifetime ($<10^{-3}$ s). However, in dried or frozen foods, or foods containing hard components such as bones or shells, free radicals may persist for longer.

The major components of foods and contaminating organisms, such as proteins, carbohydrates, fats and nucleic acids, as well as minor components such as vitamins, are all chemically altered to some extent following irradiation. This can be through direct effects of the incident electrons or Compton electrons. In aqueous solutions, reactions occur through secondary effects by interaction with the radiolytic products of water. The relative importance of primary and secondary effects depends on the concentration of the component in question.

The chemical changes are important in terms of their effects on living food contaminants whose elimination is a major objective of food irradiation. However, it is also essential to consider effects on the components of foods inasmuch as this may affect their quality, e.g. nutritional status, texture, off flavours. A further consideration is their effects on living foods such as fruits and vegetables, where the goal is to delay ripening or senescence.

5.2.3 Biological Effects

The major purpose of irradiating food is to cause changes in living cells. These can either be contaminating organisms such as bacteria or insects, or cells of living foods such as raw fruits and vegetables. Ionising radiation is lethal to all forms of life, the lethal dose being inversely related to the size and complexity of the organism (see Table 5.1). The exact mechanism of action on cells is not fully understood, but the chemical changes described above are known to alter cell membrane structure, reduce enzyme activity, reduce nucleic acid synthesis, affect energy metabolism through phosphorylation and produce compositional changes in cellular DNA. The latter is believed to be far and away the most important component of activity but membrane effects may play an additional role.

The DNA damage may be caused by *direct effects* whereby ionisations and excitations occur in the nucleic acid molecules themselves. Alternatively, the radiations may produce free radicals from other molecules, especially water, which diffuse towards and cause damage to the DNA through *indirect effects*. Direct effects predominate under dry conditions, such as when dry spores are irradiated, while indirect effects are more important under wetter conditions, such as within the cell structures of fruits or vegetative bacterial cells.

It should be noted that remarkably little chemical damage to the system is required to cause cell lethality, because DNA molecules are enormous compared to other molecules in the cell and thus present a large target. Yet, only a very small amount of damage to the molecule is required to render the cell irreparably damaged. This is clearly illustrated in two separate studies reported by Diehl [3], which calculated that during irradiation of cells:

- A radiation dose of 0.1 kGy would damage 2.8% of DNA molecules, which would be lethal to most cells, while enzyme activity would only fall by 0.14%, which is barely detectable, and only 0.005% of amino acid molecules would be affected, which is completely undetectable [5].
- A dose of 10.0 kGy would affect 0.0072% of water molecules and 0.072% of glucose molecules, but a single DNA molecule (10^9 Da) would be damaged at about 4000 locations, including 70 doublestrand breaks [6].

Table 5.1 Approximate lethal doses of radiation for different organisms; data from [4].

Organism	Lethal dose (kGy)
Mammals	0.005–0.01
Insects	0.01–1.0
Vegetative bacteria	0.5–10.0
Sporulating bacteria	10–50
Viruses	10–200

This is fortuitous for food irradiation, as the low doses required for efficacy of the process result in only minimal chemical damage to the food.

The phenomenon also explains the variation in lethal doses between organisms (see Table 5.1). The target presented by the DNA in mammalian cells is considerably larger than that of insect cells, which in turn are much larger than the genome of bacterial cells; and, consequently, radiation sensitivity follows the same pattern. In contrast, viruses have a much smaller nucleic acid content; and the high doses required for their elimination make irradiation an unlikely treatment procedure. Fortunately, it is unusual for foodborne viruses to cause health problems in humans.

5.3 Equipment

5.3.1 Isotope Sources

Co 60 is the major isotope source for commercial irradiation. It is manufactured in specific reactors and over 80% of the world supply is produced in Canada. Co 60 is produced from nonradioactive Co 59, which is compressed into small pellets and fitted into stainless steel tubes or rods a little larger than pencils. These are bombarded with neutrons in a nuclear reactor over a period of about 1 year to produce highly purified Co 60, which decays in a controlled manner to stable Ni 60, with the emission of γ rays with energies of 1.17 MeV and 1.33 MeV and a halflife of about 5.2 years. Co 60 is water-insoluble and thus presents minimal risk for environmental contamination.

Cs 137 is an alternative possibility, but is much less widely used than Co 60. It decays to stable Ba 137, emitting a γ photon (0.662 MeV) with a halflife of 30 years. It is unlikely to gain more importance and future discussion of radioisotopes will be confined to Co 60.

In practice, a radioactive 'source' is comprised of a number of Co 60 tubes arranged into the appropriate geometric pattern (e.g. placques or cylinders). The short halflife of Co 60 means that the source will be depleted by approximately 1% per month. This depletion must be considered in calculations of dose and requires that tubes be replaced periodically. In practice, individual rods are replaced in rotation at intervals maintaining the total source energy at a fairly constant level. An individual rod may be utilised for three or four halflives when activity has decayed to about 10% of its original level. When not in use, the source is normally held under a sufficient depth of water to completely absorb radiation to the surface so that personnel may safely enter the radiation cell area to load and unload radionuclides.

Examples of irradiation plants are shown in Figs. 5.4 and 5.5. Designs can be batch or continuous, the latter being more appropriate for large-scale processing.

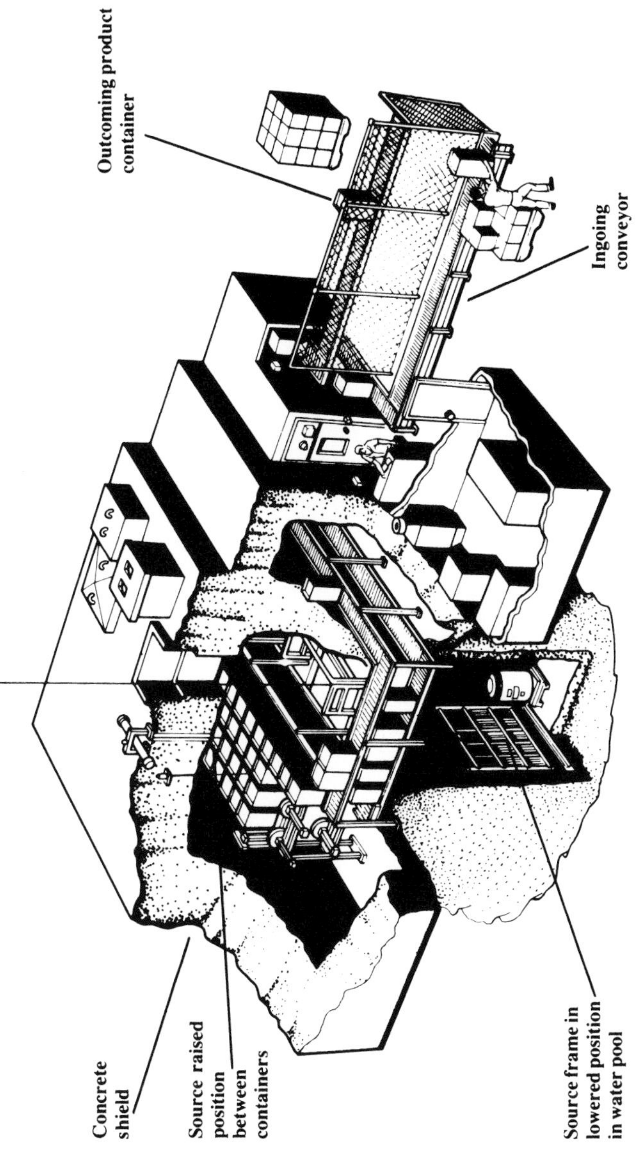

Fig. 5.4 Commercial automatic tote box γ (^{60}Co) irradiator; by courtesy of Atomic Energy of Canada Ltd.

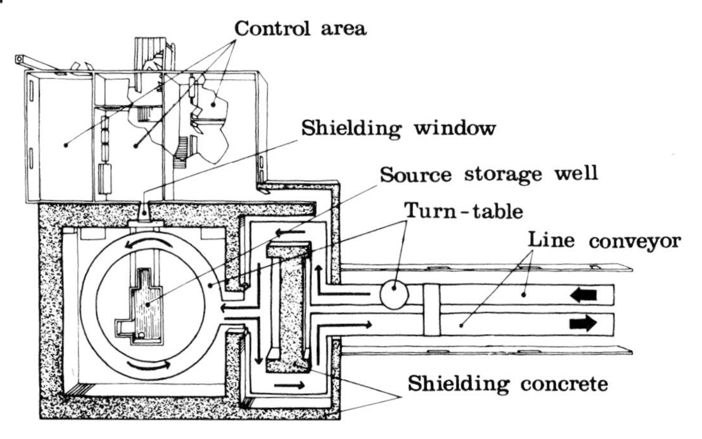

Fig. 5.5 Plan of commercial potato irradiator, by courtesy of Kawasaki Heavy Industries.

As the source emits radiation in all directions and the rate of emission cannot be controlled, it is essential to control product movement past the source in the most efficient way possible, to make best use of the radiations. Another aim is to achieve the lowest possible dose uniformity ratio (see Section 5.3.3). For this reason, containerised products may follow a complex path around the source, often two or more product units in depth, the units being turned to effect two-sided irradiation. An example of product progress around a source is shown in Fig. 5.6.

The dose rates provided by isotope sources are generally low, so that irradiation may take around 1 h to complete. Therefore, product movement is often sequential with a finite time allowed in each position without movement, to allow absorption of sufficient radiation.

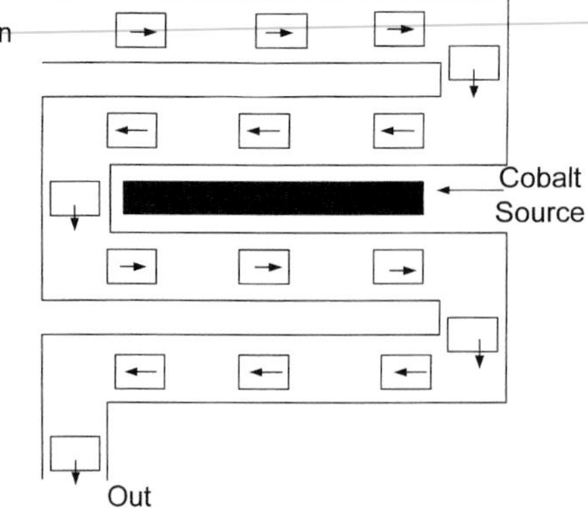

Fig. 5.6 Example of progress of containerised product around a γ irradiation source.

5.3.2
Machine Sources

Both electron and X-ray machines use electrons, which are accelerated to speeds approaching the speed of light by the application of energy from electric fields in an evacuated tube. The resulting electron beams possess a considerable amount of kinetic energy.

The main designs of electron irradiator available are the Dynamitron, which will produce electron energies up to 4.5 MeV, or linear accelerators for higher energies. In either case, the resulting beam diameter is only a few millimetres or centimetres. To allow an even dose distribution in the product, it is necessary to scan the beam using a scanning magnet, which creates an alternating mag-

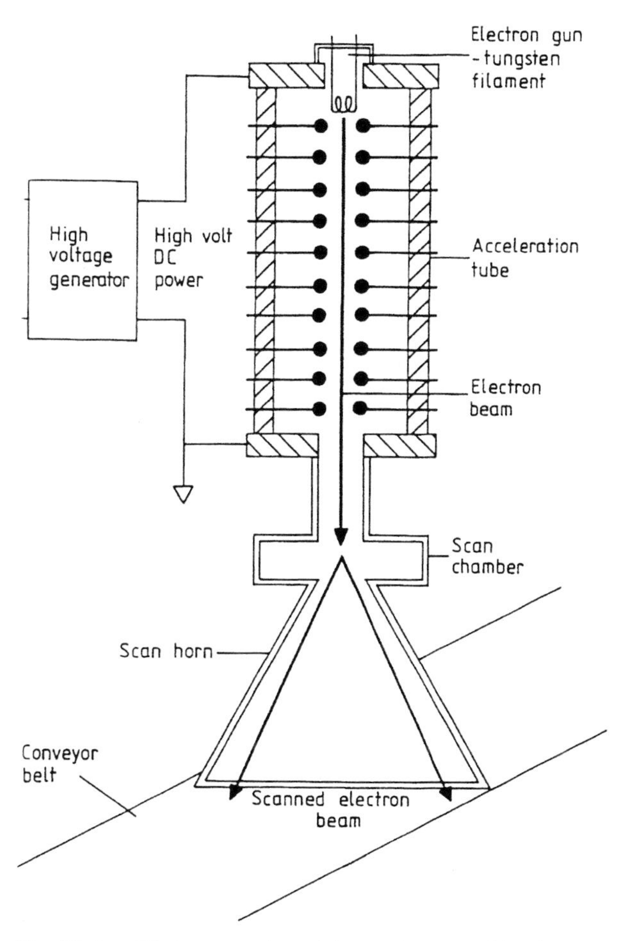

Fig. 5.7 Simplified construction of electron beam machine; by courtesy of Leatherhead Food RA, UK.

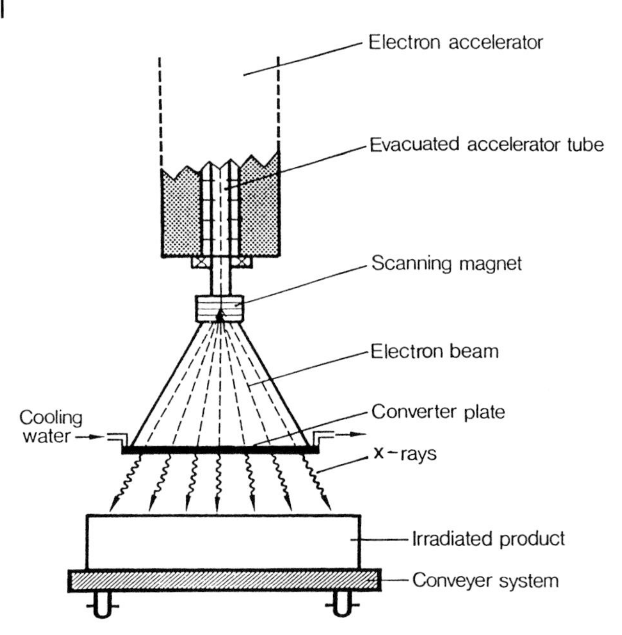

Fig. 5.8 Electron accelerator with X-ray convertor; from Diehl [3], with permission.

netic field (analogous to the horizontal scan of a television tube) which moves the beam back and forwards at 100–200 Hz. An even, fan-shaped field of emitted electrons is created, through which the food is conveyed. A simplified diagram of an electron beam machine is shown in Fig. 5.7.

As described in Section 5.2.1, the low penetration of electrons requires that this technology is limited either to surface treatment of food or to foods of limited thickness (8 cm max., with two-sided irradiation).

When electrons strike a target, they produce X-rays which can be utilised to give greater penetration depth, but which suffer from the disadvantage of a low conversion efficiency. The efficiency of conversion depends on the energy of the electrons and the atomic number of the target material. In practice, therefore, X-rays are produced by firing high-energy electrons at a heavy metal target plate, as shown schematically in Fig. 5.8. Even with 10 MeV electrons and a tungsten (atomic number 74) plate, the efficiency of conversion is only 32%, hence cooling water must be applied to the converter plate.

Both electrons and X-rays deliver much higher dose rates than isotope sources, so that processing is complete in a matter of seconds. Also the beams may be directed, so that the complex transport of packages around the source is not required.

5.3.3
Control and Dosimetry

As with any processing technology, radiation treatment must be controlled and validated; and hence the level of treatment must be monitored in a quantifiable form. The subject is considered in detail by Ehlermann [7]. The basic parameter of importance is the absorbed dose, although information on dose rate may be important in some circumstances. A knowledge of absorbed dose is necessary, both to conform to legal limits and to ensure that the advisory technological considerations have been met. In other words, to know whether the food has received sufficient treatment without exceeding either the legal limit or the threshold dose for sensory impairment. The dose absorbed by a food depends on the magnitude of the radiation field, the absorption characteristics of the food and the exposure time.

A further consideration is the dose uniformity ratio, i.e. the ratio between the maximum and minimum dose absorbed by different parts of a food piece or within a food container. Radiation processing results in a range of absorbed doses in the product, in the same way as heating results in a range of temperatures. Dose is normally expressed in terms of average dose absorbed by the food during the treatment. Dose distribution depends partly on the type of radiation used, but is affected by the geometry of individual food units and the way in which the food is packaged and loaded into containers for processing. The ideal uniformity ratio of 1 is not achievable in any type of plant, a value of about 1.5–2.0 being a rough guideline for practical applications.

Validation of dose and quality assurance (as well as optimising plant performance) can be carried out using dosimeters. Routine dosimetry is carried out by attaching dosimeters to packages and then reading on completion of treatment. The dosimeters are usually in the form of plastic strips whose absorption characteristics change in a linear fashion in relation to a given irradiation dose. Radiation-induced changes are measured with a spectrophotometer at the appropriate wavelength. It is essential that routine dosimeters be ultimately related to a primary standard at a specialised national standards laboratory, e.g. the National Physical Laboratory in the UK. The primary standard is the actual energy absorbed by water as determined by calorimetry, i.e. measurement of the temperature rise and hence heat absorbed during irradiation.

Chemical dosimetry systems such as the Fricke system can be used as reference systems to ensure the reliability of routine systems. This system is based on the conversion of ferrous to ferric ions in acidic solutions, measured with a spectrophotometer, which is highly accurate but too complex for routine use.

5.4
Safety Aspects

The safety of consuming irradiated foods is no longer in serious question; and this is discussed in other parts of this chapter. However, the safe operation of irradiation facilities warrants further consideration. In particular, precautions must be taken to protect workers, the public and the environment from accidental exposure. These considerations must encompass both the irradiation facilities and the transport and disposal of the radioactive materials associated with γ plants. There are more than 170 γ plants and 600 electron beam facilities in operation worldwide [3], but only a small fraction of these are used for treatment of food. As with any industrial process, there is the potential for accidents, but radiation processing is one of the most strictly regulated and controlled industries, with an excellent safety record [8].

Any design of plant must incorporate a 'cell' to contain the radiation source. Commercial plants usually have walls constructed of concrete, while lead may be employed in smaller plants. Gamma sources are held under water when not in use. Irradiators are designed with overlapping protection to prevent leaking of radiation to workers or the public; and examples are shown in Figs. 5.4 and 5.5. It is of paramount importance that personnel cannot enter the cell when irradiation is taking place, as exposure for even a few seconds could deliver a lethal dose. Precautions against uncontrolled entry, such as interlocks, are therefore an essential feature. Transportation of radioactive source materials is carried out in special casks, which are designed to survive the most severe accidents and disasters. Waste radioactive materials are returned to the manufacturer and either reused or stored until harmless. The quantity of waste is, in fact, very small. The North American Food Irradiation Processing Alliance [9] report that, in 40 years of transporting radioactive materials, there has never been a problem with the materials and, during 35 years of operating irradiation facilities, there has never been a fatality. This is an excellent record compared to other industries, such as those involving shipping toxic materials or crude oil and petroleum products.

It should be remembered that electron and X-ray equipment carries no problem of transporting, storing or disposal of radioactive materials.

5.5
Effects on the Properties of Food

The radiation doses suitable for food systems cause very little physicochemical damage to the actual food, but potential effects on the major components will be considered.

The mineral content of foods is unaffected by irradiation, although the status of minerals could be changed, e.g. extent of binding to other chemicals. This could feasibly affect their bioavailability.

The major effect of radiation on carbohydrates concerns the breakdown of the glycosidic bond in starch, pectin or cellulose to give smaller carbohydrates. This can lead to a reduction in the viscosity of starches, or a loss of texture in some foods, for example leading to the softening of some fruits.

Irradiation of proteins at doses up to 35 kGy causes no discernible reduction in amino acid content [4] and hence no reduction in their protein nutritional quality. Changes in secondary and tertiary structures are also minimal, although there may be isolated examples where functionality is impaired. For example, the whipping quality of egg whites is impaired following irradiation. As stated previously, effects on enzyme action are very limited and are probably undetectable at the doses likely to be used for foods.

Insignificant changes in the physical (viscosity, melting point) and many chemical properties (iodine value, peroxide number etc.) of lipids are produced by irradiation up to 50 kGy [4]. The major concern with fat-containing foods is the acceleration of autoxidation when irradiated in the presence of oxygen. Unsaturated fatty acids are converted to hydroperoxyl radicals, which form unstable hydroperoxides, which break down to a range of mainly carbonyl compounds. Many of the latter have low odour thresholds and lead to rancid off flavours. It should be noted that the same endproducts are found following longterm storage of unirradiated lipids. This is obviously a concern when irradiating foods which contain significant quantities of unsaturated fats; and the irradiation and subsequent storage of these foods in the absence of oxygen is advisable.

Loss of vitamins during processing is an obvious concern and has been studied in great detail in a variety of foods [10, 11]. Inactivation results mainly from their reaction with radiolytic products of water and is dependent on the chemical structure of the vitamin. The degree of inactivation is also determined by the composition of the food, as other food components can act as 'quenchers', in competition for the reactive products. This has led to an overestimation of vitamin losses in the literature where irradiation of pure solutions of vitamins has been studied. Another consideration is the role of postirradiation storage where, in common with other processing methods, further breakdown of some vitamins may occur. Hence, losses of vitamins during irradiation vary from vitamin to vitamin and from food to food and are obviously dose-dependent. In addition, subsequent storage or processing of food must be considered.

According to Stewart [1], the sensitivity of water-soluble vitamins to irradiation follows the order: vitamin B_1 > vitamin C > vitamin B_6 > vitamin B_2 > folate = niacin > vitamin B_{12}, while the sensitivity of fat-soluble vitamins follows the order: vitamin E > carotene > vitamin A > vitamin D > vitamin K.

As a general rule, vitamin losses during food irradiation are quite modest compared to other forms of processing. In fact, many studies have demonstrated 100% retention of individual vitamins following low or medium dose treatments.

5.6
Detection Methods for Irradiated Foods

There is a clear need to be able to distinguish between irradiated and nonirradiated foods. This would permit proof of authenticity of products labelled as irradiated, or conversely, of unlabelled products which had been irradiated. In addition, it would be useful to be able to estimate accurately the dose to which a food had been treated. Solving these issues would improve consumer confidence in the technology and would benefit international trade in irradiated foods, especially where legislation differs between countries. However, as discussed in Section 5.5, the physicochemical changes occurring during food irradiation are minimal and thus detection is difficult. The changes measured may not be specific to radiation processing and may alter during subsequent storage. Also the compositional and structural differences between foods and the different doses required for different purposes mean that it is unlikely that a universal detection method is possible. Various approaches have been studied in many different foods with some success; and the subject has been reviewed in full by Stewart [12] and Diehl [3]. Methods currently available depend on physical, chemical, biological and microbiological changes occurring during irradiation and are summarised only briefly below.

Electron Spin Resonance (ESR) can detect free radicals produced by radiation, but these are very short-lived in high moisture foods (see Section 5.2.2). However, in foods containing components with high dry matter, such as bones, shells, seeds or crystalline sugars, the free radicals remain stable and may be detected. ESR has been successfully demonstrated in some meats and poultry, fish and shellfish, berries, nuts, stone fruit spices and dried products.

Luminescence techniques are also very promising and are based on the fact that excited electrons become trapped in some materials during irradiation. The trapped energy can be released and measured as emitted light, either by heat in thermoluminescence (TL), or by light in photostimulated luminescence (PSL). In fact the luminescence is produced in trapped mineral grains rather than in the actual food, but successful tests have been demonstrated in a wide range of foods where mineral grains can be physically separated.

Other physical principles, including viscosity changes to starch, changes in electrical impedance of living tissues and near infrared reflectance, hold some promise but suffer from errors due to variation between unprocessed samples, dose thresholds and elapsed time from processing.

The most promising chemical methods are the detection of long-chain hydrocarbons and 2-alkylcyclobutanones which are formed by radiolysis of lipids and are hence limited to foods with quite high fat contents. Numerous other potentially useful chemical changes have been studied with limited success.

DNA is the main target for ionising radiation; and hence it is logical that DNA damage should be an index for detection. The 'comet assay' is used to detect the presence of tails of fragmented DNA produced by irradiation, on electrophoresis gels, as opposed to more distinct nuclei from nonirradiated samples.

It is limited by the fact that cooking and other processing damages DNA; but it is nonetheless promising, as it is applicable to most foods.

Microbiological methods involve looking at changes in populations of microorganisms which may have resulted from irradiation. They are nonspecific, but may act as useful screening procedures for large numbers of samples.

A further development is the use of antibody assays in which antibodies are raised to products of radiolysis and incorporated into enzyme-linked immunosorbent assays. These tests are rapid and specific, but not yet in routine use.

5.7 Applications and Potential Applications

In practice, the application of irradiation is limited by legal requirements. Approximately 40 countries have cleared food irradiation within specified dose limits for specific foods. This does not mean, however, that the process is carried out in all these countries, or indeed that irradiated foods are freely available within them, the UK being a prime example of a country where irradiation of many foods is permitted, but none is actually carried out. The picture is further complicated by trade agreements, labelling requirements, etc.; and legal questions will not be pursued here. Legislation generally requires that irradiated foods, or foods containing irradiated ingredients, be labelled appropriately. The 'radura' symbol (see Fig. 5.9) is accepted by many countries.

One general principle has been to limit the overall average dose to 10 kGy, which essentially means that radappertisation of foods is not an option. This figure was adopted by the WHO/FAO Codex Alimentarius Commission in 1983 [13, 14] as a result of masses of research studying the nutritional, toxicological and microbiological properties of foods irradiated up to this level. There was never any implication, however, that larger doses were unsafe; and there have been exceptions to this generalisation. South Africa has permitted average doses up to 45 kGy to be used in the production of shelf-stable meat products and several countries have permitted doses greater than 10 kGy for the treatment of spices. A recent joint FAO/IAEA/WHO study group [15] concluded that doses greater than 10 kGy can be considered safe and nutritionally adequate and it would not be surprising to see more widespread national clearances of higher doses in the future.

Fig. 5.9 Radura symbol indicates that a food has been irradiated.

Two basic purposes can be achieved by food irradiation:

- extension of storage life
- prevention of foodborne illness.

It is difficult to classify the applications of food irradiation, as the process may be acting through different mechanisms in different foods or at different doses. Some foods are much more suitable for irradiation than others and the factors determining shelf life vary between foods. This section will therefore consider the general effects and mechanisms of irradiation, followed by a brief overview of applications in the major food classes.

5.7.1 General Effects and Mechanisms of Irradiation

5.7.1.1 Inactivation of Microorganisms

When a population of microorganisms is irradiated, a proportion of the cells will be damaged or killed, depending on the dose. In a similar way to heat treatment, the number of surviving organisms decreases exponentially as dose is increased. A common measure of the radiation sensitivity in bacteria is the D_{10} value, which is the dose required to kill 90% of the population. Fig. 5.10 shows

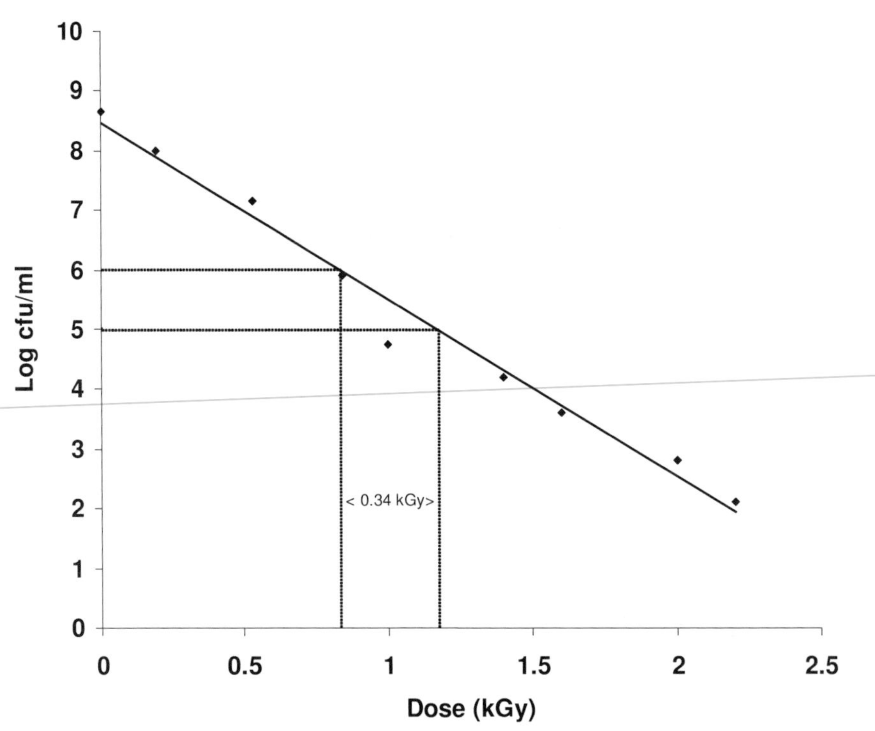

Fig. 5.10 Survival curve for electron-irradiated *Escherichia coli*.

a typical survival curve for *Escherichia coli* (data from Fielding et al. [16]) showing a D_{10} value of 0.34 kGy.

Radiation resistance varies widely among different species of bacteria, yeasts and moulds. Bacterial spores are generally more resistant than vegetative cells (see Table 5.1), which is at least partly due to their lower moisture content. Vegetative cells may contain 70% water, whereas spores contain less than 10% water. Hence indirect damage to DNA through the radiolytic products of water is much less likely in spores. An extremely important concept determining the radiation sensitivity of different species and genera of bacteria is their widely varying ability to repair DNA. D_{10} values in the range 0.03–10.0 kGy have been reported. Some organisms have developed highly efficient repair mechanisms, but fortunately these species are not pathogenic and have no role in food spoilage.

The sensitivity of microorganisms to radiation is also related to environmental conditions. Temperature of irradiation, a_w, pH and the presence of salts, nutrients or toxins such as organic acids all exert a great effect on microbial populations after irradiation. The mechanisms of effect may be through modification of the lethality of the applied dose, for example in dry or frozen environments the effectiveness may be reduced because of suppression of indirect effects caused by radiolytic products of water. Alternatively, the environmental conditions may undoubtedly affect the ability of the organisms to repair themselves and their ability to reproduce subsequent to treatment. Generally, reducing a_w by freezing, water removal or the addition of osmotically active substances increases the resistance of microorganisms, as the secondary effects due to radiolytic products of water are reduced. The presence of food components generally reduces sensitivity, as they can be considered to compete with microorganisms for interaction with the radiolytic products of water. pH also affects lethality in two ways: (a) radiolysis of water is pH-sensitive, and (b) pH changes affect the general functioning of cells, including the efficiency of DNA repair mechanisms. In addition, the recovery of surviving cells following irradiation is pH-dependent [16].

Radiation treatments aimed at the inactivation of microorganisms are conveniently classified as:

- Radappertisation: a treatment which aims to reduce the number and/or activity of microorganisms to such a level that they are undetectable. Properly packaged radappertised foods should keep indefinitely, without refrigeration. Doses in the range 25–50 kGy are normally required.
- Radicidation: this aims to reduce the number of viable spore-forming pathogenic bacteria to an undetectable level. Doses of 2–8 kGy are normally required.
- Radurisation: a treatment sufficient to enhance the keeping quality of foods through a substantial reduction in the numbers of viable specific spoilage organisms. Doses vary with the type of food and level of contamination, but are often in the range 1–5 kGy.

5.7.1.2 **Inhibition of Sprouting**

The shelf life of tuber and bulb crops, such as potatoes, yams, garlic and onions, may be extended by irradiation at low dose levels. Sprouting is the major sign of deterioration during storage of these products and occurs after a time lag (dormant period) after harvest. The duration of the dormant period differs between different crops, different agricultural practices and different storage conditions, but is usually a number of weeks.

It is believed that the inhibitory effect of irradiation on sprouting results from a combination of two metabolic effects [17]. Firstly, irradiation impairs the synthesis of endogenous growth hormones such as gibberellin and indolyl-3-acetic acid, which are known to control dormancy and sprouting. Secondly, nucleic acid synthesis in the bud tissues, which form the sprouts, is thought to be suppressed. Treatments in the range 0.03–0.25 kGy are effective, depending on the commodity, while higher doses may cause deterioration of the tissue.

5.7.1.3 **Delay of Ripening and Senescence**

Living fruits and vegetables may be irradiated to extend shelf life by delaying the physiological and biochemical processes leading to ripening. The mechanisms involved are complex and not well understood; and it is probable that different mechanisms predominate in different cultivars.

In some fruits, ripening is associated with a rapid increase in the rate of respiration and associated quality changes (flavour, colour, texture, etc.) known as the 'climacteric', which more or less coincides with eating ripeness. This represents the completion of maturation and is followed by senescence. Ripening is triggered by ethylene, which is then produced autocatalytically by the fruit. Climacteric fruit such as tomatoes, bananas or mangoes can either be harvested at full ripeness for immediate consumption, or harvested before the climacteric for storage and transport before ripening (either through endogenous or exogenous ethylene). Irradiation can be used either to delay senescence in fully ripe fruit, or to extend the preclimacteric life of unripe fruit. Applied doses of radiation are usually limited to 2 kGy and often much less, due to radiation injury to the fruit, leading to discolouration or textural damage.

Vegetables and nonclimacteric fruit, e.g. citrus, strawberries, cherries, are usually fully mature at harvest and respiration rate declines steadily thereafter. Irradiation can be used to delay the rate of senescence in these products in some instances, although other purposes such as control of fungi or sprout inhibition may be more important.

5.7.1.4 **Insect Disinfestation**

Insects can cause damage to food as well as leading to consumer objections. Fortunately insects are sensitive to irradiation (see Table 5.1) and can be controlled by doses of 0.1–1.0 kGy. These doses may not necessarily cause immediate lethality, but will effectively stop reproduction and egg development.

5.7.1.5 Elimination of Parasites

Relatively few foodborne parasites afflict humans. The two major groups are singlecelled protozoa and intestinal worms (helminths) which can occur in meats, fish, fruit and vegetables. Irradiation treatment is feasible although not widespread.

5.7.1.6 Miscellaneous Effects on Food Properties and Processing

Although chemical changes to food resulting from irradiation are very limited, it is possible that irradiation could produce beneficial changes to the eating or processing quality of certain foods. There have been reports of improvements in the flavour of some foods following processing, but these are not particularly well substantiated. Chemical changes could result in textural changes in the food. The most likely examples are depolymerisation of macromolecules such as starch, which could lead to altered baking performance or changes in drying characteristics. Irradiation may cause cellular injury in some fruits, giving rise to easier release of cell contents and hence increased juice recovery from berry fruits.

5.7.1.7 Combination Treatments

Combining food processes is a strategy that permits effective processing while minimising the severity of treatment. The benefits of combining low dose irradiation with heat, low temperatures, high pressures, modified atmosphere packaging or chemical preservatives have been described [18, 19]. In some cases, an additive effect of combining processes may be obtained, but synergistic combinations are sometimes observed.

The major drawback of these strategies is the economics of such complex processing regimes.

5.7.2 Applications to Particular Food Classes

A brief outline of the potential application of radiation processing of the major food classes is given. It should be reiterated that the actual practices adopted by different countries usually result from local legislation and public attitudes rather than the technical feasibility and scientific evaluation of product quality and safety.

5.7.2.1 Meat and Meat Products

The application of irradiation for control of bacteria and parasites in meat and poultry is discussed in detail by Molins [20]. A major problem with these products is the development of undesirable flavours and odours, which have variously been described as 'wet dog' or 'goaty'. The precise chemical nature of the irradiated flavour is unclear despite much research, but lipid oxidation is a contributing factor. The phenomenon is dose-dependent, species-dependent and can be minimised by irradiating at low (preferably sub-zero) temperatures or under vacuum or in an oxygen-free atmosphere. It is notable that the flavours and odours are transient and have been shown in chicken and different meats to be reduced or disappear following storage for a period of days. The sensory problems may also be masked by subsequent cooking. The phenomenon clearly limits applications and applied doses even though there is no suggestion of toxicological problems. Appropriate dose-temperature combinations that yield acceptable flavour, while achieving the intended purpose, must be evaluated for any application.

Radiation sterilisation of meat and meat products was the objective of much research in the 1950s and 1960s, largely aimed at military applications. The high doses required, e.g. 25–75 kGy, are usually prohibitive on legislative grounds, but the process is technically feasible if combined with vacuum and very low temperatures (say –40 °C) during processing [21]. Radiation-sterilised, shelf-stable precooked meals incorporating meat and poultry have been produced for astronauts and military purposes and are produced commercially in South Africa for those taking part in outdoor pursuits.

Effective shelf life extensions of many fresh, cured or processed meats and poultry have been demonstrated with doses in the range 0.5–2.5 kGy. Increases in shelf life of 2–3 times or more, without detriment to the sensory quality, are reported [20]. Irradiation cannot, however, make up for poor manufacturing practice. For example, Roberts and Weese [22] demonstrated that quality and shelf life extension of excellent initial quality ground beef patties [$<10^2$ aerobic colony-forming units (CFU) g^{-1}] was much better than lower initial quality (10^4 CFU g^{-1}) materials when irradiated at the same dose. The former were microbiologically acceptable for up to 42 days at 4 °C. Also, subsequent handling of irradiated meats or poultry, which may give rise to recontamination following treatment, should be avoided. It is preferable to irradiate products already packaged for retail. Packaging considerations are important for both preventing recontamination and inhibiting growth of contaminating microorganisms. Vacuum or modified atmosphere packaging may be useful options. It is important to note that such products continue to require refrigeration following treatment.

Elimination of pathogens in vegetative cell form is a further aim of irradiating meat and poultry products and may go hand in hand with shelf life extension. Many studies have reported effective destruction of different pathogens in meat – reviewed by many authors including Farkas [23] and Lee et al. [24]. Much work has focused on *Campylobacter* spp in pork and *Salmonella* spp in chicken, but recent outbreaks of food poisoning attributed to *Escherichia coli* 0157:H7 in hamburgers and other meats have gained attention and accelerated FDA approval of treatment of red meat in the USA.

5.7.2.2 **Fish and Shellfish**

Both finfish and shellfish are highly perishable unless frozen on board fishing vessels or very shortly after harvesting. If unfrozen, their quality depends on rapid ice cooling. Irradiation can play a role in preservation and distribution of unfrozen fish and shellfish, with minimal sensory problems. The subject is reviewed extensively by Kilgen [25] and Nickerson et al. [26]. The mechanism of action involves reducing the microbial load (hence extending refrigerated shelf life), inactivating parasites and reducing pathogenic organisms. Low and medium dose irradiation (up to 5 kGy) has been studied as a means of preserving many varieties of finfish and commercially important shellfish, and shelf life extensions of up to 1 month under refrigerated storage have been reported in many species, without product deterioration. On board irradiation of eviscerated, fresh, iced fish has been suggested.

Problems have been noted with fatty fish, such as flounder and sole, which must be irradiated and stored in the absence of oxygen to avoid rancidity. Salmon and trout exhibit bleaching of the carotenoid pigments following irradiation, which may limit the application [4]. Combination treatments involving low dose irradiation (1 kGy) with either chemical dips, e.g. potassium sorbate or sodium tripolyphosphate, or elevated CO_2 atmospheres may be effective for the treatment of fresh fish [27].

Strategies for irradiation treatment of frozen fish products, dried fish, fish paste and other fish products have been investigated [27].

5.7.2.3 **Fruits and Vegetables**

There are a number of purposes for irradiating fresh fruits and vegetables, including the extension of shelf life by the delay of ripening and senescence, the control of fungal pathogens which lead to rotting and insect disinfestation [28]. Radurisation and radicidation of processed fruits and vegetables have also been studied. The unusual property of fruits and vegetables is that they consist of living tissue, hence irradiation may effect life changes on the product and such changes may not be immediately apparent, but result in delayed effects.

The main problems associated with these products are textural problems which result from radiation-induced depolymerisation of cellulose, hemicellulose, starch and pectin, leading to softening of the tissue. The effect is dose-related and hence severely limits the doses applied. Nutritional losses at such doses are minimal. Other disorders include discolouration of skin, internal browning and increased susceptibility to chilling injury. Tolerance to radiation varies markedly, for example some citrus fruits can withstand doses of 7.5 kGy, whereas avocados may be sensitive to 0.1 kGy.

The conditions used for irradiating any fruit or vegetable cultivars are very specific, depending on the intended purpose and the susceptibility of the tissue to irradiation damage; and specific examples are discussed in detail by Urbain [4] and Thomas [17]. Combination treatments involving low dose irradiation and hot water or chemical dips, or modified atmospheres, may be effective.

5.7.2.4 Bulbs and Tubers

Potatoes, sweet potatoes, yams, ginger, onions, shallots and garlic may be treated by irradiation to produce effective storage life extensions due to inhibition of sprouting as described in Section 5.6.1.2. A general guideline for dose requirements is 0.02–0.09 kGy for bulb crops and 0.05–0.15 kGy for tubers [29]. These low doses have no measurable detrimental effect on nutritional quality and are too low to produce significant reductions in microbiological contaminants.

5.7.2.5 Spices and Herbs

Spices and herbs are dry materials which may contain large numbers of bacterial and fungal species, including organisms of public health significance. Small quantities of contaminated herbs and spices could inoculate large numbers of food portions and hence decontamination is essential. Chemical fumigation with ethylene oxide is now banned in many countries, on account of its toxic and potentially carcinogenic properties, and radiation treatment offers a viable alternative [27]. Fortunately, being dry products, herbs and spices are resistant to ionising radiation and can usually tolerate doses up to 10 kGy. In general, doses in the range 3–10 kGy are employed, which gives a reduction in the aerobic viable count to below 10^3 CFU g^{-1} or 10^4 CFU g^{-1}, which is considered equivalent to chemical fumigation. Individual treatments are discussed in detail by Farkas [30].

5.7.2.6 Cereals and Cereal Products

Lorenz [31] reviewed the application of radiation to cereals and cereal products. Insects are the major problem during the storage of grains and seeds. Disinfestation is therefore the main purpose of the irradiation of cereal grain. e.g. wheat, maize, rice, barley. This can be achieved with doses of 0.2–0.5 kGy, with minimal change to the properties of flour or other cereal products [27].

Chemical fumigants, such as methyl bromide gas, are considered a health risk and could be phased out in favour of the irradiation disinfestation of grain.

Radurisation of flour for bread making, at a dose of 0.75 kGy, to control the 'rope' defect caused by *Bacillus subtilis,* gives rise to a 50% increase in the shelf life of the resulting bread. However, higher doses lead to reduced bread quality. Alternatively, finished loaves and other baked goods may be irradiated to increase storage life by suppression of mould growth, with a dose of 5 kGy.

5.7.2.7 Other Miscellaneous Foods

Radiation processing of most foods has been investigated at some point. Some foods which do not appear in the above categories are described below.

Milk and dairy products are very susceptible to radiation-induced flavour changes, even at very low doses. Significant application to dairy products is therefore unlikely, with the possible exception of Camembert cheese produced from raw milk.

Control of *Salmonella* in eggs would be a very beneficial application. However, irradiation of whole eggs causes weakening of yolk membrane and loss of yolk appearance. There may be potential for the irradiation of liquid egg white, yolk or other egg products.

Nuts contain high levels of oil and again are susceptible to lipid oxidation. Low doses to control insects, or sprouting are feasible, but higher doses required to control microbial growth are less likely.

Ready meals, such as those served on airlines or in hospitals and other institutions, offer a potential application for radiation treatments. The major problem is the variety of ingredients, each with different characteristics and radiation sensitivities. However, careful selection of ingredients, the use of low doses and combination treatments may lead to some practical applications.

References

1 Stewart, E.M. **2001**, Food Irradiation Chemistry, in *Food Irradiation: Principles and Applications*, ed. R. Molins, Wiley Interscience, Chichester, pp. 37–76.

2 Hall, K.L., Bolt, R.O., Carroll, J.G. **1963** Radiation Chemistry of Pure Compounds, in *Radiation Effects on Organic Materials*, ed. R.O. Bolt, J.G. Carroll, Academic Press, New York, pp. 63–125.

3 Diehl, J.F. **1995**, *Safety of Irradiated Foods*, Marcel Dekker, New York.

4 Urbain, W.M. **1986**, *Food Irradiation*, Academic Press, London.

5 Pollard, E.C. **1966**, Phenomenology of Radiation Effects on Microorganisms, in *Encyclopedia of Medical Radiology*, vol. 2, 2nd edn, ed. A. Zuppinger, Springer, New York, pp. 1–34.

6 Brynjolffson, A. **1981**, Chemiclearance of Food Irradiation: its Scientific Basis, in *Combination Processes in Food Irradiation*, Proceedings Series, International Atomic Energy Agency, Vienna, pp. 367–373.

7 Ehlermann, D. **2001**, Process Control and Dosimetry in Food Irradiation, in *Food Irradiation: Principles and Applications*, ed. R. Molins, Wiley Interscience, Chichester, pp. 387–413.

8 IAEA **1992**, *Radiation Safety of Gamma and Electron Irradiation Facilities*, Safety Series 107, International Atomic Energy Agency, Vienna.

9 Anon. **2001**, *Food Irradiation: Questions and Answers*, Technical Document, Food Irradiation Processing Alliance.

10 WHO **1994**, *Safety and Nutritional Adequacy of Irradiated Food*, World Health Organization, Geneva.

11 Thayer, D.W., Fox, J.B., Lakrotz, L. **1991**, Effects of Ionising Radiation on Vitamins, in *Food Irradiation*, ed. S. Thorne, Elsevier Applied Science, London, pp. 285–325.

12 Stewart, E.M. **2001**, Detection Methods for Irradiated Foods, in *Food Irradiation: Principles and Applications*, ed. R. Molins, Wiley Interscience, Chichester, pp. 347–386.

13 CAC **1984**, *Codex General Standard for Irradiated Foods*, Vol. XV, E-1, CODEX STAN 106-1983, Joint FAO/WHO Food Standards Programme, Rome.

14 CAC **1984**, *Recommended International Code of Practice for the Operation of Radiation Facilities Used for the Treatment of Foods*, Vol. XV, E-1, CAC/RCP 19-1979 (rev. 1), Joint FAO/WHO Food Standards Programme, Rome.

15 WHO **1999**, *High Dose Irradiation: Wholesomeness of Food Irradiated with Doses Above 10 kGy*, Technical Report Series No. 890, World Health Organisation, Geneva.

16 Fielding, L. M., Cook, P. E., Grandison, A. S. **1994**, The Effect of Electron Beam Irradiation and Modified pH on the Survival and Recovery of *Escherichia coli*, *J. Appl. Bacteriol.* 76, 412–416.

17 Thomas, P. **1990**, Irradiation of Fruits and Vegetables, in *Food Irradiation: Principles and Applications*, ed. R. Molins, Wiley Interscience, Chichester, pp. 213–240.

18 Campbell-Platt, G., Grandison, A. S. **1990**, Food Irradiation and Combination Processes, *Radiat. Phys. Chem.* 35, 253–257.

19 Patterson, M. **2001**, Combination Treatments Involving Food Irradiation, in *Food Irradiation: Principles and Applications*, ed. R. Molins, Wiley Interscience, Chichester, pp. 313–327.

20 Molins, R. A. **2001**, Irradiation of Meats and Poultry, in *Food Irradiation: Principles and Applications*, ed. R. Molins, Wiley Interscience, Chichester, pp. 131–191.

21 Thayer, D. W. **2001**, Development of Irradiated Shelf-Stable Meat and Poultry Products, in *Food Irradiation: Principles and Applications*, ed. R. Molins, Wiley Interscience, Chichester, pp. 329–345.

22 Roberts, W. T., Weese, J. O. **1998**, Shelf-Life of Ground Beef Patties Treated by Gamma Radiation, *J. Food Prot.* 61, 1387–1389.

23 Farkas, J. **1987**, Decontamination, Including Parasite Control, of Dried, Chilled and Frozen Foods by Irradiation, *Acta Aliment.* 16, 351–384.

24 Lee, M., Sebranek, J. G., Olson, D. G., Dickson, J. S. **1996**, Irradiation and Packaging of Fresh Meat and Poultry, *J. Food Prot.* 59, 62–72.

25 Kilgen, M. B. **2001**, Irradiation Processing of Fish and Shellfish Products, in *Food Irradiation: Principles and Applications*, ed. R. Molins, Wiley Interscience, Chichester, pp. 193–211.

26 Nickerson, J. F. R, Licciardello, J. J., Ronsivalli, L. J. **1983**, Radurization and Radicidation: Fish and Shellfish, in *Preservation of Food by Ionizing Radiation*, ed. E. S. Josephson, M. S. Peterson, CRC Press, Boca Raton, pp. 12–82.

27 Wilkinson, V. M., Gould, G. W. **1996**, *Food Irradiation: A Reference Guide*, Butterworth-Heinemann, Oxford.

28 Kader, A. A. **1986**, Potential Applications of Ionising Radiation in Postharvest Handling of Fresh Fruits and Vegetables, *Food Technol.* 40, 117–121.

29 Thomas, P. **2001**, Radiation Treatment for Control of Sprouting, in *Food Irradiation: Principles and Applications*, ed. R. Molins, Wiley Interscience, Chichester, pp. 241–271.

30 Farkas, J. **2001**, Radiation Decontamination of Spices, Herbs, Condiments and Other Dried Food Ingredients, in *Food Irradiation: Principles and Applications*, ed. R. Molins, Wiley Interscience, Chichester, pp. 291–312.

31 Lorenz, K. **1975**, Irradiation of Cereal Grains and Cereal Grain Products, *CRC Crit. Rev. Food Sci. Nutr.* 6, 317–382.

6
High Pressure Processing

Margaret F. Patterson, Dave A. Ledward and Nigel Rogers

6.1
Introduction

In recent years there has been a consumer-driven trend towards better tasting and additive-free foods with a longer shelf life. Some of the scientific solutions, such as genetic modification and gamma irradiation, have met with consumer resistance. However, high pressure processing (HPP) is one technology that has the potential to fulfil both consumer and scientific requirements.

The use of HPP has, for over 50 years, found applications in diverse, nonfood industries. For example, HPP has made a significant contribution to aircraft safety and reliability where it is currently used to treat turbine blades to eliminate minute flaws in their structure. These microscopic imperfections can lead to cracks and catastrophic failure in highly stressed aero engines. HPP now gives a several-fold increase in component reliability, leading to a longer engine life and, ultimately, lower flight cost to the public.

Although the effect of high pressure on food has been known for about 100 years [1], the technology has remained within the R&D environment until recent times (see Table 6.1).

Early R&D pressure vessels were generally regarded as unreliable, costly and had a very small usable vessel volume. Hence the prospect of 'scaling up' the R&D design to a full production system was commercially difficult.

Equally the food industry, particularly in Europe, has mainly focussed upon cost reductions, restructuring and other programmes, often to the neglect of emerging technologies.

A situation had to develop where a need was created that could not be fully satisfied by current technology. This occurred on a small scale in Japan with the desire to produce delicate, fresh, quality, long shelf life fruit-based products for a niche market. The HPP products were, and are today, produced on small machines at a premium price to satisfy that particular market need.

Food Processing Handbook. Edited by James G. Brennan

ISBN: 3-527-30719-2

Table 6.1 The history of HPP for food products.

Year	Event(s)
1895	Royer (France) used high pressure to kill bacteria experimentally
1899	Hite (USA) used high pressure for food preservation
1980s	Japan started producing high-pressure jams and fruit products
1990s	Avomex (USA) began to produce high-pressure guacamole from avocados with a fresh taste and extended shelf-life
2000	Mainland Europe began producing and marketing fresh fruit juices (mainly citrus) and delicatessen-style cooked meats. High-pressure self-shucking oysters, poultry products, fruit juices and other products were marketed in the USA
2001	HPP fruit pieces given approval for sale in the UK. Launch of the first HPP fruit juices in the UK

In the USA, issues relating to food poisoning outbreaks, notably with unprocessed foods such as fruit juices and oysters led to action by the Food and Drug Administration. They attempted to regulate the situation by requiring a significant reduction in the natural microbial levels of a fresh food. Those foods which did not achieve this reduction were required to be labelled with a warning notice. This meant that producers of freshly squeezed orange juice, for example, marketing their product as wholesome and healthy were faced with a dilemma. If they reverted to the established method of reducing microbial counts, i.e. by heating the juice, then the product became of inferior quality in all respects. If they continued to sell the juice 'untreated', they faced market challenges with a product essentially labelled to say that it could be harmful to the consumer.

Another example involved the producers of avocado-based products who knew that there was a huge demand for fresh quality product. However, the very short shelf life of the product coupled with the opportunity for harmful microbes to flourish with time effectively restricted this market opportunity.

Oysters are another case in point. They have traditionally been a niche market with romantic connotations but food poisoning outbreaks associated with *Vibrio* spp in the oyster population caused serious illness and even death, which effectively devastated the market.

In all these cases, the producers needed to look for a process that reduced the numbers of spoilage or harmful microorganisms but left the food in its natural, fresh state. HPP has been shown to be successful in achieving this aim. HPP guacamole is now produced in Mexico and sells very successfully in the US, while HPP oysters are being produced in Louisiana.

High pressure processing of food is the application of high pressure to a food product in an isostatic manner. This implies that all atoms and molecules in the food are subjected to the same pressure at exactly the same time, unlike heat processing where temperature gradients are established.

The second key feature of high pressure processing, arising from Le Chatelier's principle, indicates that any phenomenon that results in a volume decrease is enhanced by an increase in pressure. Thus, hydrogen bond formation is favoured by the application of pressure while some of the other weak linkages found in proteins are destabilised. However, covalent bonds are unaffected.

For food applications, 'high pressure' can be generally considered to be up to 600 MPa for most food products (600 MPa = 6000 bar = 6000 atmospheres = 87 000 psi). With increasing pressure, the food reduces in overall size in proportion to the pressure applied but retains its original shape. Hence, a delicate food such as a grape can be subjected to 600 MPa of isostatic pressure and emerge apparently unchanged although the different rates and extents of compressibility of the gaseous (air), liquid and solid phases may lead to some physical damage. Pressure kills microorganisms, including pathogens and spoilage organisms, leading to a high quality food with a significantly longer and safer chilled shelf life. The conventional way to do this is to process the food by heat, but this may also damage the organoleptic and visual quality of the food, whereas HPP does not.

In summary, the advantages of HPP are:

- the retention of fresh taste and texture in products, such as fruit juices, shellfish, cooked meats, dips, sauces and guacamole;
- the increase in microbiological safety and shelf-life by inactivation of pathogens and spoilage organisms and also some enzymes;
- the production of novel foods, for example gelled products and modification of the properties of existing foods, e.g. milk with improved foaming properties;
- the savings in labour which some HPP processes bring about, compared to more traditional techniques, e.g. self-shucking oysters;
- low energy consumption;
- minimal heat input, thus retaining fresh-like quality in many foods;
- minimal effluent;
- uniform isostatic pressure and adiabatic temperature distribution throughout the product, unlike thermal processing.

The current disadvantages of HPP are:

- Initial outlay on equipment, which remains high (in the region of $1.8 million for a typical production system). However, numerous companies have justified this cost by offsetting it against new product opportunities, supported by the relatively low running cost of the HPP equipment.
- Uncertainty introduced by the European 'novel foods' directive (May 1997) about the ease of gaining approval for new products has made some companies reluctant to apply. Better understanding of the regulations is making this easier and the fact that approval has now been granted for some HPP products has renewed confidence in the marketplace. Furthermore, many HPP products are in fact exempt from the regulations as they are 'substantially equivalent' to nonHPP products on the market.

High pressure has many effects on the properties of the food ingredients themselves as well as on the spoilage organisms, food poisoning organisms and enzymes. In addition to preserving a fresher taste than most other processing technologies, HPP can affect the texture of foods such as cheese and the foaming properties of milk. This chapter looks at how these effects can make HPP foods more marketable, less labour-intensive to produce and generally more attractive to the producer, retailer and consumer alike.

6.2 Effect of High Pressure on Microorganisms

The lethal effect of high pressure on bacteria is due to a number of different processes taking place simultaneously. In particular, damage to the cell membrane and inactivation of key enzymes, including those involved in DNA replication and transcription are thought to play a key role in inactivation [2]. The cell membranes are generally regarded as a primary target for damage by pressure [3]. The membranes consist of a bilayer of phospholipids with a hydrophilic outer surface (composed of fatty acids) and an inner hydrophobic surface (composed of glycerol). Pressure causes a reduction in the volume of the membrane bilayers and the cross-sectional area per phospholipid molecule [4]. This change affects the permeability of the membrane, which can result in cell damage or death. The extent of the pressure inactivation achieved depends on a number of interacting factors as discussed below. These factors have to be considered when designing process conditions to ensure the microbiological safety and quality of HPP foods.

6.2.1 Bacterial Spores

Bacterial spores can be extremely resistant to high pressure, just as they are resistant to other physical treatments, such as heat and irradiation. However, low/moderate pressures are more effective than higher pressures. It was concluded that inactivation of spores is a two-step process involving pressure-induced germination [5]. This has led to the suggestion that spores could be killed by applying pressure in two stages. The first pressure treatment would germinate or activate the spores while the second treatment would kill the germinated spores [6].

Temperature has a profound effect on pressure-induced germination. In general, the initiation of germination increases with increasing temperature over a specified temperature range [7, 8]. Applying a heat treatment before or after pressurisation can also enhance spore kill [9]. In recent years the application of high temperatures along with high pressures has shown promise as a way of producing shelf-stable foods [10]. It is claimed these products have superior sensory and nutritional quality compared to those produced by conventional thermal processing [11].

6.2.2
Vegetative Bacteria

Gram positive bacteria, especially cocci such as *Staphylococccus aureus*, tend to be more pressure-resistant than Gram negative rods, such as *Salmonella* spp. However, there are exceptions to this general rule. For example, certain strains of *Escherichia coli* O157:H7 are relatively resistant to pressure [12].

One proposed explanation for the difference in pressure response between Gram positive and Gram negative bacteria is that the more complex cell membrane structure of Gram negative bacteria makes them more susceptible to environmental changes brought about by pressure [13]. The cell walls of Gram negative bacteria consist of an inner and outer membrane with a thin layer of peptidoglycan sandwiched between. The cell walls of Gram positive bacteria are less complex, with only an inner plasma membrane and a thick peptidoglycan outer layer, which can constitute up to 90% of the cell wall.

6.2.3
Yeasts and Moulds

Yeasts are generally not associated with foodborne disease. They are, however, important in food spoilage due to their ability to grow in low water activity (a_w) products and to tolerate relatively high concentrations of organic acid preservatives. Yeasts are thought to be relatively sensitive to pressure, although ascospores are more resistant than vegetative bacteria. Unlike yeasts, certain moulds are toxigenic and may present a safety problem in foods. There is relatively little information on the pressure sensitivity of these moulds. In one study, a range of moulds including *Byssochlamys nivea*, *B. fulva*, *Eupenicillium* sp. and *Paecilomyces* sp. were exposed to a range of pressures (300–800 MPa) in combination with a range of temperatures (10–70 °C) [14]. The vegetative forms were inactivated within a few minutes using 300 MPa at 25 °C. However, ascospores were more resistant. A treatment of 800 MPa at 70 °C for 10 min was required to reduce a starting inoculum of $<10^6$ ascospores ml^{-1} of *B. nivea* to undetectable levels. A treatment of 600 MPa at 10 °C for 10 min was sufficient to reduce a starting inoculum of 10^7 ascospores ml^{-1} of *Eupenicillium* to undetectable levels.

Information on the effect of pressure on preformed mycotoxins is limited. Brâna [15] reported that patulin, a mycotoxin produced by several species of *Aspergillus*, *Penicillium* and *Byssochlamys*, could be degraded by pressure. The patulin content in apple juice decreased by 42%, 53% and 62% after 1 h treatment at 300, 500, and 800 MPa, respectively, at 20 °C.

6.2.4
Viruses

There is relatively little information on high-pressure inactivation of viruses, compared to the information available on vegetative bacteria. However, some reports suggest that Polio virus suspended in tissue culture medium is relatively resistant to pressure, with 450 MPa for 5 minutes at 21 °C giving no reduction in plaque-forming units [16]. However, Feline calicivirus, a Norwalk-virus surrogate, hepatitis A [16] and human rotavirus [17] were more pressure sensitive. These results would suggest that the pressure treatments necessary to kill vegetative bacterial pathogens would also be sufficient to cause significant inactivation of human virus particles.

There is also some evidence emerging that pressure combined with heat may have some effect on prions, but further work still needs to be completed in this area [18].

6.2.5
Strain Variation Within a Species

Pressure resistance varies not only between species but also within a species. For example, Linton et al. [19] reported a 4 log difference in resistance of pathogenic *E. coli* strains to treatment at 600 MPa for 15 min at 20 °C. Alpas et al. [20] also demonstrated variability in pressure resistance within strains of *Listeria monocytogenes*, *Salmonella* spp, *S. aureus* and *E. coli* O157:H7. However, they found that the range of pressure differences within a species decreased when the temperature during the pressure treatment was increased from 25 °C to 50 °C. This finding would be helpful in a commercial situation, where the combination of pressure and mild heat could be used to enhance the lethal effect of the treatment.

6.2.6
Stage of Growth of Microorganisms

Vegetative bacteria tend to be most sensitive to pressure when treated in the exponential phase of growth and most resistant in the stationary phase of growth [21]. When bacteria enter the stationary phase they can synthesise new proteins that protect the cells against a variety of adverse conditions, such as high temperature, high salt concentrations and oxidative stress. It is not known if these proteins can also protect bacteria against high pressure but this may explain the increase in resistance in the stationary phase.

6.2.7
Magnitude and Duration of the Pressure Treatment

Pressure is similar to heat, in that there is a threshold below which no inactivation occurs. This threshold varies depending on the microorganism. Above the threshold, the lethal effect of the process tends to increase as the pressure increases but not necessarily as the time increases. This can lead to inactivation curves with very definite 'tails'. These nonlinear inactivation curves have been reported by many workers [12, 13, 22]. Metrick et al. [23] reported tailing effects for S. Typhimurium and S. Senftenberg. When the resistant tail populations were isolated, grown and again exposed to pressure, there was no significant difference in the pressure resistance between them and the original cultures.

Several theories to explain the tailing effect have been proposed. The phenomenon may be independent of the mechanisms of inactivation but be due to population heterogeneity such as clumping or genetic variation. Alternatively, tailing may be a normal feature of the mechanism of resistance (adaptation and recovery) [24]. In practice, the nonlogarithmic inactivation curves make it difficult to calculate accurate D and z values.

6.2.8
Effect of Temperature on Pressure Resistance

The temperature during pressure treatment can have a significant effect on microbial resistance. As a general rule, pressure treatments carried out below 20 °C or above the growth range for the microorganism result in greater inactivation. Takahaski et al. [25] reported that the pressure inactivation of S. Bareilly, *V. parahaemolyticus* and *S. aureus* in 2 mM sodium phosphate buffer, pH 7.0, was greater at −20 °C than at +20 °C. The simultaneous application of pressure (up to 700 MPa) with mild heating (up to 60 °C) was more lethal than either treatment alone in inactivating pathogens such as *E. coli* O157:H7 and *S. aureus* in milk and poultry meat [26].

6.2.9
Substrate

The composition of the substrate can significantly affect the response of microorganisms to pressure and there can be significant differences in the levels of kill achieved with the same organism on different substrates. For example, *E. coli* O157:H7 treated under the same conditions of 700 MPa for 30 min at 20 °C resulted in a 6 log reduction in numbers in phosphate-buffered saline, a 4 log reduction in poultry meat and a <2 log reduction in UHT milk [12]. The reasons for these effects are not clear but it may be that certain food constituents like proteins and carbohydrates can have a protective effect on the bacteria and may even allow damaged cells to recover more readily.

There is evidence that the a_w and pH of foods can significantly affect the inactivation of microorganisms by pressure. A reduction in a_w appears to protect microorganisms from pressure inactivation. Oxen et al. [27] reported that in sucrose ($a_w \sim 0.98$), the pressure inactivation (at 200–400 MPa) of *Rhodotorula rubra* was independent of pH at pH 3.0–8.0. However, at a_w values below 0.94 there was a protective effect, irrespective of solute (glucose, sucrose, fructose or sodium chloride). Most microorganisms are more susceptible to pressure at lower pH values and the survival of pressure-damaged cells is less in acidic environments. This can be of commercial value, such as in the pressure treatment of fruit juices where, in the high acid conditions, pathogens, such as *E. coli* O157:H7, which may survive the initial pressure treatment will die within a relatively short time during cold storage [28].

6.2.10 Combination Treatments Involving Pressure

High pressure processing can be successfully used in combination with other techniques to enhance its preservative action and/or reduce the severity of one or all of the treatments. It is possible that this hurdle approach will be used in many of the commercial applications of HPP technology. Beneficial combination treatments include the use of pressure (usually < 15 MPa) with carbon dioxide to improve the microbial quality of chicken, egg yolk, shrimp, orange juice [29] and fermented vegetables [30]. Pressure combined with irradiation has been proposed to improve the microbial quality of lamb meat [31] and poultry meat [32]. Pressure combined with heat and ultrasound has been successful in inactivating *B. subtilis* spores [33]. Various antimicrobial compounds have been used in combination with pressure. Nisin and pressure caused a significant reduction in numbers of *B. coagulans* spores [7]. Other antimicrobials such as pediocin AcH [34] and the monoterpenes [35] have also been combined with pressure treatment to enhance microbial inactivation with mixed success. The commercial value of these combinations still has to be assessed, given that one of the advantages of HPP is that it can be regarded as a 'natural' minimal processing technology.

6.2.11 Effect of High Pressure on the Microbiological Quality of Foods

High pressure processing is already used commercially to enhance the microbiological quality of certain food products. A number of potential applications have also been reported in the scientific literature and there is a significant amount of ongoing research in this area. Fruit products have been most extensively studied. These products are acidic so, in terms of their microbiology, pathogens are generally not so important but spoilage microorganisms, particularly yeasts and moulds are of concern. The limiting factor for shelf life of such products is often the action of enzymes, particularly those which can cause browning, although the problem may be at least partially overcome by blanching or adding an oxygen scavenger such as ascorbic acid. Most fruit products are given a treatment of around 400 MPa for up to a few minutes. This can give a shelf life of up to several months provided the products are stored at 4 °C [36]. Pressure treatment of vegetables tends to be less successful due to their higher pH and the potential presence of spores, which can be very pressure-resistant. In addition, the quality of some vegetables deteriorates as a result of pressure processing, which can also be a limiting factor [37]. However, it should be noted that one of the most successful commercial products available to date is HPP guacamole. Research has shown that clostridial spores cannot outgrow in this product and it can have a shelf life of around 1 month at 4 °C without modification of colour, texture or taste.

The use of HPP to improve the microbiological quality of meat, fish and dairy products has been investigated by a number of workers. These products tend to have a more neutral pH and provide a rich growth medium for most microorganisms, with pathogens being of particular concern. The need to ensure microbiological safety is one of the reasons why, to date, there are relatively few commercial applications of HPP meat and dairy products available. However, research has shown that pressure processing can be successful, in terms of improving microbiological safety and quality, for the treatment of pork [38], minced beef [39], duck Foie Gras (liver pate) [40], fish [41], ovine milk [42] and liquid whole egg [43]. In many cases, the authors also comment on the ability of the pressure treatment to maintain or enhance sensory, nutritional or functional quality compared to conventional processing methods. In all cases, the optimum treatment conditions need to be carefully defined and thoroughly tested to ensure food safety is not compromised. For example, this may include extensive inoculation studies, under standardised conditions and using the most resistant strains of pathogens to ensure that the product will be microbiologically safe during its shelf life.

6.3 Ingredient Functionality

It is now well established [44, 45] that changes in protein structure and functionality occur during high pressure treatment. Studies carried out on volume changes in proteins, have shown that the main targets of pressure are hydrophobic and electrostatic interactions [46, 47]. Hydrogen bonding, which stabilizes the α helical and β pleated sheet forms of proteins, is almost pressure-insensitive. At the pressures used in food processing, covalent bonds are unaffected [48] but at pressures of about 300 MPa sulfhydril groups may oxidise to S-S bonds in the presence of oxygen.

It is readily apparent from the above that pressure and temperature do not normally work synergistically with respect to protein unfolding, since the weak linkages that are most labile to heat, i.e. hydrogen bonds, are stabilised or only marginally affected by pressure whilst the bonds most labile to pressure (electrostatic and hydrophobic interactions) are far less temperature-sensitive. How-

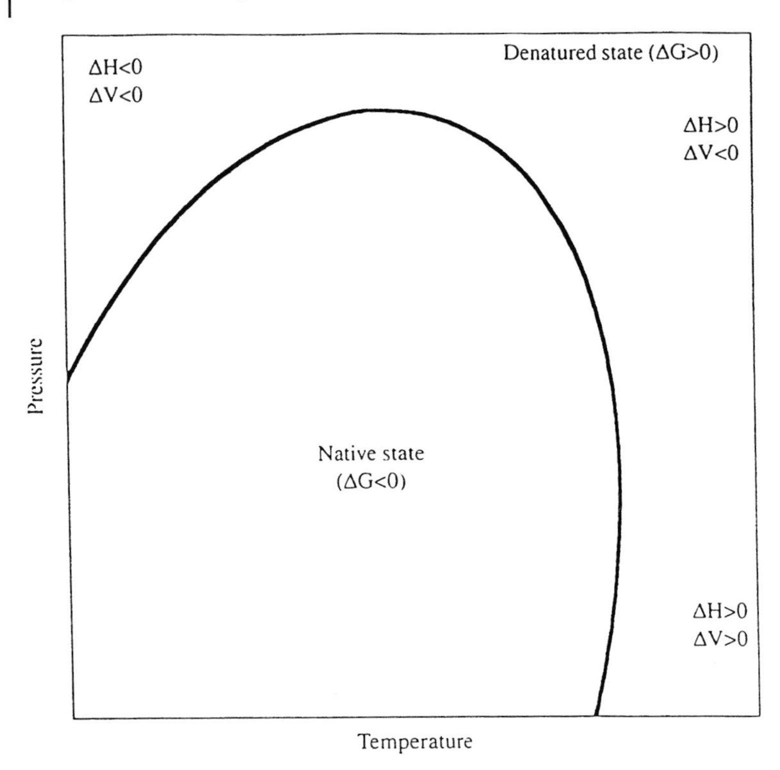

Fig. 6.1 Phase diagram for native/denatured proteins. Most proteins denature in the range 40–80 °C.

ever, in the presence of oxygen, both increasing temperature and pressure encourage disulphide bond formation and/or interchange. Such considerations help to explain the effects of pressure and temperature on the phase diagram for most, if not all, native/denatured protein systems (see Fig. 6.1) where for example it is seen that, up to a certain temperature, pressure stabilises the protein against heat denaturation. On removal from the denaturing environments, the proteins will, if free to do so, refold to a native-like structure. However, because the pressure/temperature dependence of many weak linkages differ so markedly, on pressure release the 'reformed' structures often differ from that of the native structure, i.e. conformational drift occurs. Thus the amount of α helix, β sheet and random coil present will vary, as also will such properties as surface hydrophobicity and charge. As these 'renatured' pressure-treated proteins yield such different structures, it is not surprising that they exhibit marked differences in behaviour to the native protein or its heat-denatured product. Thus their functional properties vary.

6.4 Enzyme Activity

Enzymes, being proteins, will at sufficiently high pressure undergo conformational changes and thus lose activity. Most enzymes of importance in food deterioration are relatively resistant to pressure and complete inactivation is difficult to achieve. Thus polyphenoloxidases, the enzymes responsible for browning in fruits and vegetables, require pressures of 800 MPa, at room temperature, or more to bring about complete inactivation. The degree of inactivation is usually dependent on pressure, temperature and time and, not unnaturally, pH is a further very important factor [49].

A recent review by Ludikhuyze et al. [50] has summarised present knowledge regarding the combined effects of pressure and temperature on those enzymes that are related to the quality of fruits and vegetables. The enzymes of importance include the polyphenoloxidases, pectin methylesterases, which induce cloud loss and consistency changes, and lipoxygenases which are responsible for the development of off flavours, loss of essential fatty acids and colour changes. These authors discuss how complete characterisation of the pressure/temperature phase diagram for inactivation of these enzymes has been achieved and also suggest how this information could be useful in generating integrated kinetic information with regard to process engineering associated with HPP treatment. Interestingly, these authors found that, for most of the enzymes so far studied, the reaction kinetics were first order, the only exception being pectin methylesterase which only gave fractional conversion and thus the kinetics were difficult to resolve. As discussed in Section 6.3, these studies showed that most enzymes (proteins) have maximum stability to pressure at temperatures around 25–40 °C.

In these model systems, the loss of activity is invariably due to a change in the conformation of the protein, i.e. unfolding/denaturation, which tends to be irreversible. However, although the loss of enzyme activity in the majority of cases is associated with major structural changes, this is not always the case. For example, the secondary and tertiary structure of papain is little affected by pressures up to 800 MPa [51], but a significant loss of activity is observed on treatment at these pressures. This loss in activity can be largely inhibited by applying pressure in the absence of oxygen, since treatment at 800 MPa in air causes a loss in activity of about 41%, but only 23% of the initial activity is lost on pressure treatment after flushing with nitrogen; and, after flushing with oxygen, the loss in activity is 78% at 800 MPa. Gomes et al. [51] suggested that, in this case, the loss in activity is related to specific thiol oxidation at the active site. The active site in papain contains both a cysteine group and a histidine group and, at pH 7, they exist as a relatively stable S^--N^+ ion pair. In an aqueous environment, pressure causes this linkage to rupture due to the associated decrease in volume due to electrostriction of the separated charges. However, steric considerations mean the S^- ion can not form a disulphide linkage and thus, in the presence of oxygen, the ion oxidises to the stable SO_3^- [52].

Since many enzymes are relatively difficult to denature, it is not surprising that, when whole foods are subjected to pressure, the effects on enzyme activity are difficult to predict. Thus many fruits and vegetables, when subjected to pressure, undergo considerable browning since at pressures below that necessary to inactivate the enzyme some change occurs which makes the substrate more available [53]. For this reason, combined with the cost of the process, high pressure is an effective means of preventing enzymic browning only if applied to the food with appropriate control. Although the application of high pressure may well accelerate the activity of polyphenoloxidase (PPO) in some fruits and vegetables, it can be controlled and thus in the commercial manufacture of avocado paste (guacamole) treatment at 500 MPa for a few minutes is adequate to extend the colour shelf life of the product so that it has a shelf life at chill temperatures of 8 weeks compared to a few hours under the same conditions if not subjected to pressure. From the previous discussion, it is apparent that 500 MPa does not fully inactivate PPO but it brings about sufficient decrease in activity to permit the extended shelf life. However, where modification of enzymic activity is required, pressure treatment may be of benefit as in optimising protease activity in meat and fish products [54] or in modifying the systems that affect meat colour stability. For example, Cheah and Ledward [55, 56] have shown that subjecting fresh beef to pressures of only 70–100 MPa leads to a significant increase in colour stability due to some, as yet unidentified, modification of an enzyme-based system that causes rapid oxidisation of the bright red oxymyoglobin to the brown oxidised metmyoglobin.

Enzymes, as being responsible for many colour changes in fruit, vegetables and meat systems, are also intimately related to flavour development in fruits and vegetables. Thus, lipoxygenase plays an important role in the genesis of volatiles [57], as this enzyme degrades linoleic and linolenic acids to volatiles such as hexanal and *cis*-3-hexenal. The latter compound transforms to *trans*-2-hexenal which is more stable. These compounds are thought to be the major volatile compounds contributing to the fresh flavour of blended tomatoes [58]. Tangwongchai et al. [59] reported that pressures of 600 MPa led to a complete and irreversible loss of lipoxygenase activity in cherry tomatoes when treated at ambient temperature. This loss of enzyme activity resulted in flavour differences between the pressure processed tomato and the fresh product. Compared to unpressurised tomatoes, treatment at 600 MPa gave significantly reduced levels of hexanal, *cis*-3-hexenal and *trans*-2-hexenal, all of which are important contributors to fresh tomato flavours. It is well established that high pressure can very satisfactorily maintain the flavour quality of fruit juices, as well as their colour quality, but obviously with regard to the fresh fruit differences become apparent and it is likely therefore that the technology will not be of benefit for some whole fruits, although its use for fruit juices cannot be disputed.

As well as being involved in flavour development in fruits and vegetables, enzymes are intimately involved in the textural changes that take place during growth and ripening. The enzymes primarily responsible are believed to be polygalacturonase and pectin methylesterase. Tangwongchai et al. [60] showed that, in whole cherry tomatoes, these two enzymes are affected very differently by pressure. Although a sample of purified commercial pectin methylesterase was partially inactivated at all pressures above 200 MPa, irrespective of pH, in whole cherry tomatoes no significant inactivation was seen even after treatment at 600 MPa, presumably because other components in the tomato offered protection, or the isoenzymes were different. Polygalacturonase was more susceptible to pressure, being almost totally inactivated after treatment at 500 MPa. It is interesting to note that these authors observed, both visually and by microscopy, that whole cherry tomatoes showed increasing textural damage with increasing pressures up to about 400 MPa. However at higher pressures (500–600 MPa) there was less apparent damage than that caused by treatment at the lower pressures, the tomatoes appearing more like the untreated samples. These authors concluded that the textural changes in tomato induced by pressure involve at least two related phenomena. Initially, damage is caused by the greater compressibility of the gaseous phase (air) compared to the liquid and solid components, giving rise to a compact structure which on pressure release is damaged as the air rapidly expands, leading to increases in membrane permeability. This permits egress of water and the damage also enables enzymic action to increase, causing further cell damage and softening. The major enzyme involved in the further softening is polygalacturonase (which is inactivated above 500 MPa) and not pectin methylesterase (which in the whole fruit is barotolerant). Thus, at pressures above 500 MPa, less damage to the texture is seen.

From the brief overview above it is apparent that high pressure has very significant effects on the quality of many foods, especially fruits and vegetables, if enzymes are in any way involved in the development of colour, flavour or texture.

Although to date pressure has largely been concerned with the preservation of quality, either by inhibiting bacteria or inactivating enzymes, it does offer potential as a processing aid in assisting reactions which are pressure-sensitive. An example that may have commercial application is that moderate pressures (300–600 MPa) cause significant increases in the activity of the amylases in wheat and barley flour, because the pressure-induced gelatinisation of the starch makes it, the starch, more susceptible to enzymic attack [61]. Higher pressures lead to significant decreases in activity due to unfolding and aggregation of the enzymes. This technology thus has potential in producing glucose syrups from starch by a more energy-efficient process than heat.

The effects of pressure on enzyme activity suggest that it may well be a very effective processing tool for some industrial applications in the future.

6.5 Foaming and Emulsification

The structural changes undergone by a protein will affect its functionality and, for example, pressure-treated β-lactoglobin, at 0.01% concentration has significantly improved foaming ability compared to its native counterpart (see Fig. 6.2). How-

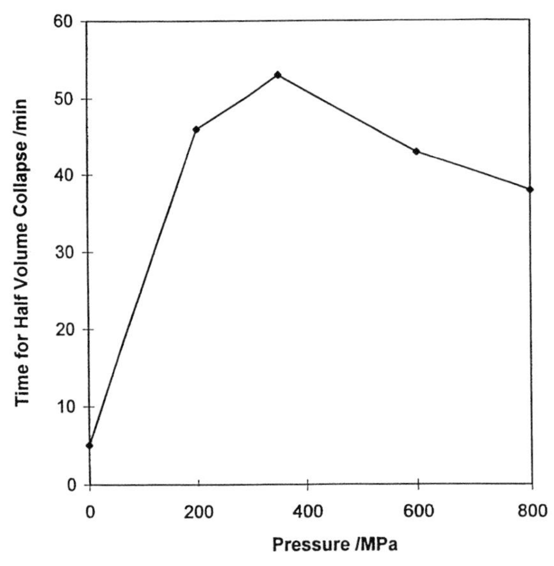

Fig. 6.2 Effects of high HPP treatment on the foam stability of β-lactoglobulin at pH 7.0. The time for half volume collapse is plotted as a function of pressure applied for 20 min; from Galazka et al. [62].

ever, if pressure is applied to a relatively concentrated solution of β-lactoglobin (0.4%), disulphide bond formation may lead to significant aggregation so that it is less useful as a functional ingredient. If disulphide bond-induced aggregation does occur on pressure treatment, then only dilute solutions should be treated, to avoid loss of functionality due to increased size. In addition, other factors that aid aggregation or disulphide bond formation, such as pressures above that necessary to cause unfolding, extended treatment times, alkaline pH and the presence of oxygen, should be avoided so as to optimise functionality.

In addition to β-lactoglobin, many studies have been reported on the effect of high pressure on the emulsifying and foaming properties of other water-soluble proteins, including ovalbumin, vegetable proteins such as soy and pea and casein. One advantage of pressure treatment on proteins to improve or modify their functional properties is that the process is invariably easier to control than thermal processing, where the effects are rather drastic and less easy to control. For example, model emulsions prepared with high pressure treated (<600 MPa) protein after homogenisation show that pressurisation induced extensive droplet flocculation which increased with protein concentration and severity of treatment [63]. However, it was also noted that moderate thermal processing (80 °C for 5 min) had a far greater effect than pressure treatment at 800 MPa for 40 min on the state of flocculation of ovalbumin-coated emulsion droplets. The increase in emulsion viscosity is due to the formation of a network from the aggregated dispersed oil droplets and denatured polymers in aqueous solution.

The level of pressure-induced modification can be controlled more efficiently by altering the intensity of high pressure treatment rather than by controlling the temperature in thermal processing. Thus, HPP can be viewed as a novel way of manipulating the microstructure of proteins such as ovalbumin while maintaining the nutritional value and natural flavour of such compounds.

As with ovalbumin, emulsions made with pressure treated 11S soy protein were found to have poorer emulsifying and stabilising ability with respect to initial droplet size and creaming behaviour than the native protein. This is probably due to the enhanced association of subunits and/or aggregation induced by the formation of intermolecular disulphide bridges via a SH/-S-S interchange. As with ovalbumin, moderate heat treatment (80 °C for 2 min) had a far greater effect than high pressure treatment on the changes in emulsion stability and droplet size distribution.

Though only in its infancy, the ability of HPP to modify the structure and surface hydrophobicity of a protein does suggest that, as well as a preservation technique, HPP may well have commercial/industrial application for modifying the functional properties of potential emulsifiers and foaming agents.

6.6 Gelation

At sufficiently high concentrations and at the appropriate pH many proteins, especially if they have potential disulphide bond forming abilities, will gel or precipitate, but the texture of the gels formed will be markedly different to their heat-set counterparts. Such pressure-set gels will normally contain a relatively high concentration of hydrogen-bonded structure(s) and thus will melt or partially melt on heating. In addition, they will be less able to hold water, i.e. they will synerese and be much softer in texture and 'glossier' in appearance. For example, myoglobin (a protein with no amino acids containing sulphur) will unfold at sufficiently high pressure on pressure release and then normally revert back to a soluble monomer or dimer; but at its isoelectric point (pH 6.9) it will precipitate. However, unlike the precipitate formed on heat denaturation, it will be relatively unstable and, because it is primarily stabilised by hydrogen bonds, will dissolve or melt on gently raising the temperature [64].

If, as suggested, pressure-set gels are stabilised, at least to some extent by hydrogen bonds, then it would be expected that heat treatment of such a system will destroy this network and enable a 'heat' set gel to form. Such effects are apparent in both whey protein concentrate gels at pH 7 and in the toughness or hardness of fish and meat flesh, i.e. myosin gels. For example, calorimetric studies on fish and meat myofibrillar proteins and whole muscle have clearly demonstrated the presence of a hydrogen-bonded network in pressure-treated muscle, myofibrillar protein and myosin [65, 66], that is destroyed on heat treatment. Just as the thermal stability of different myosins reflects the body temperature of the species, that from cod being less stable than that from beef or pork, so their relative pressure sensitivities also vary, with cod myosin denaturing at about 100–200 MPa at 20 °C, whilst that from turkey and pork only unfolds at pressures above 200 MPa.

Thus, the simultaneous or sequential treatment of proteins with heat and pressure does raise the possibility of generating gels with interesting and novel textures. The likely mechanisms are discussed in more detail by Ledward [63].

As well as being able to modify the functional properties of water soluble proteins such as β-lactoglobulin and myoglobin and generate heat-sensitive gels in proteinaceous foods such as meat and fish, high pressures can also be used to texturise many insoluble plant proteins such as gluten and soya. An extensive

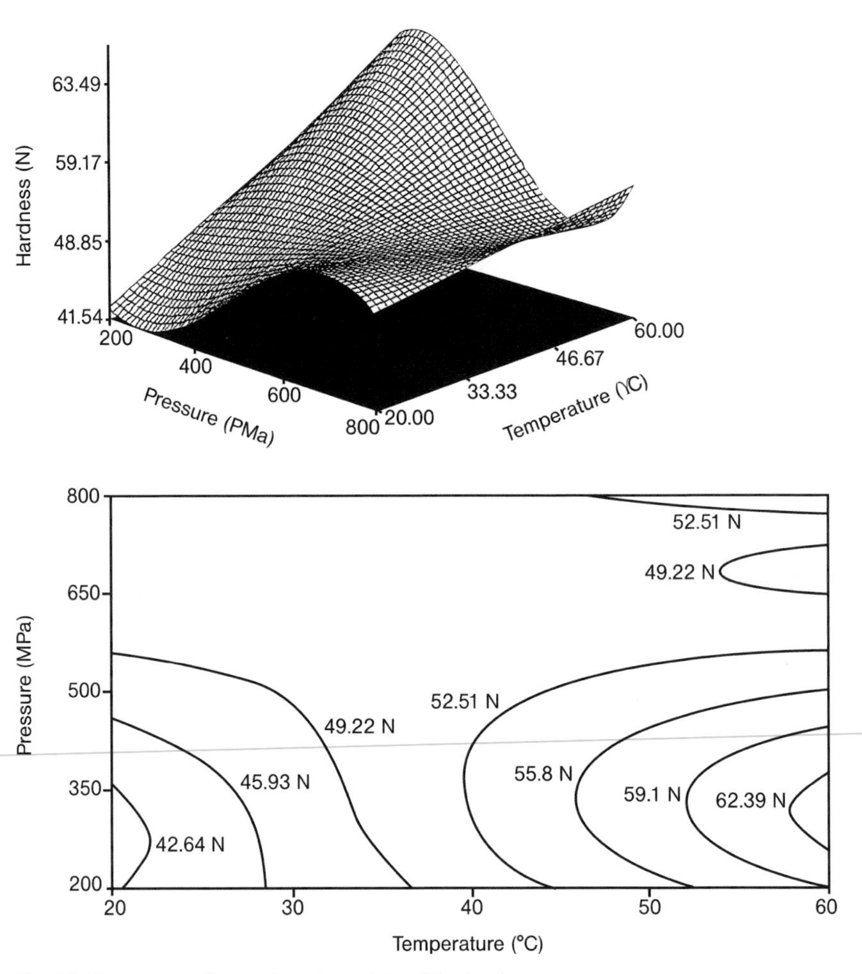

Fig. 6.3 Response surface and contour plots of the hardness of soy protein gels prepared from a commercial soy concentrate mixed with 3.75 times its weight of water after treatment at temperatures of 20–60 °C and pressures up to 800 MPa for 50 min; from Apichartsrangkoon [69].

study has been carried out by Apichartsrangkoon et al. [67, 68] on the use of various pressure/temperature treatments to texturise wheat gluten and soy protein and it can be seen from Fig. 6.3 how pressure/temperature and time can be used to generate a range of soya gels of different rheological properties.

The above are but a few examples of the rapidly expanding literature on the use of pressure and/or temperature to generate gels with very different rheological properties. The potential of such technology in the development of new proteinaceous foods is very exciting.

From the above brief review it is readily apparent that, although our knowledge of the effects of pressure on proteins has advanced in recent years, a great deal more research is needed before our understanding approaches that of the effect of other parameters on proteins, such as temperature and pH.

6.7 Organoleptic Considerations

Since covalent bonds are unaffected by pressure, many of the small molecules that contribute to the colour, flavour or nutritional quality of a food are unchanged by pressure. This is a major advantage of the process and has led to its successful application to such products as guacamole, fruit juices and many other fruit-based jams and desserts.

However, if the organoleptic quality of a food depends upon the structural or functional macromolecules, especially proteins, pressure may affect the quality. The most obvious case of this is with meat and fish products where, at pressures above ~100 MPa for coldwater fish and 300–400 MPa for meats and poultry, the myosin will unfold/denature and the meat or fish will take on a cooked appearance as these proteins gel. In addition, in red meats such as beef and lamb, myoglobin, the major haem protein responsible for the red colour of the fresh product, will denature at pressures around 400 MPa and give rise to the brown haemichrome, further contributing to the cooked meat appearance. However, since the smaller molecules are not affected, the flavour of the fish or meat will be that of the uncooked product. On subsequent heat treatment, these foods will develop a typical cooked flavour.

Although in many circumstances the flavour of a food will remain unchanged after pressure treatment, Cheah and Ledward [55] and Angsupanich and Ledward [65] have shown that, under pressure, inorganic transition metals (especially iron) can be released from the transition metal compounds in both meat and fish. These may catalyse lipid oxidation and thus limit shelf life and may also contribute to the flavour when the product is subsequently cooked. This release of inorganic iron takes place at pressures above 400 MPa and does not appear to be a problem with cured meats or shellfish. This phenomenon may though limit the usefulness of HPP for many uncured fish and meat products.

Since enzyme activity is affected by pressure, many foods whose flavour, texture or colour is dependant on enzymic reactions, may have their sensory properties modified on pressure treatment. Also, in multiphase foods, such as fruits and vegetables, pressure can give rise to significant textural changes due both to modification of the enzyme activity and because of the different rates and extents of compression and decompression of the aqueous, solid and gaseous phases. This may lead to physical damage, as described above for cherry tomatoes.

6.8 Equipment for HPP

The key features of any HPP system used for food processing include: a pressure vessel, the pressure transmission medium and a means of generating the pressure. The emergence of production-sized HPP equipment for food occurred more or less simultaneously in the USA and Europe in the mid-1990s.

For convenience, potable water is used as the pressurising medium. The water is usually separated from the food by means of a flexible barrier or package. Products such as shellfish, however, are usually immersed directly into the water as beneficial effects upon yield can be achieved. Up to the minute details of HPP technology and applications can be found on www.avure.com and www.freshunderpressure.com.

6.8.1 'Continuous' System

'Continuous' output of pumpable foods such as fruit juice is achieved by arranging three or more small pressure vessels (isolators) in parallel. Each vessel is automatically operated 'out of phase' with the others so as to give an effective 'continuous' output of treated food. Each 'isolator' typically comprises a 25-l pressure vessel incorporating a floating piston and valve block.

Juice is pumped into the vessel under low pressure, forcing the piston down. With valves closed, a high-pressure pump is used to pump water beneath the piston, so forcing it up and thereby pressurising the juice. After the set pressure and hold time conditions have been satisfied, the high-pressure water is vented and the treated juice is pumped out to the production line, by means of low-pressure water pushing the piston to the top of the vessel.

An illustration of a 'continuous' system is shown in Fig. 6.4.

The 'continuous' system has the advantage of treating juices in their unpackaged condition as an integral part of the production process. If juices were previously thermally treated, the HPP system simply replaces the thermal system, with few changes to the remainder of the production process.

The HPP juice may be packed into any suitable package. If a very long chilled shelf life is required however, then the producer should consider the use of aseptic packaging conditions (see Chapter 9). Packaging materials or methods to inhibit the oxidation of the juice by ingress of oxygen through the packaging material over time may also need to be considered.

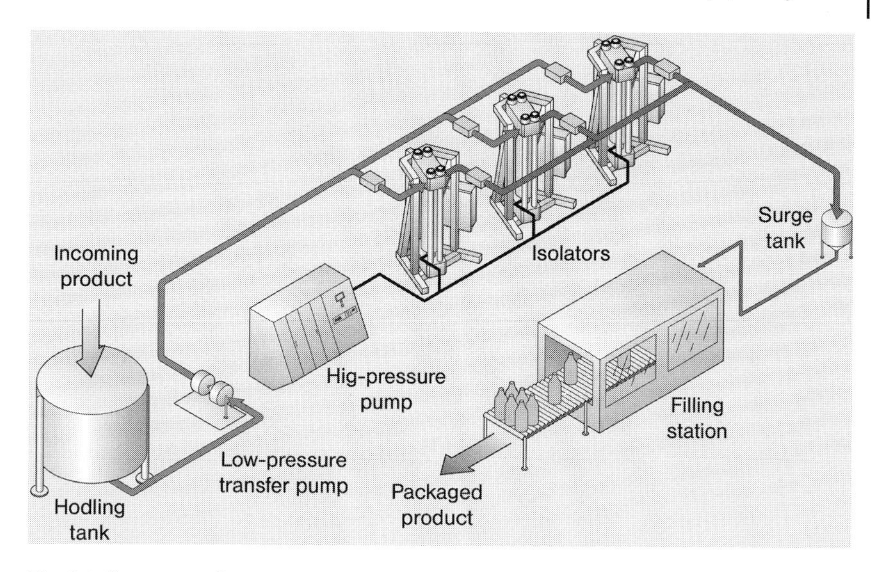

Fig. 6.4 Illustration of a 'continuous' HPP system; courtesy Avure Technologies AB.

The HPP production line can effectively be run for extended periods without stopping, as the juice now has a long shelf life and can be directed to chilled storage prior to distribution. The cost savings by adopting this strategy can be very considerable compared to the 'normal' production strategy of producing small quantities and shipping immediately.

As the HPP system incorporates its own 'clean in place' (CIP) system, the whole production system can be cleaned and sanitised together. Risk assessment under HACCP (see Chapter 11) includes the HPP system in conjunction with the remainder of the production equipment, methods and materials.

The 'continuous' system is suitable for most 'pumpable' products, such as fruit juices, purees, smooth sauces and soups, vegetable juices and smoothies.

6.8.2 'Batch' System

These vessels, typically 35–680 l in capacity, use water as the pressurising medium and accept prepackaged food products (see Fig. 6.5).

When a high water content food is subjected to isostatic pressure up to 600 MPa, it compresses in volume by 10–15%, depending on the food type and structure. As the pressurising medium is water, the food packaging must be capable of preventing water ingress and accepting the volume reduction.

The packaging has to withstand the pressure applied but it should be noted that standard plastic bottles, vacuum packs and plastic pouches are usually suitable. These flexible materials have essentially no resistance to volume under the

Fig. 6.5 Illustration of a 'batch' HPP system; courtesy Avure Technologies AB.

effects of high pressure. Hence, a plastic bottle of fruit juice, having a small (standard) headspace, will compress by the volume of the headspace plus the compressibility of the juice. Provided that the pack is resilient enough to withstand the volume reduction imposed, then the pressurising water will be kept out of the bottle. Once depressurised, the bottle and contents will revert to their original state, minus the spoilage organisms.

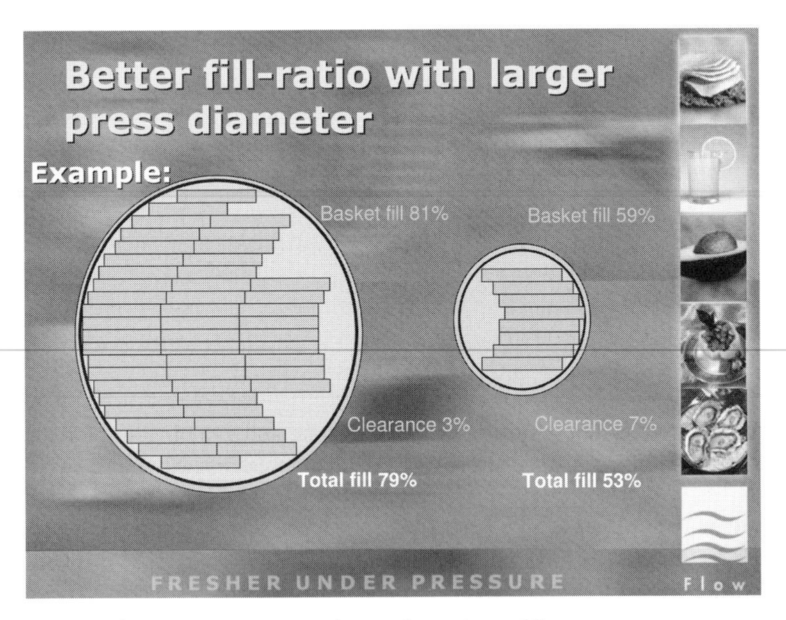

Fig. 6.6 Schematic representation showing how a better fill ratio is attainable with a larger pressure vessel; courtesy Avure Technologies AB.

A large vessel size gives better utilisation of the available space; and therefore the greatest production output per batch cycle. A 320-l vessel could be filled with a 320-l plastic bag of food, therefore obtaining the maximum use of available volume. Normally however, the batch vessel will be used for a variety of prepacked consumer food products of different size and styles. Fruit juice bottles, for example, will not normally pack together without gaps and therefore the effective output of the vessel will reduce accordingly (see Fig. 6.6).

Batch systems are ideal where wide ranges of products are required and the products comprise solids and liquids combined. Cooked meats, stews, guacamole, fruits in juice, shellfish and ready meals are typical batch HPP examples. Batch systems filled with consumer type packs of food benefit from both the packaging and the food being HPP-treated. This considerably aids quality assurance through distribution and shelf life and may reduce or simplify some production processes.

6.9 Pressure Vessel Considerations

In a commercial environment, a HPP system is likely to be run at least 8 h day^{-1}, possibly continuously, and so the design criteria is critical.

'Standard' pressure vessels use a thick steel wall construction, where the strength of the steel parts alone is used to contain the pressure within. This is fine for a vessel expected to perform a low number of cycles. However, every time the pressure is applied, the steel vessel expands slightly as it takes up the strain. This imposes expansion stresses within the steel structure and, with time, can lead to the creation of microscopic cracks in the steel itself. If undetected, it is possible for the vessel to fail and rapidly depressurise, causing a safety hazard and destroying the vessel.

The technique used to create a long life and guaranteed 'leak before break' vessel is known as 'Quintus™ wire winding'.

In this case, a relatively thin wall pressure vessel (too thin to contain the working pressure of its own) is wrapped in steel wire. The wire wrapping, which can amount to several hundred kilometres of wire for a large vessel, is stretched and wrapped onto the vessel by a special machine and powered turntable.

The wire attempts to return to its unstretched state and in doing so exerts a compressive force upon the pressure vessel. The compressive force of the wire wrapping is engineered to slightly exceed the expansion force of the HPP water pressure. Hence, the thin-walled pressure vessel, even with 600 MPa water pressure within it, is actually still under compression from the wire windings. This means that the only forces within the steel of the pressure vessel are compressive and therefore cracks cannot propagate.

A wire-wound frame is made to hold in place 'floating' top and bottom plugs, so as to allow access to the vessel for loading and unloading. The frame con-

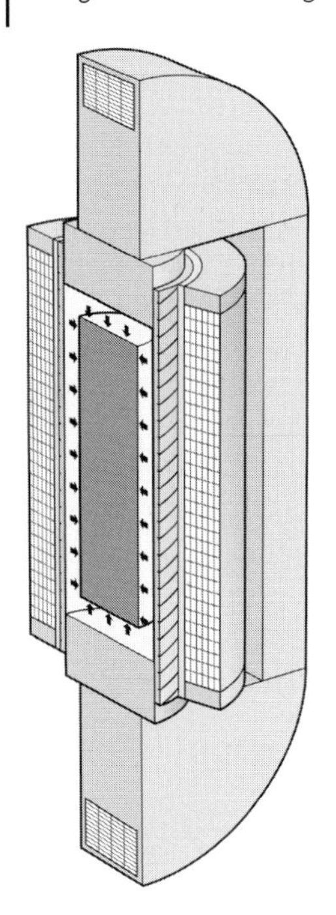

Fig. 6.7 A Quintus^{TM} wire-wound pressure vessel; copyright Avure Technologies AB.

struction is similar to that described for the vessel (see Fig. 6.7). A wire-wound vessel has theoretically unlimited life expectancy.

High-grade forged pressure vessel steel is not the preferred material for contact with foods and so the wire-wound vessel incorporates a replaceable inner liner of stainless steel.

6.9.1
HP Pumps

Normal pumps cannot achieve the required pressures for HPP food applications. Therefore, a standard hydraulic pump is used to drive an intensifier pump, comprising a large piston driven back and forth by the hydraulic oil in a low-pressure pump cylinder. The large piston has two smaller pistons connected to it, one each side, running in high-pressure cylinders. The ratio of large and small piston areas and hydraulic pump pressure gives a multiplication of the pressure seen at the output of the high-pressure cylinders. The small, high-pressure pistons are pumping the potable water used as the pressurising medium in the HPP food process.

The pressure and volume output from the intensifier is dependent upon the overall sizes of the pistons and hydraulic pressure. However, it is often more convenient to use several small intensifiers working 'out of phase' to give a smooth pressure delivery and some degree of redundancy, rather than use a single, large intensifier.

6.9.2
Control Systems

Both the batch and continuous HPP systems are mainly sequentially operated machines and so are controlled by standard PLCs. Verification of each sequence before proceeding to the next is essential in some cases, due to the very high pressures involved and the materials used for the sanitisation of food processes. Many critical sequences will include 'fail safe' instrumentation and logic.

The user interface forms an important part of any process and the demands today are for even the most complex machine or system to be operable by non-engineering operatives. The operator is therefore presented with a screen or monitor with the process parameters and operator requirements presented in a format similar to that of a home PC running the most popular software. System logic design and password protection means that any suitably trained person can carry out normal production and maintenance.

Fault conditions are automatically segregated into those that the operator can deal with and those that dictates either the automatic safety shutdown of the system (or part thereof) or a sequence halt. In every case, the fault and probable solution is available to the user.

6.10
Current and Potential Applications of HPP for Foods

France was the first country in the European Community to have HPP food products commercially available. Since 1994, Ulti have been processing citrus juices at 400 MPa. Orange juice is the main product, although some lemon and grapefruit juice is also produced. Their motivation for moving into HPP was a desire to extend the shelf life of their fresh fruit juice, then only 6 days at chilled temperature. High pressure treatment allowed a shelf life of up to 16 days. This reduced logistical problems and transportation costs, without harming the sensory quality and vitamin content of the juice. HPP-processed fruit juice and fruit smoothies have also been available in the UK since 2002.

At present, a number of products are currently in development, including fruit, delicatessen and duck fat liver products. Before commercialisation is approved, convincing physicochemical, microbial and toxicity analyses must be carried out.

Elsewhere in Europe, Espuña of Spain process sliced ham and delicatessen meat products, in flexible pouches, using an industrial 'cold pasteuriser' unit. Throughputs of 600 kg h^{-1} are achievable, using operating pressures of 400–500 MPa and hold times of a few minutes. The organoleptic properties of fresh ham are preserved; and an extended shelf life of 60 days under chilled storage has been reported.

As discussed above, HPP guacamole has been commercially available for some time in the USA. It is actually manufactured across the border in Mexico, to take full advantage of low raw material costs and avoid the import costs associated with the import of raw avocados to the USA. Its market share continues to grow and is reportedly based on the consumer preference for the 'fresher' taste of guacamole processed in this manner. The same company now produces HPP meal kits, consisting of pressure-treated cooked meat, salsa, guacamole, peppers and onion. Only the flour tortilla is not pressure-treated. The products have a chilled shelf life of at least 35 days.

'Gold Band' oysters from Motivatit Seafoods Inc. www.the perfectoyster.com are now achieving top national awards in America for HPP shucked oysters. They report up to 75% yield increase with HPP compared to manual shucking processes and now 'contract shuck' oysters for other famous oyster companies in the region. Shellfish shucking using HPP has also recently started in Australia and is expected in Europe within the near future.

High Pressure Research Inc. (Corvallis, Ore.) is another USA-based company to produce HPP products. Their range includes oysters, salmon, yoghurt, spreads, fruit and fruit juices. These products are currently appearing in major grocery chains and supermarkets; and a refrigerated shelf life of 60 days is possible. Meanwhile, pilot plant work is underway on the production of Spanish rice, oriental chicken, vegetarian pasta salad and seafood creole. Evaluation of these products after storage at room temperature for various times is to be carried out by the US Army at Natick Laboratories.

The Japanese market for HPP foods is better developed than in the West. A range of fruit products, jams and yoghurts has been available for a number of years, and the most recent innovations have been rice products. Rice cakes and convenience packs of boiled rice are now available commercially. A recent Japanese patent described a HPP technique to improve the perceived 'freshness' of fruit flavours. The process described involves adding a small amount of fruit juice to a fruit flavour and then processing at 100–400 MPa for 1–30 min. The processing temperature is maintained between freezing and room temperature.

Markets demanding better food quality, extended chilled shelf life and higher food safety can now be commercially accessed using high pressure. These new market opportunities and new products can offset the capital cost of the equipment, as has been seen in many commercial application successes. The textural changes possible using high pressure have started to lead to the development of new, or novel, foodstuffs. Food producers and retailers should carefully consider the currently accepted high costs associated with short production runs of fresh quality, limited shelf life foods. Compare those to the savings associated with longer production runs of longer shelf life and safer HPP foods, while still benefiting from the original organoleptic food quality. HPP offers foods with quality and convenience, in keeping with today's consumer trends.

References

1 Hite, B.H. **1989**, The Effect of Pressure in the Preservation of Milk, *Bull. West Virginia Univ. Agric. Exp. Stn* 58, 15–35.

2 Hoover, D.G., Metrick, K., Papineau, A.M., Farkas, D.F., Knorr, D. **1989**, Biological Effects of High Hydrostatic Pressure on Food Microorganisms, *Food Technol.* 43, 99–107.

3 Morita, R.Y. **1975**, Psychrophilic Bacteria, *Bacteriol. Rev.* 39, 144–167.

4 Chong, G., Cossins, A.R. **1983**, A Differential Polarized Fluorometric Study of the Effects of High Hydrostatic Pressure upon the Fluidity of Cellular Membranes, *Biochemistry* 22, 409–415.

5 Sale, A.J.H., Gould, G.W., Hamilton, W.A. **1970**, Inactivation of Bacterial Spores by High Hydrostatic Pressure, *J. Gen. Microbiol.* 60, 323–334.

6 Ananth, E., Heinz, V., Schlter, O., Knorr, D. **2001**, Kinetic Studies on High Pressure Inactivation of *Bacillus stearothermophilus* Spores Suspended in Food Matrices, *Innov. Food Sci. Emerging Technol.* 2, 261–272.

7 Roberts, C.M., Hoover, D.G. **1996**, Sensitivity of *Bacillus coagulans* Spores to Combinations of High Hydrostatic Pressure, Heat, Acidity and Nisin, *J. Appl. Bacteriol.* 81, 363–368.

8 Mills, G., Earnshaw, R., Patterson, M.F. **1998**, Effects of High Hydrostatic Pressure on *Clostridium sporogenes* Spores, *Lett. Appl. Microbiol.* 26, 227–230.

9 Gould, G.W. **1973**, Inactivation of Spores in Food by Combined Heat and Hydrostatic Pressure, *Acta Aliment.* 2, 377–383.

10 Meyer, R.S. **2000**, *Ultra High Pressure, High Temperature Food Preservation Process.* US patent No: 6,177,115, BI (Richard S. Meyer, Tacoma, WA, USA).

11 Master, A.M., Krebbers, B., Van den Berg, R.W., Bartels, P.V. **2004**, Advantages of high pressure sterilisation on the quality of food products. *Trends Food Sci Technol.* 15, 79–85.

12 Patterson, M.F., Quinn, M., Simpson, R., Gilmour, A. **1995**, Sensitivity of Vegetative Pathogens to High Hydrostatic Pressure Treatment in Phosphate-Buffered Saline and Foods, *J. Food Prot.* 58, 524–529.

13 Shigehisa, T., Ohmori, T., Saito, A., Taji, S., Hayashi, R. **1991**, Effects of High Pressure on Characteristics of Pork Slurries and Inactivation of Microorganisms Associated with Meat and Meat Products, *Int. J. Food Microbiol.* 12, 207–216.

14 Butz, P., Funtenberger, S., Haberditzl, T., Tauscher, B. **1996**, High Pressure Inactivation of *Byssochlamys nivea* Ascospores and Other Heat-Resistant Moulds, *Lebensm.-Wiss. Technol.* 29, 404–410.

15 Brâna, D., Voldrich, M., Marek, M., Kamarád, J. **1997**, Effect of High Pressure Treatment on Patulin Content in Apple Concentrate, in *High Pressure Research in the Biosciences*, ed. K. Heremans, Leuven University Press, Leuven, pp. 335–338.

16 Kingsley, D.H., Hoover, D.G., Papafragkou, E., Richards, G.P. **2002**, Inactivation of hepatitis A and a calicivirus by high hydrostatic pressure. *J. Food Potect.* 65, 1605–1609.

17 Khadre, M.A., Yousef, A.E. **2002**, Susceptibility of human rotavirus to ozone, high pressure and pulsed electric field. *J. Food Protect.* 65, 1441–1446.

18 Garcia, A.F., Heindl, P., Voight, H., Büttner, M., Wienhold, D., Butz, P., Stärke, J., Tausher, B. et al., **2004**. Reduced proteinase K resistance and infectivity of prions after pressure treatment at 60 °C. *J. Gen. Virol.*, 85, 261–264.

19 Linton, M., McClements, J.M.J., Patterson, M.F. **2001**, Inactivation of Pathogenic *Escherichia coli* in Skimmed Milk Using High Hydrostatic Pressure, *Int. J. Food Sci. Technol.* 2, 99–104.

20 Alpas, H., Kalchayanand, N., Bozoglu, F., Sikes, T., Dunne, C.P., Ray, B. **1999**, Variation in Resistance to Hydrostatic Pressure Among Strains of Foodborne Pathogens, *Appl. Environ. Microbiol.* 65, 4248–4251.

21 Mackey, B.M., Forestiere, K., Isaacs, N. **1995**, Factors Affecting the Resistance of *Listeria monocytogenes* to High Hydrostatic Pressure, *Food Biotechnol.* 9, 1–11.

22 Styles, M. F., Hoover, D. G., Farkas, D. F. **1991**, Response of *Listeria monocytogenes* and *Vibrio parahaemolyticus* to High Hydrostatic Pressure. *J. Food Sci.* 56, 1404–1497.
23 Metrick, C., Hoover, D. G., Farkas, D. F. **1989**, Effects of High Hydrostatic Pressure on Heat Resistant and Heat Sensitive Strains of *Salmonella*, *J. Food Sci.* 54, 1547–1564.
24 Earnshaw, R. G. **1995**, High Pressure Microbial Inactivation Kinetics, in *High Pressure Processing of Foods*, ed. D. A. Ledward, D. E. Johnston, R. G. Earnshaw, A. P. M. Hastings, Nottingham University Press, Nottingham, pp. 37–46.
25 Takahashi, K., Ishii, H., Ishikawa, H. **1991**, Sterilisation of Microorganisms by Hydrostatic Pressure at Low Temperatures, in *High Pressure Science of Food*, ed. R. Hayashi, San-Ei Publishing, Kyoto, pp. 225–232.
26 Patterson, M. F., Kilpatrick, D. J. **1998**, The Combined Effect of High Hydrostatic Pressure and Mild Heat on Inactivation of Pathogens in Milk and Poultry, *J. Food Prot.* 61, 432–436.
27 Oxen, P., Knorr, D. **1993**, Baroprotective Effects of High Solute Concentrations Against Inactivation of *Rhodotorula rubra*, *Lebensm.-Wiss. Technol.* 26, 220–223.
28 Linton, M., McClements, J. M. J., Patterson, M. F. **1999**, Survival of *Escherichia coli* O157,H7 During Storage in Pressure-Treated Orange Juice. *J. Food Prot* 62, 1038–1040.
29 Wei, C. I., Balaban, M. O., Fernando, S. Y., Peplow, A. J. **1991**, Bacterial Effect of High Pressure CO_2 Treatment of Foods Spiked with *Listeria* or *Salmonella*, *J. Food Prot.* 54, 189–193.
30 Hong, S. I., Park, W. S., Pyun, Y. R. **1997**, Inactivation of *Lactobacillus* sp. from Kimchi by High Pressure Carbon Dioxide, *Lebensm.-Wiss. Technol.* 30, 681–685
31 Paul, P., Chawala, S. P., Thomas, P., Kesavan, P. C. **1997**, Effect of High Hydrostatic Pressure, Gamma-Irradiation and Combined Treatments on the Microbiological Quality of Lamb Meat During Chilled Storage, *J. Food Saf.* 16, 263–271.
32 Crawford, Y. J., Murano, E. A., Olson, D. G., Shenoy, K. **1996**, Use of High Hydrostatic Pressure and Irradiation to Eliminate *Clostridium sporogenes* in Chicken Breast, *J. Food Prot.* 59, 711–715.
33 Raso, J., Palop, A., Pagan, R., Condon, S. **1998**, Inactivation of *Bacillus subtilis* Spores by Combining Ultrasonic Waves Under Pressure and Mild Heat Treatment. *J. Appl. Microbiol.* 85, 849–854.
34 Kalchayanand, N., Sikes, T., Dunne, C. P., Ray, B. **1998**, Interaction of Hydrostatic Pressure, Time and Temperature of Pressurization and Pediocin AcH on Inactivation of Foodborne Bacteria, *J. Food Prot.* 61, 425–431.
35 Adegoke, G. O., Iwahashi, H., Komatsu, Y. **1997**, Inhibition of *Saccharomyces cerevisiae* by Combination of Hydrostatic Pressure and Monoterpenes, *J. Food Sci.* 62, 404–405.
36 Ogawa, H., Fukuhisa, K., Fukumoto, H. **1992**, Effect of Hydrostatic Pressure on Sterilization and Preservation of Citrus Juice, in *High Pressure and Biotechnology*, ed. C. Balny, R. Hayashi, K. Heremans, P. Masson, John Libbey Eurotext, London, pp. 269–278.
37 Arroyo, G., Sanz, P. D., Prestamo, G. **1997**, Effects of High Pressure on the Reduction of Microbial Populations in Vegetables, *J. Appl. Microbiol.* 82, 735–742.
38 Ananth, V., Dickson, J. S. Olson, D. G., Murano, E. A. **1998**, Shelf–Life Extension, Safety and Quality of Fresh Pork Loin Treated with High Hydrostatic Pressure, *J. Food Prot.* 61, 1649–1656.
39 Carlez, A., Rosec, J. P., Richard, N., Cheftel, J. C. **1994**, Bacterial Growth During Chilled Storage of Pressure-Treated Minced Meat. *Lebensm.-Wiss. Technol.* 27, 48–54.
40 El Moueffak, A. C., Antoine, M., Cruz, C., Demazeau, G., Largeteau, A., Montury, M., Roy, B., Zuber, F. **1995**, High Pressure and Pasteurisation Effect on Duck Foie Gras, *Int. J. Food Sci. Technol.* 30, 737–743.
41 Carpi, G., Buzzoni, M. M., Gola, S., Maggi, A., Rovere, P. **1995**, Microbial and Chemical Shelf-Life of High-Pressure Treated Salmon Cream at Refrigeration Temperatures, *Ind. Conserve* 70, 386–397.

42 Gervilla, R., Felipe, X., Ferragut, V., Guamis, B. **1997**, Effect of High Hydrostatic Pressure on *Escherichia coli* and *Pseudomonas fluorescens* Strains in Ovine Milk, *J. Dairy Sci.* 80, 2297–2303.

43 Ponce, E., Pla, R., Mor-Mur, M., Gervilla, R., Guamis, B. **1998**, Inactivation of *Listeria innocua* Inoculated in Liquid Whole Egg by High Hydrostatic Pressure, *J. Food Prot.* 72, 119–122.

44 Balny, C., Masson, P., Travers, F. **1989**, Some Recent Aspects of the Use of High Pressure for Protein Investigations in Solutions, *High Pressure Res.* 2, 1–28.

45 Heremans, K. **1899**, From Living Systems to Biomolecules, in *High Pressure and Biotechnology*, vol. 224, ed. C. Balny, R. Hayashi, K. Heremans, P. Masson, Colloque INSERM, Paris

46 Balny, C., Masson, P. **1993**, Effects of High Pressure on Proteins, *Food Rev. Int.* 9, 611–628.

47 Heremans, K. **1982**, High Pressure Effects on Proteins and Other Biomolecules, *Annu. Rev. Biophys. Bioeng.* 11, 1–21.

48 Mozhaev, V.V., Heremanks, K., Frunk, J. **1994**, Exploiting the Effects of High Hydrostatic Pressure in Biotechnological Applications, *Trends Biotechnol.* 12, 493–501.

49 Galazka, V.B., Ledward, D.A. **1998**, High Pressure Effects on Biopolymers, in *Functional Properties of Food Macromolecules*, 2nd edn, ed. S. E. Hill, D. A. Ledward, J. R. Mitchell, Aspen Press, Aspen, 278–301.

50 Ludikhuyze, L., Van Loey, A., Indrawati, Denys, S., Hendrickx, M. **2002**, The Effect of Pressure Processing on Food Quality Related Enzymes, from Kinetic Information to Process Engineering, in *Trends in High Pressure Bioscience and Biotechnology*, ed. R. Hayashi, Elsevier, London, pp. 517–524.

51 Gomes, M.R.A., Sumner, I.G., Ledward, D.A. **1997**, High Pressure Treatment of Papain, *J. Sci. Food Agric.* 75, 67–75.

52 Baker, E.N., Drenth, J. **1987**, The Thiol Proteases, Structure and Mechanism, in *Biological Macromolecules and Assemblies*, ed. F. A. Jurnak, A. McPherson, J. Wiley and Sons, New York, pp. 313–368.

53 Gomes, M.R.A., Ledward, D.A. **1996**, High Pressure Effects on Some Polyphenoloxidases, *Food Chem.* 56, 1–5.

54 Cheftel, J.C., Culioli, J. **1997**, Effects of High Pressure on Meat, a Review, *Meat Sci.* 46, 211–236.

55 Cheah, P.B., Ledward, D.A. **1997a**, Catalytic Mechanisms of Lipid Oxidation in High Pressure Treated Pork Fat and Meat, *J. Food Sci.* 62, 1135–1141.

56 Cheah, P.B., Ledward, D.A. **1997b**, Inhibition of Metmyoglobin Formation in Fresh Beef by Pressure Treatment, *Meat Sci.* 45, 411–418.

57 Eskin, N.A.M., Grossman, S., Pinsky, A. **1977**, Biochemistry of Lipoxygenase in Relation to Food Quality, *Crit. Rev. Food Sci. Nutr.* 22, 1–33.

58 Kazeniac, S.J., Hall, R.M. **1970**, Flavour Chemistry of Tomato Volatiles, *J. Food Sci.* 35, 519–530.

59 Tangwongchai, R., Ledward, D.A., Ames, J.M. **2000**, Effect of High Pressure Treatment on Lipoxygenase Activity, *J. Agric. Food Chem.* 48, 2896–2902.

60 Tangwongchai, R., Ledward, D.A., Ames, J.M. **2000**, Effect of High Pressure on the Texture of Cherry Tomato, *J. Agric. Food Chem.* 48, 1434–1441.

61 Gomes, M.R.A., Clark, R., Ledward, D.A. **1998**, Effects of High Pressure on Amylases and Starch in Wheat and Barley Flours, *Food Chem.* 63, 363–372.

62 Galazka, V.B., Ledward, D.A., Varley, J. **1997**, High Pressure Processing of β-Lactoglobulin and Bovine Serum Albumen, in *Food Colloids: Proteins, Lipids and Polysaccharides*, ed. E. Dickenson, B. Bergenstahl, Royal Society of Chemistry, Cambridge, 127–136.

63 Ledward, D.A. **2000**, Effects of pressure on protein structure, *High Pressure Research*, 19, 1–10.

64 Defaye, A.B., Ledward, D.A., MacDougall, D.B., Tester, R.F. **1995**, Renaturation of subjected to high isostatic pressure, *Food Chemistry*, 52, 19–22.

65 Angsupanich, K., Ledward, D.A. **1998**, High pressure treatment effects on cod muscle, *Food Chemistry*, 63, 39–50.

66 Angsupanich K., Edde, M., Ledward, D.A. **1999**, The effects of high pressure on the myofibrillar proteins of cod and

turkey, *Journal of Agriculture and Food Chemistry*, 47, 92–99.

67 Apichartsrangkoon, A., Ledward, D. A., Bell, A. E., Brennan, J. G. **1998**, Physio-chemical properties of high pressure treated wheat gluten, *Food Chemistry*, 63, 215–220.

68 Apichartsrangkoon, A., Ledward, D. A., Bell, A. E., Schofield, J. D. **1999**, Dynamic viscoelastic behaviour of high pressure treated wheat gluten, *Cereal Chemistry*, 76, 777–782.

69 Apichartsrangkoon, A. **1999**, *PhD thesis*, University of Reading.

7
Pulsed Electric Field Processing, Power Ultrasound and Other Emerging Technologies

Craig E. Leadley and Alan Williams

7.1 Introduction

The extended preservation of food has been a challenge for mankind throughout the ages. Preservation processes such as drying, curing, pickling and fermenting have been carried out for generations and examples of these processes can be found throughout the world. The products resulting from these processes, almost without exception, are radically changed in comparison to the fresh counterpart. Nevertheless these products have become established in the diet throughout the world and are important in their own right.

The advent of rapid freezing technologies fundamentally shifted the consumer's expectations upward in terms of the 'quality' of extended shelf life products. In the western world, freezing remains an excellent method for preserving foods in a cost-effective manner and in some cases the frozen product is the closest to fresh that the consumer is ever likely to get. Peas for example rapidly deteriorate after picking and the frozen product is better, in terms of sensory and nutritional value, than 'fresh' as purchased over the counter.

The application of heat for food preservation is a long established production process for which we must thank pioneers such as Nicolas Appert and Louis Pasteur. Thermal preservation has an excellent safety record and near universal consumer acceptance. However, even relatively mild thermal processes can result in a food product that is substantially different in terms of colour, flavour and texture to the unprocessed food.

A wide range of novel preservation processes have been studied over the last 100 years. A general mistrust of 'artificial' preservation and consumer demand for convenient 'fresh like' products has given fresh impetus to research into so called 'nonthermal' preservation methods. A selection of the most prominent emerging preservation technologies of interest to the food industry is summarised in Table 7.1. Many of these technologies remain very much in the research arena, some are on the brink of commercialisation. This chapter provides an overview of some of the main nonthermal emerging technologies showing promise for commercial food processing. It is not intended to be ex-

Food Processing Handbook. Edited by James G. Brennan

ISBN: 3-527-30719-2

Table 7.1 Summary of emerging preservation technologies attracting interest worldwide.

High pressure processing
Pulsed electric field processing
Power ultrasound
High intensity pulsed light
Oscillating magnetic fields
Irradiation (X-ray, electron beam, γ ray)
High voltage arc discharge
Plasma processing
Microwave and radio frequency heating
Ohmic and inductive heating
Ultraviolet

haustive, merely illustrative of the range of techniques that could be available to food manufacturers in the coming years.

High pressure processing (HPP) has been the subject of intense research effort over the last 15–20 years. Food products pasteurised by high pressure are now commercially available in a number of countries including Japan, France, Spain, North America and the UK. Since HPP is covered in Chapter 6, it will not be discussed here.

Irradiation, which is covered in detail in Chapter 5, has perhaps been more widely investigated than any other novel preservation method. Its torturous route to commercialisation and the arguments for and against the technique have been widely debated. Today, despite the availability of commercial systems and the proven efficacy of the process, industrial use is on a limited scale. In the UK, this is largely due to strong consumer resistance. Since 1999, no food has been irradiated in the UK for commercial purposes [1].

There are around 15 facilities in the EU that are approved for food irradiation [2]. The exact amount of food irradiated per year in the EU is not certain, but an estimate for 2001 is around 22 000 t [1]. Herbs, spices and poultry products account for the major proportion of this total.

In the USA, there appears to have been a softening of consumer attitudes towards irradiation, in part due to concerns over *Escherichia coli* in ground meat products and the need for an effective intervention method. Electron beam irradiated beef burgers and ground meat 'chubs' have been successfully introduced in the US market. Over 5000 US retail stores in 48 states now carry products that have been pasteurised using electron beam irradiation [3]. The US supermarket chain Wegmans is reported to offer irradiated ground beef products at stores throughout the chain and Dairy Queen is similarly reported to offer irradiated ground beef products in all of its Minneapolis stores.

Irradiation has been so comprehensively studied that it can almost be considered as a 'conventional' nonthermal preservation method. Consumer confidence is the main barrier to irradiation being considered as a credible preservation method in the UK. Any review of nonthermal preservation would be incomplete without a discussion of irradiation and so it is briefly mentioned here for completeness (see also Chapter 5).

The main focus of this chapter will be on Pulsed Electric Field (PEF) and power ultrasound. Intensive research into PEF has brought the technology to the brink of commercial uptake. Power ultrasound has interesting potential as a novel preservation method but is some way from being utilised commercially for this purpose. It does, however, have numerous nonpreservation applications, some of which are already being used commercially. The remainder of the chapter will provide an overview of a range of technologies that are attracting research interest, but that are some years away from being viable commercial processes.

7.2 Pulsed Electric Field Processing

7.2.1 Definition of Pulsed Electric Fields

Pulsed electric field processing is a technique in which a food is placed between two electrodes and exposed to a pulsed high voltage field (typically 20–80 kV cm^{-1}) [4]. For preservation applications, treatment times are of the order of less than 1 s, achieved by multiple short duration pulses typically less than 5 μs. This process reduces levels of microorganisms whilst minimising undesirable changes in the sensory properties of the food. It is important to stress that although heat may be generated in the food product (and may need to be controlled by cooling), microbiological inactivation is achieved by nonthermal means, that is, due to the electrical field not just due to any induced thermal effects. However, there is a clear synergy between a moderate degree of heating (for example 40–45 °C) and the applied PEF [5].

7.2.2 Pulsed Electric Field Processing – A Brief History

The inactivation of microorganisms and enzymes using electric discharges started as early as the 1920s with the 'ElectroPure' process for milk production. This process consisted of heating the milk to 70 °C by passing it through carbon electrodes in an electric heating chamber to inactivate *Mycobacterium tuberculosis* and *E. coli* [6, 7]. The electric field was small, only 220 V AC, and was not pulsed; and the inactivation mechanism was purely thermal [7]. There were around 50 plants using the ElectroPure system in the USA up until the 1950s [8].

An 'electrohydraulic' process was developed in the 1950s as a method for inactivating microorganisms in liquid food products. A shock wave generated by an electric arc and the formation of highly reactive free radicals was thought to be the main mechanism for microbiological inactivation [6]. The process did not find widespread use in the food industry because particulates within the food were damaged by the shock waves and there were issues surrounding electrode erosion and the potential for contaminating the food [8].

The roots of pulsed electric field processing can be traced back to Germany and a patent by Doevenspeck [5, 9]. This inventor pioneered the design of pulsed electric field equipment and until the time of his death remained involved in collaborative projects in the PEF field, a career spanning over 50 years [5]. Unilever scientists Sale and Hamilton made valuable contributions to the field in the 1960s, studying the mechanisms of PEF inactivation of microorganisms [5, 10, 11]. PEF has been used for a number of years, in nonpreservation applications, at relatively low field strengths (5–15 $kV\,cm^{-1}$) as a means of inducing pores in cell membranes. This application finds widespread use in biotechnology for the insertion of foreign DNA into living cells to modify their characteristics [5, 12]. PEF as a preservation method differs from these reversible electroporation techniques in the equipment design, field strengths used (20–80 $kV\,cm^{-1}$ typically) and the pulse duration (typically $<5\ \mu s$ versus tenths or 100ths of microseconds) [5].

7.2.3 Effects of PEF on Microorganisms

Two mechanisms have been proposed to explain the inactivation of microorganisms using pulsed electric fields, 'electrical breakdown' and 'electroporation' [4].

7.2.3.1 Electrical Breakdown

According to Zimmermann [4, 8], the bacterial cell membrane can be considered to be a capacitor that is filled with a dielectric material. The normal resisting potential difference across the membrane (the transmembrane potential) is around 10 mV [4]. If an external electric field is applied, this increases the potential difference across the cell membrane. This increase in potential difference causes a reduction in the membrane thickness. When the potential difference across the cell reaches a critical level (normally considered to be around 1 V), pores are formed in the membrane. This leads to an immediate discharge at the membrane pore and, consequently, membrane damage [4]. Breakdown of the membrane is reversible if the pores are small in relation to the total membrane surface, but when pores are formed across large areas of the membrane then destruction of the cell membrane results. Figure 7.1 shows a *Bacillus cereus* cell before and after PEF treatment.

The transmembrane potential developed in the direction of an applied electric field is given by [8]:

$$U(t) = 1.5\,rE \tag{7.1}$$

where $U(t)$ is the transmembrane potential in the direction of the applied field (V), r is the radius of the cell (μm) and E is the applied electric field strength ($kV\,mm^{-1}$)

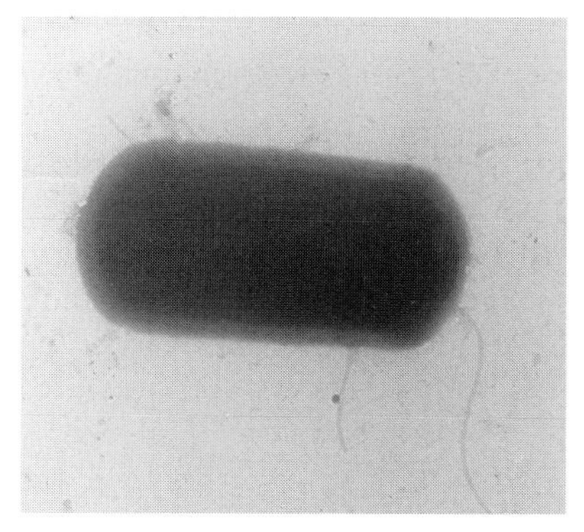
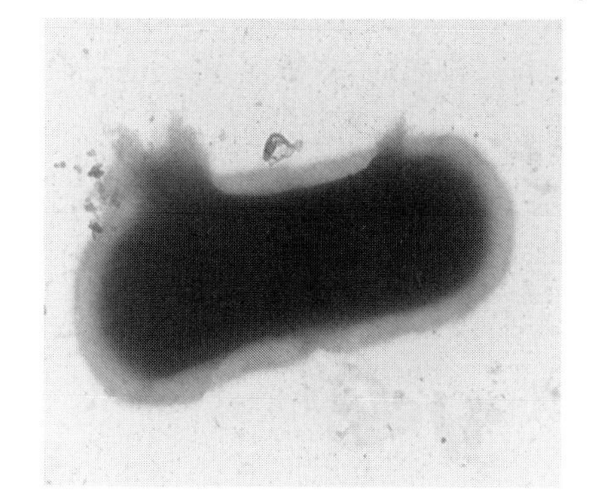

Fig. 7.1 PEF treatment of *Bacillus cereus* cell.

For a typical cell radius of 0.5 μm, the electrical field strength (*E*) required to induce poration would be 13.33 $kV\,cm^{-1}$.

7.2.3.2 Electroporation

A second proposed explanation for microbiological inactivation using PEF is that of electroporation. When a microorganism is subjected to a high voltage electric field, the lipid bilayer and proteins of the cell membrane are temporarily destabilised [4]. Changes in the conformation of lipid molecules are induced, existing pores are expanded and structurally stable hydrophobic pores are formed which can conduct current. This leads to localised heating that changes the lipid bilayer from a rigid gel to a liquid crystalline form [8]. Once the semipermeable nature of the membrane is impaired, swelling and eventual rupture of the cell is induced [4, 8].

7.2.4 Critical Factors in the Inactivation of Microorganisms Using PEF

Three key areas – process factors, product factors and microbial factors – determine the effectiveness of PEF for microbiological inactivation using PEF [4].

7.2.4.1 Process Factors

The intensity of the electric field will affect the transmembrane potential of the microbial cell (as described earlier) and therefore an increase in inactivation can be expected with an increase in electric field intensity. The pulse width used affects the level of electric field intensity that is required to achieve inactivation. Larger pulse widths reduce the field intensity that is required to produce a transmembrane potential large enough to initiate pore formation. Unfortunately, longer pulses also increase the degree of heating observed in the food so a careful balance must be established to maximise inactivation whilst minimising product heating. In general, an increase in treatment time (number of pulses multiplied by pulse duration) also increases the level of inactivation. The pulse waveshape also influences the degree of inactivation achievable with PEF. Square wave pulses are more energy efficient and more lethal than exponentially decaying waveforms. Bipolar pulses cause additional stress to the cell membrane, enhance microbial inactivation and are energy efficient. Finally the process temperature has an impact on the lethality of PEF. Moderately elevated temperatures have a synergistic effect when combined with PEF. This may be due to changes in membrane fluidity and permeability, or an increase in the conductivity of the liquid being treated.

7.2.4.2 Product Factors

The electrical conductivity of the product to be treated is a very important parameter for PEF processing. Foods with a large electrical conductivity are not suitable for processing with PEF because the peak electric field across the chamber is reduced. The ionic strength of a food material directly influences its conductivity and as the conductivity rises, the lethality of a process decreases. Reducing the pH of the product is thought to increase the inactivation achievable for a given field strength. However, work by Berlin University of Technology suggested that pH modifications down to pH 5.5 had minimal effect on the lethality of PEF processing of *B. subtilis* [13].

Particulates in the liquid also pose processing problems because high energy inputs may be needed to inactivate microorganisms in the particulates and there is a risk of dielectric breakdown of the food.

7.2.4.3 Microbial Factors

In general, the order of resistance of microorganisms to PEF (lowest to highest) is considered to be yeasts, Gram negative bacteria and Gram positive bacteria. The lifecycle stage of the microorganisms affects the lethality of the process: organisms in the log phase of growth are generally more sensitive to PEF than those in the lag or stationary phases of growth. There is also some evidence to suggest that higher initial concentrations of microorganisms can impact on the lethality of the process [4].

7.2.5 Effects of PEF on Food Enzymes

Work on PEF effects on food enzymes has been relatively limited to date and variable results have been obtained [14]. Work using simulated milk ultrafiltrate in a continuous flow unit (45 ml min^{-1}) resulted in a 20–90% reduction of plasmin (bovine milk). The process conditions used varied between 15–45 kV cm^{-1}, with a 2-μs pulse, a frequency of 0.1 Hz and between 10–50 pulses [14].

Studies on a protease (*Pseudomonas flourescens* M3/6) highlighted the effect of substrate on the levels of achieved inactivation. A 60% reduction in activity was found in skimmed milk (15 kV cm^{-1}, 2-μs pulse, frequency of 2 Hz, 98 pulses, 50 °C), whereas no effect was found using the same processing parameters in casein tris buffer [14].

Results on the inactivation of alkaline phosphatase in raw milk have been variable. Washington State University demonstrated a 96% reduction in activity using 13.2 kV cm^{-1} and 70 pulses, whereas unpublished data by Verachtert and others showed no inactivation using 13.3 kV cm^{-1}, a frequency of 1 Hz and 200 pulses of 2 μs [14].

Ho et al. examined the effects of PEF on a number of enzymes in model systems [14, 15]. They found that the activity of a wheat germ derived lipase in deionised water (pH 7) could be reduced by up to 85%, using a treatment of 87 kV cm^{-1} at a frequency of 0.5 Hz using 30 pulses of 2 μs duration. The level of inactivation increased with increasing electric field strength. For example, at 20 kV cm^{-1} only a 20% reduction in activity was achievable. Glucose oxidase activity in pH 5.1 buffer could be reduced by 20–75%, using 17–63 kV cm^{-1}, a frequency of 0.5 Hz and 30 pulses of 2 μs at 20 °C. Inactivation of α-amylase (from *B. licheniformis*) in deionised water (pH 7) varied between less than 5% to around 85% using 20–80 kV cm^{-1}, a frequency of 0.5 Hz and 30 pulses of 2 μs at 20 °C.

Work at the Katholieke Universiteit Leuven (KUL) in Belgium [14] has increased the complexity of the materials used in enzyme studies, moving from initial trials in simple model systems towards real food products. In distilled water, KUL studies found no better than a 10% reduction in activity of a range of commercial enzymes: lipoxgenase (soyabean), pectinmethylesterase (tomato), α-amylase (*B. subtilis*), polyphenoloxidase (mushroom) and peroxidase (horseradish). Processing conditions evaluated were 10, 20 and 30 kV cm^{-1}, frequencies of 1–100 Hz, pulse widths of 5–40 μs and 1–1000 pulses [14]. In raw milk, no inactivation of alkaline phosphatase was observed using field strengths of up to 20 kV cm^{-1} at a frequency of 1 Hz with 200 pulses of 2 μs duration. The only treatment that brought about a reduction in activity (74%) was one using an extended pulse duration (40 μs). This resulted in a temperature rise within the milk of up to 70 °C and this is likely to have been responsible for the inactivation. Similarly, lactoperoxidase in milk proved resistant to PEF processing. A maximum of 13% inactivation was achieved using 13 kV cm^{-1}, a frequency of 1 Hz and 200 pulses of 10 μs duration. In this experiment, the milk reached a temperature of 52 °C. PEF had very little effect on lipoxygenase in pea juice, the maximum reduction in activity that could be achieved being 9%. In apple juice, polyphenoloxidase was similarly resistant with only a 10% reduction in activity being achievable using 31 kV cm^{-1}, a frequency of 1 Hz and 1000 pulses of 1 μs duration.

In a joint piece of work by the University of Lleida, Spain, and Washington State University, the activity of endopolygalacturonase (endoPG) was reduced by 98% after a treatment for 32 ms at 10 kV cm^{-1} [16].

Debate continues regarding the effects of pulsed electric field processing on enzymes. Much of the research conducted in Europe would suggest that PEF has minimal effects, whereas research in the USA more frequently suggests that a significant degree of inactivation is achievable. Varying experimental approaches and equipment design features cloud the issue. Historically, it has proved difficult to compare results between laboratories because of these differences in equipment design and the processing parameters selected. Separating temperature effects from PEF effects with respect to enzyme inactivation is critical. Some researchers firmly believe that it is temperature increases that give rise to any observed enzyme inactivation; others maintain that inactivation is a nonthermal effect.

7.2.6 Basic Engineering Aspects of PEF

A number of excellent reviews have been published on the engineering aspects of pulsed electric field processing e.g. [4, 8, 12, 17]. To generate a high voltage pulsed electric field of several kV cm^{-1} within a food, a large flux of electrical current must flow through the food within a treatment chamber, for a very short period of time (μs) [12]. This process involves the slow charging of a capacitor followed by a rapid discharge. The typical components of PEF processing equipment include [4, 12, 17]:

- A power supply: this may be an ordinary direct current power supply or a capacitor charging power supply (this latter option can provide higher repetition rates).
- An energy storage element: either electric (capacitive) or magnetic (inductive).
- A switch which may be either closing or opening. Devices suitable for use as the discharge switch include a mercury ignitron spark gap, a gas spark gap, a thyratron, a series of SCRs, a magnetic switch or a mechanical rotary switch [12].
- A pulse shaping and triggering circuit in some cases.
- A treatment chamber (a wide variety of designs have been developed by individual laboratories).
- A pump to supply a feed of product to the chamber.
- A cooling system to control the temperature of the feed and/or output material.

7.2.6.1 Pulse Shapes

PEF can be applied in a variety of forms, including exponential, square wave, instant charge reversal, bipolar or oscillatory pulses [4]. Figure 7.2 shows a simple circuit for the generation of an exponentially decaying pulse. Exponential waveforms are characterised by a rapid rise to the target voltage followed by a slow decay towards zero volts [4]. This waveform has been widely used by researchers in the field, one reason for this being that they are relatively simple to generate and modify.

Square waveforms are also quite widely used in PEF studies. The generation of square pulses is more complex than the generation of exponential decaying

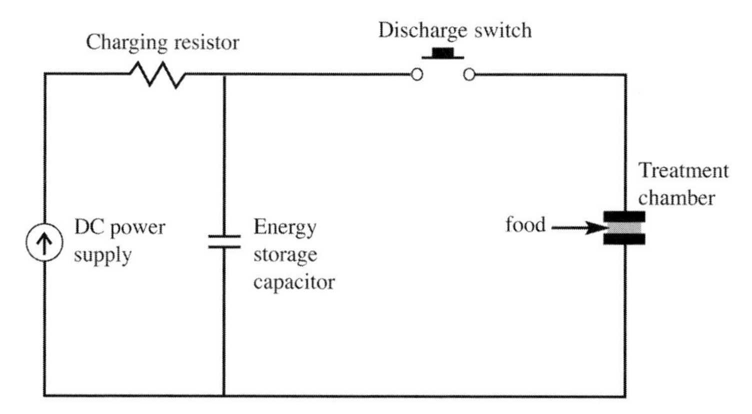

Fig. 7.2 Basic circuit for exponential waveform.

pulses and requires a pulse forming network (an array of capacitors, inductors and solid state switching devices [4]). To generate the square waveform, the pulse forming network and treatment chamber must have matching impedance and, practically, this is difficult to achieve [12].

Although both exponential and square waveforms are effective for inactivating microorganisms, the square waveform is more lethal [4] and is generally considered to be the better option of the two because it maintains peak voltage for longer than the exponential form and is more energy efficient. The prolonged tailing associated with the exponential waveform can lead to excessive heat generation in the food and additional cooling is required in comparison with the square wave [12].

Oscillatory pulses have been found to be the least efficient for microbial inactivation because although the microbial cell is subjected to multiple exposures to the high intensity field, each exposure is only for a short duration of time and irreversible breakdown of the membrane over a large area is prevented [4, 18].

Bipolar pulses, in which the polarity of the applied electric field reverses after a relaxation time, are more effective than monopolar pulses because additional stress is thought to be induced in the cell membrane. Rapidly reversing the electric field orientation changes the movement direction of charged groups in the cell membrane, causing structural fatigue and enhanced electrical breakdown [4, 8]. Biopolar pulses also minimise the deposition of solids at the electrode surfaces and the consequent detrimental effect on field uniformity within the chamber [4, 12].

Instant charge reversal pulses are characterised by having a positive and negative component with various pulse widths and peak field strengths [18]. This type of pulse can significantly reduce the energy requirements for the PEF process to as low as 1.3 J ml^{-1} [4]. Instant charge reversal differs from standard bipolar pulsing because there is no relaxation time between the changes in polarity. There is evidence to suggest that instant charge reversal pulses reduce the critical electric field strength that needs to be applied in order to induce poration of the microbial membrane [18].

7.2.6.2 Chamber Designs

A wide range of experimental PEF treatment chambers have been designed and built by researchers active in this field. Chambers can be broadly categorised as batch or continuous in design. Early chambers were designed for batch processing of static volumes and used parallel plate electrodes separated by an insulating spacer [5, 19]. An alternative batch design is a U-shaped unit, which comprises two electrodes supported on brass blocks in a U-shaped polystyrene spacer [4, 19].

Perhaps of most interest from the point of view of a future commercial process are the continuous chamber designs. Coaxial chambers consist of an inner and outer electrode with the product flowing between them [5]. The electrical current flow is perpendicular to the fluid flow [19]. In co-field designs, electrical current flow is parallel to the fluid flow [19]. The co-field chamber consists of two hollow cylindrical electrodes, separated by an insulator, forming a tube through which the product flows [5].

Fig. 7.3 Continuous-flow PEF processing: 'OSU-6' at the Ohio State University (picture courtesy of Prof. Howard Zhang, The Ohio State University).

The main hurdle restricting commercialisation of PEF has, until recently, been problems associated with scale-up of the equipment. The development of solid state switching systems in recent years has opened up the possibility of full scale-up to almost any throughput desired by the manufacturer. The work of a PEF consortium funded by the US 'DUST' programme (Dual Use of Science and Technology) has led to the manufacture of a large-scale processing unit that has been installed and commissioned at Ohio State University (Fig. 7.3). The 'OSU-6' consists of four treatment chambers with a cooling system before and after each one to control the temperature of the product. A throughput of up to 2000 l h^{-1} is achievable. The university has successfully conducted trials on the pasteurisation of products such as orange juice, tomato juice, salsas and yoghurt.

The capital cost for a commercial PEF unit capable of processing 5000 l h^{-1} has been estimated to be around £ 460 000 (€ 663 000). Depreciating the equipment over 5 years, the cost per litre of juice pasteurised using PEF has been estimated at around £ 0.02 l^{-1} (€ 0.03 l^{-1}). This includes all personnel, maintenance and utility costs. This is broadly in line with the costs associated with conventional thermal pasteurisation [20].

7.2.7
Potential Applications for PEF

7.2.7.1 Preservation Applications

A substantial body of research has demonstrated the microbiological effects of PEF. Many studies have been conducted in model food systems, but relatively few publications have related to real food products. From those that have, a range of pumpable food products have been identified that potentially could be preserved using PEF.

Juice The shelf life of apple juice (from concentrate) has been successfully extended from 21 days to 28 days using 50 kV cm^{-1}, ten pulses, a pulse width of 2 μs and a maximum process temperature of 45 °C [4]. Sensory panellists could determine no significant differences between treated and untreated juice. Work at Washington State University demonstrated that PEF could extend the shelf life of fresh apple juice and apple juice from concentrate to over 56 days and 32 days respectively when stored at 22–25 °C [4].

Rodrigo and others [21] demonstrated that, with process conditions of 28.6, 32.0 and 35.8 kV cm^{-1} for 10.3–46.3 μs, a 2.5 log reduction of *Lactobacillus plantarum* was achieved in orange/carrot juice (70% orange, 30% carrot).

Sitzmann [7], summarising work by Grahl, indicated that a five-fold reduction of *Saccharomyces cerevisiae* was achievable in orange juice using five pulses at a field strength of around 6.5 kV cm^{-1}. Work conducted by Zang and others [4] showed a 3–4 log reduction of total aerobic plate counts in orange juice processed under 32 kV cm^{-1}. The shelf life of this product at 4 °C was 5 months, vitamin losses were lower than heat-processed controls and colour was better preserved.

Milk A number of studies have demonstrated PEF inactivation of microorganisms in milk. Examples [22] include a 3 log reduction of *E. coli* using 21 kV cm^{-1}, a 4 log reduction of *Salmonella dublin* using 18 kV cm^{-1}, a 2.5 log reduction of *Streptomyces thermophilus* using 25 kV cm^{-1} and a 4.5 log reduction of *L. brevis* using 23 kV cm^{-1}. More recently, the shelf life of skimmed milk treated with PEF was reported to be 2 weeks at 4 °C using a process of 40 kV cm^{-1}, 30 pulses and a 2-µs treatment time [4].

Speaking at a CCFRA conference [23] Professor Barbosa-Cánovas of Washington State University discussed the potential of PEF processing for skimmed milk. In his experience of PEF skimmed milk processing, coliforms were inactivated with minimal treatment but naturally occurring flora survived. He also found that Gram positive species were not always more resistant to PEF than Gram negative, as is usually proposed. A relatively short shelf life was achieved in his studies on skimmed milk, primarily due to the surviving population of natural flora. From an engineering viewpoint, significant high voltage electrode damage was observed, along with solids deposition on the electrodes. A hurdle approach was recommended for effective PEF processing using a combination of mild heat and PEF.

Liquid Whole Egg Washington State University has conducted trials over a number of years, assessing the feasibility of PEF processing for liquid whole egg pasteurisation. Results have been promising. PEF has been shown to have minimal effect on the colour of liquid whole egg. Although undetectable levels of microbial populations were present in PEF-treated liquid whole egg, spoilage was found to occur within 25–28 days of storage at 4 °C according to Professor Barbosa-Cánovas [23]. Aseptic packaging of PEF-treated liquid whole egg was therefore recommended to optimise the product shelf life.

Other Liquid Products Research at Washington State University [24] showed that a 6.5 log reduction of *E. coli* (alone) and a 5.3 log reduction of *B. subtilis* (alone) were achievable in pea soup treated with 33 kV cm^{-1}, 0.5 l min^{-1}, 4.3 Hz and 30 pulses. Inactivation was reduced however when a combination of the two organisms were treated in pea soup: a 4.8 log reduction was observed using 30 kV cm^{-1}, 6.7 Hz and a flow rate of 0.75 l min^{-1}. The effects of PEF were limited if the bulk temperature of the liquid during processing was below 53 °C. The physical, chemical and sensory properties of the product did not appear to be changed following PEF treatment and 4 weeks of storage at chilled temperatures [4].

7.2.7.2 Nonpreservation Applications

Whilst the vast majority of research in the field of PEF processing has focused on microbiological inactivation and food preservation, a number of nonpreservation applications using PEF could also prove useful for the food industry.

Baking Applications PEF-treated wheat dough (50 kV, 20 min) is reported to have decreased water loss during baking and the shelf life of the bread subsequently baked from the dough is reported to be increased [25]. Another potential application is the treatment of brewer's yeast to convert nonflocculent yeast to a flocculent form [25].

Extraction/Cell Permeabilisation The irreversible permeabilisation of plant cell membranes and tissues using PEF has been demonstrated [26]. This offers interesting possibilities for improving expression, extraction and diffusion processes [26]. Potential applications include extraction processes such as those found in starch production, sugarbeet processing and juice extraction.

Pretreating vegetables with pulsed electric fields can dramatically reduce drying times. In trials on potato cubes which were dried in a fluidised bed with and without a PEF pretreatment, a one-third reduction in drying time was demonstrated [26].

Taiwo and others [27] studied the effects of a range of pretreatments (PEF, pressure treatment, freezing and blanching) prior to osmotic dehydration of apple slices. PEF pretreatments were conducted in a static, parallel plate chamber using an exponentially decaying wave shape and a field strength of 1.4 kV cm^{-1}. Twenty pulses (at 1 Hz) were administered, each with a duration of 800 μs. In the initial period of dehydration (up to 3 h) PEF processing was not the most effective treatment (in terms of enhancing water loss from the samples). However, beyond this point, the rate of change of moisture loss from PEF-treated samples stabilised at around 0.43 whereas in the other pretreatments investigated, the rate of change of moisture loss continued to fall and in some cases tended towards zero. As a result, after 6 h, the water loss from the PEF pretreated apples was around 79.7% versus 72.6, 40.5, 72.2 and 60.3% for pressure-treated, frozen, blanched and untreated samples, respectively. Uptake of sugar during dehydration was more pronounced in pressure-treated samples than PEF-treated samples. At the start of the dehydration process, PEF processing had a minimal effect on vitamin C levels (10.3±0.8 mg $(100\ g)^{-1}$ fruit versus 10.8±0.5 mg $(100\ g)^{-1}$ fruit in the control). As dehydration progressed, vitamin C levels dropped considerably regardless of pretreatment.

7.2.8 The Future for PEF

Significant steps forward have been made for pulsed electric field processing and it has reached a point where it is very close to commercial realisation. Large-scale equipment is now not only feasible but can be built to specification by companies such as Diversified Technologies Inc. (Mass., USA). It seems likely that a PEF-processed product will be launched in the not too distant future.

There are of course still some issues to be resolved, most notably that of establishing exactly how effective PEF really is with respect to microbial and enzyme inactivation. Numerous laboratory and pilot-scale trials have been conducted using a range of custom-built PEF equipment. Unfortunately, this makes it extremely difficult to compare data obtained from different laboratories and assess exactly what levels of inactivation are truly achievable in a commercial system. Harmonisation of equipment and research protocols is beginning to take place and this will greatly help the situation. As research progress is made and knowledge increases regarding the most effective design parameters for maximising microbiological and enzyme inactivation whilst minimising product deterioration, then the full potential of PEF may be realised.

7.3 Power Ultrasound

7.3.1 Definition of Power Ultrasound

Ultrasonic techniques are finding increasing use in the food industry for both the analysis and processing of foods [28]. Normal human hearing will detect sound frequencies ranging from 0.016 kHz to 18.0 kHz and the power intensity of normal quiet conversation is of the order of 1 W cm^{-2}. Low intensity ultrasound uses very high frequencies, typically 2–20 MHz with low power levels from 100 mW cm^{-2} to less than 1 W cm^{-2}. This type of ultrasound is readily used for noninvasive imaging, sensing and analysis and is fairly well established in certain industrial and analytical sectors for measuring factors such as composition, ripeness, the efficiency of emulsification and the concentration or dispersion of particulate matter within a fluid [29].

Power ultrasound, in contrast, uses lower frequencies, normally in the range of 20–100 kHz (generally less than 1 MHz), and can produce much higher power levels, in the order of 10–1000 W cm^{-2}. Low frequency high power ultrasound has sufficient energy to break intermolecular bonds, and energy intensities greater than 10 W cm^{-2} will generate cavitation effects, which are known to alter some physical properties as well as enhance or modify many chemical reactions [28, 30, 31].

It has long been known that ultrasound is able to disrupt biological structures and produce permanent effects in the medium to which it is applied, but the proposed use of power ultrasound for microbial inactivation in foods is still in its infancy. Much work has been done to investigate the mechanism by which ultrasonic disruption of biological systems occurs, but cavitation effects are thought to be a major factor [32, 33].

Cavitation occurs when ultrasound passes through a liquid medium [34], causing alternate rarefactions and compressions. If the ultrasound waves are of sufficiently high amplitude, bubbles are produced. These bubbles collapse with differing intensities and it is this that is thought to be a major contribution to cellular disruption. The mechanisms involved in cellular disruption are multifactorial and may include shear forces generated during movement (subcellular turbulance) of the bubbles or sudden localised temperature and pressure changes caused by bubble collapse.

A characteristic of ultrasonic waves is the ability to produce different effects in different media in such a way that sometimes these effects seem contradictory. For example, power ultrasound in liquid suspensions has the ability to break aggregated particles whereas using ultrasound in air or gas suspensions tends to produce particle agglomeration.

7.3.2 Generation of Power Ultrasound

Whatever type of system is used to apply power ultrasound to foods, it will consist of three basic parts [35]:

1. Generator: this is an electronic or mechanical oscillator that needs to be rugged, robust, reliable and able to operate with and without load.
2. Transducer: this is a device for converting mechanical or electrical energy into sound energy at ultrasonic frequencies.
3. Coupler: the working end of the system that helps transfer the ultrasonic vibrations to the substance being treated (usually liquid).

There are three main types of transducer: *liquid driven*, *magnetostrictive* and *piezoelectric*. Liquid driven transducers are effectively a liquid whistle where a liquid is forced across a thin metal blade causing it to vibrate at ultrasonic frequencies: rapidly alternating pressure and cavitation effects in the liquid generate a high degree of mixing. This is a simple and robust device but, because it involves pumping a liquid through an orifice and across a blade, processing applications are restricted to mixing and homogenisation (see Chapter 15).

Magnetostrictive transducers are electromechanical devices that use magnetostriction, an effect found in some ferromagnetic materials which change dimension in response to the application of a magnetic field. The dimensions of the transducer must be accurately designed so that the whole unit resonates at the correct frequency. The frequency range is normally restricted to below 100 kHz and the system is not the most efficient (60% transfer from electrical to acoustic energy, with losses mainly due to heat). The main advantages are that these transducers are rugged and able to withstand long exposure to high temperatures.

Piezoelectric transducers are electrostrictive devices that utilise ceramic materials such as lead zirconate titanate (PZT) or barium titanate and lead metaniobate. This piezoceramic element is the most common of the transducers and is more efficient (80–95% transfer to acoustic energy) but less rugged than magnetostrictive devices; piezoelectric transducers are not able to withstand long exposure to high temperatures (normally not >85 °C).

7.3.3 System Types

The design, geometry and method by which the ultrasonic transducer is inserted or attached to the reaction vessel is essential to its effectiveness and efficiency – this is an important variable and any differences between laboratory and pilot plant design and application of ultrasound can often lead to very different results. For example, with ultrasonic baths, the transducer is bonded to the base or sides of the tank and the ultrasonic energy is delivered directly to the liquid in the tank. However, with probes, the high power acoustic vibration is amplified and conducted into the media by the use of a shaped metal horn; and the shape of the horn will determine the amount of signal amplification.

There are several ultrasonic systems available, which differ mainly in the design of the power generator, the type of transducer used and the reactor to which it is coupled. Typical ultrasonic systems are.

7.3.3.1 Ultrasonic Baths

Transducers are normally fixed to the underside of the vessel, operate at around 40 kHz and produce high intensities at fixed levels due to the development of standing waves created by reflection of the sound waves at the liquid/air interface. The depth of the liquid is important for maintaining these high intensities and should not be less than half the wavelength of the ultrasound in the liquid. Frequency sweeping is often used to produce a more uniform cavitation field and reduce standing wave zones.

7.3.3.2 Ultrasonic Probes

These systems use detachable 'horns' or shapes to amplify the signal; the horns or probes are usually a half wavelength (or multiples) in length. The amount of gain in amplitude depends upon the shape and difference in diameter of the horn be-

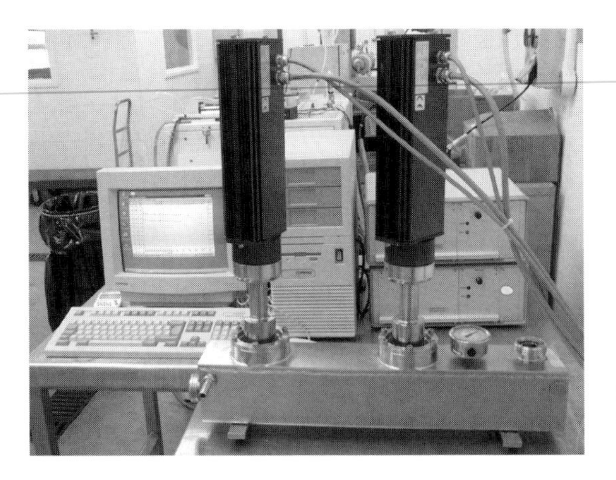

Fig. 7.4 Example of a probe type system from Dr Hielscher GmbH.

tween one face (the driven face) and the other (the emitting face). If the probe is the same diameter along its length then no gain in amplitude will occur but the acoustic energy will simply be transferred to the media (see Fig. 7.4).

7.3.3.3 **Parallel Vibrating Plates**

Opposing vibrating plates offer a better design for maximising the mechanical effect of ultrasound than a single vibrating surface. Often plates vibrate at different frequencies (for example 20 kHz and 16 kHz) to set up beat frequencies and create a larger number of different cavitation bubbles.

7.3.3.4 **Radial Vibrating Systems**

This is perhaps the ideal way of delivering ultrasound to fluids flowing in a pipe. The transducers are bonded to the outside surface of the pipe and use the pipe itself as a part of the delivery system (see Fig. 7.5). This system is very good for handling high flow rates and high viscosity fluids. A cylindrical resonating pipe will help focus ultrasound at the central region of the tube, resulting in high energy in the centre for low power emission at the surface; and this can reduce erosion problems at the surface of the emitter.

7.3.3.5 **Airborne Power Ultrasound Technology**

Air and gaseous media present problems for the efficient generation and transmission of ultrasonic energy. Due to the low density and high acoustic absorption

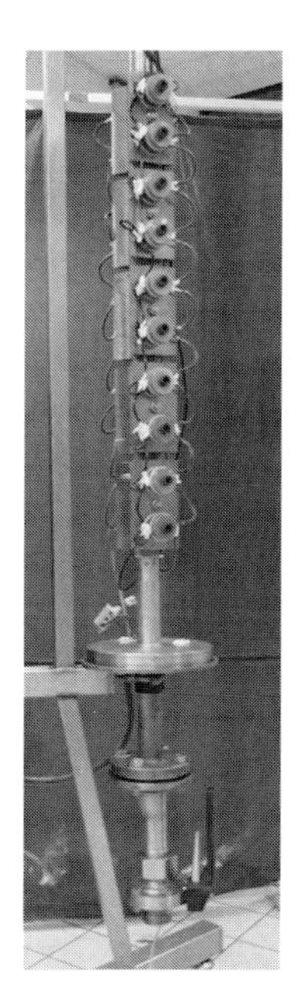

Fig. 7.5 Example of a radially vibrating system from Sonic Process Technologies.

of air and gases, ultrasonic generators require good impedance matching with the air or gas concerned, a large amplitude of vibration, highly directional or focused radiation and a high power capacity. Whistles and sirens were the most common type of ultrasonic device for use in air until the 'stepped-plate transducer' was developed in the late 1980s. In this case, a large diameter flexible radiating plate with a stepped profile is driven at its centre by a piezoelectric device. The extensive surface of the plate produces a high radiation resistance and power capacity and offers a good impedance matching with air. The special profile of the vibrating plate permits good control of the vibrating amplitude and focused radiation pattern that is very directional. Special power generators (1–2 kW macrosonic generators) are used to drive the transducer at resonance during operation and produce acoustic energy in the frequency range 10–50 kHz. These devices have found applications in food dehydration and defoaming of liquid in cans or batch tanks and could be applied to the agglomeration of particles in the air of a room or filling environment as well as gas sterilisation and mass transfer.

7.3.4
Applications for Power Ultrasound in the Food Industry

Power ultrasound is already used for the processing of food materials in a variety of ways such as mixing, emulsification, cutting, tenderising and ageing [30, 36]. Potential applications for high power ultrasound in the food industry are wide-ranging and include enzyme inhibition, hydrogenation of oils, crystallisation control, extraction of proteins and enzymes, the inactivation of microorganisms and improved heat and mass transfer [30, 31, 35]. A list summarising current and potential applications is shown in Table 7.2.

7.3.4.1 Ultrasonically Enhanced Oxidation

Ageing of fermented products and inducing rapid oxidation in alcoholic drinks for flavour development and early maturation has been developed, using higher frequency lower energy power ultrasound. In 1981, in Japan, the use of 1 MHz ultrasound was shown to alter the alcohol/ester balance [37] with possible applications for accelerating whisky maturation through the barrel wall being tested [38].

7.3.4.2 Ultrasonic Stimulation of Living Cells

Lower power sonication can be used to enhance the efficiency of whole cells without cell wall disruption. In this case, ultrasound appears to be increasing the transport of nutrients or affecting membrane/seed permeability by microstreaming. For example, work done in Belgium, at Undatim Ultrasonics, demonstrated that the use of ultrasound as a processing aid in yoghurt manufacture led to a reduction in production time of up to 40%. Sonication reduced the dependency of the process on the origin of milk and improved the consistency and texture of the product. Russian studies have shown that the application of

Table 7.2 List of current and potential applications for ultrasound in the food industry.

Application	Comments
Crystallisation of fats and sugars	Enhances the rate and uniformity of seeding
Degassing	Carbon dioxide removal from fermentation liquors
Foam breaking	Foam control in pumped liquids and during container filling
Extraction of solutes	Acceleration of extraction rate and efficacy; research on coffee, tea, brewing; scale-up issues
Ultrasonically aided drying	Increased drying efficiency when applied in warm air resulting in lower drying temperatures, lower air velocities or increased product throughput
Mixing and emulsification	Online commercial use often using 'liquid whistle'. Can also be used to break emulsions
Spirit maturation and oxidation processes	Inducing rapid oxidation in alcoholic drinks; 1 MHz ultrasound has possible applications for accelerating whisky maturation through the barrel wall
Meat tenderisation	Alternative to pounding or massaging; evidence for enhanced myofibrillar protein extraction and binding in reformed and cured meats
Humidifying and fogging	Ultrasonic nebulisers for humidifying air with precision and control; possible applications in disinfectant fogging
Cleaning and surface decontamination	Online commercial use for cleaning poultry processing equipment; possible pipe fouling and fresh produce cleaning applications; can reach crevices not easily reached by conventional cleaning methods
Cutting	Commercial units available capable of cutting difficult products – very soft/hard/fragile with less wastage, more hygienically and at high speeds
Effluent treatment	Potential to break down pesticide residues
Precipitation of airborne powders	Potential for wall transducers to help precipitate dust in the atmosphere; also removal of smoke from waste gases
Inhibiting enzyme activity	Can inhibit sucrose inversion, and pepsin activity; generally oxidases are inactivated by sonication, but catalases are only affected at low concentrations; reductases and amylases appear to be highly resistant to sonication
Stimulating living cells	Lower-power sonication can be used to enhance the efficiency of whole cells without cell wall disruption, for example in yoghurt, action of *Lactobacillus* was improved by 40%; also improved seed germination and hatching of fish eggs
Ultrasonically assisted freezing	Control of crystal size and reduced freezing time through zone of ice crystal formation
Ultrasonically aided filtration	Rate of flow through the filter medium can be increased substantially
Enhanced preservation (thermal and chemical)	Sonication in combination with heat and pressure has the potential to enhance microbial inactivation; this could result in reduced process times and/or temperatures to achieve the same lethality

20 kHz low amplitude/energy (40 μm/0.7 W cm^{-2}) ultrasound in an aqueous environment for 10 min improved the germination of lotus seeds by 30%.

7.3.4.3 Ultrasonic Emulsification

The most common use of power ultrasound in the food industry is the online commercial use of the 'liquid whistle' to emulsify a range of products. It can also be used to break emulsions. Trials have shown that fat globules undergo a substantial reduction in size (up to 80%) following ultrasonication and emulsions produced are often more stable than those produced conventionally, requiring little, if any surfactant. In trials, when temperatures of 70–75 °C were reached during ultrasonic homogenisation, a better particle distribution was achieved in comparison with lower temperature treatments. Ultrasound has been used industrially in the manufacture of salad cream, tomato ketchup, peanut butter and some cream soups and fruit juices (see Chapter 15) [31].

7.3.4.4 Ultrasonic Extraction

The mechanical effects of power ultrasound provide a greater penetration of solvent into cellular materials and also improve mass transfer [31]. Additionally, biological cells can be disrupted by power ultrasound, facilitating the release of cell contents. The benefits are acceleration of extraction rate and improved efficacy. Laboratory trials have taken place with coffee, tea, soya bean protein, sugar from sugar beet and rennin from calf stomachs and shown the benefits of improved yield. Scale-up issues could be a problem. However, a pilot-plant extraction process for soya protein was developed in the 1980s [39].

7.3.4.5 Ultrasound and Meat Processing

Power ultrasound has been used as an alternative to pounding, tumbling or massaging. There is evidence that enhanced myofibrillar protein extraction occurs and that binding in reformed and cured meats is improved following ultrasound application. The binding strength, water-holding capacity, product colour and yields of processed meats were evaluated after treating with either salt tumbling, sonication in an aqueous liquor or both. The samples that received both treatments were judged superior in all qualities [40]. Pilot studies in the early 1990s showed that sirloin steak connective tissue could be reduced when subjected to sonication at 40 kHz (2 W cm^{-2}) for 2 h [41].

7.3.4.6 Crystallisation

Controlled crystallisation of sugar solutions, hardening of fats, and the manufacture of chocolate and margarine are examples of food processes where crystallisation plays a vital part that can be improved by the application of power ultrasound (see Chapter 14). Power ultrasound acts in a number of ways during crystallisation. It initiates seeding because the cavitation bubbles tend to act like crystal nuclei and so enhances the rate and uniformity of seeding. In addition, the ultrasound can break up any large crystalline agglomerates and can also effectively remove any encrustation from heat exchanger surfaces [35].

A successful, patented, full-scale ultrasonically assisted crystallisation operation has been used in the production of a crystalline drug for some years [42]. Examples of more recent applications involving ultrasonically assisted crystallisation include a patent by Kraft Jacobs Suchard issued in the late 1990s for the transformation of unstable to stable polymorphic crystals in edible fat manufacture [43] and a patent using sonication to retard fat bloom development in chocolate confectionery products [44]. Another area where ultrasound can be used to control crystal size is in the freezing operation where, under the influence of ultrasound, food materials such as soft fruits can be frozen with a reduced freezing time through the zone of ice crystal formation. In addition, it has been reported that more rapid and uniform seeding of ice crystals and reduced cellular damage can be achieved [45].

7.3.4.7 **Degassing**

Ultrasound has been used successfully for the control or removal of carbon dioxide from fermentation liquors. Japanese brewers have shown that nitrogen gas bubbling and ultrasonic vibrations can decrease dissolved carbon dioxide when brewing in cylindroconical tanks and can help control yeast metabolism, foam separation and froth height [46].

Another area of commercial application of power ultrasound is in foam control in pumped liquids and during container filling. Trials by NIZO Food Research in The Netherlands [47] showed that following the application of 1-s pulses of high-intensity ultrasound (20 kHz) for 3 min at 20 °C, up to 80% reduction in the foaming potential of supersaturated milk could be achieved. Low energy consumption was also reported. However, longer applications were required before any noticeable change in dissolved oxygen was observed.

7.3.4.8 **Filtration**

Ultrasound in filtration has two specific effects (see Chapter 14). Sonication will (a) cause the agglomeration of fine particles and (b) supply sufficient vibration energy to the filter to keep particles suspended above the medium and prevent clogging. This has been successfully applied to the reduction of water in coal slurry. Acoustic filtration can increase the rate of flow through the filter medium substantially and, when applied to fruit extracts and drinks, this technique has been used to increase the fruit juice extracted from pulp. Applying a potential difference across the pulp bed whilst at the same time applying ultrasound (electroacoustic filtration) can enhance the removal of juice from the pulp. In a pilot study, vacuum belt filtration reduced the moisture content of apple pulp from 85% to 50% but electroacoustic filtration reduced this further to 38% [22].

7.3.4.9 Drying

Applying power ultrasound to particles in a warm air convective drier can lead to increased drying efficiency, allowing lower drying temperatures, lower air velocities or increased product throughput (see Chapter 3). It is thought that initially sonication reduces the pressure above the particles and encourages water loss into the warm air passing over the bed of the drier. This approach has been evaluated for the hot air drying of carrots and results showed dramatic reductions in treatment times with a final moisture content of less than 1% attained easily. In addition, product quality was maintained, sample rehydration was greater than 70% and energy consumption was low. The technique may only be useful to specific food applications but scale-up and potential industrial applications seem promising [48].

7.3.4.10 Effect of Ultrasound on Heat Transfer

Increasing the rate of heat transfer from liquids to solid food particles by the use of high intensity 'power ultrasound' has been described [49]. Using ultrasonic power inputs of 0.14–0.05 W g^{-1}, Sastry and colleagues were able to demonstrate an increase in the convective heat transfer coefficient (h_{fp}) from about 500 W m^{-2} K^{-1} to 1200 W m^{-2} K^{-1}: this was measured for aluminium particles in heated water. Later work [50] showed that, in a simple ultrasonic bath, this effect was dependent upon the position of the particle as well as the viscosity of the fluid medium. It was suggested that higher ultrasonication power levels would be required to ensure the enhanced heat transfer effects persisted in higher viscosity fluids. Studies at CCFRA have confirmed this effect and have also shown that, in simple batch heating trials, high intensity ultrasound has the potential both to significantly increase the rate of heat transfer and to reduce the thermal resistance of yeasts, moulds and bacteria. Heating trials have been conducted at CCFRA using water flowing over aluminium cylinders and 4% starch flowing through a packed bed of potato cubes. These trials demonstrated that, each time ultrasonic energy was applied to the carrier fluid during the heating in a mock holding tube, the centre of probed solids in the carrier fluid heated up more rapidly and more uniformly across the range of particles tested [51, 52].

7.3.5 Inactivation of Microorganisms Using Power Ultrasound

7.3.5.1 Mechanism of Ultrasound Action

Microbial cell inactivation is generally thought to occur due to three different mechanisms: cavitation, localised heating and free radical formation. There are two sorts of cavitation (transient and stable) which have been reported to have different effects.

Stable cavitation occurs due to oscillations of the ultrasound waves, which causes tiny bubbles to be produced in the liquid. It takes thousands of oscillatory cycles of the ultrasound waves to allow the bubbles to increase in size. As the ultrasonic wave passes through the medium, it causes the bubbles to vibrate, causing strong currents to be produced in the surrounding liquid. Other small bubbles are attracted into the sonic field and this adds to the creation of microcurrents. This effect, which is known as microstreaming, provides a substantial force, which rubs against the surface of cells, causing them to shear and break down without any collapse of the bubbles. This shear force is one of the modes of action, which leads to disruption of the microbial cells. The pressures produced on the cell membrane disrupt its structure and cause the cell wall to break down.

During transient cavitation, the bubbles rapidly increase in size within a few oscillatory cycles. The larger bubbles eventually collapse, causing localised high pressures and temperatures (up to 100 MPa and 5000 K) to be momentarily produced. It is widely believed that cellular stress is caused by the cavitation effect, which occurs when bubbles collapse. The pressures produced during bubble collapse are sufficient to disrupt cell wall structures, eventually causing them to break, leading to cell leakage and cell disruption.

Additionally, the localised high temperatures can lead to thermal damage, such as denaturation of proteins and enzymes. However, as these temperature changes occur only momentarily and in the immediate vicinity of the cells, it is likely that only a small number of cells are affected.

The intensity of bubble collapse can also be sufficient to dislodge particles, for example, bacteria from surfaces, and could displace weakly bound ATPase from the cell membrane, another possible mechanism for cell inactivation [53].

Free radical formation is the final proposed mode of action of microbial inactivation. Applying ultrasound to a liquid can lead to the formation of free radicals, which may or may not be beneficial. In the sonolysis of water, OH^- and H^+ ions and hydrogen peroxide can be produced, which have important bactericidal effects [54, 55]. The primary target site of these free radicals is the DNA in the bacterial cell. The action of the free radicals causes breakages along the length of the DNA and fragmentation occurs where small fragments of DNA are produced. These fragments are susceptible to attack by the free radicals produced during the ultrasound treatment and it is thought that the hydroxyl radicals attack the hydrogen bonding, leading to further fragmentation effects [32]. The chemical environment plays an important part in determining the effectiveness of the ultrasound treatment and it may be possible to manipulate or exploit these conditions in order to achieve a greater level of inactivation.

7.3.5.2 **Factors Affecting Cavitation**

The frequency of ultrasound is an important parameter and influences the bubble size [56]. At lower frequencies such as 20 kHz, the bubbles produced are larger in size and when they collapse higher energies are produced. At higher frequencies, bubble formation becomes more difficult and, at frequencies above 2.5 MHz, cavitation does not occur [33]. The amplitude of the ultrasound also influences the intensity of cavitation. If a high intensity is required then a high amplitude is necessary.

The intensity of bubble collapse also depends on factors such as temperature of the treatment medium, viscosity and frequency of ultrasound. As temperature increases, cavitation bubbles develop more rapidly, but the intensity of collapse is reduced. This is thought to be due to an increase in the vapour pressure, which is offset by a decrease in the tensile strength. This results in cavitation becoming less intense and therefore less effective as temperature increases. This effect can be overcome if required, by the application of an overpressure (200–600 kPa) to the treatment system. Combining pressure with ultrasound and heat increases the amplitude of the ultrasonic wave and it has been shown that this can increase the effectiveness of microbial inactivation. Pressures of 200 kPa (2 bar) combined with ultrasound of frequency 20 kHz and a temperature of 30 °C produced a decrease in the decimal reduction time (*D* value: the time taken to achieve a 1 log reduction in cell levels) by up to 90% for a range of microorganisms [57].

7.3.5.3 Factors Affecting Microbiological Sensitivity to Ultrasound

Bacterial cells differ in their sensitivity to ultrasound treatment [33]. In general, larger cells are more sensitive to ultrasound [58]. This may be due to the fact that larger cells have an increased surface area, making them more vulnerable to the high pressures produced during ultrasonication. The effects of ultrasound have been studied using a range of organisms such as *Staphylococcus aureus* and *Bacillus subtilis* and the Gram negative *Pseudomonas aeruginosa* and *Escherichia coli* [59]. Gram positive cells appear to be more resistant to ultrasound than Gram negative cells; and this may be due to the structure of the cell walls. Gram positive cells have thicker cell walls that provide the cells with some protection against sonication treatment. Other studies have indicated that there is no significant difference between the percentage of Gram positive and Gram negative cells killed by ultrasound [59]. Cell shape has been investigated and it has been found that spherical-shaped cells (cocci) are more resistant to ultrasound than rod-shaped cells [33]. *Bacillus* and *Clostridium* spores have been found to be more resistant to sonication than vegetative bacteria and many of the bacteria known to be resistant to heat are similarly resistant to ultrasound [60].

7.3.5.4 Effect of Treatment Medium

The characteristics of the food or substrate can influence the effectiveness of the ultrasound treatment applied. For instance, it has been found that the resistance of bacteria is different when treated in real food systems than when treated in microbiological broths [61]. In general, foods that contain a high fat content reduce the killing effect of the ultrasound treatment. Differences in effectiveness may be due to intrinsic effects of the environment on the ultrasound action (e.g. cavitation) or due to changes in ultrasound penetration and energy distribution. In a low viscosity liquid, ultrasound waves will pass through relatively easily, causing cavitation to occur, but in a more viscous solution the ultrasound waves have to be of a higher intensity to enable the same level of penetration to be achieved. Low frequency, high power ultrasound is better at penetrating viscous products than higher frequency ultrasound. This is because ultrasound waves with higher frequency are more easily dispersed within the solution, causing a reduction in the overall intensity of the energy delivered.

7.3.5.5 **Combination Treatments**

With normal laboratory probes or cleaning baths, ultrasound applied on its own does not appear to significantly reduce bacterial levels. However, if it is combined with other preservation treatments such as heat or chemicals, the vital processes and structures of bacterial cells undergo a synergistic attack.

A varying response of microorganisms to ultrasound treatment depending on the pH of the surrounding medium has been observed [62, 63]. In particular, it has been found that, if the microorganisms are placed in acidic conditions, this leads to a reduction in the resistance of the organisms to the ultrasound treatment [63]. This may be due to the effects of the ultrasound on the bacterial membranes, making them more susceptible to the antimicrobial effects of the acid or unable to maintain essential internal pH conditions.

The most commonly used combination treatment is the use of heat with ultrasound, known as *thermosonication* (TS) or *manothermosonication* (MTS) if pressure is also included as a variable [34, 64, 65]. Several studies have shown that bacteria become more sensitive to heat treatment if they have undergone an ultrasound treatment either just before (presonication) or at the same time as the heat application. Increased cell death has been demonstrated in cells that have been subjected to a combined ultrasound and heat treatment compared with cells that were exposed to ultrasound treatment only or heat treatment only [64, 65].

Studies at CCFRA used a pilot-scale continuous-flow ultrasonic system manufactured by Sonic Process Technologies (SPT) Ltd, Shrewsbury. This system comprised a 1 kW unit operating at 30 kHz with radially mounted PZT ceramic transducers arranged in opposing pairs around the outside of a 32 mm diameter stainless steel pipe. Microbiological inactivation trials with this system in batch mode showed that in the case of *Zygosaccharomyces bailii* in orange juice there was a seven-fold increase in the inactivation of organisms using ultrasound in combination with heat, when compared with heating alone at 55 °C. With *Listeria monocytogenes* in milk there was a 20-fold increase in the inactivation achieved using ultrasound and heat, compared with heat alone at 60 °C. The amount of heat used in the ultrasound trials was about one-quarter of that used in the isothermal laboratory trials [51].

Spore formers have also been shown to have some degree of reduced resistance to subsequent heat treatments if they are exposed to power ultrasound at temperatures of 70–95 °C. The increased heat sensitivity caused by sonication can be quantified in terms of changes in the decimal reduction or *D* value; and

Table 7.3 shows the synergistic effect of heat and ultrasound for a range of bacterial species. Whilst these data show up to a 43% reduction in the heat resistance of the spore formers tested [66], other studies have shown no effect or a limited effect for other spore formers. Similar reductions have also been reported in the heat resistance data of spore formers such as *Bacillus cereus* and *B. licheniformis* after treatment with ultrasound (20 kHz) [67]. It is not fully understood why there is a limited effect of ultrasound on spores but it has been attributed to the fact that spores contain a highly protective outer coat, which prevents the ultrasound passing through, thus limiting the amount of perturbation that occurs within the spore.

During treatment with a combination of pressure and thermosonication (MTS), it has been shown that chemicals such as dipicolinic acid and low molecular weight peptides are released from spores of *B. stearothermophilus* [68]. In these combined treatments, spores are subjected to violent and intense vibrations due to increased cavitation effects [57]. The loss of substances from spores during this combination of pressure, heat and ultrasound suggests that spore cortex damage and protoplast rehydration may account for the subsequent reduction in heat resistance.

As previously discussed, the frequency of ultrasound used affects the type of cavitation response observed. Data from trials treating *L. monocytogenes* at 20, 38 and 800 kHz in whole milk indicate that 20 kHz was the most effective frequency, whilst 800 kHz had very little effect and resulted in a survivor tail. These data also suggested that the order in which heat and ultrasound are applied could have affected the inactivation observed.

Studies have also been conducted on the use of ultrasound in combination with chemical treatments. Ultrasound is able to disperse bacterial cells in suspensions, making them more susceptible to treatment with sanitising agents. One of the advantages of this type of combined treatment is that it could enable large reductions in the concentrations that are required when chemical treatments are used in isolation for sanitation and disinfection. This has additional

Table 7.3 Inactivation (*D* values) of a range of bacteria using heat and high-power ultrasound [34].

Organism	Heating temperature (°C)	*D* value (min)		
		Heat only	Heat + ultrasound	Ultrasound only
Bacillus subtilis	81.5	257.0	149.0	Not tested
Bacillus subtilis	89	39.2	22.9	Not tested
Bacillus licheniformis	99	5.0	2.2	No effect seen
Bacillus cereus	110	12.0	1.0	No effect seen
Enterococcus faecium	62	11.2	1.8	30
Salmonella typhimurium	50	50.0	30.0	No effect seen
Staphylococcus aureus	50.5	19.7	7.3	Not tested

advantages in that there is less likelihood of residual cleaning agents contaminating equipment after cleaning. Chemicals such as chlorine are often used to decontaminate food products or processing surfaces and it has been demonstrated that chlorine combined with ultrasound enhances the effectiveness of the treatment [69]. Trials were conducted using *Salmonella* attached to the surface of broiler carcasses. Treatment with ultrasound caused the cells to become detached from the surfaces, making it easier for the chlorine to penetrate the cells and exert an antimicrobial effect. For example, immersion in 0.5 ppm chlorine solution for 30 min reduced *Salmonella* by 0.89 log cycles, sonication for 30 min reduced the count by 1.4 log cycles, but a combination process reduced the count by 2.88 log cycles.

Similar results have been obtained in fresh produce cleaning and decontamination in trials at CCFRA using lettuce inoculated with *S. typhimurium*. A 2-l ultrasonic cleaning bath (30–40kHz) operating both with and without the addition of chlorine showed that the combination of ultrasound with chlorine resulted in just under 3 log reductions of *S. typhimurium* compared with about 1.5–2.0 log reductions for ultrasound or chlorine applications separately [70].

7.3.6
Effect of Power Ultrasound on Enzymes

Over 60 years ago, it was reported that pure pepsin was inactivated by sonication [71]. High-power ultrasound has been reported to inhibit various enzymes and work is still ongoing in this area. Recent examples have included the inactivation of enzymes involved in the inversion of sucrose [71] and inactivation of lipases and proteinases [72, 73, 74]. Wiltshire [75] reported that power ultrasound (20 kHz with a power intensity of 371 W cm^{-2}), when applied to peroxidase dissolved in a buffered (pH 7) solution of 0.1 M potassium phosphate at 20 °C, progressively reduced the original activity of the enzyme by 90% (1 log reduction) over a 3-h period. The main mechanisms by which enzyme inactivation is thought to occur are the same as those associated with the destruction of microorganisms (cavitation, localised heating and free radical formation).

7.3.7
Effects of Ultrasound on Food Quality

Power ultrasound has numerous nonpreservation applications, as previously discussed. There is a growing body of evidence to suggest that power ultrasound can be used in combination with heating to reduce bacterial populations and to bring about a substantial reduction in enzyme activity. However, what is uncertain, especially in relation to enzymes, is whether power ultrasound is effective when using processing conditions representative of typical commercial operations. For example, if power ultrasound were used for microbial inactivation in a continuous flow system the residence time would be of the order of seconds; it is uncertain as to whether this would be a sufficient exposure time to inactivate enzymes. There is also a substantial knowledge gap regarding the extent to which the sensory qualities of foods are affected by power ultrasound. For example, there is surprisingly little published information regarding the effects of power ultrasound on food texture and nutritional composition. In order for power ultrasound to be commercially viable as a preservation technique, it must not only produce an acceptable level of food safety, it must also have a minimal effect on the sensory and nutritional qualities of the food being treated. This does not necessarily mean that there is no tolerance for minor effects. Thermal processing is known to result in vitamin C losses relative to a fresh counterpart. For example, Davidek [76] and others reported a 10–20% reduction in ascorbic acid resulting from pasteurisation. Sterilisation at 110–140 °C for 3.5 s was reported to result in a 17–30% loss. For ultrasonic processing to have any chance of commercial success, the process impact on nutritional composition should be comparable or reduced relative to conventional thermal processing. Trials have been carried out at CCFRA (unpublished) with the aim of investigating the effects of power ultrasound on the sensory qualities of foods to establish if possible undesirable effects of ultrasonication outweighed the potential benefits of the technology. In batch trials using a probe type system, no significant reduction in total vitamin C was observed in a sprout pureé after treatments of around 13–28 W m^{-2} at 20 °C and 40 °C with holding times of between 1 min and 5 min. At 40 °C there did appear to be some reduction in total vitamin C, but a statistically significant correlation could not be established due to variation in the raw material vitamin C levels. Within the same project, a substantial reduction in viscosity was observed when a 5% solution of a modified waxy maize starch was sonicated for 1 min in a flow cell with a probe assembly. In this trial, the apparent viscosity of the solution at a shear rate of 10 s^{-1} was reduced from 2360 cP to 13 cP. However, it should be stressed that, in this type of system, the ultrasonic energy is transmitted to the food in a particularly intense and localised manner and alternative equipment designs could minimise this detrimental effect.

7.3.8 The Future for Power Ultrasound

Ultrasound currently finds numerous applications in the food industry, including emulsification of fats and oils, mixing, blending, cutting and accelerating the ageing processes in meats and wines. In the laboratory, it has the potential to be applied to the pasteurisation of a range of low viscosity liquid products as well as enhance the effectiveness of other processing methods. However, the process remains some way from being a viable preservation technology. This is in part due to a lack of knowledge regarding full-scale design and scale-up, but also in part due to a considerable knowledge gap relating to the optimisation of process conditions for food processing. Much more research is required to gain a greater understanding of issues such as:

- equipment design to optimise microbial and enzyme inactivation;

- ultrasonic enhancement of heat transfer to augment existing thermal processes;
- accurate mapping of field intensity variations within a treatment chamber to develop reliable scheduled processes using ultrasound;
- inactivation mechanisms for vegetative cells, spores and enzymes, which need to be clearly identified, especially when combination technologies are used;
- development of mathematical models for the inactivation of microorganisms and enzymes involving ultrasound;
- identifying the influence of food properties such as viscosity and particle size on process lethality as well as the implication of process deviations when using ultrasound.

7.4 Other Technologies with Potential

Several other technologies have been the subject of research interest for food preservation and processing, including oscillating magnetic fields, arc discharge, pulsed broad-spectrum light and UV light for preservation. Although not well advanced, they are considered here in outline as some of them, at least, are likely to develop further.

7.4.1 Pulsed Light

Pulsed light will not penetrate deeply into foods but has potential for the treatment of surfaces – on the product, on packaging and on surfaces used for food preparation. The light in question is broad spectrum white light – which can include light from the ultraviolet and infrared regions – with an energy density of 0.01–50.0 J cm^{-2} [77]. It is generally applied as a single pulse or a short series (up to 20) of short pulses (milliseconds duration). Although the light spectrum generated has a similar composition to sunlight, the main company involved in developing this technology suggest that the effects arise because the intensity involved is roughly 20 000–90 000 times that of sunlight at the earth's surface.

According to publicity materials from equipment manufacturers, pulsed light treatments have been reported to reduce spoilage of baked products by inactivation of moulds, *Salmonella* on egg shell and chicken surfaces, and *Pseudomonas* on cottage cheese. It has been suggested that up to 9 log reduction in viable vegetative bacteria can be achieved on nonporous smooth surfaces. The mechanisms of action that have been proposed include short-term, thin-layer temperature effects (the treated surface briefly reaching 300–700 °C), photochemical effects (formation of free radicals) and DNA damage.

Significant research and independent evaluation are still needed to determine the true potential of this approach, but the scale of the potential effects – at least for certain types of product and/or material – suggest that this is warranted.

In 2002, PurePulse Technologies (the main manufacturer of this technology) ceased trading, which has made the future of the use of this technology for food preservation a little uncertain.

7.4.2 High Voltage Arc Discharge

High voltage arc discharge processing involves rapidly discharging voltages through an electrode gap immersed in aqueous suspensions [78]. The discharge is believed to generate intense physical waves and chemical changes (through electrolytic effects) which can inactivate microorganisms and enzymes without any significant rise in temperature. However, the approach is not without its problems – for example, the shock waves can cause disintegration of food particles – and relatively little has been published to support or refute its use for food preservation. That having been said, the initial findings are likely to be explored further and it might emerge as having some specialist applications, if not a wider use.

7.4.3 Oscillating Magnetic Fields

Speculation surrounding the potential use of high intensity magnetic fields is largely based on a US patent [79] issued to Maxwell Laboratories (Pure Pulse). The patent suggested that a single pulse of an oscillating magnetic field with a strength of 5–100 Tesla and a frequency of 5–500 kHz, could bring about a 2 log reduction in the number of viable microbes in the food within the field. Multiple pulses could result in a commercially sterile product.

However, other studies have not universally corroborated these findings, with some suggesting that oscillating magnetic fields do not affect the microbial population or that the treatment can even stimulate microbial growth. As with some of the other technologies discussed here, such variation may be due either to differences in the treatment intensity or means of delivery, to differences in the media/food in which the microbes are treated, or to the target microbes. It does seem, however, that further research is needed to clarify the extent and mechanisms of any such effect before the approach can be fully assessed as a tool for food preservation.

7.4.4 Plasma Processing

The use of cold gas phase plasmas has been proposed for the inactivation of microorganisms [80]. The term 'cold plasma' refers to partially ionised or activated gases existing at temperatures in the region of 30–60 °C. Cold plasma irradiation could be used to inactivate microorganisms on the surface of a range of materials, including packaging and food surfaces such as fruit, vegetables and meat. Cold plasma irradiation has the advantage that it can be readily switched on and off, making it much more controllable than something like irradiation using a radioactive source. To date, only highly exploratory studies have been carried out on plasma processing for food preservation [80]. In trials conducted by ATO in the Netherlands, structural changes were observed in microorganisms irradiated by a cold gas phase plasma. Gram negative organisms were completely fragmented after treatment. Gram positive organisms had some cell leakage. Using exposure times of 25 s to several minutes, a 5–6 log reduction was demonstrated. Factors which appeared to influence the efficacy of the process included the density of bacterial loading, exposure time and spatial location. Organisms located central to the plasma source were inactivated most readily. Interestingly, after treatment, not all of the nonviable microorganisms were structurally damaged and, conversely, some structurally damaged cells were still viable. This suggests that the mechanisms for inactivation may not simply be due to structural damage. The work at ATO has shown that cold gas phase plasmas can inactivate microorganisms, but a great deal more fundamental research is still required before the technique can be applied commercially.

7.4.5 Pasteurisation Using Carbon Dioxide

Praxair Technologies Inc. has developed a continuous nonthermal pasteurisation process for juice products [81]. The plant operates at around 34.5 MPa, which is required to solubilise the CO_2. The components of the system are relatively simple. The juice is supplied from a raw juice tank and is mixed with CO_2 under pressure. The conditions are maintained such that the CO_2 maintains a liquid state and does not freeze the product. After moving through a holding coil, the juice is de-aerated to strip off the CO_2. According to Praxair, at least a 5 log reduction for a range of pathogens is readily achievable using 5–20% w/w carbon dioxide. Inactivation of *E. coli* O157.H7, *S. muenchen*, *S. agona* and *L. monocytogenes* has been demonstrated using this technique. The process has been shown to be effective for processing orange juice concentrate and orange juice with and without pulp. In studies reported by Praxair, vitamin C and folic acid levels, brix, pH, titratable acidity and cloud stability were virtually unaffected when comparing freshly squeezed juice with a CO_2 pasteurised juice. Sensory trials by two independent laboratories found no significant differences between freshly squeezed and CO_2 pasteurised orange juice. Commercial systems are now available from Praxair operating at 40 US gallons min^{-1} (151.4 l min^{-1}).

7.5 Conclusions

This chapter has provided an overview of some of the main nonthermal preservation technologies attracting both academic and industrial interest. Some of these technologies may come to nothing while some, such as PEF processing, are likely to succeed and find, at least, niche applications in the food industry. The road to commercialisation of a new technology can be rocky and history has shown that timing, industrial need and a little good fortune all play an important role in the successful adoption of a new technology. HPP, for example, was discovered over 100 years ago but it took 80 years before it could realistically be used for commercial food processing. The success of pressure-processed products, particularly in the USA, has relied heavily on the entrepreneurial spirit of relatively small companies and the careful selection of niche products with high added value. It is highly unlikely that any of the technologies discussed in this chapter will one day replace thermal preservation. They will, however, find niche applications for products where they can provide solutions that simply cannot be delivered by conventional technologies.

References

1 EC Official Journal **2002**, EC 23/10/2002 C255/2.
2 EC Official Journal **2002**, EC 18/6/2002 C145/4.
3 Olson, D. **2002**, Electron Beam Processing of Case Ready Ground Beef, in *IFT/EFFoST Non-thermal Processing Workshop*, Ohio State University.
4 Barbosa-Cánovas, G.V., Pierson, M.D., Zhang, Q.H. and Schaffner, D.W. **2000**, Pulsed Electric Fields, *J. Food Sci. [Suppl]* 2000, 65–81.
5 Dunn, J.T. **2001**, Pulsed Electric Field Processing: An Overview, in *Pulsed Electric Fields in Food Processing; Fundamental Aspects and Applications*, ed. G.V. Barbosa-Canovas, Q.H. Zhang, Technomic Publishing Co., Lancaster, pp 1–30.
6 Vega Mercado, H., Martín-Bellosa, O., Quin, B., Chang, F.J., Góngora-Nieto, M.M., Barbosa-Cánovas, G.V., Swanson, B.G. **1997**, Non-Thermal Food Preservation: Pulsed Electric Fields, *Trends Food Sci. Technol.* 8, 151–156.
7 Sitzmann, W. **1995**, High Voltage Pulse Techniques for Food Preservation, in *New Methods for Food Preservation*, ed. G.W. Gould, Blackie Academic and Professional, London, pp 236–252.
8 Jeyamkondan, S., Jayas, D.S., Holley, R.A. **1999**, Pulsed Electric Field Processing: A Review, *J. Food Prot.* 62, 1088–1096.
9 Dovenspeck, H. **1960**, Verfahren und Vorrichtung zur Gewinnung der Einzelnen Phasen aus dispersen Systemen, German patent DE 1237541.
10 Sale, A.J.H., Hamilton, W.A. **1967**, Effects of High Electric Fields on Microorganisms I. Killing of Bacteria and Yeast, *Biochim. Biophys. Acta* 1967, 781–788.
11 Hamilton, W.A., Sale, A.J.H. **1967**, Effects of High Electric Fields on Microorganisms II. Mechanism of Action of the Lethal Effect, *Biochim. Biophys. Acta* 1967, 789–800.
12 Zang, Q., Barbosa-Cánovas, G.V., Swanson, B.G. **1995**, Engineering Aspects of Pulsed Electric Field Pasteurisation, *J. Food Eng.* 25, 261–281.
13 Heinz, V., Knorr, D. **2000**, Effect of pH, Ethanol Addition and High Hydrostatic Pressure on the Inactivation of *Bacillus subtilis* by Pulsed Electric Fields, *Innov. Food Sci. Emerg. Technol.* 1, 151–159.
14 Van Loey, A., Verachtert, B., Hendrickx, M. **2001**, Pulsed Electric Field and Enzyme Inactivation? in *International Seminar on Electric Field Processing – the Potential to Make a Difference*, Campden & Chorleywood Food Research Association, Chipping Campden.
15 Ho, S.Y., Mittal, G.S. and Cross, J.D. **1997**, Effects of High Field Electric Pulses on the Activity of Selected Enzymes, *J. Food Eng.* 31, 69–84

16 Giner, J, Gimeno, V, Palomes, M., Barbosa-Canovas, G.V., Martín O. **2001**, Effects of High Intensity Pulsed Electric Fields on Endopolygalacturonase Activity in a Commercial Enzyme Formulation, *Abstr. Eur. Conf. Adv. Technol. Safe High Quality Foods*, poster number 3.11

17 Ho S., Mittal, G.S. **2000**, High Voltage Pulsed Electric Field for Liquid Food Pasteurisation, *Food Rev. Int.* 16, 395–434.

18 Barbosa-Cánovas, G.V., Fernández-Molina, J.J., Swanson, B.G. **2001**, Pulsed Electric Fields: A Novel Technology for Food Preservation, *Agro Food Ind. Hi-Tech.* 12, 9–14.

19 Yeom, H.W., Mccann, K.T., Streaker, C.B., Zhang, Q.H. **2002**, Pulsed Electric Field Processing of High Acid Liquid Foods: A Review, *Adv. Food Nutr. Res.* 44, 1–32.

20 Kempkes, M. **2002**, Pulsed Electric Field Systems, in *IFT/EFFoST Non-Thermal Processing Workshop*, Ohio State University.

21 Rodrigo, D., Martínez, A., Harte, F., Barbosa-Cánovas, G.V., Rodrigo, M. **2001**, Study of Inactivation of *Lactobacillus plantarum* in Orange-Carrot Juice by Means of Pulsed Electric Fields: Comparison of Inactivation Kinetics Models, *J. Food Prot.* 64, 259–263.

22 Zang, Q., Chang, F.J., Barbosa-Cánovas, G.V., Swanson, B.G. **1994**, Inactivation of Microorganisms in a Semisolid Model Food using High Voltage Pulsed Electric Fields, *Lebensm. Wiss. Tech.* 27, 538–543.

23 Barbosa-Cánovas, G.V. **2001**, Developments in Pulsed Electric Fields – USA Research and Consortium Activities, in *International Seminar on Electric Field Processing – the Potential to Make a Difference*, ed. CCFRA, Campden & Chorleywood Food Research Association, Chipping Campden.

24 Vega-Mercado, H., Martín-Bellosa, O., Chang, F.J., Barbosa-Cánovas, G.V., Swanson, B.G. **1996**, Inactivation of *Escherichia coli* and *Bacillus subtilis* Suspended in Pea Soup using Pulsed Electric Fields, *J. Food Process. Preserv.* 20, 501–510.

25 Knorr, D., Geulen, M., Grahl, T., Sitzman, W. **1994**, Food Applications of High Electric Field Pulses, *Trends Food Sci. Technol.* 5, 71–75.

26 Knorr, D., Angersbach, A. **1998**, Impact of High Intensity Electric Field Pulses on Plant Membrane Permeabilization, *Trends Food Sci. Techno.* 9, 185–191.

27 Taiwo, K.A., Angersbach, A., Ade-Omowaye, B.I.O., Knorr, D. **2001**, Effects of Pre-Treatments on the Diffusion Kinetics and Some Quality Parameters of Osmotically Dehydrated Apple Slices, *J. Agric. Food Chem.* 49, 2804–2811.

28 McClements, D.J. **1995**, Advances in the Application of Ultrasound in Food Analysis and Processing, *Trends Food Sci. Technol.* 6, 293–299.

29 Povey, M.J.W., McClements, D.J. **1988**, Ultrasonics in Food Engineering. Part 1: Introduction and Experimental Methods, *J. Food Eng.* 8, 217–245.

30 Roberts, R.T., Wiltshire, M.P. **1990**, High Intensity Ultrasound in Food Processing, in *Food Technology International Europe*, ed. A. Turner, Sterling Publications International, London, pp 83–87.

31 Mason, T.J., Paniwnyk, L., Lorimer, J.P. **1996**, The Uses of Ultrasound on Food Technology, *Ultrasonics Sonochem.* 3, S253–S260.

32 Hughes, D.E., Nyborg, W.L. **1962**, Cell Disruption by Ultrasound, *Science* 138, 108–144.

33 Alliger, H. **1975**, Ultrasonic Disruption, *Am. Lab.* 10, 75–85.

34 Sala, F.J., Burgos, J., Condón, S., Lopez, P., Raso, J. **1995**, Manothermosonication, in *New Methods of Food Preservation by Combined Processes*, ed. G.W. Gould, Blackie, London, pp 176–204.

35 Mason, T.J. **1998**, Power Ultrasound in Food Processing – The Way Forward, in *Ultrasound in Food Processing*, ed. M.J.W. Povey, T.J. Mason, Blackie Academic and Professional, London, pp 105–126.

36 Roberts, R.T. **1993**, High Intensity Ultrasound in Food Processing, *Chem. Ind.* 3, 119–121.

37 Ishimori, Y., Karube, I., Suzuki, S. **1981**, Acceleration of Immobilised Alpha-Chymotrypsin Activity with Ultrasonic Irradiation, *J. Mol. Catal.* 12, 253.

38 Rosenfeld, E., Schmidt, P. **1984**, *Arch. Acoust.*. 9, 105.

39 Moulton, K.J., Wang, L.C. **1982**, A Pilot-Plant Study of Continuous Ultrasonic Extraction of Soy Bean Protein, *J. Food Sci.* 47, 1127.

40 Vimini, R.J., Kemp, J.D., Fox, J. **1983**, Effects of Low Frequency Ultrasound on Properties of Restructured Beef Rolls, *J. Food Sc.* 48, 1572.

41 Roberts, T. **1991**, Sound for Processing Food, *Nutr. Food Sci.* 130, 17.

42 Midler, M. **1970**, Production of Crystals in a Fluidised Bed with Ultrasonic Vibrations, US patent 3,510,266.

43 Baxter, J.F., Morris, G.J., Gaim-Marsoner, G. **1997**, Process for Accelerating the Polymorphic Transformation of Edible Fats Using Ultrasonication, European patent 95–30683.

44 Baxter, J.F., Morris, G.J., Gaim-Marsoner, G. **1997**, Process for Retarding Fat Bloom in Fat-Based Confectionery Masses, European patent 95–306833.

45 Acton, E., Morris, G.J.Y. **1992**, Method and Apparatus for the Control of Solidification in Liquids, Worldwide patent WO 92/20420.

46 Morikawa, T., Oka, K., Kojima, K. **1996**, Fluidisation and Foam Separation in Brewing, *Tech. Q. Master Brew. Assoc. Am.* 33, 54–58.

47 Villamiel, M., Verdurmen, R., De Jong, P. **2000**, Degassing of Milk by High Intensity Ultrasound, *Milchwissenschaft* 55, 123–125.

48 Gallego-Juarez, J.A., Rodriguez-Corral, G., Galvez-Maraleda, J.C., Yang, T.S. **1999**, A New High Intensity Ultrasonic Technology for Food Dehydration, *Dry. Technol.* 17, 597–608.

49 Sastry, S.K., Shen, G.Q., Blaisdell, J.L. **1989**, Effect of Ultrasonic Vibration on Fluid-to-Particle Convective Heat Transfer Coefficients – A Research Note, *J. Food Sci.* 54, 229–230.

50 Lima, M., Sastry, S.K. **1990**, Influence of Fluid Rheological Properties and Particle Location on Ultrasound-Assisted Heat Transfer Between Liquid and Particles, *J. Food Sci.* 55, 1112–1119.

51 Williams, A., Leadley, C.E., Lloyd, E., Betts, G., Oakley, R., Gonzalez, M. **1998**, Ultrasonically Enhanced Heat Transfer and Microbiological Inactivation (*Research Summary Sheet 43*), Campden & Chorleywood Research Association, Chipping Campden.

52 Williams, A., Leadley, C.E., Lloyd, E., Betts, G., Oakley, R., Gonzalez, M. **1999**, Ultrasonically Enhanced Heat Transfer and Microbiological Inactivation, *Abstr. Eur. Conf. Emerg. Food Sci. Technol.* Tampere, Finland.

53 Schuett-Abraham, I., Trommer, E., Levetzow, R. **1992**, Ultrasonics in Sterilisation Sinks. Applications of Ultrasonics on Equipment for Cleaning and Disinfection of Knives at the Workplace in Slaughter and Meat Cutting Plants, *Fleischwirtschaft* 72, 864–867.

54 Mason, T.J., Newman, A.P., Phull, S.S., Charter, C. **1994**, Sonochemistry in Water Treatment: A Sound Solution to Traditional Problems, *World Water Environ. Eng.* 1994, 16.

55 Suslick, K.S. **1988**, Homogenous Sonochemistry, in *Ultrasound. Its Chemical, Physical and Biological Effects*, ed. K.S. Suslick, VCH Publishers, New York, pp 122–163.

56 Suslick, K.S. **1989**, The Chemical Effects of Ultrasound. *Sci. Am.* 2, 62–68.

57 Raso, J., Condón, S., Sala, F.J. **1994**, Manothermosonication – a New Method of Food Preservation?, in *Food Preservation by Combined Processes. (Final Report for FLAIR Concerted Action No. 7, Subgroup B)*, ed. FLAIR, pp 37–41.

58 Ahmed, F.I.K., Russell, C. **1975**, Synergism Between Ultrasonic Waves and Hydrogen Peroxide in the Killing of Microorganisms, *J. Appl. Bacteriol.* 39, 1–40.

59 Scherba, G., Weigel, R.M., O'Brien, J.R. **1991**, Quantitative Assessment of the Germicidal Efficiency of Ultrasonic Energy, *Appl. Environ. Microbiol.* 57, 2079–2084.

60 Sanz, P., Palacios, P., Lopez, P., Ordonez, J.A. **1985**, Effect of Ultrasonic Waves on the Heat Resistance of *Bacillus stearothermophilus* Spores, in *Fundamental and Applied Aspects of Bacterial Spores*, ed. D.J.E. Dring, G.W. Gould, Academic Press, New York, pp 251–259.

61 Lee, B.H., Kermala, S., Baker, B.E. **1989**, Thermal, Ultrasonic and Ultraviolet Inactivation of *Salmonella* in Films of

Aqueous Media and Chocolate, *Food Microbiol.* 6, 143–152.

62 Kinsloe, H., Ackerman, E., Reid, J. J. **1954**, Exposure of Microorganisms to Measured Sound Fields, *J. Bacteriol.* 68, 373–380.

63 Utsunomyia, Y., Kosaka, Y. **1979**, Application of Supersonic Waves to Foods, *J. Fac. Appl. Biol. Sci. Univ. Tokyo* 18, 225–231.

64 Hurst, R. M., Betts, G. D., Earnshaw, R. G. **1995**, The Antimicrobial Effect of Power Ultrasound (*CCFRA R&D Report 4*), Campden & Chorleywood Research Association, Chipping Campden.

65 Earnshaw, R. G. **1998**, Ultrasound: a New Opportunity for Food Preservation, in *Ultrasound in Food Processing*, ed. M. J. W Povey, T. J. Mason, Blackie Academic and Professional, London, p. 183.

66 Garcia, M. L., Burgos, J., Sanz, B., Ordonez, J. A. **1989**, Effect of Heat and Ultrasonic Waves on the Survival of Two Strains of *Bacillus subtilis*, *J. Appl. Bacteriol.* 67, 619–628.

67 Burgos, J., Ardennes, J. A., Sala, F. J. **1972**, Effect of Ultrasonic Waves on the Heat Resistance of *Bacillus cereus* and *Bacillus licheniformis* Spores, *Appl. Microbiol.* 24, 497–478.

68 Palacios, P., Borgos, J., Hoz, L., Sanz, B., Ordonez, J. A. **1991**, Study of Substances Released by Ultrasonic Treatment from *Bacillus stearothermophilus* Spores, *J. Appl. Bacteriol.* 71, 445–451.

69 Lillard, H. S. **1993**, Bactericidal Effect of Chlorine on Attached Salmonellae with and Without Sonification, *J. Food Prot.* 56, 716–717.

70 Seymour, I. **1999**, Novel Techniques for Cleaning and Decontaminating Raw Vegetables and Fruits (*Research Summary Sheet 14*), Campden & Chorleywood Research Association, Chipping Campden.

71 Chambers, L. A. **1937**, The Influence of Intense Mechanical Vibration on the Proteolytic Activity of Pepsin, *J. Biol. Chem.* 117, 639.

72 Vercet, A., Lopez, P., Burgos, J. **1997**, Inactivation of Heat-Resistant Lipase and Protease from *Pseudomonas fluorescens* by Manothermosonication, *J. Dairy Sci.* 80, 29–36.

73 Lu, A. T., Whitaker, J. R. **1974**, Some Factors Affecting Rates of Heat Inactivation and Reactivation of Horseradish Peroxidase, *J. Food Sci.* 39, 1173–1178.

74 Villamiel, M., De Jong, P. **2000**, Influence of High-Intensity Ultrasound and Heat Treatment in Continuous Flow on Fat, Proteins, and Native Enzymes of Milk, *J. Agric. Food Chem.* 48, 472–478.

75 Wiltshire, M. **1992**, Presentation at Sonochemistry Symposium, *R. Soc. Chem. Annu. Congr*, 1992.

76 Davidek, J., Velisek, J. **1990**, Chemical Changes During Food Processing, *Dev. Food Sci.* 1990, 21.

77 Barbosa-Canovas, G. V., Schaffner, D. W., Pierson, M. D., Zhang, H. Q. **2000**, Pulsed Light Technology, *J. Food Sci. [Suppl]* 2000, 82–85.

78 Barbosa-Canovas, G. V., Schaffner, D. W., Pierson, M. D., Zhang, H. Q. **2000**, High Voltage Arc Discharge, *J. Food Sci. [Suppl]* 2000, 80–81.

79 Hofmann, G. A. **1985**, Deactivation of Microorganisms by an Oscillating Magnetic Field, US patent 4524079.

80 Mastwijk, H. **2002**, Inactivation of Microorganisms by Cold Gas Phase Plasmas, *IFT/EFFoST Non-Thermal Processing Workshop*, Ohio State University.

81 Ho, G. **2002**, Carbon Dioxide Pressure Processing, *IFT/EFFoST Non-Thermal Processing Workshop*, Ohio State University.

8
Baking, Extrusion and Frying

Bogdan J. Dobraszczyk, Paul Ainsworth, Senol Ibanoglu and Pedro Bouchon

8.1
Baking Bread

Bogdan J. Dobraszczyk

8.1.1
General Principles

Baking is a term commonly applied to the production of cereal-based products such as bread, biscuits, cakes, pizzas, etc., and in its English usage baking is generally applied to the production of fermented bread. Baking is at heart a process: the conversion of some relatively unpalatable ingredients (starch, gluten, bran, in the case of most cereals) into the aerated, open cell sponge structure we know as bread has taken millennia to develop. Bread is one of the oldest 'functional' or engineered foods, where the addition of bubbles of air by mixing and fermentation has created an entirely different product with considerably enhanced sensory properties. Historical records have been found in ancient Egyptian tomb carvings dating from 3000 BC, which show fermented bread being made from wheat flour and baked in clay ovens (see Fig. 8.1). The cultivation of wheat for breadmaking has played a key role in the development of modern civilization; and bread is one of the principal sources of nutrition for mankind.

Nutritionally, the production of aerated bread is of little benefit. During fermentation, the yeast converts some of the starch sugars into CO_2, which escapes during baking, decreasing the nutritional benefit of the flour. The production of aerated bread is also energetically wasteful – requiring considerable effort and expertise. It is the sensory and textural benefits of the aerated structure of fermented bread that outweigh the losses which encourage us to spend the time and effort in baking bread.

Each country has its own particular methods of baking, but in essence bread is made by simply mixing flour, water, yeast (and air) into a dough, allowing the

Food Processing Handbook. Edited by James G. Brennan

ISBN: 3-527-30719-2

Fig. 8.1 Ancient Egyptian baking.

yeast to ferment for some time to produce an expanding aerated foam and then setting the structure at high temperature in an oven to produce bread. A critical feature of baking is the creation of a stable, aerated structure in the baked bread. Air is a key ingredient in bread, which is added during mixing, manipulated and controlled during proof, and transformed into an interconnected sponge structure during baking. It is the unique viscoelastic properties of gluten, the major protein in wheat flour, which allows the continued expansion and stability of bubbles during breadmaking. No other cereal is capable of retaining gas to the same extent as wheat during fermentation and baking. The polymer structure and rheological properties of gluten are fundamental in maintaining bubble stability during fermentation and in determining the subsequent texture and quality of baked bread. Baking is a deceptively simple-looking process that hides a complex set of chemical and physical processes that are still not completely understood.

8.1.2
Methods of Bread Production

A large number of baking methods exist around the world, with each country having its own traditions and practices handed down over generations, producing a wide variety of types of bread. In practice, these methods can be classified into two main processes: (a) bulk fermentation (BF) and (b) mechanical dough development, or the Chorleywood bread process (CBP).

8.1.2.1 Bulk Fermentation

The major traditional breadmaking process is the bulk fermentation process, in which the dough mass is allowed to ferment over a lengthy period of time after mixing. Cauvain [1] lists the three essential features of bulk fermentation as:

- mixing the ingredients, usually at slow speed, to form a homogeneous dough;
- resting the dough in bulk for a prescribed period of time (typically 3–24 h), during which the yeast ferments to inflate the dough;
- remixing the dough partway through the bulk fermentation, to remove most of the larger gas cells produced and subdivide the smaller gas cells to give an effective increase in the number of gas cells (known as 'punching' or 'knock-back').

The purpose of this bulk fermentation period is partly for flavour development, but primarily for developing the dough protein structure so that it is better able to retain gas and retard coalescence of gas cells during subsequent proving. The slow inflation during bulk fermentation stretches the dough proteins biaxially, so that they align into a cooperative network that imparts superior elasticity and extensibility to the dough. Historical experience taught bakers that this structure development produced a larger loaf volume and finer crumb structure, and bulk fermentation has thus been a feature of breadmaking for centuries. Baker and Mize [2] showed that gas cells must be incorporated into the dough during mixing, to act as nucleation sites for the carbon dioxide produced during proving; yeast is unable to generate bubbles from nothing. Without these sites, the CO_2 has nowhere within the dough to diffuse and is largely lost from the dough, resulting in a loaf of low volume and a coarse, unattractive structure. The operations of punching and moulding (in which the dough is sheeted between rollers, then rolled up into a 'Swiss roll' to orient subsequent bubble growth and give a firmer crumb structure) serve to redistribute the gas bubbles, so that the relationship between aeration during mixing and final baked loaf structure is weak.

8.1.2.2 Chorleywood Bread Process

The modern commercial process used in commercial bakeries in the UK and many other countries is known as the Chorleywood bread process (CBP), which was developed in 1961 by the Flour Milling and Baking Research Association at Chorleywood [3]. This method produces bread and other fermented bakery goods without the need for a long fermentation period. Dough development in CBP is achieved during high-speed mixing by intense mechanical working of the dough in a few minutes. Not only does the CBP save considerable time in the baking process, which helps keep down the production costs and delivers cheaper bread, but it also produces bread which is better in respect of volume, colour and keeping qualities. CBP is now by far the most common method used throughout all sectors of the bread baking industry. The main features of the CBP process are shown in Fig. 8.2.

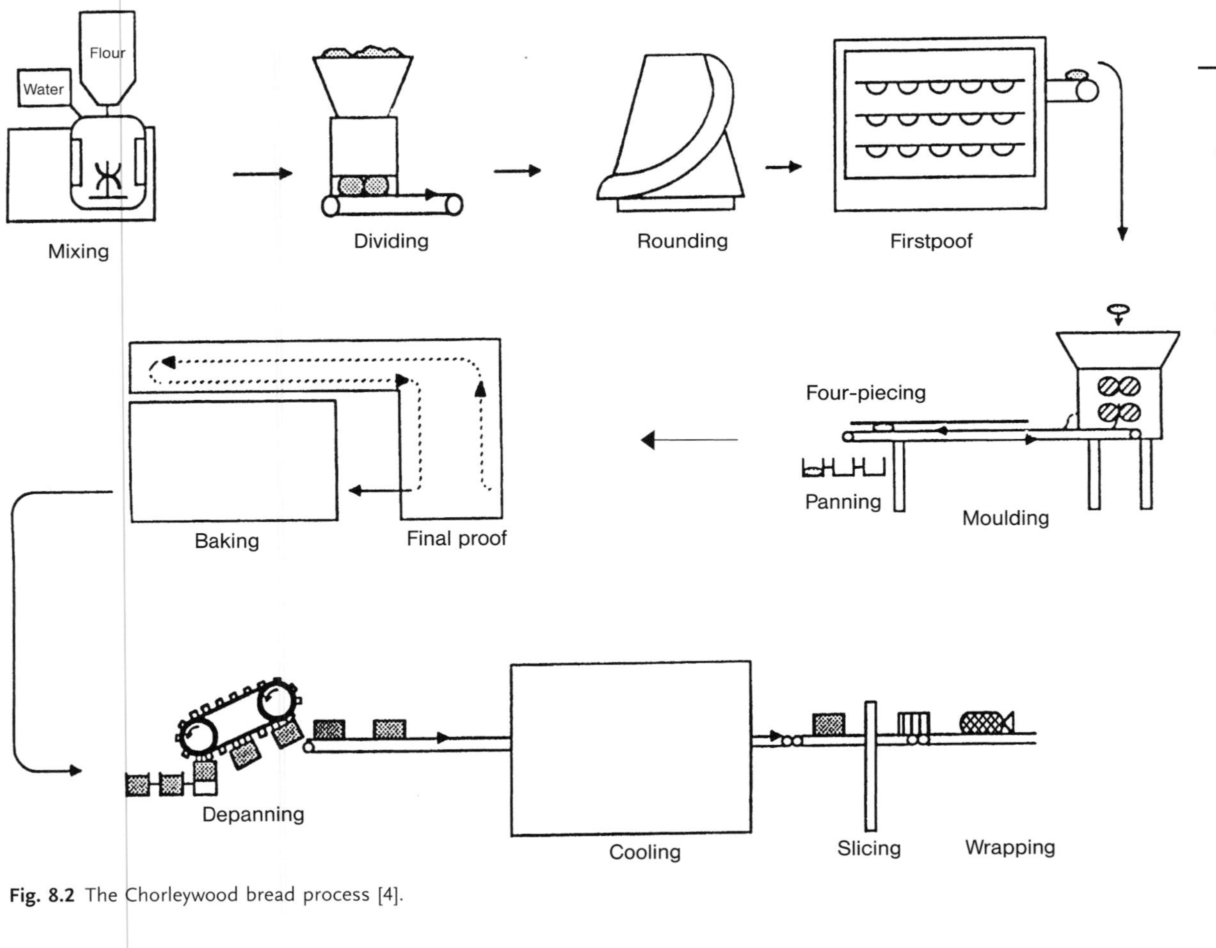

Fig. 8.2 The Chorleywood bread process [4].

Table 8.1 describes the major operations in the CBP and contrasts these with bulk fermentation processes. In the CBP, mixing and dough development are combined into a single operation of a few minutes, in contrast to the much slower, low-energy mixing for BF (8–15 min). Oxidants such as ascorbic acid are added in the CBP method to facilitate rapid dough development. Increased yeast levels and higher proof temperatures are required in the CBP to ensure adequate CO_2 production, as the yeast has less time to multiply during the much shorter proof period. The yeast metabolises less of the flour, therefore extra water is required in the dough, thereby increasing yield. The CBP can also use flour with a lower protein content than that required for BF, thereby increasing the use of more homegrown UK wheat which traditionally has a lower protein content than imported wheats from North America. High melting point fats and emulsifiers are required in the dough formulation for CBP bread, to aid bubble stabilisation during baking. Partial vacuum or pressure/vacuum is applied during mixing to regulate the bubble sizes and structure in the final bread texture. Because the CBP eliminates punching and performs moulding before much yeast activity has occurred, the bubble size distribution at the end of mixing has a much greater influence on baked loaf structure than for traditional BF processes. Due to the rapid development of the dough structure during mixing in the CBP process, it is considered that flavour is not developed as strongly in comparison with the BF process. This is counteracted with the addition of various ingredients such as enzymes and the use of prefermented extracts obtained from BF doughs.

Table 8.1 Comparison of bulk fermentation and Chorleywood bread process (CBP).

	Bulk fermentation	**CBP**
Ingredients	Flour Water Yeast ~1% Salt	Flour Water Yeast ~2.5% Salt Oxidants (ascorbic acid) Fat Emulsifiers Lower protein content and more water relative to BF
Mixing	Slow mixing (8–15 min) Low energy (5–15 kJ kg^{-1})	Rapid mixing (<3 min) High energy (40 kJ kg^{-1}) Partial vacuum or pressure-vacuum
Proof	Fermentation for 3–24 h at 26 °C Remixing during proof to expel excess gas	First proof 10–15 min at room temperature Second proof 45 min at ~40 °C
Baking	20–25 min at ~200 °C	As for BF

8.1.3 The Baking Process

The baking process can be divided into three main processes: mixing, fermentation (or proof), and baking. These three processes are reviewed in the following sections.

8.1.3.1 Mixing

Mixing of bread dough has three main functions: (a) to blend and hydrate the dough ingredients, (b) to develop the dough and (c) to incorporate air into the dough. In the production of doughs, the nature of the mixing action develops the viscoelastic properties of gluten and also incorporates air, which has a major effect on their rheology and texture. There is an intimate relationship between mixing, aeration and rheology: the design and operation of the mixer develops texture, aeration and rheology to different extents; and conversely the rheology of the food affects the time and energy input required to achieve optimal development. This is seen in the great variety of mixers used in the baking industry, where certain mixers are required to produce a desired texture or rheology in the bread.

Extensive work on dough mixing has shown that mixing speed and energy (work input) must be above a certain value to develop the gluten network and to produce satisfactory breadmaking [5]; and an optimum work input or mixing time has been related to optimum breadmaking performance, which varies depending on mixer type, flour composition and ingredients [6]. Mixing beyond the optimum (overmixing) is thought to damage the dough, causing the gluten network to break down, resulting in more fragile bubble walls and less gas retention and lower baking volume. Overmixing can also result in sticky, difficult to handle doughs which causes production problems. The strong relationship between mixing and handling and baking properties has resulted in a large number of commercial force-recording dough mixers, such as the Farinograph and Mixograph [7, 8], which are used to determine the optimum in mixing speed and energy. Mixing doughs by elongational flow in sheeting to achieve optimum development requires only 10–15% of the energy normally used in conventional high-speed shear mixers [9], suggesting that much higher rates of work input can be achieved due to the enhanced strain hardening of doughs under extension.

8.1.3.2 Fermentation (Proof)

Fermentation, or proof, is the critical step in the breadmaking process, where the expansion of air bubbles previously incorporated during mixing provides the characteristic aerated structure of bread, which is central to its appeal. During proof, the gas content within the dough increases from around 4–8% to approximately 80%. The original bubble structure formed during mixing and sub-

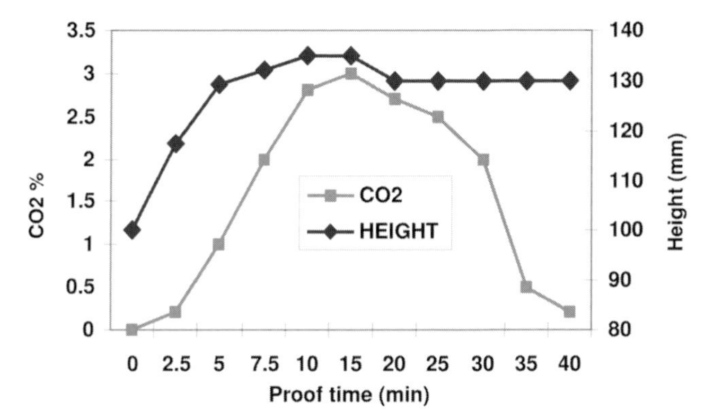

Fig. 8.3 CO_2 release and change in dough height during proof.

sequently altered during punching and moulding is slowly expanded by the diffusion of CO_2 which is dissolved in the surrounding liquid phase within the dough [10]. This causes the steady increase in the volume of the dough known as proof, since at this stage the bubbles are discrete and no gas can escape. Eventually the bubbles start to interconnect or coalesce, CO_2 gas begins to escape and volume expansion ceases (see Fig. 8.3). The growth and stability of the gas bubbles during proof determine the ultimate expansion of the dough and its final baked texture and volume [11].

The coalescence of gas bubbles is the key event which determines the extent of bubble expansion and volume increase during proof, and also the final loaf texture. If coalescence is delayed, bubbles can expand for much longer, giving a larger proof volume and a finer texture, i.e. smaller gas cells. If coalescence occurs early, then gas escape occurs much earlier and proof volume is decreased and the texture is much coarser. The rheological properties of the bubble cell walls are very important in controlling coalescence in doughs, in particular the elongational strain hardening properties of the gluten polymers [12, 13].

8.1.3.3 **Baking**

In baking the foam structure of the dough, created by the gas cells, is turned into a loaf of bread with an interconnected sponge structure supported by the starch-protein matrix. Within the first few minutes of baking, the volume of the dough increases rapidly and reaches the maximum size of the loaf; and this period is called oven rise or oven spring [14]. At the end of oven spring, there is a sharp increase in the rate of gas lost from the loaf, which is due to the rapid coalescence or rupture of the bubble cell walls. At the temperature of starch gelatinisation ($\sim 65\,^{\circ}C$), there is a transfer of water from the protein to the starch, leading to a swelling of the starch granules and a rapid increase in viscosity of the dough, which sets the sponge structure. These physical changes lead to a change from a closed cell foam structure to an open cell sponge structure (see Fig. 8.4).

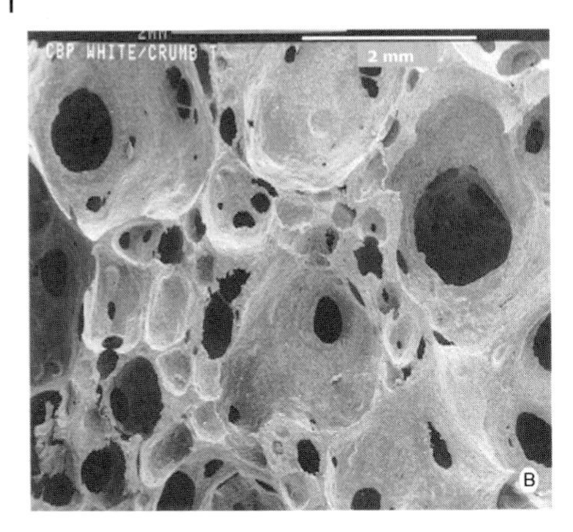

Fig. 8.4 Open cell sponge structure of bread.

During baking, there is an evaporation of water from the loaf, which is particularly marked near the surface of the loaf; and this evaporation plus the occurrence of the Maillard reaction cause a characteristic dark brown crust to be formed on the exterior of the loaf. The Maillard reaction is a complex set of chemical reactions in which the amino acids in proteins react with reducing sugars such as glucose and fructose and which are very important to our perception of flavour in baked bread.

8.1.4
Gluten Polymer Structure, Rheology and Baking

Gluten is the major protein in wheat flour doughs, responsible for their unique viscoelastic behaviour during deformation. It is now widely accepted that gluten proteins are responsible for variations in baking quality; and in particular it is the insoluble fraction of the high molecular weight (HMW) glutenin polymer which is best related to differences in dough strength and baking quality amongst different wheat varieties [15, 16]. For example, the rheological properties of gluten are known to be important in the entrainment, retention and stability of gas bubbles within dough during mixing, fermentation and baking [13, 17], which ultimately are responsible for the texture and volume of the final baked product.

Gluten proteins comprise a highly polydisperse system of polymers, classically divided into two groups based on their extractability in alcohols: gliadins and glutenins. The gliadins are single-chain polypeptides with MW ranging from 2×10^4 to 7×10^4, whilst the glutenins are multiple-chain polymeric proteins in which individual polypeptides are thought to be linked by interchain disulphide and hydrogen bonds to give a wide MW distribution, ranging between 10^5 and 10^9. Gluten has a bimodal MW distribution which roughly parallels the classic division based on solubility into gliadins and glutenins (see Fig. 8.5) [18, 19].

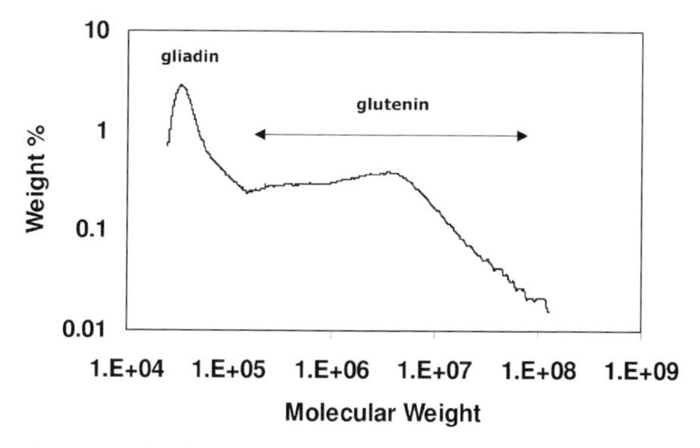

Fig. 8.5 Molecular weight distribution of gluten, showing bimodal distribution.

Individual glutenin polymers have an extended rod-like structure ca. 50–60 nm in length, made up of a central repetitive region, which is known to adopt a regular spiral structure (see Fig. 8.6) [20] and terminal regions which contain cysteine residues which are associated with intermolecular crosslinking by disulphide bonds.

The rheological properties of gluten are known to depend on the three-dimensional organisation of the polymer network, where individual linear HMW glutenin polymers (subunits) are linked by disulphide and hydrogen bonding and entanglements to form a network structure. The microscopic structure of gluten is seen clearly in confocal laser scanning microscopy (see Fig. 8.7), where gluten strands form a three-dimensional network in which are embedded rigid starch granules.

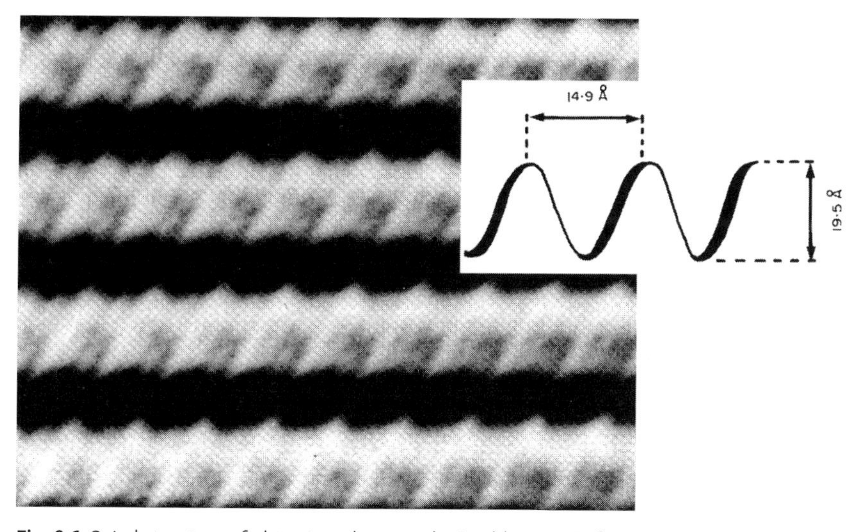

Fig. 8.6 Spiral structure of glutenin polymers; obtained by atomic force microscopy.

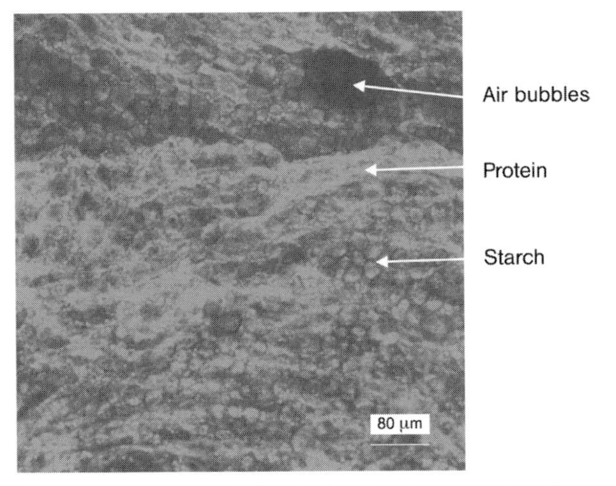

Fig. 8.7 Microstructure of dough, showing gluten, starch and air bubbles during the early stages of proof; obtained by confocal laser scanning microscopy.

Recent studies in polymer physics have shown that molecular size, structure and MW distributions of polymers are intimately linked to their rheological properties and ultimately to their performance in various end use applications [21, 22]. Beyond a critical molecular weight (MW_c), characteristic for each polymer, rheological properties such as viscosity, relaxation time and strain hardening start to increase rapidly with increasing MW. Above MW_c, the polymers start to entangle, giving rise to the observed rapid increase in viscosity with MW (see Fig. 8.8).

Entanglements can be viewed as physical constraints between segments of the polymer chain, rather like knots (see Fig. 8.9).

A relatively small variation in the highest end of the MW distribution can give rise to a large increase in viscosity and strain hardening and is likely to have a large effect on baking performance. If the polymers are branched, viscosity rises even more rapidly. The effect of polymer chain branching on shear and extensional viscosity of polymers and doughs is shown in Fig. 8.10. At low strains, shear and extensional viscosities are very similar, but as deformation increases, the effect of polymer branching and entanglement becomes apparent, with a steep rise in viscosity with strain (known as strain hardening) apparent for extensional deformation and a decrease in viscosity with shear deformation (known as shear thinning).

An increase in the number of branches increases strain hardening and extensional viscosity and decreases shear viscosity. Extensional rheological properties appear to be more sensitive to changes in MW, polymer entanglements and branching than dynamic shear properties. This indicates that it is more likely to be the physical interactions of the secondary molecular structure of the insoluble HMW glutenin (such as branching and entanglements) that are responsi-

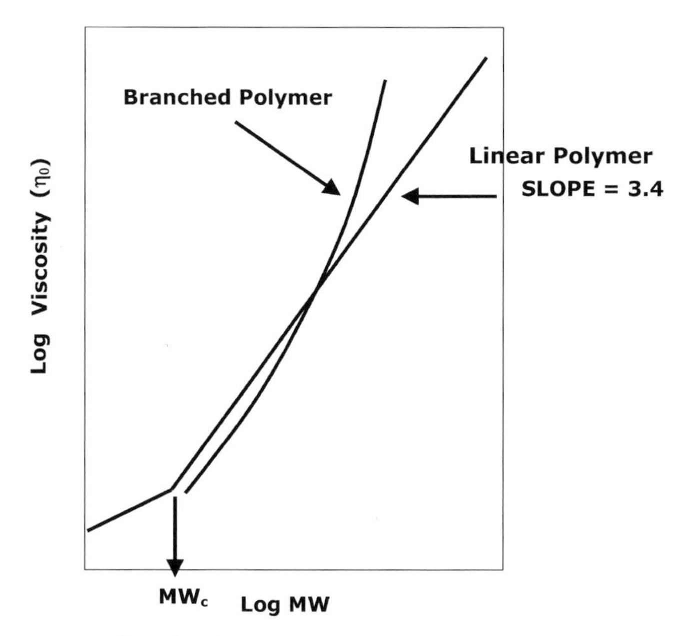

Fig. 8.8 Effect of molecular weight (MW) and branching on zero-shear viscosity for polymer melts. Beyond a critical molecular weight for entanglements (MW_c), viscosity (η_0) increases rapidly for linear polymer as $MW^{3.4}$.

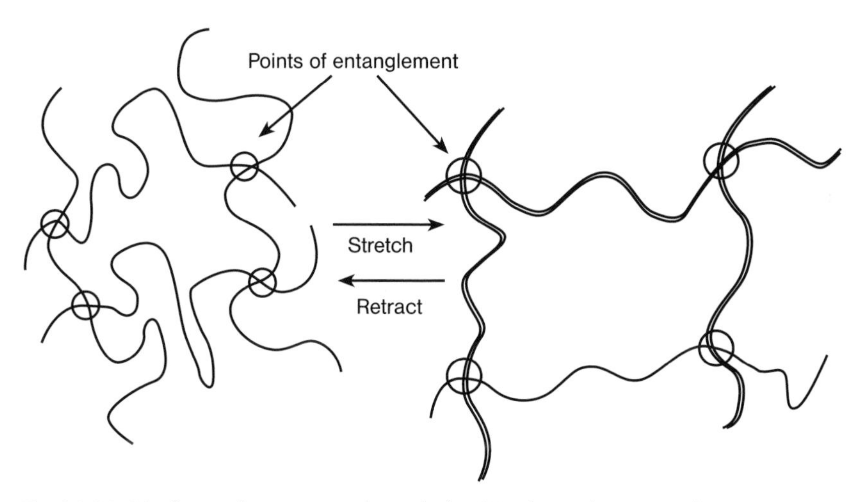

Fig. 8.9 Model of entanglement network in a high MW polymer during stretching.

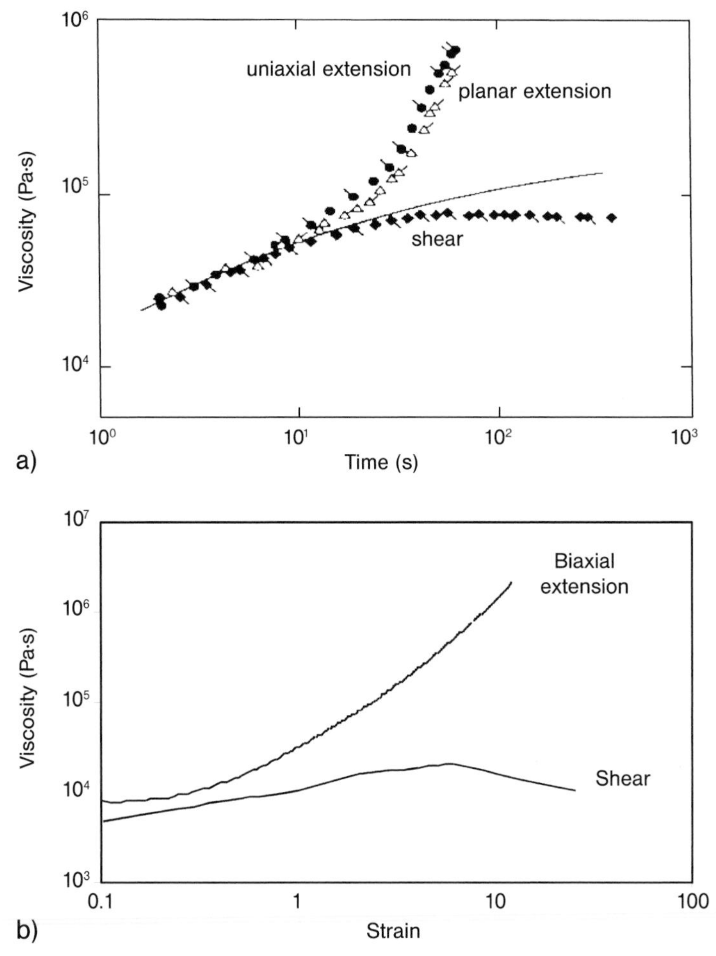

Fig. 8.10 Effect of polymer chain branching on shear and extensional viscosities of (a) branched low-density polyethylene at 125 °C and (b) dough.

ble for the rheological properties of dough and its baking performance than the primary chemical structure or size of individual glutenin molecules. Recent evidence suggests that (a) these insoluble HMW polymers are entangled with a corresponding long relaxation time, (b) they are branched and form extensive intermolecular secondary structures held together by hydrogen bonding and (c) differences in these secondary structures are likely to be strongly related to extensional rheology and baking performance [13, 23, 24, 25].

8.1.5 Baking Quality and Rheology

During proof and baking, the growth and stability of gas bubbles within the dough determines the expansion of the dough and therefore the ultimate volume and texture of the baked product. The limit of expansion of these bubbles is related directly to their stability, due to retardation of coalescence and loss of gas when the bubbles fail. The rheological properties of the expanding bubble walls are therefore important in maintaining stability in the bubble wall and promoting gas retention. The relevant rheological conditions around an expanding gas cell during proof and baking are biaxial extension, large extensional strain and low strain rates. Any rheological tests which seek to relate to baking performance should be performed under conditions similar to those of baking expansion. Methods such as bubble inflation and lubricated compression offer the most appropriate method for measuring extensional rheological properties of doughs. The major advantage of these tests is that the deformation closely resembles practical conditions experienced by the cell walls around the expanding gas cells within the dough during proof and oven rise, i.e. large deformation biaxial extension and can be carried out at the low strain rates and elevated temperatures relevant to baking.

The failure of gas cell walls in doughs has been shown to be directly related to the elongational strain hardening properties of the dough measured under large deformation biaxial extension [26]. Strain hardening is shown as an increase in the slope of the stress-Hencky strain curve with increasing extension, giving rise to the typical J-shaped stress-strain curve observed for highly extensible materials (see Fig. 8.11).

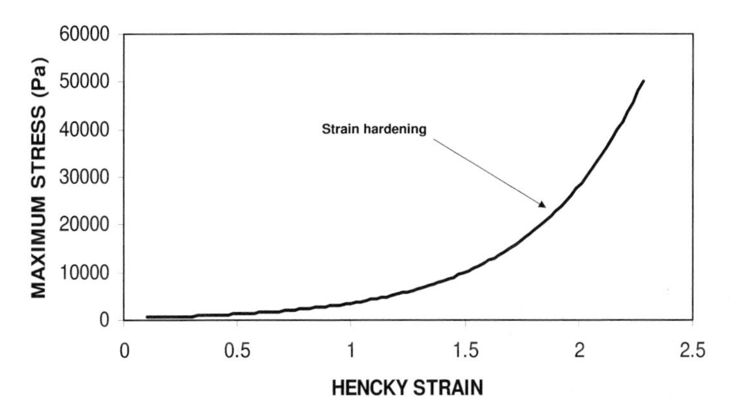

Fig. 8.11 Typical J-shaped stress-Hencky strain curve in biaxial extension for a dough bubble, indicating strain hardening as a rapid increase in stiffness with increasing inflation (Hencky strain). Bubble inflation using SMS Dough Inflation System, maximum stress and Hencky strain calculated for bubble wall polar region.

Strain hardening in doughs is thought to arise mainly from stretching of polymer chains between points of entanglement in the larger glutenin molecules, which gives rise to the increasing stiffness observed at large strains. Under extensional flow, entangled polymers exhibit strain hardening which is enhanced for polymers with a broad MW distribution, particularly a bimodal distribution and branching. It is therefore expected that the broad bimodal MW distribution and branched structure typical of gluten will result in enhanced strain hardening and a bimodal distribution of relaxation times. Recent work has shown that bread doughs exhibit strain hardening under large extensional deformations and that these extensional rheological properties are important in baking performance [12, 13, 27]. Strain hardening allows the expanding gas cell walls to resist failure by locally increasing resistance to extension as the bubble walls become thinner and provides the bubble walls with greater stability against early coalescence and better gas retention. It is expected therefore that doughs with good strain hardening characteristics should result in a finer crumb texture, e.g. smaller gas cells, thinner cell walls, an even distribution of bubble sizes and a larger baked volume than doughs with poor strain hardening properties. Good breadmaking doughs have been shown to have good strain hardening properties and inflate to larger individual bubble volumes before rupture, whilst poor breadmaking doughs inflate to lower volumes and have much lower strain hardening (see Fig. 8.12).

Loaf volume and crumb score for a number of doughs from flour varieties with varying baking performance has been related directly to the strain hardening properties of single dough bubbles measured at elevated temperatures in

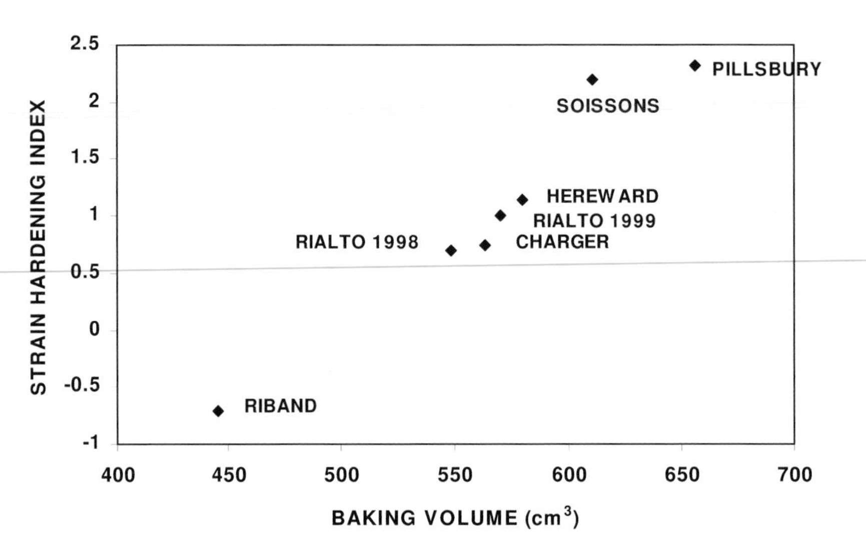

Fig. 8.12 Relationship between strain hardening and baking volume for a number of flour varieties with varying baking performance.

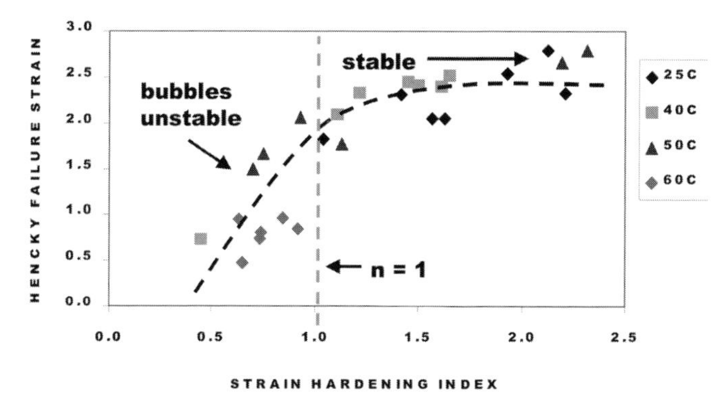

Fig. 8.13 Dough bubble wall stability (measured as Hencky failure strain) as a function of strain hardening, showing instability at strain hardening <1.

biaxial extension. Once strain hardening drops below a value of around 1, bubble wall stability decreases rapidly (see Fig. 8.13). Bubble wall stability (as indicated by a strain hardening value of 1) is increased to progressively higher temperatures with increasing baking volume, allowing the bubbles to resist coalescence and retain gas for much longer during the baking process. Bubble wall instability in poorer breadmaking varieties occurs at much lower temperatures, giving earlier bubble coalescence and release of gas, resulting in lower loaf volumes and poorer texture [28].

8.2 Extrusion

Paul Ainsworth and Senol Ibanoglu

8.2.1 General Principles

Extrusion can be defined as the process of forcing a pumpable material through a restricted opening. It involves compressing and working a material to form a semisolid mass under a variety of controlled conditions and then forcing it, at a predetermined rate, to pass through a hole.

The origins of extrusion are in the metallurgical industry, where in 1797 a piston driven device was used to produce seamless lead pipes [29]. The current understanding of extrusion technology and the developments in machine design are largely due to research carried out by the plastics industry.

Extrusion technology was first applied to food materials in the mid-1800s, when chopped meat was stuffed into casings using a piston type extruder. In the 1930s, a single-screw extruder was introduced to the pasta industry, to both mix the ingredients (semolina and water) and to shape the resulting dough into macaroni in one continuous operation.

Today, a wide variety of intermediate or food products are produced by extrusion.

8.2.1.1 **The Extrusion Process**

Extrusion is predominantly a thermomechanical processing operation that combines several unit operations, including mixing, kneading, shearing, conveying, heating, cooling, forming, partial drying or puffing, depending on the material and equipment used. During extrusion processing, food materials are generally subjected to a combination of high temperature, high pressure and high shear. This can lead to a variety of reactions with corresponding changes in the functional properties of the extruded material.

In the extrusion process, there are generally two main energy inputs to the system. Firstly there is the energy transferred from the rotation of the screws and secondly the energy transferred from the heaters through the barrel walls. The thermal energy that is generated by viscous dissipation and/or transferred through the barrel wall results in an increase in the temperature of the material being extruded. As a result of this, there may be phase changes, such as melting of solid material, and/or the evaporation of moisture.

The ingredients used in extrusion are predominantly dry powdered materials, with the most commonly used being wheat, maize and rice flours. The conditions in the extruder transform the dry powdered materials into fluids and therefore, characteristics such as surface friction, hardness and cohesiveness of particles become important. In the high solids concentration of the extruder melt, the presence of other ingredients, such as lipids and sugars, can cause significant changes in the final product characteristics.

In addition to starch-based products, a range of protein-rich products can be manufactured by extrusion, using raw materials such as soya or sunflower, fava beans, field bean and isolated cereal proteins. Fig. 8.14 shows a schematic diagram of an extruder.

In order to convey the dry raw material to the extruder barrel, volumetric and gravimetric feeders tend to be used. Volumetric devices include single- and twin-screw feeders, rotary airlock feeders, disk feeders, vibratory feeders and volumetric belt feeders. In all of these feed mechanisms, it is assumed that the density of the feed material remains constant over time and hence a constant volume of feed will result in a constant mass flow rate [30]. Gravimetric feeders are more expensive and more complex than volumetric feeders. They are usually microprocessor-controlled to monitor the mass flow rate and adjust the feeder speed as required. The most commonly used gravimetric feeders are the weight-belt and the 'loss in weight' feeders [31].

Addition of liquid feed ingredients to the extruder can be achieved using a variety of devices, including rotameters, fluid displacement meters, differential

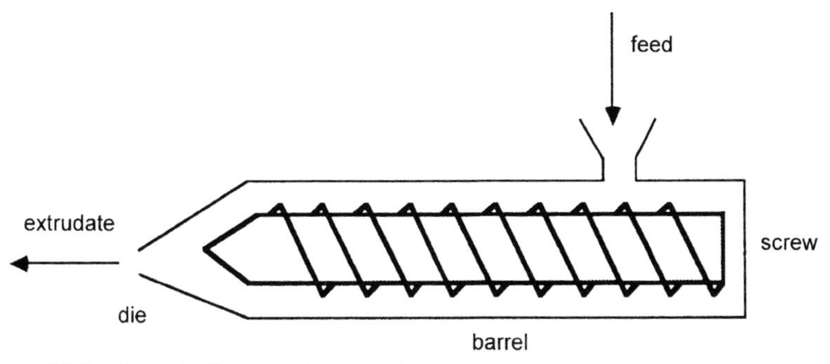

Fig. 8.14 Schematic diagram of an extruder.

pressure meters, mass flow meters, velocity flow meters and positive displacement pumps [30].

8.2.1.2 **Advantages of the Extrusion Process**

There are many benefits to using extruders to process food materials. They are capable of producing a wide variety of different product types and shapes, often with only small changes to the extruder, its operational settings or the raw materials used. From an engineering perspective, extruders can be described as a combination of a pump and a scraped surface heat exchanger for which operating conditions are relatively insensitive to material viscosity [32]. Thus extrusion systems are able to process highly viscous materials that are difficult or impossible to handle using conventional methods.

The ability of extruders to process biopolymers and ingredient mixes at relatively high temperatures (250 °C) and pressures (e.g. 25 MPa) with high shear forces and low moisture contents (10–40%) leads to a variety of fast and comparatively efficient chemical reactions and functional changes of the extruded material [33].

The ability of extrusion systems to carry out a series of unit operations simultaneously and continuously gives rise to savings in labour costs, floor space costs and energy costs whilst increasing productivity [34]. These production efficiencies, combined with the ability to produce shapes not easily formed with other production methods, have led to extensive use of extrusion in the food industry. An indication of the range of applications is given in (Table 8.2) [33].

Table 8.2 Extrusion cooking applications [33].

Bread crumbs	Degermination of spices
Precooked starches	Flavour encapsulation
Anhydrous decrystallisation of sugars to make confectioneries	Enzymic liquefaction of starch for fermentation into ethanol
Chocolate conching	Quick-cooking pasta products
Pretreated malt and starch for brewing	Oilseed treatment for subsequent oil extraction
Stabilisation of rice bran	Preparation of specific doughs
Gelatin gel confectioneries	Destruction of aflatoxins or gossypol in peanut meal
Caramel, liquorice, chewing gum	Precooked soy flours
Corn and potato snack	Gelation of vegetable proteins
Co-extruded snacks with internal fillings	Restructuring of minced meat
Flat crispbread, biscuits, crackers, cookies	Preparation of sterile baby foods
Precooked flours, instant rice puddings	Oilseed meals
Cereal-based instant dried soup mixes or drink bases	Sterile cheese processes
Transformation of casein into caseinate	Animal feeds
Precooked instant weaning foods or gruels	Texturised vegetable proteins

8.2.2 Extrusion Equipment

Extruders come in a wide variety of shapes, sizes and methods of operation, but can be categorised into one of three main types: piston, roller and screw extruders [32, 35]. The simplest of these is the piston extruder, which consists of a single piston or a battery of pistons that force the material through a nozzle onto a wide conveyor. The pistons can deliver very precise quantities of materials and are often used in the confectionery industry to deposit the centre fillings of chocolates. Roller extruders consist of two counter-rotating drums placed close together. The material is fed into the gap between the rollers that rotate at similar or different speeds and have smooth or profiled surfaces. A variety of product characteristics can be obtained by altering the rotation speeds of the rollers and the gap between them. This process is used primarily with sticky materials that do not require high-pressure forming. Screw extruders are the most complex of the three categories of extruder and employ single-, twin-, or multiple-screws rotating in a stationary barrel to convey material forward through a specially designed die. Amongst the screw type extruders, it is common to classify machines on the basis of the amount of mechanical energy they can generate. A low-shear extruder is designed to minimise the mechanical energy produced and is used primarily to mix and form products. Conversely, high-shear extruders aim to maximise mechanical energy input and are used in applications where heating is required.

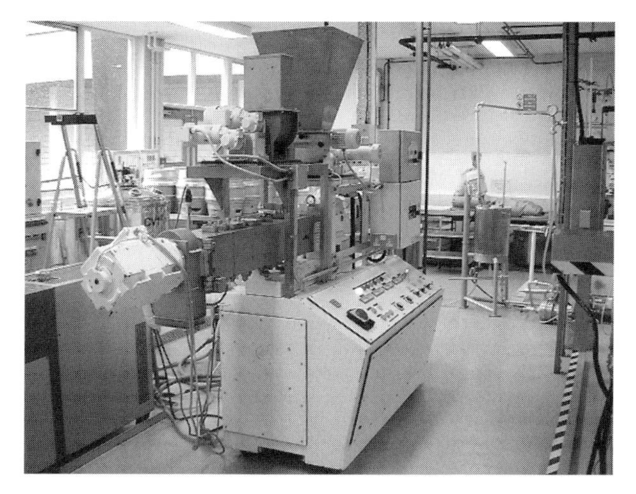

Fig. 8.15 A co-rotating twin-screw extruder (Continua 37; Werner and Pfleiderer, Stuttgart, Germany).

In the food industry, single- and twin-screw extruders predominate and hence are discussed in more detail.

Fig. 8.15 shows a co-rotating twin-screw extruder (Continua 37; Werner and Pfleiderer, Stuttgart, Germany).

8.2.2.1 **Single-Screw Extruders**

Single-screw extruders were first used in the 1940s to produce puffed snacks from cereal flours and grits. During transport through the extruder, mechanical energy from the rotation of the screw is converted to heat, raising the temperature of the mixture to over 150 °C. The resulting plasticised mixture is then forced through the die. The sudden reduction in pressure at the die causes moisture to flash off rapidly as steam, puffing the product. From the late 1950s, extrusion processes were developed to manufacture products such as dry expanded pet food, dry expanded 'ready to eat' breakfast cereals and textured vegetable protein.

Because single-screw extruders have relatively poor mixing ability, they are often used with materials that have been either premixed or preconditioned. Preconditioning is used to increase residence time, to reduce mechanical power consumption and/or to increase capacity. The preconditioner is an atmospheric or pressurised chamber in which raw granular food ingredients are uniformly moistened or heated or both by contact with live steam or water before entering the extruder.

The single-screw extruder relies upon drag flow to convey the feed material through the barrel of the extruder and to develop pressure at the die. In order for the product to advance along the barrel, it must not rotate with the screw.

The frictional force between the material and the barrel wall is the only force that can keep the material from turning with the screw and hence many single-screw machines have grooves cut in the barrel to promote adhesion to the barrel wall.

The rotation of the screw in the barrel gives rise to a second flow, called the cross-channel flow. This flow does not contribute to the net movement of material along the barrel but instead recirculates it within the screw flights and hence is responsible for some of the mixing action of the extruder.

In forcing the feed material along the barrel of the extruder and through the restricted opening of the die, a third flow known as the pressure flow is generated. The pressure flow causes movement backwards down the extruder barrel, causing further mixing of the product. The three flows combine to give the net flow of material out of the die.

Single-screw extrusion operation depends upon the pressure requirements of the die, the slip at the barrel wall (controlled by the barrel wall temperature, the presence of barrel wall grooves, or both) and the degree to which the screw is filled. Screw fill is dictated by feed rate, screw speed, melt characteristics and viscosity of the material extruded. The coupling of these variables limits the operating range and flexibility of single-screw extruders [36].

In the compression section of the screw, the compression ratio increases rapidly causing most of the mechanical energy used to turn the screw to be dissipated causing an increase in the temperature of the material. This results in the plasticisation of the dry feed ingredients. Energy input to the extrusion system may also arise from heat transfer through the jacket and latent heat from steam injected into the ingredients in the preconditioner.

8.2.2.2 Twin-Screw Extruders

Twin-screw extruders were introduced to the food industry in the 1970s and are now extensively used in food production.

In addition to manufacturing foods similar to those produced by single-screw extruders, twin-screw extruders have found a wide application in the food industry due to their better process control and versatility, their flexible design permitting easy cleaning and rapid product changeover and their ability to handle a wide variety of formulations.

Twin-screw extruders differ from the single-screw extruder in terms of their processing capability and mechanical characteristics and are largely responsible for the increasing popularity of extrusion processing. The screws in a twin-screw extruder are positioned adjacent to each other and are retained in position by a profiled barrel housing, having a horizontal ‘figure of eight’ appearance. The position of the screws in relation to one another and their direction of rotation, can be used to categorise twin-screw machines.

Twin-screw extruders can have intermeshing screws in which the flights of one screw engage the other or they can have nonintermeshing screws in which the threads of the screws do not engage one another, allowing one screw to turn without interfering with the other. Nonintermeshing screw extruders function like single-screw extruders but have a higher capacity.

Twin-screw extruders may have co-rotating or counter-rotating screws. Both co-rotating and counter-rotating extruders can have fully, partially or nonintermeshing screws. Co- and counter-rotating screws differ in their transport characteristics and are therefore suited to different technological applications [35, 37].

Intermeshing twin-screw extruders generally act like positive displacement pumps, forcing the material within the flights to move towards the die by rotation of the screws. The movement of the material is dependant upon the screw geometry and occurs independently of the operating conditions. Twin-screw intermeshing extruders have found wide applications due to their positive pumping action, efficient mixing and self-cleaning characteristics.

Intermeshing counter-rotating machines are particularly suited to the processing of relatively low-viscosity materials that require low screw speeds and a long residence time. Examples of products suited to this type of extruder include chewing gum, jelly and liquorice confections [35]. These extruders exhibit poor mixing characteristics as each half of the chamber housing the screws acts independently and thus two streams of material that have little interaction are generated. Hence the only mixing that is done is due to the recirculation within the chamber itself.

Although high pressures can be achieved in counter-rotating extruders if the screw speed is increased, large separating forces are generated at the interface between the screws, giving a calendaring effect which can cause excessive wear. Hence, the production of expanded products with counter-rotating extruders is considered uneconomic [30, 35].

Intermeshing co-rotating extruders are particularly suited to applications where a high degree of heat transfer is required but not forced conveyance and thus are widely used for the production of expanded products. In this type of extruder, the material being extruded is transferred from one screw to the other. The flow mechanism is a combination of both drag and positive displacement flow [38]. The self-wiping style of co-rotating extruders is most commonly used due to their high capacity and enhanced mixing ability [30]. Co-rotating extruders can be operated at higher speeds than counter-rotating extruders because the radial forces generated are more evenly distributed. The conveying ability of twin-screw extruders allows them to handle sticky and other difficult to handle ingredients [35].

In order to improve the mixing, heat transfer and viscous dissipation of mechanical energy, sections of the extruder are completely filled with material. To create these filled sections, some type of restriction is placed on the screw configuration. The addition of forward- or reverse-conveying discs into the screw configuration alters the pressure profile within the barrel. The forward-conveying discs push the material towards the die increasing the pressure in the barrel. Reverse-conveying discs reduce the pressure by delaying the passage of material through the extruder, increasing the barrel fill prior to the restriction, allowing additional processing and improved efficiency of heat transfer through the barrel wall. Restrictions in the screw configuration are placed under greater stresses and hence they tend to wear, requiring more frequent replacement than other elements of the screw.

Nonintermeshing twin-screw extruders can be described as two single-screw extruders sitting side by side with only a small portion of the barrels in common [39]. Like the single-screw extruder, these extruders rely on friction for extrusion.

8.2.2.3 Comparison of Single- and Twin-Screw Extruders

A comparison of single- and twin-screw extruders is shown in Table 8.3 [37].

For a given throughput, twin-screw extruders are 1.5 to 2.0 times as expensive as single-screw extruders, primarily due to the complexity of the screw, drive and heat-transfer jackets [40].

The preconditioning of feed ingredients with live steam is widely used in conjunction with single-screw extrusion processes and provides around half of the heat necessary for cooking/processing, the remainder of the heat being derived from the mechanical energy inputs. The rate of heat transfer to a material using the direct injection of live steam is very high and thus represents the lowest-cost method of heating the product. For this reason, single-screw extruders usually have lower energy costs than twin-screw extruders.

The cost of twin-screw extruders can, however, be offset by their ability to process at lower moisture levels, thus reducing or eliminating the need for additional postprocess drying [40].

The geometry and characteristics of the screws used in twin-screw extrusion present some advantages over single-screw extruders and enables the twin-screw extruder to process a wide range of ingredients that cause feeding problems for single-screw machines [39]. The conveying angle, combined with the self-wiping feature, results in an extruder that is less prone to surging. Increased uniformity of processing also occurs in twin-screw extruders due to the consistency of shear rate across the channel depth, leading to a narrower residence time distribution and increased mixing within the screw channel [30].

A single-screw extruder is relatively ineffective in transferring heat from barrel jackets because convective heat transfer is limited by poor mixing within the channel. Instead, the jackets in single-screw extruders control barrel wall temperature to regulate slip between the materials and the wall. Twin-screw extruders have considerably more heat exchange capability that expands their application to heating and cooling viscous pastes, solutions and slurries [40].

Single- and twin-screw extruders have found many different applications in the food industry. For example, single-screw extruders are considered an economic and effective method for the thermal processing and forming of pet foods, whilst twin-screw extruders have been widely used in snack food production, where better control and flexibility is required [37].

Table 8.3 Relative comparison of single- and twin-screw extruders [37].

Item	Single-screw	Twin-screw
Relative cost/unit capacity		
Capital cost of extruder	1.0	1.5–2.5
Capital cost of system	1.0	0.9–1.3
Relative maintenance	1.0	1.0–2.0
Energy		
With preconditioner	Half from steam	Generally not used
Without preconditioner	Mechanical energy	Mix of mechanical and heat exchange
Screw		
Conveying angle	10°	30°
Wear	Highest at discharge and transition sections	Highest at restrictions and kneading discs
Positive displacement	No	No, but approached by fully intermeshing screws
Self-cleaning	No	Self-wiping
Variable flight height	Yes	No
L/D	4–25	10–25
Mixing	Poor	Good
Uniformity of shear	Poor	Good
Relative RTD spread	1.2	1.0
Venting	Requires two extruders	Yes
Drive		
Relative screw speed	1.0–3.0	1.0
Relative thrust-bearing capacity	Up to 2.5	1.0
Relative torque/pressure	Up to 5.0	1.0
Heat transfer	Poor; jackets control barrel wall temperature and slip at wall	Good in filled sections
Operations		
Moisture	12–35%	6% to very high
Ingredients	Flowing granular materials	Wide range
Flexibility	Narrow operating range	Greater operating range

8.2.3
Effects of Extrusion on the Properties of Foods

8.2.3.1 Extrusion of Starch-Based Products

Cereal flours and other starchy materials are widely used as raw materials in the production of many extruded products. The physical characteristics of cereal fluids developed within an extruder and their extrudates are predominantly due to the starch component present, which usually represents between 50% and 80% of the dry solids in the mixture. Accordingly, many of the studies relating to cereal extrusion focus on changes occurring in the starch component of the product under varying extrusion conditions and the resulting effects on its physical, chemical and organoleptic properties. The type of extruder, feed moisture, feed rate, barrel temperature, screw speed, screw profile and die size are all important in developing the characteristics of the extruded product.

Cereals and starch-based products have been processed by extrusion since the initial introduction of the process to the food industry.

A study of the effect of feed moisture and barrel temperature on the extrusion of commercial yellow corn grits found that an increase in feed moisture or barrel temperature, up to 177 °C, causes an increase in WAI [41]. At temperatures above 177 °C WAI was found to decrease. Similarly, WSI was found to gradually increase with an increase in barrel temperature up to a value of 177 °C, with a more pronounced increase above this temperature. For any given set of operating conditions, a decrease in feed moisture content resulted in reduced WAI and increased WSI. Determination of the final cooked-paste viscosities showed maximum viscosity at 25% feed moisture.

A study carried out by Mercier and Feillet [42] looked at the modification of carbohydrate components in a range of cereal products, namely corn grits, corn, waxy corn, Amylon 5 and 7 (52% and 61% amylose), wheat and rice. The barrel temperature was varied from 70 °C to 225 °C with a moisture content of 22%. Their study showed a consistent increase in WSI with increasing barrel temperature, agreeing with earlier work [43]. Analysis of WAI for waxy corn showed a decrease with increasing barrel temperature from 70 °C to 225 °C, whilst the Amylon 5 and 7 samples showed little change until 200 °C, after which there was a sharp increase. WAI data for corn, wheat and rice products showed a gradual increase with barrel temperature, reaching a maximum at around 180 °C. At an extrusion temperature of 135 °C, the cooked viscosity of corn and rice starch was similar, while wheat starch gave a higher final cooked viscosity. Data obtained from the extrusion of starch with different amylose contents at different temperatures showed that the WSI and water-soluble carbohydrate increased less with increasing amylose content.

Research on the effect of several extrusion variables such as moisture, barrel temperature, screw geometry and screw speed on the gelatinisation of corn starch indicated that barrel temperature and moisture had the greatest effect on the gelatinisation of starch [43]. The results showed that the maximum gelatinisation occurred at high moisture and low barrel temperatures or vice versa. Higher screw speed reduced gelatinisation and was related to lower residence times. During the extrusion of cereal starches, it was found that starch was solubilised in a macromolecular form and no small oligo- or monosaccharides were formed [42]. Research on potato starch [44] showed that extrusion preferentially broke the α-1–4 linkages of amylose and not the outer chains of amylopectin. Linear oligosaccharides having no α-1-6 linkages were found.

The effect of feed moisture, barrel temperature, screw speed and die size on the gelatinisation of wheat flour during extrusion showed that the degree of starch gelatinisation increased sharply with increasing temperature, when the feed moisture contents were 24–27%, but increased more gradually when moisture contents were 18–21% [45]. At lower temperatures (65–80 °C) an increase in feed moisture content was found to cause a slight decrease in gelatinisation, whilst at higher temperatures (95–110 °C), an increase in moisture resulted in a significantly increased degree of gelatinisation. Increase in the screw speed was found to decrease the degree of starch gelatinisation, despite the increased shear. This was explained by the decrease in retention time of the sample in the extruder. Increasing the die size was found to reduce the degree of starch gelatinisation. This was explained by the possible decrease in the retention time of the sample in the extruder due to lower pressures and decreased surface shear.

At a constant moisture content, the effect of barrel temperature, feed rate and screw speed on starch gelatinisation during the extrusion of a yoghurt-wheat mix showed that barrel temperature had the most pronounced effect, followed by feed rate and screw speed [46].

A comparison of the appearance of starch granules from wheat semolina prior to and after extrusion at 60 °C showed little difference in the shape of the starch granules [47]. Increasing the barrel temperature resulted in a flattening of the granule although the original shape (unextruded) was still recognisable. A complete destruction of the granule was not seen until the barrel temperature reached 125 °C.

The physiochemical properties of several blends of raw, gelatinised and dextrinised commercial yellow corn flour were evaluated [48] and the results compared to those of raw commercial yellow corn flour extruded at a range of moisture contents. The extruded samples had properties similar to blends containing gelatinised and dextrinised corn. Reduction of the extrusion moisture level resulted in an increase in the relative proportion of dextrinised corn from 10% to 60%. It was suggested that dextrinisation is the dominant mechanism for starch degradation during extrusion, especially at low moisture contents.

Colonna et al. [49] extruded modified wheat starch and found this led to a macromolecular degradation of amylose and amylopectin, by random chain splitting. The water soluble fractions were composed of partly depolymerised amylose and amylopectin. It was also concluded that shear in the extruder completely disperses starch components by decreasing molecular entanglement.

A study [50] to evaluate the structural modifications occurring during extrusion cooking of wheat starch revealed that the amylopectin fraction of the starch was significantly degraded during extrusion and the degradation products were also macromolecules. It was felt that the structural modifications occurring during the extrusion was limited debranching, attributed to mechanical rupture of covalent bonds.

A study [51] of the effect of moisture content, screw speed and barrel temperature on starch fragmentation of corn starch found that carbohydrate from the extruded samples dissolved at a significantly faster rate than from native corn starch, with a pronounced difference in the amount of material solubilised within the first 2 h. It was suggest that this was indicative of an increase in the amount of linear polysaccharide and thus an increase in the degree of fragmentation. The amount of large molecular weight material decreased from 68% to a range of 24–58%, depending upon the extrusion conditions applied. Reductions in moisture content or temperature resulted in an increase in fragmentation, whilst decreasing the screw speed resulted in a decrease in fragmentation.

Bhattacharya and Hanna [52] studied the effect of moisture content, barrel temperature and screw speed on the textural properties of extrusion-cooked corn starch. Their results showed that the feed moisture content and barrel temperature affected the expansion of the extruded products predominantly. Increasing the moisture content reduced expansion in both waxy and nonwaxy corn starch samples. They suggest that this is due to a reduction in the dough/melt temperature that in turn reduces the degree of gelatinisation. An increase in the barrel temperature was found to increase the degree of expansion reflecting an increase in gelatinisation. Similar results were reported on the extrusion of potato flakes [53]. Screw speed was shown to be insignificant in its effect on expansion of the product. Although an increase in the screw speed increased the rate of shear and hence the degree of starch modification, this was accompanied by a reduction of residence time which cancelled out the effect of the additional shear. The shear strength of the products was found to increase with increasing moisture content, reflecting the decrease in expansion and increase in density.

The effect of screw speed on the extrusion of mixtures of yellow corn meal with wheat and oat fibres has been reported [54]. An increase in the screw speed was found to reduce the torque and die pressure, whilst increasing the specific mechanical energy and resulting in a decrease in radial expansion, with increases in axial expansion, bulk density and breaking force. It is suggested that the reduction in torque and die pressure is the result of a decrease in the screw fill. The reduction in radial expansion and the increases in axial expansion, bulk density and breaking force are reported to be due to the reduced die pressure and resistance to the flow of extrudate at the die. These results are contrary to those of Fletcher et al. [55], who observed an increase in die pressure and radial expansion with increasing screw speed. The conflicting results in these studies suggest that the textural characteristics may or may not be affected by screw speed, depending upon the feed material and the geometry and design of the screw used.

Although the extrusion characteristics of cereals are dominated by the physical and chemical changes occurring in the starch component, cereals typically contain 6–16% protein and 0.8–7.0% lipid, which can significantly affect the properties of the extruded product [56]. In addition to the components of the cereal itself, a wide variety of materials are incorporated into cereal extrusion mixes to modify the characteristics of the final product.

Protein Typically protein acts as a 'filler' in cereal extrudates and is dispersed in the continuous phase of the extrusion melt, modifying the flow behaviour and characteristics of the cooled extrudates. Proteinaceous materials hydrate in the mixing stage of the process and become soft viscoelastic doughs during formation of the extrusion melt. The shearing forces generated in the extruder cause breakage of the protein into small particles of roughly cylindrical and globular shapes. At levels of around 5–15%, they tend to reduce the extensibility of the starch polymer foam during its expansion at the die exit, reducing the degree of expansion [56].

Extrudates show a reduction in cell size with the addition of protein proportional to the amount of protein added. At higher levels of protein, severely torn regions in the cell walls of the extrudate are noted, indicating a loss of elasticity in the extrusion melt.

Martinez-Serna et al. [57] investigated the effects of whey protein isolates on the extrusion of corn starch. The isolates were blended with corn starch at concentrations between 0% and 20% and then extruded at varying barrel temperatures, screw speeds and moisture contents. Their results showed that an increase in barrel temperature increased the level of starch-protein interaction. Extrusion of the blend showed a 30% reduction in expansion when compared to extruded corn starch alone. They suggest that this is due to modification in the viscoelastic properties of the dough, as a result of competition for the available water between the starch and protein fractions, leading to a delay in starch gelatinisation. Increasing the screw speed and hence shear rate led to an increase in viscosity due to unfolding of the protein, involving rupture of covalent bonds or interaction with the starch. A similar trend was also observed when the protein concentration was increased.

Fat Fats and oils have two functions in starch extrusion processes. They act as a lubricant in the extrusion melt and modify the eating qualities of the final extruded product. The action of the extruder screws causes the oils to be either dispersed into small droplets or smeared on the polymers [56].

Extrusion of starches with low lipid contents, such as potato or pea starch, at low moisture contents (<25%) is extremely difficult, due to degradative dehydration of the starch polymers. This results in the formation of a very sticky melt, which tends to cause blockages. The addition of 0.5–1.0% of an oil to the starch reduces the degradation of the starch and enables extrusion without blockages [56].

The macromolecular modifications occurring in manioc starch extruded without and with a range of lipids (oleic acid, dimodan, copra and soya lecithin) have been investigated [58]. The results showed that an increase in extrusion temperature or screw speed increased the degree of macromolecular degradation of the native manioc starch. Addition of all lipids at 2% was shown to reduce the degree of macromolecular degradation, with all samples having higher intrinsic viscosities than the native starch extruded under the same conditions. Whilst all of the added lipids gave rise to an increase in the intrinsic viscosity, differences were apparent between each of the lipid-starch extrudates, suggesting different modes of action for each of the lipids studied.

A highly expanded oat cereal product is difficult to achieve due to its high fat content. The effects of the process conditions on the physical and sensory properties of an extruded oat-corn puff showed that increasing the level of oat flour (and hence fat content) caused an increase in the extrudate bulk density and a reduction in specific length and expansion [59].

An investigation into the improvement of extruded rice products showed that defatting of the flour resulted in improved expansion and lower bulk densities [60].

Sugars Sucrose and other sugars are commonly added to extruded products, particularly breakfast cereals. The level of sugar added to a product varies but is typically within the range 6–25 wt% on a final dry product basis. Sugar contributes to binding, flavour and browning characteristics and is important in controlling texture and mouthfeel. In addition, it can act as a carrier and potentiator of other flavours.

The effects of sucrose on the structure and texture of extrudates has been studied extensively.

When extrusion is carried out at moisture contents above 16%, addition of sucrose progressively reduces extrudate expansion, with an accompanying increase in product density. This effect was noted at sucrose concentrations as low as 2% when extruding with a feed moisture of 20% [61]. In addition to the reduction in expansion and increase in density, an increase in mechanical strength and the number of cells formed per unit area has been observed [62]. The structural changes in the extrudates brought about by the addition of sucrose have been attributed to competition for moisture, inhibition of gelatinisation and plasticisation of starch-based systems by low molecular weight constituents during extrusion.

The effect of sucrose and fructose on the extrusion of maize grits indicated inhibition of starch conversion due to a reduction in specific mechanical energy input [63]. In addition, a change in the packing of amylose-lipid complexes in the extrudate was noted. It is suggested that this rearrangement process is accelerated by the addition of the sugars due to the enhanced molecular mobility.

The effect of sucrose on both maize and wheat flour extrudates showed that, in contrast to maize extrudates, sucrose addition had little effect on the degree of starch conversion and sectional expansion of wheat flour extrudates [64]. It was postulated that the observed differences between the wheat-sucrose and maize-sucrose extrudates may be a result of particle size and the presence of gluten.

8.2.3.2 Nutritional Changes

Protein Extrusion cooking, like other food processing, may have both beneficial and undesirable effects on the nutritional value of proteins. During extrusion, chemical constituents of the feed material are exposed to high temperature, high shear and/or high pressure that may improve or damage the nutritional quality of proteins in the extruded materials by various mechanisms. These changes depend on temperature, moisture, pH, shear rate, residence time, their interactions, the nature of the proteins themselves and the presence of materials such as carbohydrates and lipids.

The proteins present in the feed material may undergo structural unfolding and/or aggregation when subjected to heat or shear during extrusion. Intact protein structures represent a significant barrier to digestive enzymes; and the combination of heat and shear is a very efficient way of disrupting such structures.

In general, denaturation of protein to random configurations improves nutritional quality by making the molecules more accessible to proteases and, thus, more digestible. This is especially important in legume-based foods that contain active enzyme inhibitors in the raw state.

Disulphide bonds are involved in stabilizing the native tertiary configurations of most proteins. Their disruption aids in protein unfolding and thus digestibility. Mild shearing can contribute to the breaking of these bonds.

Partial hydrolysis of proteins during extrusion increases their digestibility by producing more open configurations and increasing the number of exopeptidase-susceptible sites. Conversely, production of an extensively isopeptide cross-linked network could interfere with protease action, reducing the digestibility [65].

The Maillard reaction may take place during extrusion cooking of protein foods containing reducing sugars. The chemical reaction between the reducing sugars and a free amino group on an amino acid has important nutritional and functional consequences for extruded products. Maillard reactions can result in a decrease in protein quality, by lowering digestibility and producing nonutilisable products. During extrusion, the Maillard reaction is favoured by conditions of high temperatures ($>180\,^{\circ}C$) and shear (>100 rpm) in combination with low moisture ($<15\%$) [66].

Loss of total lysine and changes in the *in vitro* availability of amino acids in protein-enriched biscuits when extruded at different mass temperatures and moisture contents have been investigated [67, 68]. It was found that digestibility of the product extruded at 170 °C was not different from that of the raw material. However, increasing the mass temperature to 210 °C decreased the *in vitro* digestibility.

In a similar study [69], it was reported that screw speed was not a major factor in the retention of available lysine, but 40% of the available lysine present in the unextruded mix was lost during extrusion conditions above 170 °C mass temperature at a moisture content of 13%. The loss of lysine decreased when the water was increased to 18%, despite an increase in barrel temperature to produce an equivalent mass temperature to the 13% moisture mixture. Higher loses in lysine were observed with increased sucrose content and reduced pH.

Extrusion of corn gluten meal and blends of corn gluten meal and whey at various screw speeds and barrel temperatures showed an increase in the *in vitro* digestibility of the extruded product when compared to the raw material [70]. It was also found that the addition of whey had no significant effect on digestibility.

The protein nutritional value of extruded wheat flours was studied by Bjorck et al. [71]. Amino acid analysis showed that lysine retention was between 63% and 100%. It was found that the retention was positively affected by an increase in feed rate and negatively by an increase in screw speed. The authors felt that the prominent lysine damage under severe conditions was probably due to the formation of reducing carbohydrates through the hydrolysis of starch. The loss of other amino acids was found to be small.

The effects of a range of extrusion process variables on the *in vitro* protein digestibility of minced fish and wheat flour mixes have been studied [72]. The extrudates showed slight increases in their *in vitro* protein digestibility values. The authors found that, of the process variables studied, only the effects of feed rate and temperature of extrusion were significant.

The effect of extrusion on the *in vitro* protein digestibility of sorghum showed that varying screw speed and moisture content did not have a significant effect on the digestibility of sorghum, but temperature was significant in improving the digestibility of sorghum [73].

Dahlin and Lorenz [74] investigated the effect of extrusion on the protein digestibility of various cereal grains (sorghum, millet, quinoa, wheat, rye and corn). When the extrusion feed moisture was decreased from 25% to 15%, an increase in protein digestibility was observed. It was found that a screw speed of 100 rpm and a product temperature of 150 °C improved the *in vitro* protein digestibility of all cereals studied.

Extrusion of a yoghurt-wheat mixture at a constant moisture content of 43% showed no decrease in the *in vitro* protein digestibility, up to barrel temperatures of 120 °C and screw speeds of 300 rpm [75].

In addition to lysine loss and decrease in protein digestibility, the colour of an extruded product is another indication of the extent of the Maillard reactions. A correlation was found between the degree of browning of extruded wheat flour and the total lysine content of the samples [71]. It was also found that there was a positive correlation between Hunter *L* values for wheat-based breakfast cereals and *in vitro* protein digestibility and available lysine [76].

Vitamins The retention of vitamins in extrusion cooking generally decreases with increasing temperature, increasing screw speed, decreasing moisture, decreasing throughput, decreasing die diameter and increasing specific energy input [77].

The stability of thiamin, riboflavin and niacin during the extrusion cooking of full-fat soy flour has been studied [78, 79]. Different barrel temperatures and water contents of feed material with a residence time of 1 min did not affect riboflavin and niacin. However, some loss of thiamin was observed at high moisture ($<15\%$), high barrel temperature (<153 °C) and long residence times (<1 min).

The retention of thiamin and riboflavin in maize grits during extrusion has been studied [80]. The average retention was 54% for thiamin and 92% for riboflavin. These workers found that moisture content (13–16%) did not have any effect on the stability of thiamin. An increased degradation was seen for thiamin with increasing temperature and screw speed. Riboflavin degradation increased with increasing moisture content and screw speed.

Thiamin losses in extruded potato flakes at a range of barrel temperatures and screw speeds, different screw compression ratios and die diameters and a moisture content of 20% have been studied [53]. It was reported that thiamin loss did not exceed 15% in all runs.

The importance of moisture content on the retention of thiamin in potato flakes during extrusion indicated that, at moisture contents of 25–59%, retention ranged from 22% to 97% [81]. The retention was poor at low moisture contents.

Thiamin losses in extruded legume products with increase in process temperature (93–165 °C), pH (6.2–7.4) and screw speed (10–200 rpm) have been reported [82]. The retention of thiamin increased with higher moisture contents (30–45%).

The effect of throughput and moisture content on the stability of thiamin, riboflavin, B_6, B_{12} and folic acid when extruding flat bread showed that vitamin stability improved with increased throughput and increased water content [83].

Guzman-Tello and Cheftel [84] studied the stability of thiamin as an indicator of the intensity of thermal processing during extrusion cooking. They found that retention of thiamin decreased from 88.5% to 57.5% when the product temperature increased from 131 °C to 176 °C. Other extrusion parameters were kept constant. More thiamin was retained by increasing the moisture content (14.0–28.5% wet basis), while higher screw speeds (125 rpm and 150 rpm) caused higher losses.

At screw speeds up to 300 rpm at a barrel temperature of 120 °C, no loss of thiamin and riboflavin was observed when extruding a yoghurt-wheat mixture at a moisture content of 43% [75].

Although extrusion systems can be successfully used with higher vitamin retentions, food manufacturers often carry out vitamin fortification postextrusion, by dusting, enrobing, spraying or coating.

8.2.3.3 Flavour Formation and Retention During Extrusion

Much of the flavour, or the volatile components of flavour are either lost to the atmosphere as the extrudate exits the die or they become bound to starch or proteins during extrusion. Because of the high losses, flavours are generally added after extrusion. Postextrusion flavouring processes have many disadvantages, including difficulty in obtaining even application, possibility of contamination and rubbing off of the coating in the package or consumer's hands. A limited range of encapsulated and extrusion stable flavours have been developed to address these problems. However, these are often restricted in application due to their cost. An alternative approach to the problem of flavouring extruded products involves the addition of reactive flavour precursors to the extrusion mix. The conditions applied during the extrusion process cause a reaction of the precursors to produce the desired flavour compounds. These precursors are usually compounds known to participate in the browning and flavour reactions that occur normally during the extrusion process.

In the extrusion of cereal-based products, flavours are predominantly generated by nonenzymic browning reactions, typified by caramelisation, the Maillard reaction and oxidative decomposition.

The temperature and shear conditions generated in the extruder provide the physical and chemical means whereby starch and protein can be partially degraded to provide the reactants that can then participate in nonenzymic browning reactions. In addition, lipids (especially unsaturated lipids) may undergo thermal degradation, thereby providing additional flavour compounds.

An evaluation of the volatile compounds and colour formation in a whey protein concentrate-corn meal extruded product showed 71 volatile compounds in the headspace of the samples, 68 of which were identifiable [85]. The compounds present included 12 aldehydes, ten ketones, six alcohols, two esters, six aromatics and ten hydrocarbons. In addition, 11 pyrazines, four furans, five heterocyclics and two sulphur compounds were isolated. They suggest that these compounds were products of the Maillard reaction and are possibly the most important in contributing to the corn flavour of the product. The concentrations of pyrazines, furans and other heterocyclics were found to increase with increasing levels of whey protein in the extrudates.

Nair et al. [86] identified 91 compounds in the condensate flash off at the die and 56 compounds from the headspace of the extrudate of corn flour. They suggest that the difference in the composition of the two samples is a reflection of the volatility with water of the compounds. Whilst their study identified a significant number of compounds formed during extrusion were retained in the extrudate, it was concluded that flash off at the die is still a major hindrance to the flavouring of extruded products.

The effects of product temperature, moisture content and residence time on the aroma volatiles generated during the extrusion of maize flour have been investigated by Bredie et al. [87]. Their results showed that low temperatures and high moisture contents favoured the production of volatiles associated with lipid degradation. Increasing both the temperature and residence time, along with a reduction in moisture content, was reported to increase the production of compounds derived from the Maillard reaction whilst reducing the levels of those compounds associated with lipid degradation.

The effect of pH on the volatiles formed in a model system, containing wheat starch, lysine and glucose, showed that both the total yield and number of compounds formed are greater at pH 7.7 than at pH 4.0 and pH 5.0, where the total yields are similar [88]. At pH 7.7, the volatile compounds from the extrudate were dominated by pyrazines that gave the extrudates a nutty, toasted and roasted character. Modification to pH 4.0 or pH 5.0 reduced the production of pyrazines and increased the production of 2-furfural and 5-methylfurfural, which accounted for over 80% of the total volatiles. It is suggested that the presence of these compounds is an indication of starch degradation, leading to an additional source of carbohydrate precursor during extrusion. They conclude that it may be possible to control flavour and colour development by careful control of the processing conditions.

Using a trained sensory panel to study the aroma profile of extrudates produced from wheat flour and wheat starch fortified with mixtures of cysteine, glucose and xylose [89], a vocabulary of 24 odour attributes was developed, with

17 of the attributes significantly differentiating the samples. The results indicated that extrusion of the wheat flour alone resulted in an extrudate aroma that was characterised by the terms biscuity, cornflakes, sweet and cooked milk. Addition of cysteine and glucose resulted in major differences in aroma, with the cereal terms popcorn, nutty/roasted and puffed wheat predominating. Cysteine/xylose mixtures gave rise to sulphur odour notes, such as garlic-like, onion-like, rubbery and sulphury, with the less desirable terms acrid/burnt, sharp/acidic and stale cooking oil becoming apparent.

A more extensive study using wheat starch, cysteine, xylose and glucose extruded at a range of pH and target die temperatures has been carried out [90]. Extrudates prepared using glucose were more frequently described as biscuity and nutty, whereas those prepared using xylose were commonly described using the terms meaty or onion-like. Analysis of the extrudates identified 80 compounds. Yields of the compounds formed were generally higher in extrudates prepared using xylose than those using glucose, under the same processing conditions. Increases in temperature and pH were reported to increase the yields of the compounds formed. Pyrazines and thiophenes were amongst the most abundant classes of compounds identified. Results from the analysis of the extrudates prepared using glucose showed aliphatic sulphur compounds and thiozoles to feature strongly in the aroma, whereas use of xylose gave rise to both nonsulphur-containing and sulphur-containing furans.

8.3 Frying

Pedro Bouchon

8.3.1 General Principles

Deep-fat or immersion frying is an old and popular process, which originated and developed around the Mediterranean area due to the influence of olive oil [91]. Today, numerous processed foods are prepared by deep-fat frying all over the world since, in addition to cooking, frying provides unique flavours and textures that improve the overall palatability. This section briefly describes the frying process from industrial and scientific perspectives. First, it introduces the process and presents some fried food quality characteristics. Thereafter, it describes the equipment and oils used in frying and some features of potato chip and potato crisp production. Finally, this section reviews main research in the field.

8.3.1.1 The Frying Process

Deep-fat frying can be defined as a process for the cooking of foods, by immersing them in an edible fluid (fat) at a temperature above the boiling point of water [92]. Frying temperatures can range from 130 °C to 190 °C, but the most common temperatures are 170–190 °C. Immersion frying is a complex process that involves simultaneous heat and mass transfer resulting in counterflow of water vapour (bubbles) and oil at the surface of the food. In addition, frying induces physicochemical alterations of major food components and significant microstructural changes [93]. In fact, most of the desirable characteristics of fried foods are derived from the formation of a composite structure: a dry, porous, crisp and oily outer layer or crust, with a moist, cooked interior or core. The crust is the result of several alterations that mainly occur at the cellular and subcellular level and are located in the outermost layers of the product. These chemical and physical changes include: physical damage produced when the product is cut and a rough surface is formed with release of intracellular material, starch gelatinisation and consequent dehydration, protein denaturation, breakdown of the cellular adhesion, water evaporation and rapid dehydration of the tissue and, finally, oil uptake itself.

Dehydration, high temperatures and oil absorption distinguish frying from simmering, which occurs in a moist medium and where the temperature does not exceed the boiling point of water. During baking, heat transfer coefficients are much lower than during frying and, although there is surface dehydration and crust formation, there is no oil uptake. In addition, the high temperatures achieved during frying (usually more than 150 °C) allow enzyme inactivation, intercellular air reduction and destruction of microorganisms, including pathogens [94].

8.3.1.2 Fried Products

A wide variety of food materials is used to make fried products, including meats, dairy, grains and vegetables. Different shapes and forms, such as chips, crisps, doughnuts, battered and breaded food, among others, can be found in the market. Therefore, frying technology is important to many sectors of the food industry: including suppliers of oils and ingredients, fast-food shop and restaurant operators, industrial producers of fully fried, parfried and snack foods and manufacturers of frying equipment [95].

The quantities of food fried and oils used at both the industrial and commercial levels are huge. For example, the USA produces, on average, over 2.3×10^6 t of sliced frozen potato and potato products every year, the majority of which are fried or partially fried [96]. Commercial deep-fat frying has been estimated to be worth £45×10^9 in the USA and at least twice this amount for the rest of the world [95]. In the UK, the crisp market is currently worth £693×10^6, while all other snack products together are worth £751×10^6 [97].

One of the most important quality parameters of fried products is the amount of fat absorbed during the frying process [98]. Per capita consumption of oils

Table 8.4 Oil absorption in fried foods.

Food item	Fat (g) absorbed in 100 g edible portion
Frozen chips	≈ 5
Fresh chips	≈ 10
Battered food (fish/chicken)	≈ 15
Low fat crisps	≈ 20
Breaded food (fish/chicken)	15–20
Potato crisps	35–40
Doughnuts [a)]	15–20

a) Doughnuts also contain about 10% fat used in preparation of the dough.

and fats was estimated to be 62.7 lb year^{-1} (28.4 kg year^{-1}) in the USA, far exceeding the recommendations found in The Surgeon General's Report on Nutrition and Health [99]. Excess consumption of fat is considered as the key dietary contributor to coronary heart disease and perhaps cancer of the breast, colon and prostate [100]. Therefore, consumer trends are moving towards healthier food and low-fat products, creating the need to reduce amount of oil in end products. Despite such market forces, the consumption of snack food is increasing in developed and developing countries, and fried products still contain large amounts of fat. An example of the total oil content of selected snack and fast foods is presented in Table 8.4 (from [101]). Most of these products have an oil content varying from 5% (frozen chips) to 40% (potato crisps).

In addition, recently, Swedish scientists sounded an alarm in April 2002, when they discovered that certain cooked foods, particularly potato crisps and potato chips, contained high levels of acrylamide, a chemical compound that is listed by the World Health Organization (WHO) as a probable human carcinogen [102]. Acrylimide is a substance used to produce plastics and dyes that can be produced when foods are heated above 120 °C, due to a reaction between amino acids and reducing sugars [103]. Today, there is no concensus about whether acrylamide in food is a danger and the WHO has not called yet for any reduction in foods containing high levels of this compound.

As mentioned, there is an increased demand for low-fat products. However, reduced and no-fat crisps varieties represent only 11% of potato crisp market sales in the USA [104]. Sales of fat-free snacks are increasing but, as these products are baked rather than fried, they have different flavour and textural characteristics to fried crisps and, therefore, consumer acceptance is low. However, low-fat snacks, such as crisps or tortilla chips, are acquiring greater acceptance. These products are usually dried prior to frying and research is focused in developing a product with enough fat to impart the desired organoleptic properties.

8.3.2 Frying Equipment

Frying equipment can be divided into two groups: (a) batch frying equipment, which is used in small plants and catering restaurants, and (b) continuous frying equipment, which is used on the industrial scale to process large amounts of product.

8.3.2.1 Batch Frying Equipment

A batch fryer consists of one or more chambers with an oil capacity in the range 5–25 l. The oil can be heated directly by means of an electrical resistance heater that may be installed a few centimetres above the bottom of the fryer. This enables a cool zone to be formed at the bottom of the vessel where debris can fall, remaining there and minimising oil damage. The fryer can also be heated by direct gas flames underneath the bottom of the vessel, although this arrangement makes the provision of a cold zone under the heaters difficult [101]. Recently developed high-efficiency fryers include turbojet infrared burners which use less than 30–40% energy than standard gas-fired fryers with the same capacity [104]. Modern batch fryers are constructed with high-grade stainless steel; and no copper or brass is used in any valve fitting or heating element, to avoid oxidation catalysis. Usually, the operators immerse and remove the baskets manually from the oil, but new equipment may include an automatic basket lift system that rises automatically when the frying time is finished. Removal of food scraps and oil filtration is an essential practice that needs to be carried out on a daily basis, to increase oil shelf life and avoid smoking, charring and off flavours. New equipment can also have an inbuilt pump filtration unit for the removal of sediments [105]. Some catering outlets, especially those devoted to chicken frying, prefer to use pressure batch fryers. These reduce frying time considerably and the fried products have a high moisture content, with a uniform colour and appearance [104].

8.3.2.2 Continuous Frying Equipment

Large-scale processing plants use continuous fryers. These are automated machines that consist of a frying vessel where oil is maintained at the desired temperature, a conveyor that displaces the product through the unit and an extraction system that eliminates the fumes, primarily made up of moisture and a fine mist of fatty acids.

The oil can be heated either directly by means of an electric heater or a battery of gas burners, or indirectly, by pumping a heated thermal fluid into the pipes immersed in the oil bath. Some fryers are equipped with external heat exchangers. In those systems, the oil is continuously removed at the discharge end of the tank, pumped through a filter unit and then through an external heat exchanger, before it is returned to the receiving end of the vessel (see

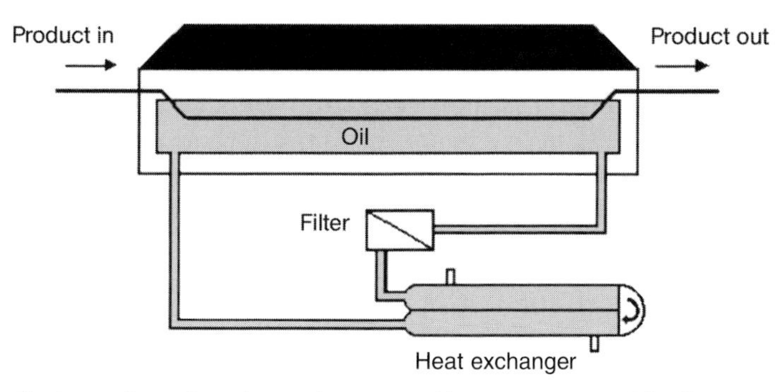

Fig. 8.16 Scheme for a fryer with an external heating system and fat filter unit.

Fig. 8.16). Some continuous fryers are designed with multiple heating zones along the fryer that can be adjusted separately, providing optimal temperature control to improve product quality.

As the oil is absorbed by the fried product, it has to be made up with fresh oil continuously. The amount of fresh oil added to the vessel defines the oil turnover [105]:

$$\text{Oil turnover} = \frac{\text{weight of oil in fryer}}{\text{weight of oil added per hour}}$$

which represents the time needed to replace all the oil contained in the fryer. A fast oil turnover is desired as it maintains satisfactorily the level of free fatty acids, preserving oil quality for longer periods of time. Normally, the oil turnover is kept at 3–8 h [105].

Industrial fryers have oil capacities ranging from 200 kg to 1000 kg and can produce up to 25 000 kg h^{-1} of chips and 2500 kg h^{-1} of crisps. Some of the main frying equipment manufacturers are Florigo B.V. in The Netherlands, Heat and Control and Stein in the USA.

In the last decade, the continuous vacuum frying system developed by Florigo B.V. during the 1960s has been reintroduced. The equipment was first created to produce high-quality chips but, due to the improvement in blanching technology and raw material quality, the use of this technology almost disappeared [104, 106]. Nowadays, vacuum frying technology is being used to maintain natural colours, flavours and nutrients in products with high added value, such as vegetables and fruits, due to the fact that much lower temperatures can be applied during frying.

8.3.2.3 **Oil-Reducing System**

Given the current concern with lowering the diet fat contents, an oil-reducing unit has been developed. The equipment is mounted at the discharge end of the fryer and removes the fat excess from the recently fried food, using a steaming and drying technology. Superheated steam at 150–160 °C, which is circulated by fan through the heat exchangers, flows through the product bed and removes nonabsorbed surface oil from the hot surface of the food. Consecutively, the oil-vapour mixture is filtered and the oil is pumped back to the fryer. The low-fat stripping system can reduce oil content in crisps by 25% [107]. Units range from batch strippers for pilot plants or product development to continuous production units.

8.3.3 Frying Oils

Food can be fried in a wide range of oils and fats, including vegetable oils, shortenings, animal fats or a mixture thereof. The main criteria used to select frying oils are: long frying stability, fluidity, bland flavour, low tendency to foam or smoke formation, low tendency to gum (polymerise), oxidative stability of the oil in the fried food during storage, good flavour stability of the product and price [101, 105]. The fat melting point is a very important parameter, as it affects the temperature of heating in tanks and pipes during storage and handling; and also it may affect the sensory attributes of fried products that are eaten at colder temperatures. Conversely, fats with a high melting point frequently have less tendency to oxidation.

Frying oils are principally from vegetable sources. Some traditional oils used for frying are: corn, cottonseed and groundnut oils, which are used as a stable source of polyunsaturated fatty acids due to their low linolenic acid content [108]. However, over recent decades, the use of groundnut oil has diminished due to its cost and also due to production problems related to naturally occurring aflatoxins [101, 105].

Nowadays, palm oils are increasingly used in industrial frying. Palm oil is a semisolid fat, which is fractionated at low temperature to give a liquid fraction called palm olein with a melting point that ranges between 19 °C and 24 °C and a fraction called palm stearin, with a high melting point, melting at no less than 44 °C. Also, a double-fractioned oil called super olein can be obtained, which has a melting point range of 13–16 °C. Palm olein is liquid enough for frying use; however, super olein is the grade used to produce fully liquid frying oil blends, together with sunflower and groundnut oils. Palm oil and palm olein have a very good frying performance due to their high resistance to oxidation and flavour reversion because of their low unsaturation; and they are now becoming virtually the standard oils for industrial frying in West Europe [109]. However, the high level of saturated (palmitic) fatty acids (38.2–42.9% in palm olein and 40.1–47.5% in palm oil) may be criticised from a nutritional view point.

Rapeseed (canola) and soyabean oils have a high level of linolenic acid (8–10%), making them vulnerable to oxidation and off flavour development; and therefore they can be slightly hydrogenated for industrial frying. This procedure can also be applied to sunflower oil and may be attractive where an oil with a high polyunsaturated to saturated ratio is needed for dietary purposes [101].

Olive oil has excellent attributes that make it suitable as a frying oil. It has a low level of polyunsaturated fatty acids and a mixture of phenolic antioxidants that make it resistant to oxidation. Extra-virgin and virgin olive oils are expensive for industrial use, although refined solvent-extracted olive oil can be satisfactory for industrial frying.

Animal fats are also used for frying in some regions, due to the characteristic flavours that they impart to the fried food, despite their high level of saturated fatty acids. Fish oils are rarely used for frying, as their high level of long-chain polyunsaturated fatty acids makes them prone to oxidation [101].

Recently some fat substitutes, such as sucrose polyesters (olestra), have been extensively studied. In fact, Olean (Procter & Gamble's brand name for olestra) has been introduced in the USA and several snacks fried in olestra are already in the market (fat-free Pringles, Frito-Lay Wow). Olestra is synthesised from sucrose and fatty acid methyl esters and has no calories because its structure prevents digestive enzymes from breaking it down. Two main negative aspects are currently discussed in relation to olestra that impair its acceptance in other countries. First, olestra can cause gastrointestinal discomfort in some people and, second, there is some debate about the ability of olestra to reduce the absorption of fat-soluble nutrients [104].

Antioxidants such as tertiary butyl hydroquinone (TBHQ), butylated hydroxyanisole (BHA) and butylated hydroxytoluene (BHT) are usually added to improve oil stability. TBHQ is regarded as the best antioxidant for protecting frying oils against oxidation and, like BHA and BHT, it provides carry-through protection to the finished fried product.

8.3.4 Potato Chip and Potato Crisp Production

Potato is the primary raw material used in the frying industry. The storage organ and hence the food part of potato is the tuber, essentially a thickened underground stem, which is made of 2% skin, 75–85% parenchyma and 14–20% pith [110]. The anatomy of the potato plant has been described in detail by Artschwager [111, 112], from Talburt [113]. In terms of their composition, potatoes are mainly made up of water, having an average of 77.5% (range 63.2–86.9%) [113]. The chemical composition of the remaining solid part can greatly vary, depending on a wide range of factors, including variety, maturity, cultural practice, environmental differences, chemical application and storage conditions. Starch represents 65–89% of the dry matter weight; and amylose and amylopectin are usually present in a 1:3 ratio [114]. Starch granules are ellipsoidal in shape, about 100×60 μm, much larger than the average starch granules of cereal grains.

For potato fried products, potatoes with high solids content (20–22%) are preferred, as they result in better finished product texture, higher yields and lower oil absorption [115]. Also, low reducing sugar contents are required to minimise colour development during processing, which is generated by the Maillard, nonenzymatic browning reaction [116].

8.3.4.1 Potato Chip Production

Potato chips are traditionally produced by cutting potato strips from fresh potatoes (parallelepiped of 1×1 cm cross-section by 4–7 cm in length), which are then deep-fat fried. Three major kinds of chip are produced at a commercial scale: (a) deep-frozen completely fried chips, which just require oven heating, (b) deep-frozen partially fried chips, which require additional frying before eating, and (c) refrigerated partially fried chips, which have a short shelf life and need additional frying [117]. A summary of the production process is described below.

The technology of chip production progressively improves as the industry develops modern equipment and entire technological lines for its manufacture [117]. Processors of frozen chips wash and peel potatoes with lye or steam, as abrasion peeling results in higher losses. Peeled potatoes are conveyed over trimmers and cut. Strip cutters orient potatoes along the long axis, in order to obtain the greatest yield of long cuts. Subsequently, chips are blanched in hot water prior to frying. The usual range of water temperatures is 60–85 °C. The positive effects of blanching include: a more uniform colour of the fried product, reduction of the frying time, since the potato is partially cooked, and improvement of texture of the fried product [118]. After blanching, excess water is removed, in order to minimise the frying time and lower the oil content of the product. This is carried out by means of dewatering screens and, subsequently, blowing warm air in continuous dryers. Afterwards, potato strips are parfried in a continuous fryer. The frying time is controlled by the rate of conveyor movement, oil temperature, dry matter content in the potato tubers, size of strips and type of processed chip (parfried or finish fried). The most common tem-

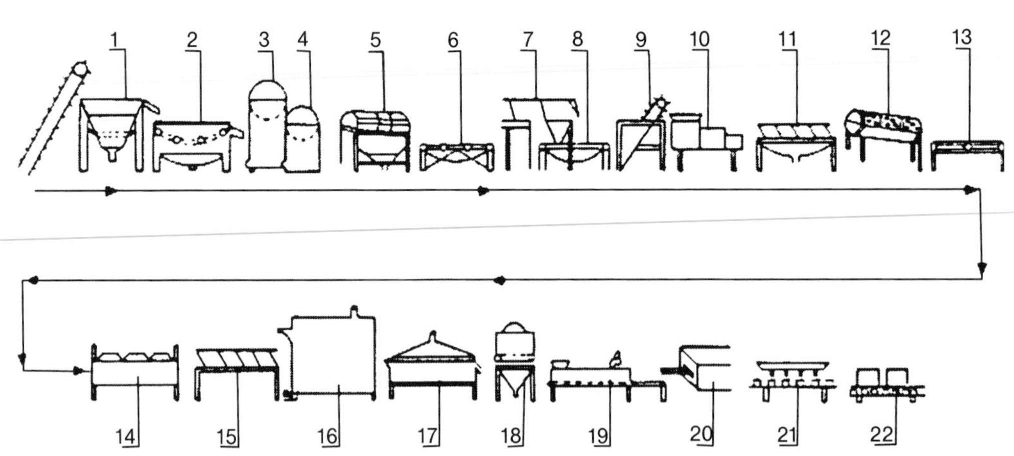

Fig. 8.17 Flow sheet for frozen chips: 1, trash remover; 2, washer; 3, preheater; 4, lye peeler; 5, washer; 6, trimming belt; 7, 8, size grader; 9, surge hopper; 10, strip cutter; 11, sliver eliminator; 12, nubbin eliminator; 13, inspection belt; 14, blancher; 15, dewatering screen; 16, dryer; 17, fryer; 18, deffater; 19, cooler; 20, freezer; 21, package filler; 22, freezing storage; from [117].

perature range falls within 160 °C and 180 °C. Temperatures above 190 °C are not used, due to the possibility of more rapid oil breakdown. The recommended frying parameters are, for finish-fried chips, 5 min at 180 °C and, for parfried, 3 min at 180 °C [117].

The excess fat is removed by passing the product immediately after emerging from the fryer over a vibrating screen, allowing the fat to drain off and, thereafter, the product is air-cooled for about 20 min while it is conveyed to the freezing tunnel. Finally, the product is packed in polyethylene/polypropylene bags or in cartons for the retail trade. A flow sheet for frozen chips is presented in Fig. 8.17.

8.3.4.2 **Potato Crisp Production**

Snack food manufacturers produce two types of potato crisps: traditional crisps made by thinly slicing fresh potatoes and crisps processed from potato dough and formed into potato crisp shapes. Processing lines for potato crisps include similar steps as in potato chip production. Potatoes are washed, peeled and sliced, commonly using a rotary slicer. Shape and thickness can be varied to meet marketing needs, but thickness is usually in the range 50–60 thou (1.27–1.52 mm). To remove excess starch, the potato slices are washed and dried on a flat wire conveyor to remove as much surface starch and water as possible [119]. Some potato processing plants use blanching prior to frying to improve the colour of the crisp. The blanching solution is heated to 65–95 °C and blanching takes around 1 min. Then excessive water is removed. Thereafter, potatoes are usually fried in a continuous fryer, where they remain from 1.5–3.0 min at 170–190 °C, until the moisture level is less than 2% of the total weight. The frying time depends on the flow of slices to the fryer, the initial moisture level of the potato and the desired browning. Subsequently, crisps are conveyed, allowing excess oil to drain off, and are salted or flavoured. Finally, crisps are cooled on a conveyor and sorted by size before packaging.

Crisp products, made from potato dough, are normally based on starch-containing ingredients such as potato flakes, flours, starches, ground slices, meals, granules or mixtures thereof. Dry ingredients are mixed, normally followed by the addition of the liquid ingredients. Thereafter, the mixture is introduced to the sheeting line, where they are formed into discrete pieces. Subsequently, the product is fried in a continuous fryer, following a similar procedure to the one described above, for potato crisp manufacture. Production lines can vary depending on the final requirements and uses of the final product. Restructured potato crisps may not have similar flavour and textural characteristics to fresh potato crisps, but they have the advantages of uniformity and absence of defects.

8.3.5
Heat and Mass Transfer During Deep-Fat Frying

Deep-fat frying is a thermal process, in which heat and mass transfer occur simultaneously. A schematic diagram of the process is shown in Fig. 8.18, where it can be observed that convective heat is transferred from the frying medium to the surface of the product and, thereafter, conductive heat transfer occurs inside the food. Mass transfer is characterised by the loss of water from the food as water vapour and by the movement of oil into the food [93].

Farkas et al. [120] observed that the temperature at any location in the core region is limited to values below the boiling point of the interstitial liquid (approximately 105 °C). When all the liquid is evaporated from the region, the moving front propagates towards the interior and the temperature begins to rise, approaching the oil temperature. On the basis of visual observations and analysis of temperature profiles and moisture data, they suggested that the frying process is composed of four distinct stages:

1. Initial heating, which lasts a few seconds and corresponds to the period of time whilst the surface temperature reaches the boiling point of the liquid. Heat transfer is by natural convection and no vaporisation of water occurs.
2. Surface boiling, which is characterised by the sudden loss of water, the beginning of the crust formation and a forced convection regime due to high turbulence, associated with nucleate boiling.
3. Falling rate, which is the longest, in which the internal moisture leaves the food, the core temperature rises to the boiling point, the crust layer increases in thickness and finally the vapour transfer at the surface decreases.
4. Bubble end point.

Natural convective heat transfer coefficients range over 250–280 W m^{-2} K^{-1} at frying temperatures according to estimations from Miller et al. [121] and Moreira et al. [122], whereas forced convective heat transfer coefficients can achieve maximum values up to two or three times bigger that those measured in the absence of bubbling [97, 123–125]. This value gradually decreases over the duration of the process.

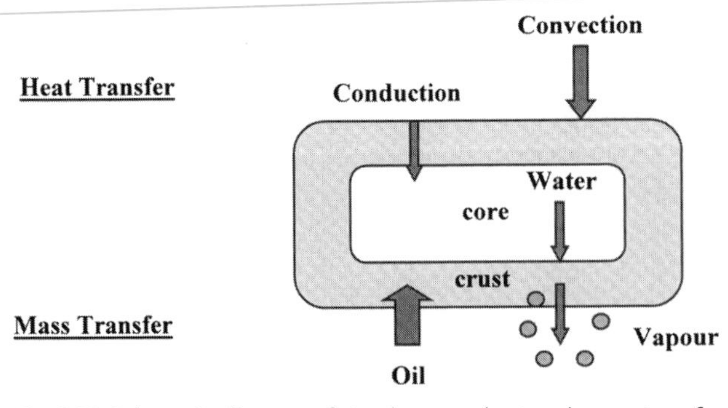

Fig. 8.18 Schematic diagram of simultaneous heat and mass transfer during frying.

8.3.6 Modelling Deep-Fat Frying

Many of the studies related to frying have been limited fundamentally to observations on the frying process. Mathematical models have been centred on the prediction of temperature profiles inside the fried food and the prediction of the kinetics of moisture loss; and very little has been carried out in relation to oil uptake. Models with different levels of complexity have been reported, which vary considerably in terms of the assumptions made. Most of the models for water evaporation during frying consider the fried product as a single phase, with no differentiation between the crust and the core regions and without the existence of a moving crust/core interface, where either energy or mass diffusion are considered to be the rate controlling mechanisms [126–135].

However, some authors have recognised the existence of two separate regions, the crust and the core, and have proposed a model considering the presence of a moving boundary [97, 120, 136, 137]. In fact, Farkas et al. [120] were the pioneers in developing a comprehensive model of thermal and moisture transport during frying. They formulated their model by analogy to freezing [138] and to the solution of the uniform retreating ice front during freeze-drying [139]. They provided different sets of equations for the two regions, separated by a moving boundary where the evaporation occurred. In their model they did not include the oil phase. They described the heat transfer in both regions using the unsteady heat transfer conduction equation, they considered water diffusional flow within the core region and they assumed that water vapour movement was pressure-driven. The final set of equations consisted of four nonlinear partial differential equations, which were solved using finite differences. The results were compared with experimental data; and they obtained a reasonable prediction for temperature profiles, water content and thickness of the crust region [140]. However, as Farid and Chen [136] pointed out, simulations were time-consuming, making it difficult to extrapolate the model to a multidimensional geometry and therefore they simplified the model. A similar approach was assumed by Bouchon [97], when modelling water evaporation in an infinite slab and in an infinite cylinder.

Ni and Datta [137] developed a multiphase porous media model to predict moisture loss, oil absorption and energy transport in a potato slab, also considering a moving front. They assumed that vapour and air transport were considered to be driven by convective and diffusive flows, while liquid water and oil were supposed to be driven by convective and capillary flows. Model predictions were compared with literature experimental data, mainly from Farkas et al. [140]. They centred their validation on temperature profiles, moisture content and crust thickness predictions, however, they did not include oil uptake absorption as part of their model validation.

8.3.7
Kinetics of Oil Uptake

The selection of a model needs to be in accordance with experimental observations that reveal how oil absorption takes place. It is not clearly understood yet when and how the oil penetrates into the structure, however it has been shown that most of the oil is confined to the surface region of the fried product [141–147] and there is strong evidence that it is mostly absorbed during the cooling period [132, 148–150].

Gamble et al. [151] gave a reasonable initial explanation of the deep-fat frying mechanism. They proposed that most of the oil is pulled into the product when it is removed from the fryer due to condensation of steam producing a vacuum effect. They suggested that oil absorption depends on the amount of water removed and on the way this moisture is lost. In accordance, Ufheil and Escher [149] suggested that oil uptake is primarily a surface phenomenon, involving an equilibrium between adhesion and drainage of oil upon retrieval of the slice from the oil.

Moreira and Barrufet [152] explained the mechanism of oil absorption during cooling in terms of capillary forces. The model considered that during frying heat transfer was controlled by convection and conduction and that during cooling mass transfer from the surface oil and the surrounding air into the structure was controlled by capillary pressure.

Recently, Bouchon and Pyle [153] developed a model for oil absorption during the cooling period, which considered the competition between drainage and suction. A key element in the model was the hypothesis that oil suction would only begin once a positive pressure driving force had developed.

8.3.8
Factors Affecting Oil Absorption

There has been much research to examine the different factors affecting oil absorption during frying and many empirical studies have correlated oil absorption measurements with process and/or product characteristics.

Gamble et al. [154] concluded that a lower oil temperature resulted in a lower oil content in potato crisps in the early stages of frying with a greater difference between 145 °C and 165 °C than between 165 °C and 185 °C. Similarly, Moreira et al. [132] determined higher differences in oil absorption in tortilla chips (wet base) between 130 °C and 160 °C than between 160 °C and 190 °C. However, results were expressed as oil uptake on a wet basis. Consequently, there was a systematic reduction in the basis, as the water content was constantly reduced due to the higher dehydration that resulted when the product was fried at a higher temperature for the same period of time. In contrast, Bouchon [97] determined that the rate of oil absorption (dry base) by potato cylinders was not significantly affected by oil temperature (155, 170, 185 °C). In addition, no significant differences were found in oil absorption between the two higher temperatures. However, when frying at 155 °C, a significant lower absorption was found, compared to the previous ones. Similarly, Moreira et al. [155] determined that the oil absorption rate was unaffected by the oil temperature when frying tortilla chips and that a frying temperature of 190 °C gave a higher oil content (3–5%) than a frying temperature of 155 °C. Nonaka et al. [156] also found that oil content increased with increasing frying temperature when frying potato chips.

The influence of oil type and quality on oil absorption and residues absorbed by fried foods is widely documented [95, 96, 156–159]. No relationship has been found between oil type and oil absorption, although it has been shown that an increase in the oil's initial interfacial tension decreases oil absorption [160]. Blumenthal [96, 157], noticed the importance of oil surface tension during deep-fat frying and developed what he called the surfactant theory of frying. He explained that several classes of surfactants are formed during frying of food, either as a result of the degradation of the frying oil itself or as a result of the reactions occurring between the food components and the oil. These compounds act as wetting agents, reducing the interfacial tension between the food and the frying oil, causing increased contact between the food and the oil and finally producing excessive oil absorption by the fried product.

An increase in additional factors such as initial solids content [117, 161, 162], slice thickness [163–165] and gel strength [166] have been shown to reduce oil uptake. Additionally, it has been found that an increase in the initial porosity of the food increases the absorption [167]. However, as explained by Saguy and Pinthus [99], crust formation plays an additional and fundamental role as soon as frying commences.

Some prefrying treatments have been shown to significantly reduce oil absorption during frying. Lowering the moisture content of the food prior to frying using microwave and hot air treatment results in a reduction in the final oil content [104, 142, 151], whereas freeze-drying increases the absorption [154]. Also, post-treatments such as hot air drying [156] have been shown to reduce oil uptake.

In addition, during the last decade much attention has been given to the use of hydrocolloids such as methylcellulose (MC), hydroxypropyl methylcellulose (HPMC), long-fibre cellulose and corn zein, to inhibit oil uptake [128, 168–171]. The hydrocolloid mixture is added to the batter or breading coating, which creates a barrier against oil absorption either prior and/or during frying. Batters and breadings are gaining more importance as they contribute significantly to the added product value [172], however, most of the information on this subject is of a proprietary nature [99].

8.3.9 Microstructural Changes During Deep-Fat Frying

The importance of microstructural changes during deep-fat frying has been recognised in modelling heat and mass transport and unravelling their mechanisms. Three of five papers by leading scientists in the field published in an overview on frying of foods (*Food Technology*, October 1995) included SEM photomicrographs of fried products, suggesting the importance of microstructure. In fact, Baumann and Escher [165] recommended that the explanation of some factors in oil absorption needs to be validated by structural analysis in relation to the location of oil deposition and to the mechanisms of oil adhesion. Since the first histological studies of deep-fat fried potatoes by Reeve and Neel [173], using light microscopy, evidence has accumulated that, except for the outermost layers damaged by cutting, the majority of the inner cells retain their individuality after frying and their interior contains dehydrated but gelatinised starch granules. The microstructural aspect of the core tissue is similar to that of cooked potatoes. In the case of potato crisps (or outer layers in potato chips) cells shrink during frying but do not rupture, while cell walls become wrinkled and convoluted around the dehydrated gelled starch [146, 173]. It is thought that the rapid dehydration occurring during deep-fat frying reduces the starch-swelling process and therefore cell walls do not break as sometimes occurs during ordinary cooking. Similar observations were determined by Costa et al. [174], when studying structural changes of potato during frying, and by Mc Donough et al. [175], when evaluating the physical changes during deep-fat frying of tortilla chips.

Keller et al. [141] and Lamberg et al. [142] determined the extent of oil penetration in fried chips, using Sudan red B, a heat-resistant and oil-soluble dye, which was added to the frying medium before frying. They concluded that oil uptake during deep-fat frying was localised on the surface of the fried product and restricted to a depth of a few cells. In fact, Bouchon et al. [147], using high spatial resolution infrared microspectroscopy, determined oil distribution profiles within fried potato cylinders. Results confirmed that oil was confined to the outer region and that oil distribution reflected the anisotropic nature of the porous network developed during the process.

Blistering in potato crisps takes place after separation of neighbouring intact cells alongside cell walls, similar to fracture observed in steam-cooked potatoes [176], and oil was found to be mainly distributed in the cell walls, intercellular spaces and blister areas [173]. Aguilera and Gloria [177] demonstrated that three distinct microstructures exist in finished fried commercial chips: (a) a thin outer layer (approx. 250 μm) formed by remnants of cell walls of broken or damaged cells by cutting, (b) an intermediate layer of shrunken intact cells which extends to the evaporation front and (c) the core with fully hydrated intact cells containing gelatinised starch.

It is thought that standard microscopic techniques may produce artefacts in samples like swelling of the interiors by solvents and smearing of oil during sectioning. In an attempt to reduce invasion and destruction of samples, Farkas et al. [143] used magnetic resonance imaging (MRI) to determine water location and oil penetration depth in immersion-fried potato cylinders. They confirmed that oil was mainly located on the surface of the product and penetrated only slightly inside the structure. Confocal laser scanning microscopy (CLSM) has been introduced recently as a new methodology for studying oil location directly in fried potato chips with minimal intrusion [145, 146]. This was achieved by frying in oil containing a heat-stable fluorochrome (Nile red) and observing the fried crust, without further preparation, under a CLSM. This technique allows optical sectioning to be carried out using a laser beam, avoiding any physical damage on the specimen. It was shown that cells seem to be quite preserved and surrounded by oil, which is not uniformly located at each depth, suggesting that at least some of the oil penetrates into the interior of a potato strip by moving between cells.

Recently, it was suggested that surface roughness increases importantly surface area, enhancing oil absorption [172]. In an effort to quantify the irregular conformation of the surface, Pedreschi et al. [178] and Rubnov and Saguy [179] used fractal geometry and confirmed the significant role of crust roughness in oil absorption.

References

1 Cauvain, S. P. **1998**, Breadmaking Processes, in *Technology of Breadmaking*, ed. S. P. Cauvain, L. S. Young, Blackie Academic and Professional, London, pp 18–44.

2 Baker, J. C., Mize, M. D. **1941**, The Origin of the Gas Cell in Bread Dough, *Cereal Chem.* 18, 19–34.

3 Chamberlain, N., Collins, T. H., Elton, G. A. H. **1962**, The Chorleywood Bread Process, *Baker's Dig.* 36, 52–53.

4 Dobraszczyk, B. J., Campbell, G. M., Gan, Z. **2001**, Bread – A Unique Food, in *Cereals & Cereal Products: Chemistry and Technology*, ed. D. A. V. Dendy, B. J. Dobraszczyk, Aspen Publishers, Gaithersburg, pp 182–232.

5 Kilborn, R. H., Tipples, K. H. **1972**, Factors Affecting Mechanical Dough Development I. Effect of Mixing Intensity and Work Input, *Cereal Chem.* 49, 4–47.

6 Mani, K., Eliasson, A.-C., Lindahl, L., Trägrdh, C. **1992**, Rheological Properties and Bread Making Quality of Wheat Flour Doughs made with Different Dough Mixers, *Cereal Chem.* 69, 222–225.

7 Dobraszczyk, B. J. **2001**, Wheat and Flour, in *Cereals & Cereal Products: Chemistry and Technology*, ed. D. A. V. Dendy, B. J. Dobraszczyk, Aspen Publishers, Gaithersburg, pp 100–139.

8 Dobraszczyk, B. J., Schofield, J. D. **2002**, Rapid Assessment and Prediction Of Wheat and Gluten Baking Quality with The Two-Gram Direct Drive Mixograph Using Multivariate Statistical Analysis, *Cereal Chem.* 79, 607–612.

9 Kilborn, R. H., Tipples, K. H. **1974**, Implications of the Mechanical Development of Bread Dough by Means of Sheeting Rolls, *Cereal Chem.* 51, 648–657.

10 Chiotellis, E., Campbell, G. M. **2003**, Proving of Bread Dough I. Modelling the Evolution of the Bubble Size Distribution, *Trans. IChemE* 81, 194–206.

11 Chiotellis, E., Campbell, G. M. **2003**, Proving of Bread Dough II. Measurement of Gas Production and Retention, *Trans. IChemE* 81, 207–216.

12 Kokelaar, J. J., van Vliet, T., Prins, A. **1996**, Strain Hardening and Extensibility of Flour and Gluten Doughs in Relation to Breadmaking Performance, *J. Cereal Sci.* 24, 199–214.

13 Dobraszczyk, B. J., Morgenstern, M. P. **2003**, Review: Rheology and the Breadmaking Process, *J. Cereal Sci.* 38, 229–245.

14 He, H., Hoseney, R. C. **1991**, Gas Retention of Different Cereal Flours, *Cereal Chem.* 68, 334–336.

15 MacRitchie, F., Lafiandra, D. **1997**, Structure-Function Relationships of Wheat Proteins, in *Food Proteins and Their Applications*, ed. S. Damodaran, A. Paraf, Marcel Dekker, New York, pp 293–323.

16 Weegels, P. L., Hamer, R. J., Schofield, J. D. **1996**, Critical Review: Functional

Properties of Wheat Glutenin, *J. Cereal Sci.* 23, 1–18.

17 Chin, N. L., Campbell, G. M. **2005**, Dough Aeration and Rheology II. Effects of Flour Type, Mixing Speed and Total Work Input on Aeration and Rheology of Bread Dough, *J. Cereal Sci.* (in press).

18 Carceller, J. L., Aussenac, T. **2001**, Size Characterisation of Glutenin Polymers by HPSEC-MALLS, *J. Cereal Sci.* 33, 131–142.

19 Arfvidsson, C., Wahlund, K.-G., Eliasson, A.-C. **2004**, Direct Molecular Weight Determination in the Evaluation of Dissolution Methods for Unreduced Glutenin, *J. Cereal Sci.* 39, 1–8.

20 Shewry, P. R., Halford, N. G., Belton, P. S., Tatham, A. S. **2002**, The Structure and Properties of Gluten: an Elastic Protein from Wheat Grain, *Phil. Trans. R. Soc. Lond. B* 357, 133–142.

21 Doi, M., Edwards, S. F. **1986**, *The Theory of Polymer Dynamics*, Oxford University Press, Oxford.

22 McLeish, T. C. B., Larson, R. G. **1998**, Molecular Constitutive Equations for a Class of Branched Polymers: The Pom-Pom Model, *J. Rheol.* 42, 81–110.

23 Li, W., Dobraszczyk, B. J., Schofield, J. D. **2003**, Stress Relaxation Behaviour of Wheat Dough and Gluten Protein Fractions, *Cereal Chem.* 80, 333–338.

24 Humphris, A. D. L., McMaster, T. J., Miles, M., Gilbert, S. M., Shewry, P. R., Tatham, A. S. **2000**, Atomic Force Microscopy (AFM) Study of Interactions of HMW Subunits of Wheat Glutenin, *Cereal Chem.* **77**, 107–110.

25 Belton, P. S. **1999**, On the Elasticity of Wheat Gluten, *J. Cereal Sci.* 29, 103–107.

26 Dobraszczyk, B. J., Roberts, C. A. **1994**, Strain Hardening and Dough Gas Cell-Wall Failure in Biaxial Extension, *J. Cereal Sci.* 20, 265–274.

27 van Vliet,T., Janssen, A. M., Bloksma, A. H., Walstra, P. **1992**, Strain Hardening of Dough as a Requirement for Gas Retention, *J. Texture Stud.* 23, 439–460.

28 Dobraszczyk, B. J., Smewing, J., Albertini, M., Maesmans, G., Schofield, J. D. **2003**, Extensional Rheology and Stability of Gas Cell Walls in Bread Doughs at Elevated Temperatures in Relation to Breadmaking Performance, *Cereal Chem.* 80, 218–224.

29 Hsieh, F. **1992**, Extrusion and Extrusion Cooking, in *Encyclopedia of Food Science and Technology*, ed. Y. H. Hui, John Wiley & Sons, New York, pp 795–800.

30 Harper, J. M. **1989**, Food Extruders and Their Applications, in *Extrusion Cooking*, ed. C. Mercier, P. Linko, J. M. Harper, AACC, St Paul, pp 1–16.

31 Lanz, R. **1983**, Successful Multi-Component, Continuous-Extruder Feeding, in *Progress in Food Engineering*, ed. C. Canterelli, C. Peri, Forster, Kisnacht, pp 625–629.

32 Janssen, L. P. B. M. **1993**, Extrusion Cooking – Principals and Practice, in *Encyclopedia of Food Science Food Technology and Nutrition*, ed. R. Macrae, R. K. Robinson, M. J. Sadler, Academic Press, London, pp 1700–1705.

33 Cheftel, J. C. **1986**, Nutritional Effects of Extrusion Cooking, *Food Chem.*, 20, 263–283.

34 Heldman, D. R., Hartel, R. W. **1997**, *Principles of Food Processing*, Chapman and Hall, New York.

35 Frame, N. D. **1994**, Operational Characteristics of the Co-Rotating Twin-Screw Extruder, in *The Technology of Extrusion Cooking*, ed. N. D. Frame, Blackie Academic and Professional, Glasgow, pp 1–51.

36 Harper, J. M. **1990**, Extrusion of Foods, in *IFT Symposium Series: Biotechnology and Food Process Engineering*, ed. H. G. Schwartzberg, M. A. Rao, Marcel Dekker, New York, pp 295–308.

37 Harper, J. M. **1986**, Processing Characteristics of Food Extruders, in *Food Engineering and Process Applications, Vol 2, Unit Operations*, ed. M. L. Mauger, P. Jelem, Elsevier Applied Science, London, pp 101–114.

38 Jager, T., van Zuilichem, D. J., Stolp, W. **1992**, Residence Time Distribution, Mass Flow, and Mixing in a Co-Rotating Twin-Screw Extruder, in *Food Extrusion Science and Technology*, ed. J. L. Kokini, C. Ho, M. V. Karwe, Marcel Dekker, New York, pp 71–88.

39 Dziezak, J. D. **1989**, Single- and Twin-Screw Extruders in Food Processing, *Food Technol.* 43, 163– 174.

40 Harper, J. M. **1992**, A Comparative Analysis of Single and Twin Screw Extruders, in *Food Extrusion Science and Technology*, ed. J. L. Kokini, C. Ho, N. W. Kaewe, Marcel Dekker, New York, pp 139–148.

41 Anderson, R. A., Conway, H. F., Pfeifer, V. F., Griffin, E. L. **1969**, Gelatinization of Corn Grits by Roll- and Extrusion-Cooking, *Cereal Sci. Today* 14, 1–12.

42 Mercier, C., Feillet, P. **1975**, Modification of Carbohydrate Components by Extrusion-Cooking of Cereal Products, *Cereal Chem.* 52, 283–297.

43 Lawton, B. T., Henderson, G. A., Derlatka, E. J. **1972**, The Effects of Extruder Variables on the Gelatinization of Corn Starch, *Can. J. Chem. Eng.* 50, 168–173.

44 Mercier, C. **1977**, Effect of Extrusion-Cooking on Potato Starch Using a Twin Screw French Extruder, *Staerke* 29, 48–52.

45 Chiang, B. Y. and Johnson, J. A., Gelatinization of starch in extruded products, *Cereal Chemistry*, **54**, 436–443, 1977.

46 Ibanoglu, S., Ainsworth, P., Hayes, G. D. **1996**, Extrusion of Tarhana: Effect of Operating Variables on Starch Gelatinizaton, *Food Chem.* 57, 541–544.

47 Kim, J. C., Rottier, W. **1980**, Modification of Aestivum Wheat Semolina by Extrusion, *Cereal Foods World* 24, 62–66.

48 Gomez, M. H., Aguilera, J. M. **1983**, Changes in the Starch Fraction During Extrusion Cooking of Corn, *J. Food Sci.* 48, 378–381.

49 Colonna, P., Doublier, J. L., Melcion, J. P., Monredon, F., Mercier, C. **1984**, Physical and Functional Properties of Wheat Starch After Extrusion Cooking and Drum Drying, in *Thermal Processing and Quality of Foods*, ed. P. Zeuthen, J. C. Cheftel, C. Eriksson, M. Jul, H. Leniger, P. Linko, G Varela, G. Vos, Elsevier Applied Science, London, pp 96–112.

50 Davidson, V. J., Paton, D., Diosady, L. L., Larocque, G. **1984**, Degradation of Wheat Starch in a Single-Screw Extruder: Characterization of Extruded Starch Polymers, *J. Food Sci.* 49, 453–459.

51 Wen, L. F., Rodis, P., Wasserman, B. P. **1990**, Starch Fragmentation and Protein Insolubilization During Twin-Screw Extrusion of Corn Meal, *Cereal Chem.* 67, 268–275.

52 Bhattacharya, M., Hanna, M. A. **1987**, Textural Properties of Extrusion-Cooked Corn Starch, *Lebensm. Wiss. Technol.* 20, 195–201.

53 Maga, J. A., Cohen, M. R. **1978**, Effect of Extrusion Parameters on Certain Sensory, Physical and Nutritional Properties of Potato Flakes, *Lebensm. Wiss. Technol.* 11, 195–197.

54 Hsieh, F., Mulvaney, S. J., Huff, H. E., Lue, L., Brent, J. **1989**, Effect of Extrusion Parameters on Certain Sensory, Physical and Nutritional Properties of Potato Flakes, *Lebensm. Wiss. Technol.* 22, 204–207.

55 Fletcher, S. I., Richmond, P., Smith, A. C. **1985**, An Experimental Study of Twin-Screw Extrusion-Cooking of Maize Grits, *J. Food Eng.* 4, 291–312.

56 Guy, R. C. E. **1994**, Raw Materials for Extrusion Cooking Processes, in *The Technology of Extrusion Cooking*, ed. N. D. Frame, Blackie Academic and Professional, Glasgow, pp 52–72.

57 Martinez-Serna, M., Hawkes, J., Villota, R. **1990**, Extrusion of Natural and Modified Whey Proteins in Starch-Based Systems, in *Engineering and Food Vol 3: Advanced Processes*, ed. W. E. L. Spiess, H. Schubert, Elsevier Applied Science, London, pp 346–365.

58 Colonna, P., Mercier, C. **1983**, Macromolecular Modifications of Manoic Starch Components by Extrusion-Cooking With and Without Lipids, *Carbohydr. Polym.* 3, 87–108.

59 Liu, Y., Hsieh, F., Heymann, H., Huff, H. E. **2000**, Effect of Process Conditions on the Physical and Sensory Properties of Extruded Oat-Corn Puff, *J. Food Sci.* 65, 1253–1259.

60 Kumagai, H., Lee, B. H., Yano, T. **1987**, Flour Treatment to Improve the Quality of Extrusion-Cooked Rice Flour Products, *J. Agric. Biol. Chem.* 51, 2067–2071.

61 Jin, Z., Hsieh, F., Huff, H. E. **1994**, Extrusion Cooking of Corn Meal with Soy Fiber, Salt and Sugar, *Cereal Chem.* 71, 227–234.

62 Ryu, G.H., Neumann, P.E., Walker, C.E. **1993**, Effects of Some Baking Ingredients on Physical and Structural Properties of Wheat Flour Extrudates, *Cereal Chem.* 70, 291–297.

63 Fan, J., Mitchell, J.R., Blanshard, J.M.V. **1996**, The Effect of Sugars on the Extrusion of Maize Grits: II. Starch Conversion, *Int. J. Food Sci. Technol.* 31, 67–76.

64 Carvalho, C.W.P., Mitchell, J.R. **2000**, Effect of Sugar on the Extrusion of Maize Grits and Wheat Flour, *Int. J. Food Sci. Technol.* 35, 569–576.

65 Phillips, R.D. **1989**, Effect of Extrusion Cooking on the Nutritional Quality of Plant Proteins, in *Protein Quality and the Effect of Processing*, ed. R.D. Phillips, J.W. Finley, Marcel Dekker, New York, pp 219–246.

66 Camire, M.E., Camire, A., Krumhar, K. **1990**, Chemical and Nutritional Changes in Foods During Extrusion, *Crit. Rev. Food Sci. Nutr.* 29, 35–57.

67 Björk, I., Asp, N.G. **1983**, The Effects of Extrusion Cooking on Nutritional Value – a Literature Review, *J. Food Eng.* 2, 281–308.

68 Björk, I., Asp, N.G. **1984**, Protein Nutritional Value of Extrusion-Cooked Wheat Flours, *Food Chem.* 15, 203–214.

69 Noguchi, A., Mosso, K., Aymard, C., Jeunink, J., Cheftel, J.C. **1982**, Protein Nutritional Value of Extrusion-Cooked Wheat Flours, *Lebensm. Wiss. Technol.* 15, 105–110.

70 Bhattacharya, M., Hanna, M.A. **1988**, Extrusion Processing to Improve Nutritional and Functional Properties of Corn Gluten, *Lebensm. Wiss. Technol.* 21, 20–24.

71 Bjorck, I., Matoba, T., Nair, B.M. **1985**, In-Vitro Enzymatic Determination of the Protein Nutritional Value and the Amount of Available Lysine in Extruded Cereal-Based Products, *Agric. Biol. Chem.* 49, 945– 951.

72 Bhattacharya, S., Das, H., Bose, A.N. **1988**, Effect of Extrusion Process Variables on In Vitro Protein Digestibility of Fish-Wheat Flour Blends, *Food Chem.* 28, 225–231.

73 Fapojuwo, O.O., Maga, J.A., Jansen, G.R. **1987**, Effect of Extrusion Cooking on In Vitro Protein Digestibility of Sorghum, *J. Food Sci.* 52, 218–219.

74 Dahlin, K., Lorenz, K. **1993**, Protein Digestibility of Extruded Cereal Grains, *Food Chem.* 48, 13–18.

75 Ibanoglu, S., Ainsworth, P., Hayes, G.D. **1997**, In Vitro Protein Digestibility and Content of Thiamin and Riboflavin in Extruded Tarhana, a Traditional Turkish Cereal Food, *Food Chem.* 58, 141–144.

76 McAuley, J.A., Kunkel, M.E., Acton, J.C. **1987**, Relationships of Available Lysine to Lignin, Color and Protein Digestibility of Selected Wheat-Based Breakfast Cereals, *J. Food Sci.* 52, 1580–1582.

77 Killeit, U. **1994**, Vitamin Retention in Extrusion Cooking, *Food Chem.* 49, 149–155.

78 Mustakas, G.C., Griffin, E.L., Alien, L.E., Smith, O.B. **1964**, Production and Nutritional Evaluation of Extrusion Cooked Full Fat Soybean Flour, *Am. Oil Chem. Soc.* 41, 607–615.

79 Mustakas, C.G., Albrecht, W.J., Bookwalter, G.N., McGhee, J.E., Kwolek, W.F., Griffin, E.L. **1970**, Extruder-Processing to Improve Nutritional Quality, Flavor, and Keeping Quality of Full-Fat Soy Flour, *Food Technol.* 24, 102–108.

80 Beetner, G., Tsao, T., Frey, A., Harper, J.M. **1974**, Degradation of Thiamine and Riboflavin During Extrusion Processing, *J. Food Sci.* 39, 207–208.

81 Maga, J.A., Sizer, C.E. **1978**, Ascorbic Acid and Thiamine Retention During Extrusion of Potato Flakes, *Lebensm. Wiss. Technol.* 11, 192–194.

82 Pham, C.B., Rosario, R.R. **1986**, Studies on the Development of Texturized Vegetable Products by the Extrusion Process. III. Effects of Processing Variables of Thiamine Retention, *J. Food Technol.* 21, 569–576.

83 Millauer, C., Wiedmann, W.M., Killeit, U. **1984**, Studies on the Development of Texturized Vegetable Products by the Extrusion Process. III. Effects of Processing Variables of Thiamine Retention, in *Thermal Processing and Quality of Foods*, ed. P. Zeuthen, J.C. Cheftel, C. Eriksson, M. Jul, H. Leniger, P. Linko, G. Varela, G. Vos, Elsevier Applied Science, London, pp 208–213.

84 Guzman-Tello, R., Cheftel, J.C. **1987**, Thiamine Destruction During Extrusion Cooking as an Indication of the Intensity of Thermal Processing, *Int. J. Food Sci. Technol.* 22, 549–562.

85 Bailey, M.E., Gutheil, R.A., Hsieh, F., Cheng, C., Gerhardt, K.O. **1994**, Maillard Reaction Volatile Compounds and Color Quality of a Whey Protein Concentrate-Corn Meal Extruded Product, in *Thermally Generated Flavours: Maillard, Microwave and Extrusion Processes* (*ACS Symposium Series 543*), ed. T.H. Parliment, M.J. Morello, R.J. McGorrin, American Chemical Society, Washington, D. C., pp 315–327.

86 Nair, M., Shi, Z., Karwe, M., Ho, C.T., Daun, H. **1994**, Collection and Characterization of Volatile Compounds Released at the Die During Twin Screw Extrusion of Corn Flour, in *Thermally Generated Flavours: Maillard, Microwave and Extrusion Processes* (*ACS Symposium Series 543*), ed. T.H. Parliment, M.J. Morello, R. J. McGorrin, American Chemical Society, Washington, D.C., pp 334–347.

87 Bredie, W.L.P., Mottram, D.S., Guy, R.C.E. **1998**, Aroma Volatiles Generated During Extrusion Cooking of Maize Flour, *J. Agric. Food Chem.* 46, 1479–1487.

88 Ames, J.M., Defaye, A.B., Bates, L. **1997**, The Effect of pH on the Volatiles Formed in an Extruded Starch-Glucose-Lysine Model System, *Food Chem.* 58, 323–327.

89 Bredie, W.L.P., Hassell, G.M., Guy, R.C.E., Mottram, D.S. **1997**, Aroma Characteristics of Extruded Wheat Flour and Wheat Starch Containing Added Cysteine and Reducing Sugars, *J. Cereal Sci.* 25, 57–63.

90 Ames, J.M., Guy, R.C.E., Kipping, G.L. **2001**, Effect of pH and Temperature on the Formation of Volatile Compounds in Cysteine/Reducing Sugar/Starch Mixtures During Extrusion Cooking, *J. Agric. Food Chem.* 49, 1885–1894.

91 Varela, G. **1988**, Current Facts About Frying of Food, in *Frying of Foods: Principles, Changes, New Approaches,* ed. G. Varela, A.E. Bender, I.D. Morton, Ellis Horwood, Chichester, pp 9–25.

92 Farkas, B.E. **1994**, Modelling Immersion Frying as a Moving Boundary Problem, *Ph.D dissertion,* University of California, Davis.

93 Singh, R.P. **1995**, Heat and Mass Transfer in Foods During Deep-Fat Frying, *Food Technol.* 49, 134– 137.

94 Aguilera, J.M. **1997**, Fritura de Alimentos, in *Temas en Tecnologia de Alimentos,* ed. J.M. Aguilera, Instituto Politecnico Nacional, Mexico D.F., pp 187–214.

95 Blumenthal, M.M. **1991**, A New Look at the Chemistry and Physics of Deep-Fat Frying, *Food Chem.* 45, 68–71.

96 Blumenthal, M.M., Stier, R.F. **1991**, Optimization of Deep-Fat Frying Operations, *Trends Food Sci. Technol.* 2, 144–148.

97 Bouchon, P. **2002**, Modelling Oil Uptake During Frying, *Ph.D. dissertation,* University of Reading, Reading.

98 Bouchon, P., Pyle, D.L. **2004**, Studying Oil Absorption in Restructured Potato Chips, *J. Food Sci.* 69, 115–122.

99 Saguy, I.S., Pinthus, E.J. **1995**, Oil Uptake During Deep-Fat Frying: Factors and Mechanism, *Food Technol.* 49, 142–145, 152.

100 Browner, W. S., Westonhouse, J., Tice, J.A. **1991**, What if Americans Ate Less Fat? A Quantitative Estimate of the Effect on Mortality, *J. Am. Med. Assoc.* 265, 3285–3291.

101 Rossel, J.B. **1998**, Industrial Frying Process, *Grasas Aceites,* 49, 282–295.

102 Mitka, M. **2002**, Fear of Frying: Is Acrylamide in Foods a Cancer Risk?, *J. Am. Med. Assoc.* 288, 2105–2106.

103 Mottram, D.S., Wedzicha, B.L., Dodson, A.T. **2002**, Food Chemistry: Acrylamide is Formed in the Maillard Reaction, *Nature* 419, 448–449.

104 Moreira, R.G., Castell-Perez, M.E., Barrufet, M.A. **1999**, *Deep-Fat Frying: Fundamentals and Applications,* Aspen Publications, Gaithersburg.

105 Kochhar, S.P. **1998**, Security in Industrial Frying Processes, *Grasas Aceites* 49, 282–302.

106 Banks, D. **1996**, Food Service Frying, in *Deep Frying,* ed. E.G. Perkins, M.D. Erickson, AOAC Press, Champaign, pp 245–257.

107 Kochhar, S.P. **1999**, Safety and Reliability During Frying Operations – Effects of Detrimental Components and Fryer Design Features, in *Frying of Food*, ed. D. Boskou, I. Elmadfa, Technomic Publishing, Lancaster, pp 253–269.

108 Pavel, J. **1995**, *Introduction to Food Processing*, Reston Publishing Co., Reston.

109 Pantzaris, T.P. **1999**, Palm Oil in Frying, in *Frying of Food*, ed. D. Boskoy, I. Elmadfa, Technomic Publishing, Lancaster, pp 223–252.

110 Montaldo, A. **1984**, *Cultivo y Mejoramiento de la Papa*, Instituto Interamericano Cooperacion para la Agricultura, San Jose.

111 Artschwager, E. **1918**, Anatomy of the Potato Plant, With Special Reference to the Ontogeny of the Vascular System, *J. Agric. Res.* 14, 221–252.

112 Artschwager, E. **1924**, Studies on the Potato Tuber, *J. Agric. Res.* 27, 809–835.

113 Talburt, W.F., Schwimmer, S., Burr, H.K. **1975**, Structural and Chemical Composition of the Potato Tuber, in *Potato Processing*, 3rd edn, ed. W. Talburt, O. Smith, AVI Publishing Co., Westport, pp 11–42.

114 Banks, W., Greenwood, C.T. **1959**, The Starch of the Tuber and Shoots of the Sprouting Potato, *Biochem. J.* 73, 237–241.

115 True, R.H., Work, T.M., Bushway, R.J., Bushway, A.A. **1983**, Sensory Quality of French Fries Prepared from Belrus and Russet Burbank Potatoes, *Am. Potato J.* 60, 933–937.

116 Mottur, G.P. **1998**, A Scientific Look at Potato Chips – The Original Savoury Snack, *Am. Assoc. Cereal Chem.* 34, 620–626.

117 Lisinska, G., Leszczynski, W. **1991**, *Potato Science and Technology*, Elsevier Applied Science, London.

118 Weaver, M.L., Reeve, R.M., Kueneman, R.W. **1975**, Frozen French Fries and Other Frozen Potato Products, in *Potato Processing*, ed. G. Varela, A.E. Bender, D. Morton, AVI Publishing Co., Westport, pp 403–442.

119 Gebhardt, B. **1996**, Oils and Fats in Snack Foods, in *Bailey's Industrial Oil & Fat Products*, 5th edn, ed. Y.I. Hui, Wiley, New York, pp 409–428.

120 Farkas, B.E., Singh, R.P., McCarthy, M.J. **1996a**, Modeling Heat and Mass Transfer in Immersion Frying, I. Model Development, *J. Food Eng.* 29, 211–226.

121 Miller, K.S., Singh, R.P., Farkas, B.E. **1994**, Viscosity and Heat Transfer Coefficients for Canola, Corn, Palm and Soybean Oil, *J. Food Process. Preserv.* 18, 461–472.

122 Moreira, R.G., Palau, J. Sweat, V., Sun, X. **1995a**, Thermal and Physical Properties of Tortilla Chips as a Function of Frying Time, *J. Food Process. Preserv.* 19, 175–189.

123 Hubbard, L.J., Farkas, B.E. **1999**, A Method for Determining the Convective Heat Transfer Coefficient During Immersion Frying, *J. Food Process Eng.* 22, 201–214.

124 Hubbard, L.J., Farkas, B.E. **2000**, Influence of Oil Temperature on Convective Heat Transfer During Immersion Frying, *J. Food Process. Preserv.* 24, 143–162.

125 Costa, R.M., Oliveira, F.A.R., Delaney, O., Gekas, V. **1999**, Analysis of the Heat Transfer Coefficient During Potato Frying, *J. Food Eng.* 39, 293–299.

126 Ashkenazi, N., Mizrahi, S., Berk, Z. **1984**, Heat and Mass Transfer in Frying, in *Engineering and Food*, ed. B.M. McKenna, Elsevier Applied Science, London, pp 109–116.

127 Rice, P., Gamble, M.H. **1989**, Technical Note: Modelling Moisture Loss During Potato Slice Frying, *Int. J. Food Sci. Technol.* 24, 183–187.

128 Kozempel, M.F., Tomasula, P.M., Craig, J.C. Jr. **1991**, Correlation of Moisture and Oil Concentration in French Fries, *Lebensm. Wiss. Technol.* 24, 445–448.

129 Ateba, P., Mittal, G.S. **1994**, Modelling the Deep-Fat Frying of Beef Meatballs, *Int. J. Food Sci. Technol.* 29, 429–440.

130 Rao, V.N.M., Delaney, R.A.M. **1995**, An Engineering Perspective on Deep-Fat Frying of Breaded Chicken Pieces, *Food Technol.* 49, 138–141.

131 Dincer, I. **1996**, Modelling for Heat and Mass Transfer Parameters in Deep Frying of Products, *Heat Mass Transfer* 32, 109–113.

132 Moreira, R.G., Sun, X., Chen, Y. **1997**, Factors Affecting Oil Uptake in Tortilla Chips in Deep-Fat Frying, *J. Food Eng.* 31, 485–498.

133 Chen, Y., Moreira, R.G. **1997**, Modelling of a Batch Deep-Fat Frying Process for Tortilla Chips, *Trans. IChemE Part C* 75, 181–190.

134 Ngadi, M.O., Watts, K.C., Correia, L.R. **1997**, Finite Element Method Modelling of Moisture Transfer in Chicken Drum During Deep-Fat Frying, *J. Food Eng.* 32, 11–20.

135 Sahin, S., Sastry, S.K., Bayindirli, B. **1999**, Heat Transfer During Frying of Potato Slices, *Lebensm. Wiss. Technol.* 32, 19–24.

136 Farid, M.M., Chen, X.D. **1998**, The Analysis of Heat and Mass Transfer During Frying of Food Using a Moving Boundary Solution, *Heat Mass Transfer* 34, 69–77.

137 Ni, H., Datta, A.K. **1999**, Moisture, Oil and Energy Transport During Deep-Fat Frying of Food Materials, *Trans. IChemE Part C* 77, 194–204.

138 Carslaw, H.S., Jaeger, J.C. **1959**, *Conduction of Heat in Solids*, 2nd edn, Oxford University Press, New York.

139 King, C.J. **1970**, Freeze Drying of Foodstuffs, *Crit. Rev. Food Technol.* 1, 379–451.

140 Farkas, B.E., Singh, R.P., Rumsey, T.R. **1996b**, Modeling Heat and Mass Transfer in Immersion Frying, II, Model Solution and Verification, *J. Food Eng.* 29, 227–248.

141 Keller, C., Escher, F., Solms, J.A. **1986**, Method of Localizing Fat Distribution in Deep-Fat Fried Potato Products, *Lebensm. Wiss. Technol.* 19, 346–348.

142 Lamberg, I., Hallstrom, B., Olsson, H. **1990**, Fat Uptake in a Potato Drying/Frying Process, *Lebensm. Wiss. Technol.* 23, 295–300.

143 Farkas, B.E., Singh, R.P., McCarthy, M.J. **1992**, Measurement of Oil/Water Interface in Foods During Frying, in *Advances in Food Engineering*, ed. R.P. Singh, A. Wirakartakusumah, CRC Press, Boca Raton, pp 237–245.

144 Saguy, I.S., Gremaud, E., Gloria, H., Turesky, R.J. **1997**, Distribution and Quantification of Oil Uptake in French Fries Utilizing Radiolabeled 14C Palmitic Acid, *J. Agric. Food Chem.* 45, 4286–4289.

145 Pedreschi, F., Aguilera, J.M., Arbildua, J.J. **1999**, CLSM Study of Oil Location in Fried Potato Slices, *Microsc. Anal.* 37, 21–22.

146 Bouchon, P., Aguirera, J.M. **2001**, Microstructural Analysis of Frying of Potatoes, *Int. J. Food Sci. Technol.* 36, 669–676.

147 Bouchon, P., Hollins, P., Pearson, M., Pyle, D.L., Tobin, M.J. **2001**, Oil Distribution in Fried Potatoes Monitored by Infrared Microspectroscopy, *J. Food Sci.* 66, 918–923.

148 Bouchon, P., Aguilera, J.M., Pyle, D.L. **2003**, Structure-Oil Absorption Kinetics Relationships During Deep-Fat Frying, *J. Food Sci.* 68, 2711–2716.

149 Ufheil, G., Escher, F. **1996**, Dynamics of Oil Uptake During Deep-Fat Frying of Potato Slices, *Lebensm. Wiss. Technol.* 29, 640–644.

150 Aguilera, J.M., Gloria, H. **2000**, Oil Absorption During Frying of Frozen Parfried Potatoes, *J. Food Sci.* 65, 476–479.

151 Gamble, M.H., Rice, P. **1987**, Effect of Pre-Fry Drying on Oil Uptake and Distribution on Potato Crisp Manufacture, *Int. J. Food Sci. Technol.* 22, 535–548.

152 Moreira, R.G., Barrufet, M.A. **1998**, A New Approach to Describe Oil Absorption in Fried Foods: A Simulation Study, *J. Food Eng.* 35, 1–22.

153 Bouchon, P., Pyle, D.L. **2002**, School of Food Biosciences, University of Reading, Reading.

154 Gamble, M.H., Rice, P., Selman, J.D. **1987**, Relationship Between Oil Uptake and Moisture Loss During Frying of Potato Slices from c.v. Record UK Tubers, *Int. J. Food Sci. Technol.* 22, 233–241.

155 Moreira, R.J., Palau, J., Sun, X. **1995b**, Simultaneous Heat and Mass Transfer During the Deep Fat Frying of Tortilla Chips, *J. Food Process Eng.* 18, 307–320.

156 Nonaka, M, Sayre, R. N., Weaver, M. L. **1977**, Oil Content of French Fries as Affected by Blanch Temperatures, Fry Temperatures and Melting Point of Frying Oils, *Am. Potato J.* 54, 151–159.
157 Blumenthal, M. M. **2001**, A New Look at Frying Science, *Cereals Foods World* 46, 352–354.
158 Pokorny, J. **1980**, Effect of Substrates on Changes of Fats and Oils During Frying, *Riv. Ital. Sostanze Grasse*, 57, 222–225.
159 Krokida, M. K., Oreopoulou, V., Maroulis, Z. B. **2000**, Water Loss and Oil Uptake as a Function of Frying Time, *J. Food Eng.* 44, 39–46.
160 Pinthus, E. J., Saguy, I. S. **1994**, Initial Interfacial Tension and Oil Uptake by Deep-Fat Fried Foods, *J. Food Sci.* 59, 804–807, 823.
161 Gamble, M. H., Rice, P. **1988 a**, Effect of Initial Solids Content on Final Oil Content of Potato Chips, *Lebensm. Wiss. Technol.* 21, 62–65, 1988 a.
162 Lulai, E. G., Orr, P. H. **1979**, Influence of Potato Specific Gravity on Yield and Oil Content of Chips, *Am. Potato J.* 56, 379–390.
163 Gamble, M. H., Rice, P. **1989**, The Effect of Slice Thickness on Potato Crisp Yield and Composition, *J. Food Eng.* 8, 31–46.
164 Selman, J. D., Hopkins, M. **1989**, Factors Affecting Oil Uptake During the Production of Fried Potato Products (*Technical Memorandum 475*), Campden Food and Drink Research Association, Chipping Campden.
165 Baumann, B., Escher, E. **1995**, Mass and Heat Transfer During Deep Fat Frying of Potato Slices – 1. Rate of Drying and Oil Uptake, *Lebensm. Wiss. Technol.* 28, 395–403.
166 Pinthus, E. J, Weinberg, P., Saguy, I. S. **1992**, Gel-Strength in Restructured Potato Products Affects Oil Uptake During Deep-Fat Frying, *J. Food Sci.* 57, 1359–1360.
167 Pinthus, E. J., Weinberg, P., Saguy, I. S. **1995**, Oil Uptake in Deep-Fat Frying as Affected by Porosity, *Journal of Food Sci.* 60, 767–769.
168 Pinthus, E. J., Weinberg, P., Saguy, I. S. **1993**, Criterion for Oil Uptake During Deep-Fat Frying, *J. Food Sci.* 58, 204–205.
169 Balasubramaniam, V. M., Chinnan, M. S., Mallikarjunan, P., Phillips, R. D. **1997**, The Effect of Edible Film on Oil Uptake and Moisture Retention of a Deep-Fat Fried Poultry Product, *J. Food Process Eng.* 20, 17–29.
170 Williams, R., Mittal, G. S. **1999**, Low-Fat Fried Foods with Edible Coatings: Modelling and Simulation, *J. Food Sci.* 64, 317–322.
171 Malikarjunan, P., Chinnan, N. S., Balasubramaniam, V. M., Phillips, R. D. **1997**, Edible Coatings for Deep-Fat Frying of Starchy Products, *Lebensm. Wiss. Technol.* 30, 709–714.
172 Saguy, I. S., Ufheil, G., Livings, S. **1998**, Oil Uptake in Deep-Fat Frying: Review, *Oleag. Corps Gras Lipids* 5, 30–35.
173 Reeve, R. M., Neel, E. M. **1960**, Microscopy Structure of Potato Chips, *Am. Potato J.* 37, 45–57.
174 Costa, R. M., Oliveira, F. A. R., Boutcheva, G. **2000**, Structural Changes and Shrinkage of Potato During Frying, *Int. J. Food Sci. Technol.* 36, 11–24.
175 McDonough, C., Gomez, M. H., Lee, J. K., Waniska, R. D., Rooney, L. W. **1993**, Environmental Scanning Electron Microscopy Evaluation of Tortilla Chip Microstructure During Deep-Fat Frying, *J. Food Sci.* 58, 199–203.
176 van Marle, J. T., Clerkx, A. C. M., Boekstein, A. **1992**, Cryo-Scanning Electron Microscopy Investigation of the Texture of Cooked Potatoes, *Am. Potato J.* 11, 209–216.
177 Aguilera, J. M., Gloria, H. **1997**, Determination of Oil in Fried Potato Products by Differential Scanning Calorimetry, *J. Agric. Food Chem.* 45, 781–785.
178 Pedreschi, F., Aguilera, J. M., Brown, C. **2000**, Characterization of Food Surfaces Using Scale-Sensitive Fractal Analysis, *J. Food Process Eng.* 23, 127–143.
179 Rubnov, M., Saguy, I. S. **1997**, Fractal Analysis and Crust Water Diffusivity of a Restructured Potato Product During Deep-Fat Frying, *J. Food Sci.* 62, 135–137, 154.

9 Packaging

James G. Brennan and Brian P. F. Day

9.1 Introduction

The main functions of a package are to contain the product and protect it against a range of hazards which might adversely affect its quality during handling, distribution and storage. The package also plays an important role in marketing and selling the product. In this chapter only the protective role of the package will be considered. In this context the following definition may apply: 'Packaging is the protection of materials by means of containers designed to isolate the contents, to some predetermined degree, from outside influences. In this way, the product is contained in a suitable environment within the package'. The qualification 'to some predetermined degree' is included in the definition as it is not always desirable to completely isolate the contents from the external environment.

Today most food materials are supplied to the consumer in a packaged form. Even foods which are sold unpackaged, such as some fruits and vegetables, will have been bagged, boxed or otherwise crudely packaged at some stage in their distribution. A wide range of packaging materials is used for packaging foods including: papers, paperboards, fibreboards, regenerated cellulose films, polymer films, semirigid and rigid containers made from polymer materials, metal foil, rigid metals, glass, timber, textiles and earthenware. Very often a combination of two or more materials is employed to package one product. It is important to look on packaging as an integral part of food processing and preservation. The success of most preservation methods depends on appropriate packaging, e.g. to prevent microbiological contamination of heat-processed foods or moisture pick up by dehydrated foods. It is essential to consider packaging at an early stage in any product development exercise, not only for its technical importance, but also because of its cost implications. Packaging should not give rise to any health hazard to the consumer. No harmful substances should leach from the packaging material into the food. Packaging should not lead to the growth of pathogenic microorganisms when anaerobic conditions are created within the package.

Food Processing Handbook. Edited by James G. Brennan

ISBN: 3-527-30719-2

Packages should be convenient to use. They should be easy to open and resealable, if appropriate. The contents should be readily dispensed from the container. Other examples of convenient packaging are 'boil in the bag' products and microwavable packaging.

There are many environmental implications to packaging. In the manufacture of packaging materials, the energy requirements and the release of undesirable compounds into the atmosphere, have environment implications. The use of multitrip containers and recycling of packaging materials can have positive influences on the environment. The disposal of waste packaging materials, particularly those that are not biodegradable, is a huge problem. These topics are outside the scope of this chapter but are covered elsewhere in the literature [1–5].

9.2 Factors Affecting the Choice of a Packaging Material and/or Container for a Particular Duty

9.2.1 Mechanical Damage

Fresh, processed and manufactured foods are susceptible to mechanical damage. The bruising of soft fruits, the break up of heat processed vegetables and the cracking of biscuits are examples. Such damage may result from sudden impacts or shocks during handling and transport, vibration during transport by road, rail and air and compression loads imposed when packages are stacked in warehouses or large transport vehicles. Appropriate packaging can reduce the incidence and extent of such mechanical damage. Packaging alone is not the whole answer. Good handling and transport procedures and equipment are also necessary.

The selection of a packaging material of sufficient strength and rigidity can reduce damage due to compression loads. Metal, glass and rigid plastic materials may be used for primary or consumer packages. Fibreboard and timber materials are used for secondary or outer packages. The incorporation of cushioning materials into the packaging can protect against impacts, shock and vibration. Corrugated papers and boards, pulpboard and foamed plastics are examples of such cushioning materials. Restricting movement of the product within the package may also reduce damage. This may be achieved by tight-wrapping or shrink-wrapping. Inserts in boxes or cases or thermoformed trays may be used to provide compartments for individual items such as eggs and fruits.

9.2.2 Permeability Characteristics

The rate of permeation of water vapour, gases (O_2, CO_2, N_2, ethylene) and volatile odour compounds into or out of the package is an important consideration, in the case of packaging films, laminates and coated papers. Foods with relatively high moisture contents tend to lose water to the atmosphere. This results in a loss of weight and deterioration in appearance and texture. Meat and cheese are typical examples of such foods. Products with relatively low moisture contents will tend to pick up moisture, particularly when exposed to a high humidity atmosphere. Dry powders such as cake mixes and custard powders may cake and lose their freeflowing characteristics. Biscuits and snack foods may lose their crispness. If the water activity of a dehydrated product is allowed to rise above a certain critical level, microbiological spoilage may occur. In such cases a packaging material with a low permeability to water vapour, effectively sealed, is required. In contrast, fresh fruit and vegetables continue to respire after harvesting. They use up oxygen and produce water vapour, carbon dioxide and ethylene. As a result, the humidity inside the package increases. If a high humidity develops, condensation may occur within the package when the temperature fluctuates. In such cases, it is necessary to allow for the passage of water vapour out of the package. A packaging material which is semipermeable to water vapour is required in this case.

The shelf life of many foods may be extended by creating an atmosphere in the package which is low in oxygen. This can be achieved by vacuum packaging or by replacing the air in the package with carbon dioxide and/or nitrogen. Cheese, cooked and cured meat products, dried meats, egg and coffee powders are examples of such foods. In such cases, the packaging material should have a low permeability to gases and be effectively sealed. This applies also when modified atmosphere packaging (MAP) is used (see Section 9.4).

If a respiring food is sealed in a gastight container, the oxygen will be used up and replaced with carbon dioxide. The rate at which this occurs depends on the rate of respiration of the food, the amount in the package and the temperature. Over a period of time, an anaerobic atmosphere will develop inside the container. If the oxygen content falls below 2%, anaerobic respiration will set in and the food will spoil rapidly. The influence of the level of carbon dioxide in the package varies from product to product. Some fruits and vegetables can tolerate, and may even benefit from, high levels of carbon dioxide while others do not. In such cases, it is necessary to select a packaging material which permits the movement of oxygen into and carbon dioxide out of the package, at a rate which is optimum for the contents. Ethylene is produced by respiring fruits. Even when present in low concentrations, this can accelerate the ripening of the fruit. The packaging material must have an adequate permeability to ethylene to avoid this problem.

To retain the pleasant odour associated with many foods, such as coffee, it is necessary to select a packaging material that is a good barrier to the volatile compounds which contribute to that odour. Such materials may also prevent the contents from developing taints due to the absorption of foreign odours. It is worth noting here that films that are good barriers to water vapour may be permeable to volatiles.

In those cases where the movement of gases and vapours is to be minimised, metal and glass containers, suitably sealed, may be used. Many flexible film materials, particularly if used in laminates, are also good barriers to vapours and gases. Where some movement of vapours and/or gases is desirable, films that are semipermeable to them may be used. For products with high respiration rates the packaging material may be perforated. A range of microperforated films is available for such applications.

In the case of an intact polymer film, the rate at which vapours and gases pass through it is specified by its 'permeability' or 'permeability constant', P, defined by the following relationship:

$$P = \frac{ql}{A(p_1 - p_2)} \tag{9.1}$$

where q is the quantity of vapour or gas passing through A, an area of the film in unit time, l is the thickness of the film and p_1, p_2 are the partial pressures of the vapour or gas in equilibrium with the film at its two faces. The permeability of a film to water vapour is usually expressed as x g m^{-2} day^{-1} (i.e. per 24 h) and is also known as the water vapour transfer rate (WVTR). Highly permeable films have values of WVTR in the range from 200 g m^{-2} day^{-1} to >800 g m^{-2} day^{-1}, while those with low permeability have values of 10 g m^{-2} day^{-1} or below. The permeability of a film to gases is usually expressed as x cm^3 m^{-2} day^{-1}. Highly permeable films have P values from 1000 cm^3 m^{-2} day^{-1} to >25 000 cm^3 m^{-2} day^{-1}, while those with low permeability have values of 10 cm^3 m^{-2} day^{-1} or below. When stating the P value of a film, the thickness of the film and the conditions under which it was measured, mainly the temperature and (p_1, p_2), must be given.

9.2.3 Greaseproofness

In the case of fatty foods, it is necessary to prevent egress of grease or oil to the outside of the package, where it would spoil its appearance and possibly interfere with the printing and decoration. Greaseproof and parchment papers (see Section 9.3.1) may give adequate protection to dry fatty foods, such as chocolate and milk powder, while hydrophilic films or laminates are used with wet foods, such as meat or fish.

9.2.4 Temperature

A package must be able to withstand the changes in temperature which it is likely to encounter, without any reduction in performance or undesirable change in appearance. This is of particular importance when foods are heated or cooled in the package. For many decades metal and glass containers were used for foods which were retorted in the package. It is only in relatively recent times that heat resistant laminates were developed for this purpose. Some packaging films become brittle when exposed to low temperatures and are not suitable for packaging frozen foods. The rate of change of temperature may be important. For example, glass containers have to be heated and cooled slowly to avoid breakage. The method of heating may influence the choice of packaging. Many new packaging materials have been developed for foods which are to be processed or heated by microwaves.

9.2.5 Light

Many food components are sensitive to light, particularly at the blue and ultraviolet end of the spectrum. Vitamins may be destroyed, colours may fade and fats may develop rancidity when exposed to such light waves. The use of packaging materials which are opaque to light will prevent these changes. If it is desirable that the contents be visible, for example to check the clarity of a liquid, coloured materials which filter out short wavelength light may be used. Amber glass bottles, commonly used for beer in the UK, perform this function. Pigmented plastic bottles are used for some health drinks.

9.2.6 Chemical Compatibility of the Packaging Material and the Contents of the Package

It is essential in food packaging that no health hazard to the consumer should arise as a result of toxic substances, present in the packaging material, leaching into the contents. In the case of flexible packaging films, such substances may be residual monomers from the polymerisation process or additives such as stabilisers, plasticisers, colouring materials etc. To establish the safety of such packaging materials two questions need to be answered: (a) are there any toxic substances present in the packaging material and (b) will they leach into the product? Toxicological testing of just one chemical compound is lengthy, complicated and expensive, usually involving extensive animal feeding trials and requiring expert interpretation of the results. Such undertakings are outside the scope of all but very large food companies. In most countries there are specialist organisations to carry out such this type of investigation, e.g. the British Industrial Biological Research Association (BIBRA) in the UK. Such work may be commissioned by governments, manufacturers of packaging materials and food companies.

To establish the extent of migration of a chemical compound from a packaging material into a food product is also quite complex. The obvious procedure would be to store the food in contact with the packaging material for a specified time under controlled conditions and then to analyse the food to determine the amount of the specific compound present in it. However, detecting a very small amount of a specific compound in a food is a difficult analytical problem. It is now common practice to use simulants instead of real foods for this purpose. These are liquids or simple solutions which represent different types of foods in migration testing. For example the EC specifies the following simulants:

- Simulant A: distilled water or equivalent (to represent low acid, aqueous foods);
- Simulant B: 3% (w/v) acetic acid in aqueous solution (to represent acid foods);
- Simulant C: 15% (w/v) ethanol in aqueous solution (to represent foods containing alcohol);
- Simulant D: rectified olive oil (to represent fatty foods).

The EC also specifies which simulants are to be used when testing specific foods. More than one simulant may be used with some foods. After been held in contact with the packaging material, under specified conditions, the simulant is analysed to determine how much of the component under test it contains. Migration testing is seldom carried out by food companies. Specialist organisations mostly do this type of work, e.g. in the UK Pira International.

Most countries have extensive legislation in place controlling the safety of flexible plastic packaging materials for food use. These include limits on the amount of monomer in the packaging material. There is particular concern over the amount of vinyl chloride monomer (VCM) in polyvinylchloride (PVC). The legislation may also include: lists of permitted additives which may be incorporated into different materials, limits on the total migration from the packaging material into the food and limits on the migration of specific substances, such as VCM. The types of simulants to be used in migration tests on different foods and the methods to be used for analysing the simulants may also be specified. While the discussion above is concerned only with flexible films, other materials used for food packaging may result in undesirable chemicals migrating into foods. These include semirigid and rigid plastic packaging materials, lacquers and sealing compounds used in metal cans, materials used in the closures for glass containers, additives and coatings applied to paper, board and regenerated cellulose films, wood, ceramics and textiles.

Apart from causing a health hazard to the consumer, interaction between the packaging material and the food may affect the quality and shelf life of the food and/or the integrity of the package; and it should be avoided. An example of this is the reaction between acid fruits and tinplate cans. This results in the solution of tin in the syrup and the production of hydrogen gas. The appearance of the syrup may deteriorate and coloured fruits may be bleached. In extreme cases, swelling of the can (hydrogen swelling) and even perforation may occur. The solution to this problem is to apply an acid resistant lacquer to the inside of the can. Packaging materials, which are likely to react adversely with the contents, should be avoided, or another barrier substance should be interposed between the packaging material and the food [6–9].

9.2.7
Protection Against Microbial Contamination

Another role of the package may be to prevent or limit the contamination of the contents by microorganisms from sources outside the package. This is most important in the case of foods that are heat-sterilised in the package, where it is essential that postprocess contamination does not occur. The metal can has dominated this field for decades and still does. The reliability of the double seam (see Section 9.3.8) in preventing contamination is one reason for this dominance. Some closures for glass containers are also effective barriers to contamination. It is only in relatively recent times that plastic containers have been developed, which not only withstand the rigours of heat processing, but also whose heat seals are effective in preventing postprocess contamination. Effective seals are also necessary on cartons, cups and other containers which are aseptically filled with UHT products. The sealing requirements for containers for pasteurised products and foods preserved by drying, freezing, curing, etc. are not so rigorous. However, they should still provide a high level of protection against microbial contamination.

9.2.8
In-Package Microflora

The permeability of the packaging material to gases and the packaging procedure employed can influence the type of microorganisms that grow within the package. Packaging foods in materials that are highly permeable to gases is not likely to bring about any significant change in the microflora, compared to unpackaged foods. However, when a fresh or mildly processed food is packaged in a material that has a low permeability to gases and when an anaerobic atmosphere is created within the package, as a result of respiration of the product or because of vacuum or gas packaging, the type of microorganisms that grow inside the package are likely to be different to those that would grow in the unpackaged food. There is a danger that pathogenic microorganisms could flourish under these conditions and result in food poisoning. Such packaging procedures should not be used without a detailed study of the microbiological implications, taking into account the type of food, the treatment it receives before packaging, the hygienic conditions under which it is packaged and the temperature at which the packaged product is to be stored, transported, displayed in the retail outlet and kept in the home of the consumer.

9.2.9
Protection Against Insect and Rodent Infestation

In temperate climates, moths, beetles and mites are the insects that mainly infest foods. Control of insect infestation is largely a question of good housekeeping. Dry, cool, clean storage conditions, good ventilation, adequate turnaround of warehouse stocks and the controlled use of fumigants or contact insecticides can all help to limit insect infestation. Packaging can also contribute, but an insectproof package is not normally economically feasible, with the exception of metal and glass containers. Some insects are classified as penetrators, as they can gnaw their way through some packaging materials. Paper, paperboard and regenerated cellulose materials offer little resistance to such insects. Packaging films vary in the resistance they offer. In general, the thicker the film the more resistant it is to penetrating insects. Oriented films are usually more resistant that unoriented forms of the same materials. Some laminates, particularly those containing foil, offer good resistance to penetrating insects. Other insects are classified as invaders as they enter through openings in the package. Good design of containers to eliminate as far as possible cracks, crevices and pinholes in corners and seals can limit the ingress of invading insects. The use of adhesive tape to seal any such openings can help. The application of insecticides to some packaging materials is practised to a limited extent, e.g. to the outer layers of multiwall paper sacks. They may be incorporated into adhesives. However, this can only be done if regulations allow it [10, 11].

Packaging does not make a significant contribution to the prevention of infestation by rodents. Only robust metal containers offer resistance to rats and mice. Good, clean storekeeping, provision of barriers to infestation and controlled use of poisons, gassing and trapping are the usual preventive measures taken to limit such infestation.

9.2.10 Taint

Many packaging materials contain volatile compounds which give rise to characteristic odours. The contents of a package may become tainted by absorption or solution of such compounds when in direct contact with the packaging materials. Food not in direct contact with the packaging material may absorb odorous compounds present in the free space within the package. Paper, paperboard and fibreboard give off odours which may contaminate food. The cheaper forms of these papers and boards, which contain recycled material, are more likely to cause tainting of the contents. Clay, wax and plastic coatings applied to such materials may also cause tainting. Storage of these packaging materials in clean, dry and well ventilated stores can reduce the problem. Some varieties of wood, such as cedar and cypress, have very strong odours which could contaminate foods. Most polymers are relatively odour-free, but care must be taken in the selection of additives used. Lacquers and sealing compounds used in metal and glass containers are possible sources of odour contamination. Some printing inks and adhesives give off volatile compounds, when drying, which may give rise to tainting of foods. Careful selection of such materials is necessary to lessen the risk of contamination of foods in this way [9].

9.2.11
Tamper-Evident/Resistant Packages

There have been many reports in recent years of food packages being deliberately contaminated with toxic substances, metal or glass fragments. The motive for this dangerous practise is often blackmail or revenge against companies. Another less serious, but none the less undesirable activity, is the opening of packages to inspect, or even taste, the contents and returning them to the shelf in the supermarket. This habit is known as grazing. There is no such thing as a tamper-proof package. However, tamper-resistant and/or tamper-evident features can be incorporated into packages. Reclosable glass or plastic bottles and jars are most vulnerable to tampering. Examples of tamper-evident features include: a membrane heat-sealed to the mouth of the container, beneath the cap, roll-on closures (see Section 9.3.9), polymer sleeves heat-shrunk over the necks and caps, breakable caps which are connected to a band by means of frangible bridges that break when the cap is opened and leave the band on the neck of the container [12–14].

9.2.12
Other Factors

There are many other factors to be considered when selecting a package for a particular duty. The package must have a size and shape which makes it easy to handle, store and display on the supermarket shelf. Equipment must be available to form, fill and seal the containers at an acceptable speed and with an adequately low failure rate. The package must be aesthetically compatible with the contents. For example, the consumer tends to associate a particular type of package with a given food or drink. Good quality wines are packaged in glass, whereas cheaper ones may be packaged in 'bag in box' containers or plastic bottles. The decoration on the package must be attractive. A look around a supermarket confirms the role of the well designed package in attracting the consumer to purchase that product. The labelling must clearly convey all the information required to the consumer and comply with relevant regulations.

Detailed discussion of these factors is not included in this chapter but further information is available in the literature [13–20].

9.3
Materials and Containers Used for Packaging Foods

9.3.1
Papers, Paperboards and Fibreboards

9.3.1.1 Papers

While paper may be manufactured from a wide range of raw materials, almost all paper used for food packaging is made from wood. Some papers and boards are made from repulped waste paper. Such materials are not used in direct contact with foods. The first stage in the manufacture of papers and boards is pulping. Groundwood pulp is produced by mechanical grinding of wood and contains all the ingredients present in the wood (cellulose, lignin, carbohydrates, resins, gums). Paper made from this type of pulp is relatively weak and dull compared to the alternative chemical pulp. Chemical pulp is produced by digesting wood chips in an alkaline (sulphate pulp) or acid (sulphite pulp) solution, followed by washing. This pulp is a purer form of cellulose, as the other ingredients are dissolved during the digestion and removed by washing. Some mechanical pulp may be added to chemical pulp for paper manufacture, but such paper is not usually used in direct contact with foods. The first step in the paper making process itself is known as beating or refining. A dilute suspension of pulp in water is subjected to controlled mechanical treatment in order to split the fibres longitudinally and produce a mass of thin fibrils. This enables them to hold together when the paper is manufactured thus increasing the strength of the paper. The structure and density of the finished paper is mainly determined by the extent of this mechanical treatment. Additives such as mineral fillers and sizing agents are included at this stage to impart particular properties to the paper. The paper pulp is subjected to a series of refining operations before being converted into paper. There are two types of equipment used to produce paper from pulp. In the *Fourdrinier machine*, a dilute suspension of the refined pulp is deposited on to a fine woven, moving and vibrating mesh belt. By a sequence of draining, vacuum filtration, pressing and drying, the water content is reduced to 4–8% and the network of fibres on the belt is formed into paper. In the alternative *cylinder machine*, six or more wire mesh cylinders rotate partly immersed in a suspension of cellulose fibres. They pick up fibres and deposit them in layers onto a moving felt blanket. Water is removed by a sequence of operations similar to those described above. This method is mainly used for the manufacture of boards where combinations of different pulps are used.

Types of papers used for packaging foods include:

- Kraft paper, which is made form sulphate pulp. It is available unbleached (brown) or bleached. It is a strong multipurpose paper used for wrapping individual items or parcelling a number of items together. It may also be fabricated into bags and multiwall sacks.
- Sulphite paper, which is made from pulp produced by acid digestion. It is again a general purpose paper, not as strong as Kraft. It is used in the form of sachets and bags.

- Greaseproof paper, which is made from sulphite pulp, which is given a severe mechanical treatment at the beating stage. It is a close-textured paper with greaseproof properties under dry conditions.
- Glassine paper, which is produced by polishing the surface of greaseproof paper. It has some resistance to moisture penetration.
- Vegetable parchment, which is produced by passing paper made from chemical pulp through a bath of sulphuric acid, after which it is washed, neutralised and dried. The acid dissolves the surface layers of the paper, decreasing its porosity. It has good greaseproof characteristics and retains its strength when wet better than greaseproof paper.
- Tissue paper, which is light and has an open structure. It is used to protect the surface of fruits and provide some cushioning.
- Wet-strength papers, which have chemicals added which are crosslinked during the manufacturing process. They retain more of their strength when wet, compared to untreated papers. They are not used in direct contact with food, but mainly for outside packaging.
- Wax-coated papers, which are heat-sealable and offer moderate resistance to water and water vapour transfer. However, the heat seals are relatively weak and the wax coating may be damaged by creasing and abrasion.
- Other coatings may be applied to papers to improve their functionality. These include many of the polymer materials discussed in Section 9.3.2. They may be used to increase the strength of paper, make it heat-sealable and/or improve its barrier properties.

These various types of papers may be used to wrap individual items or portions. Examples include waxed paper wraps for toffees and vegetable parchment paper wraps for butter and margarine. They may be made into small sachets or bags. Examples include sulphite papers sachets for custard powders or cake mixes and bags for sugar and flour. Kraft papers may be fabricated into multiwall papers sacks containing from two to six plies. They are used for fruits, vegetables, grains, sugar and salt in quantities up to 25 kg. Where extra protection is required against water vapour, one or more plies maybe wax- or polyethylene-coated. The outer layer may consist of wet-strength paper.

9.3.1.2 **Paperboards**

Paperboards are made from the same raw materials as papers. They normally are made on the cylinder machine and consist of two or more layers of different quality pulps with a total thickness in the range 300–1100 μm. The types of paperboard used in food packaging include:

- Chipboard, which is made from a mixture of repulped waste with chemical and mechanical pulp. It is dull grey in colour and relatively weak. It is available lined on one side with unbleached, semi or fully bleached chemical pulp. A range of such paperboards are available, with different quality liners. Chipboards are seldom used in direct contact with foods, but are used as outer car-

tons when the food is already contained in a film pouch or bag e.g. breakfast cereals.
- Duplex board, which is made from a mixture of chemical and mechanical pulp, usually lined on both sides with chemical pulp. It is used for some frozen foods, biscuits and similar products.
- Solid white board, in which all plies are made from fully, bleached chemical pulp. It is used for some frozen foods, food liquids and other products requiring special protection.
- Paperboards are available which are coated with wax or polymer materials such as polyethylene, polyvinylidene chloride and polyamides. These are mainly used for packaging wet or fatty foods.

Paperboards are mainly used in the form of cartons. Cartons are fed to the filling machine in a flat or collapsed form where they are erected, filled and sealed. The thicker grades of paperboards are used for set-up boxes which come to the filling machine already erected. These are more rigid than cartons and provide additional mechanical protection.

9.3.1.3 **Moulded Pulp**

Moulded pulp containers are made from a waterborne suspension of mechanical, chemical or waste pulps or mixtures of same. The suspension is moulded into shape either under pressure (pressure injection moulding) or vacuum (suction moulding) and the resulting containers are dried. Such containers have good cushioning properties and limit in-pack movement, thus providing good mechanical protection to the contents. Trays for eggs and fruits are typical examples.

9.3.1.4 **Fibreboards**

Fibreboard is available in solid or corrugated forms. Solid fibreboard consists of a layer of paperboard, usually chipboard, lined on one or both faces with Kraft paper. Solid fibreboard is rigid and resistant to puncturing. Corrugated fibreboard consists of one or more layers of corrugated material (medium) sandwiched between flat sheets of paperboard (linerboard), held in place by adhesive. The medium may be chipboard, strawboard or board made from mixtures of chemical and mechanical pulp. The completed board may have one (single wall) two (double wall) or three (triple wall) layers of corrugations with linerboard in between. Four different flute sizes are available:
- A (104–125 flutes m^{-1}) is described as coarse and has good cushioning characteristics and rigidity.
- B (150–184 flutes m^{-1}) is designated as fine and has good crush resistance.
- C (120–145 flutes m^{-1}) is a compromise between these properties.
- E (275–310 flutes m^{-1}) is classed as very fine and is used for small boxes and cartons, when some cushioning is required.

Wax and polymer coated fibreboards are available. Fibreboards are usually fabricated into cases which are used as outer containers, to provide mechanical protection to the contents. Unpackaged products such as fruits, vegetables and eggs are packaged in such containers. Inserts within the case reduce in-pack movement. Fibreboard cases are also used for goods already packaged in pouches, cartons, cans and glass containers.

9.3.1.5 **Composite Containers**

So called composite containers usually consist of cylindrical bodies made of paperboard or fibreboard with metal or plastic ends. Where good barrier properties are required, coated or laminated board may be used for the body or aluminium foil may be incorporated into it. Small containers, less than 200 mm in diameter, are referred to as tubes or cans and are used for foods such as salt, pepper, spices, custard powders, chocolate beverages and frozen fruit juices. Larger containers, known as fibreboard drums, are used as alternatives to paper or plastic sacks or metal drums for products such as milk powder, emulsifying agents and cooking fats [13, 16–22].

9.3.2 Wooden Containers

Outer wooden containers are used when a high degree of mechanical protection is required during storage and transport. They take the form of crates and cases. Wooden drums and barrels are used for liquid products. The role of crates has largely been replaced by shipping containers. Open cases find limited use for fish, fruits and vegetables, although plastic cases are now widely used. Casks, kegs and barrels are used for storage of wines and spirits. Oak casks are used for high quality wines and spirits. Lower quality wines and spirits are stored in chestnut casks [13].

9.3.3 Textiles

Jute and cotton are woven materials which have been used for packaging foods. Sacks made of jute are used, to a limited extent, for fresh fruit and vegetables, grains and dried legumes. However, multiwall paper sacks and plastic sacks have largely replaced them for such products. Cotton bags have been used in the past for flour, sugar, salt and similar products. Again, paper and plastic bags are now mainly used for these foods. Cotton scrims are used to pack fresh meat. However, synthetic materials are increasingly used for this purpose [13].

9.3.4 Flexible Films

Nonfibrous materials in continuous sheet form, up to 0.25 mm thick, are termed packaging films. They are flexible, usually transparent, unless deliberately pigmented and, with the exception of regenerated cellulose, thermoplastic to some extent. This latter property enables many of them to be heat-sealed. With the exception of regenerated cellulose, most films consist of a polymer, or a mixture of two or more polymers, to which are added other materials to give them particular functional properties, alter their appearance or improve their handling characteristics. Such additives may include plasticisers, stabilisers, colouring materials, antioxidants, antiblocking and slip agents.

- Extrusion is the method most commonly used to produce polymer films. The mixture of polymer and additives is fed into the extruder, which consists of a screw revolving inside a close-fitting, heated barrel. The combination of the heat applied to the barrel and that generated by friction, melts the mixture, which is then forced through a die in the form of a tube or flat film. The extrudate is stretched to control the thickness of the film and rapidly cooled. By using special adaptors, it is possible to extrude two or more different polymers, simultaneously. They fuse together to form a single web. This is known as coextrusion.
- Calendering is another techniques used to produce polymer films and sheets. The heated mixture of polymer and additives is squeezed between a series of heated rollers with a progressively decreasing clearance. The film formed then passes over cooled rollers. Some polyvinylchloride, ethylene-vinyl acetate and ethylene-propylene copolymers are calendered.
- Solution casting is also used to a limited extent. The plastic material, with additives, is dissolved in a solvent, filtered and the solution cast through a slot onto a stainless steel belt. The solvent is driven off by heating. The resulting film is removed from the belt. Films produced in this way have a clear, sparkling appearance. Cellulose acetate and ethyl cellulose films are among those that are produced by solvent casting.
- Orientation is a process applied to some films in order to increase their strength and durability. It involves stretching the film in one (uniaxial orientation) or two directions at right angles to each other (biaxial orientation). This causes the polymer chains to line up in a particular direction. In addition to their improved strength, oriented films have better flexibility and clarity and, in some cases, lower permeability to water vapour and gases, compared to nonoriented forms of the same polymer film. Oriented films tear easily and are difficult to heat-seal. The process involves heating the film to a temperature at which it is soft before stretching it. Flat films are passed between heated rollers and then stretched on a machine known as a tenter, after which they are passed over a cooling roller. Films in the form of tubes are flattened by passing through nip rollers, heated to the appropriate temperature and stretched by increasing the air pressure within the tube. When stretched to

the correct extent they are cooled on rollers. Polyester, polypropylene, low-density polyethylene and polyamide are the films that are mainly available in oriented form.
- Irradiation of some thermoplastic films can bring about crosslinking of the C-C bonds, which can increase their tensile strength, broaden their heating-sealing range and improve their shrink characteristics. Polyethylene is the film most widely irradiated, using an electron beam accelerator.

The following are brief details of the packaging films which are commonly used to package food.

9.3.4.1 **Regenerated Cellulose**

Regenerated cellulose (cellophane) differs from the polymer films in that it is made from wood pulp. Good quality, bleached sulphite pulp is treated with sodium hydroxide and carbon disulphide to produce sodium cellulose xanthate. This is dispersed in sodium hydroxide to produce viscose. The viscose is passed through an acid-salt bath which salts out the viscose and neutralises the alkali. The continuous sheet of cellulose hydrate produced in this way is desulphured, bleached and passed through a bath of softener solution to give it flexibility. It is then dried in an oven. This is known as plain (P) regenerated cellulose. It is clear, transparent, not heat-sealable and has been described as a transparent paper. It provides general protect against dust and dirt, some mechanical protection and is greaseproof. When dry it is a good barrier to gases, but becomes highly permeable when wet. Plain cellulose is little used in food packaging. Plain regenerated cellulose is mainly used coated with various materials which improve its functional properties. The most common coating material is referred to as 'nitrocellulose' but is actually a mixture of nitrocellulose, waxes, resins, plasticisers and some other agents. The following code letters are used to reflect the properties of coated regenerated cellulose films:
- A: anchored coating i.e. lacquer coating
- D: coated on one side only
- M: moistureproof
- P: uncoated
- Q: semimoistureproof
- S: heat-sealable
- T: transparent
- X or XD: copolymer coated on one side only
- XX: copolymer coated on both sides.

The types of film most often used for food packaging include:
- MSAT: nitrocellulose-coated on both sides, a good barrier to water vapour, gases and volatiles and heat-sealable;
- QSAT: nitrocellulose-coated on both sides, more permeable to water vapour than MSAT and heat-sealable;

- DMS: nitrocellulose-coated on one side only;
- MXXT: copolymer coated on both sides, very good barrier to water vapour, gases and volatiles, strong heat-seal;
- MXDT: copolymer coated on one side only.

The copolymer used in the X and XX films is a mixture of polyvinyl chloride (PVC) and polyvinylidene chloride (PVdC). The various coated films are used in the form of pouches and bags and as a component in laminates.

9.3.4.2 Cellulose Acetate

Cellulose acetate is made from waste cotton fibres which are acetylated and partially hydrolysed. The film is made by casting from a solvent or extrusion. It is clear, transparent and has a sparkling appearance. It is highly permeable to water vapour, gases and volatiles. It is not much used in food packaging except as window material in cartons. It can be thermoformed into semirigid containers or as blister packaging.

9.3.4.3 Polyethylene

Polyethylene (PE), commonly called polythene, is made in one of two ways. Ethylene is polymerised at high temperature and pressure, in the presence of a little oxygen and the polymer converted into a film by extrusion. Alternatively, lower temperatures and pressures may be used to produce the polymer if certain alkyl metals are used as catalysts. The film is available in low (LDPE), medium (MDPE) and high (HDPE) density grades. The lower density grades are most widely used in food packaging. The main functional properties of LDPE are its strength, low permeability to water vapour and it forms a very strong heat seal. It is not a good barrier to gases, oils or volatiles. It is used on its own in the form of pouches, bags and sacks. It is also used for coating papers, boards and plain regenerated cellulose and as a component in laminates. HDPE has a higher tensile strength and stiffness than LDPE. Its permeability to gases is lower and it can withstand higher temperatures. It is used for foods which are heated in the package, so called 'boil in the bag' items.

9.3.4.4 Polyvinyl Chloride

Polyvinyl chloride (PVC) is made by chlorination of acetylene or ethylene followed by polymerisation under pressure in the presence of a catalyst. The film can be formed by extrusion or calendering. It is a clear, transparent film which on its own is brittle. The addition of plasticizers and stabilisers to the polymer are necessary to give it flexibility. It is essential that PVC film used in food packaging contains only permitted additives to avoid any hazard to the consumer (see Section 9.2.6). It has good mechanical properties. Its permeability to water vapour, gases and volatiles depends on the type and amount of plasticizers added to the polymer. The most common grade used for food packaging is slightly more permeable to water vapour than LDPE, but less permeable to gases and volatiles. It is a good grease barrier. It can be sealed by high-frequency welding. It can be orientated and as such is heat-shrinkable. Highly plasticised grades are available with stretch and cling properties. This is one form of 'cling film' which is used for stretch-wrapping foods in industry and in the home.

9.3.4.5 **Polyvinylidene Chloride**

Polyvinylidene chloride (PVdC) is made by further chlorination of vinyl chloride in the presence of a catalyst, followed by polymerisation. The polymer itself is stiff and brittle and unsuitable for use as a flexible film. Consequently, it is a copolymer of PVdC with PVC that is used for food packaging. Typically, 20% of VC is used in the copolymer, although other ratios are available. The film is usually produced by extrusion of the copolymer. The properties of the copolymer film depend on the degree of polymerisation, the properties of the monomers used and the proportion of each one used. The copolymer film most widely used for food packaging has good mechanical properties, is a very good barrier to the passage of water vapour, gases and volatiles and is greaseproof and heat-sealable. It can withstand relatively high temperatures such as those encountered during hot filling and retorting. It is available in oriented form which has improved strength and barrier properties and is highly heat-shrinkable. PVdC/PVC copolymer film is used for shrink-wrapping foods such as meat and poultry products and as a component in laminates.

9.3.4.6 **Polypropylene**

Polypropylene (PP) is produced by low-pressure polymerisation of propylene in the presence of a catalyst. The film is normally extruded onto chilled rollers and is known as cast polypropylene. Its mechanical properties are good except at low temperature, when it becomes brittle. The permeability of cast PP to water vapour and gases is relatively low, comparable with high-density polyethylene. It is heat-sealable, but at a very high temperature, 170 °C. It is usually coated with PE or PVdC/PVC copolymer to facilitate heat-sealing. Cast PP is used in the form of bags or overwraps for applications similar to PE. Oriented polypropylene (OPP) has better mechanical properties than cast PP, particularly at low temperature, and is used in thinner gauges. It is a good barrier to water vapour but not gases. It is often coated with PP or PVdC/PVC copolymer to improve its barrier properties and to make it heat-sealable. It is normally heat-shrinkable. It is used in coated or laminated form to package a wide range of food products, including biscuits, cheese, meat and coffee. It is stable at relatively high temperature and is used for in-package heat processing. A white opaque form of OPP, known as pearlised film, is also available. Copolymers of PP and PE are also available. Their functional properties tend to be in a range between PP and HDPE.

9.3.4.7 Polyester

Polyester (PET) film used in food packaging is polyethylene terephthalate, which is usually produced by a condensation reaction between terephthalic acid and ethylene glycol and extruded. There is little use of the nonoriented form of PET but it widely used in the biaxially oriented form. Oriented PET has good tensile strength and can be used in relatively thin gauges. It is often used coated with PE or PVdC/PVC copolymer to increase its barrier properties and facilitate heat-sealing. It is stable over a wide temperature range and can be used for 'boil in the bag' applications. Metallised PET is also available and has a very low permeability to gases and volatiles. Metallised, coextruded PE/PET is used for packaging snack foods.

9.3.4.8 Polystyrene

Polystyrene (PS) is produced by reacting ethylene with benzene to form ethylbenzene. This is dehydrogenated to give styrene which is polymerised at a relatively low temperature, in the presence of catalysts, to form polystyrene. PS film is produced by extrusion. It is stiff and brittle with a clear sparkling appearance. In this form it is not useful as a food packaging film. Biaxially oriented polystyrene (BOPS) is less brittle and has an increased tensile strength, compared to the non-oriented film. BOPS has a relatively high permeability to vapours and gases and is greaseproof. It softens at ca. 80–85 °C, but is stable at low temperature, below 0 °C. It shrinks on heating and may be heat-sealed by impulse sealers. The film has few applications in food packaging, apart from wrapping of fresh produce. PS is widely used in the form of thermoformed semirigid containers and blow-moulded bottle. For these applications it is coextruded with ethylene-vinyl alcohol (EVOH) or PVdC/PVC copolymer. PS is also used in the form of a foam for containers such as egg cartons, fruit trays and containers for takeaway meals.

9.3.4.9 Polyamides

Polyamides (PAs) known generally as Nylons, are produced by two different reactions. Nylon 6,6 and 6,10 are formed by condensation of diamines and dibasic acids. The numbers indicate the number of carbon atoms in the diamine molecules followed by the number in the acid. Nylon 11 and 12 are formed by condensation of ω-amino acids. Here, the numbers indicate the total number of carbon atoms involved. The film may be extruded or solution cast. PA films are clear and attractive in appearance. As a group they are mechanically strong, but the different types do vary in strength. The permeability to water vapour varies from high, Nylon 6, to low, Nylon 12. They are good barriers to gases, particularly under dry conditions, volatiles and greases. They are stable over a very wide temperature range. They can be heat-sealed but at a high temperature, 240 °C. They do absorb moisture and their dimensions can change by 1–2% as a result. Nylon films may be combined with other materials, by coating, coextrusion or lamination, in order to facilitate heat-sealing and/or improve their mechanical and barrier properties. Polyethylene, ionomers, EVA and EAA (see below) are among such other materials. Different types of Nylon may be combined as copolymers e.g. Nylon 6/6,6 or 6/12. Biaxially oriented Nylon films are also available. Their functional properties may be further modified by vacuum-metallising. Applications for Nylon films include packaging of meat products, cheese and condiments.

9.3.4.10 **Polycarbonate**

Polycarbonate (PC) is made by the reaction of phosgene or diphenyl carbonate with bisphenol A. The film is produced by extrusion or casting. It is mechanically strong and grease-resistant. It has a relatively high permeability to vapours and gases. It is stable over a wide temperature range, from –70 °C to 130 °C. It is not widely used for food packaging but could be used for 'boil in the bag' packages, retortable pouches and frozen foods.

9.3.4.11 **Polytetrafluoroethylene**

Polytetrafluoroethylene (PTFE) is made by the reaction of hydrofluoric acid with chloroform followed by pyrolysis and polymerisation. The film is usually produced by extrusion. It is strong, has a relatively low permeability to vapours and gases and is grease-resistant. It is stable over a wide temperature range, –190 °C to 190 °C and has a very low coefficient of friction. It is not widely used in film form for packaging of foods but could be used for retortable packages and where a high resistance to the transfer of vapours and gases is required e.g. for freeze-dried foods. It is best known for its nonstick property and is used on heat sealers and for coating cooking utensils.

9.3.4.12 **Ionomers**

'Ionomers' are formed by introducing ionic bonds as well as the covalent bonds normally present in polymers such as PE. This is achieved by reacting with metal ions. Compared to LDPE, they are stiffer and more resistant to puncturing and have a higher permeability to water vapour and good grease resistance. They are most widely used as components in laminates with other films, such as PC or PET, for packaging cheese and meat products.

9.3.4.13 **Ethylene-vinyl Acetate Copolymers**

Ethylene-vinyl acetate copolymers (EVAs) are made by the polymerisation of polyethylene with vinyl acetate. Compared to LDPE, they have higher impact strength, higher permeability to water vapour and gases and are heat-sealable over a wider temperature range. EVA with other polymers such as ethylene-ethyl acrylate (EEA) and ethylene-acrylic acid (EAA) form a family of materials that may be used, usually in laminates with PE, PP and other films, for food packaging. Care must be taken in selecting these materials as there are limitations on the quantity of the minor components which should be used for particular food applications. EVA itself has very good stretch and cling characteristics and can be used, as an alternative to PVC for cling-wrap applications [13, 16, 17, 23, 24].

9.3.5 Metallised Films

Many flexible packaging films can have a thin coating, less than 1 μm thick, of metal applied to them. This was originally introduced for decorative purposes. However, it emerged that metallising certain films increased their resistance to the passage of water vapour and gases, by up to 100%. Today metallised films are used extensively to package snack foods. The process involves heating the metal, usually aluminium, to temperatures of 1500–1800 °C in a vacuum chamber maintained at a very low pressure, ca. 10^{-4} Torr (0.13 Pa). The metal vaporises and deposits onto the film which passes through the vapour on a chilled roller. PET, PP, PA, PS, PVC, PVdC and regenerated cellulose are available in metallised form [17, 24, 25].

9.3.6 Flexible Laminates

When a single paper or film does not provide adequate protection to the product, two or more flexible materials may be combined together in the form of a laminate. In this form the functional properties of the individual components complement each other to suit the requirements of a particular food product. The materials involved may include papers or paperboards, films and aluminium foil (see Section 10.3.8). The paper or paperboard provides stiffness, protects the foil against mechanical damage and has a surface suitable for printing. The film(s) contributes to the barrier properties of the laminate, provides a heat-sealable surface and strengthens the laminate. The foil acts as a barrier material and has an attractive appearance. Laminates may be formed from paper-paper, paper-film, film-film, paper-foil, film-foil and paper-film- foil combinations. The layers of a laminate may be bonded together by adhesive. When one or more of the layers is permeable to water vapour, an aqueous adhesive may be used. Otherwise, nonaqueous adhesives must be used. If one or more of the components is thermoplastic, it may be bonded to the other layer by passing them between heated rollers. A freshly extruded thermoplastic material, still in molten form, may be applied directly to another layer and thus bonded to it. Two or more thermoplastic materials may be combined to together by coextrusion (see Section 9.3.4). There are hundreds of combinations of different materials available. Examples include:

- vegetable parchment-foil for wrapping butter and margarine;
- MXXT regenerated cellulose-PE for vacuum packed cheese, cooked and cured meats;
- PET-PE for coffee, paperboard-foil-PE for milk and fruit juice cartons.

Retortable pouches may be made of a threeply laminate typically consisting of PET-foil-PP or PET- foil-HDPE [13, 16, 17, 24].

9.3.7 Heat-Sealing Equipment

Many flexible polymer films are thermoplastic and heat sealable. Nonthermoplastic materials may be coated with or laminated to thermoplastic material to facilitate heat-sealing. Heat-sealing equipment must be selected to suit the type of material being sealed. Nonthermoplastic materials such as papers, regenerated cellulose and foil, which are coated with heat-sealable material, are best sealed with a *hot bar* or *resistance* sealer. The two layers of material are clamped between two electrically heated metal bars. The temperature of the bars, the pressure exerted by them and the contact time all influence the sealing. Metal jaws with matching serrations are often used for coated regenerated cellulose films. The serrations stretch out wrinkles and improve the seal. In the case of laminates, smooth jaws meeting uniformly along their length are used. Alternatively, one of the jaws is made of resilient material, often silicone elastomer, which is not heated.

For continuous heat-sealing of coated material, heated rollers are used. Heated plates are used to seal wrapped items. For unsupported, thermoplastic materials, *impulse sealers* are used. In such sealers, the layers of film are clamped between jaws of resilient material, one or both of which has a narrow metal strip running the length of the jaw. An accurately timed pulse of low-voltage electricity is passed through the strip(s), heating it and fusing the two layers of material together. The jaws are held apart by unmelted material each side of the strip. This minimises the thinning of the sealed area, which would weaken the seal. The jaws remain closed until the melted material solidifies. The jaws are coated with PTFE to prevent the film sticking to them. For continuous sealing of unsupported thermoplastic material a *band sealer* may be used. A pair of moving, endless metal belts or bands is heated by stationary, heated shoes. The shoes are so shaped that they touch the centre of the bands only. This minimises thinning of the seal. After passing between the heated shoes, the layers of material pass between pressure rollers and then between cooled shoes to solidify the melted film. A heated wire may be used to simultaneously cut and seal unsupported thermoplastic films. *Electronic sealing* is used on relatively thick layers of polymer material with suitable electrical properties, mainly PVC and PVC/PVdC copolymers. The layers of film are placed between shaped electrodes and subjected to a high-frequency electric field. This welds the layers of material together. *Ultrasonic sealing* may be used to seal layers of film or foil together. This is particularly suited to uncoated, oriented materials that are difficult to seal by other methods [15–17, 24].

9.3.8
Packaging in Flexible Films and Laminates

Flexible films may be used to overwrap items of food such as portions of meat. The meat is usually positioned on a tray made of paperboard or foamed plastic, with an absorbent pad between it and the tray. The film is stretched over the meat and under the tray. It may be heat sealed on a heated plate or held in position by clinging to itself. Films may also be made into preformed bags which are filled by hand or machine and sealed by heat or other means. Heavy gauge material, such as PE, may be made into shipping sacks for handling large amounts, 25–50 kg, of foods such as grains or milk powder.

However, films and laminates are most widely used in the form of sachets or pillow packs. A sachet is a small square or rectangular pouch heat-sealed on all four edges (see Figs. 9.1, 9.3). A pillow pack is a pouch with a longitudinal heat seal and two end seals (see Figs. 9.2, 9.4). These are formed, filled and sealed by a sequential operation, known as a form-fill-seal (FFS) system. Form-fill-seal machines may operate vertically or horizontally. The principle of one vertical FFS machine, for making sachets, is shown in Fig. 9.1.

Vertical FFS machines can also make pillow packs (see Fig. 9.2).

Vertically formed pillow and sachet packs may be used for liquids and solids. The principle of one horizontal FFS for sachets is shown in Fig. 9.3.

Such a system is used for both solid and liquid products. Horizontal FFS machines can also produce pillow packs (see Fig. 9.4).

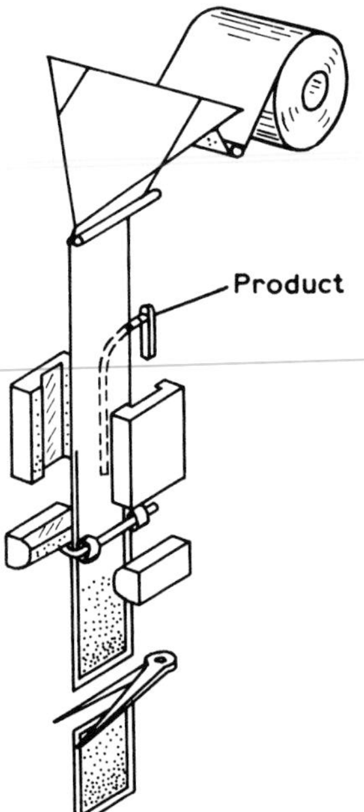

Fig. 9.1 Vertical form-fill-seal machine for sachets; adapted from [16] with permission.

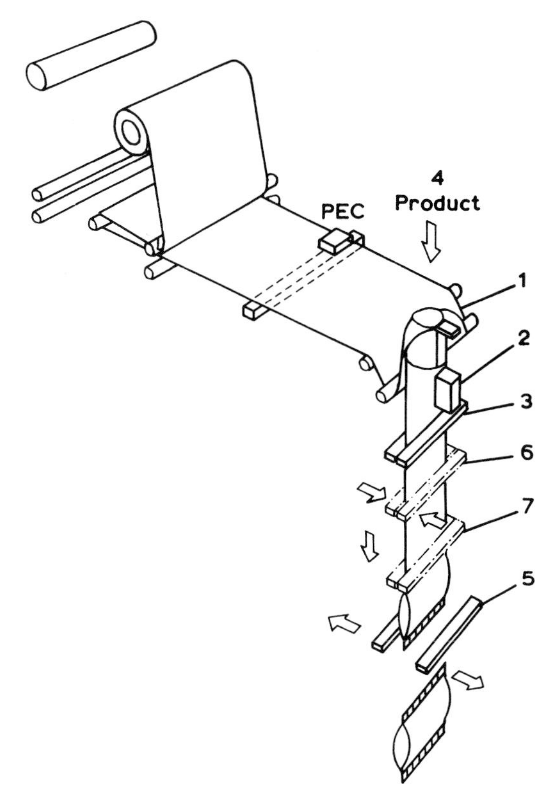

Fig. 9.2 Vertical form-fill-seal machine for pillow packs. 1. Film from reel made into a tube over forming shoulder. 2. Longitudinal seal made. 3. Bottom of tube closed by heat crimped jaws which move downwards drawing film from reel. 4. Predetermined quantity of product falls through collar into pouch. 5. Jaws open and return on top of stroke. 6. Jaws partially close and 'scrape' product into pouch out of seal area. 7. Jaws close, crimp seal top of previous pouch and bottom of new one. Crimp-sealed container cut-off with knife. Adapted from [16] with permission.

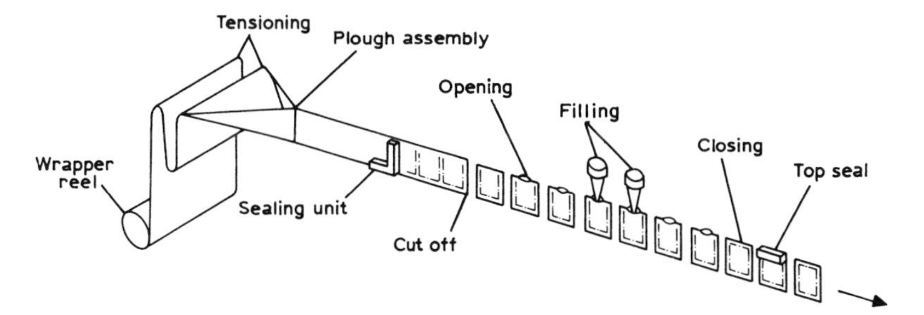

Fig. 9.3 Horizontal form-fill-seal machine for sachets; adapted from [16] with permission.

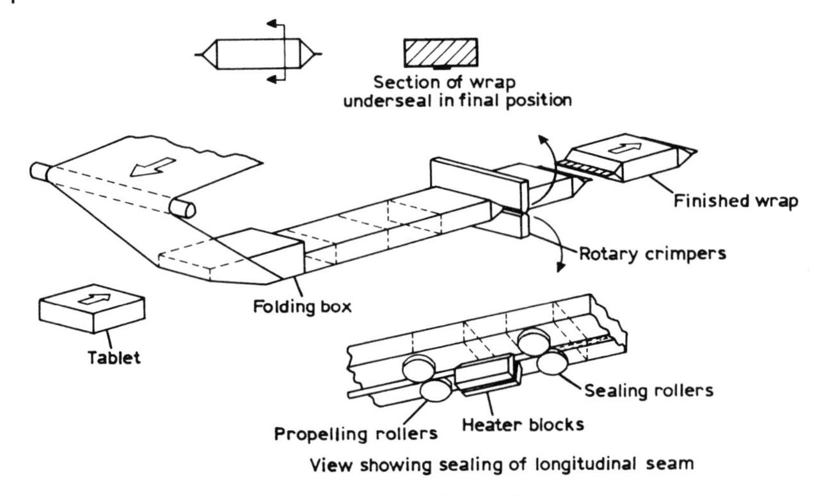

Fig. 9.4 Horizontal form-fill-seal machine for pillow packs. Film is drawn from reel and formed into horizontal tube around product with continuous seal underneath formed by heater blocks and crimping rollers. Then, rotary heaters make the crimped end seals and cut-off produces individual packs. Adapted from [16] with permission.

Systems like these are used for solid items such as candy bars or biscuits.

Pillow packs are more economical than sachets in the use of packaging material. The packaging materials must be thin and flexible, have good slip characteristics and form a strong seal, even before cooling. Sachets are made from stiffer material and can be used for a wider range of product types. They are usually used in relatively small sizes, e.g. for individual portions of sauce or salad dressing [13, 15–17, 24, 26].

9.3.9 Rigid and Semirigid Plastic Containers

Many of the thermoplastic materials described above can be formed into rigid and semirigid containers, the most common being LDPE, HDPE, PVC, PP, PET and PS, singly or in combinations. Acrylic plastics are also used for this purpose, including polyacrylonitrile and acrylonitrile-butadiene-styrene (ABS). Urea formaldehyde, a thermosetting material, is used to make screwcap closures for glass and plastic containers.

The following methods are used to convert these materials into containers:

9.3.9.1 Thermoforming

In thermoforming, a plastic sheet is clamped in position above a mould. The sheet is heated until it softens and then made to take up the shape of the mould by either (a) having an air pressure greater than atmospheric applied above the sheet, (b) having a vacuum created below the sheet or (c) sandwiching the sheet between a male and female mould. The sheet cools through contact with the mould, hardens and is ejected from the mould. Plastic materials that are thermoformed include PS, PP, PVC, HDPE and ABS. Thermoforming is used to produce opentopped or widemouthed containers such as cups and tubs for yoghurt, cottage cheese or margarine, trays for eggs or fresh fruit and inserts in biscuit tins or chocolate boxes.

9.3.9.2 **Blow Moulding**

In blow moulding, a mass of molten plastic is introduced into a mould and compressed air is used to make it take up the shape of the mould. The plastic cools, hardens and is ejected from the mould. Blow moulding is mainly used to produce narrownecked containers. LDPE is the main material used for blow moulding, but PVC, PS and PP may also be processed in this way. Food applications include bottles for oils, fruit juices and milk and squeezable bottles for sauces and syrups.

9.3.9.3 **Injection Moulding**

In injection moulding, the molten plastic from an extruder is injected directly into a mould, taking up the shape of the mould. On cooling, the material hardens and is ejected from the mould. Injection moulding is mainly used to produce widemouthed containers, but, narrownecked containers can be injection-moulded in two parts which are joined together by a solvent or welding. PS is the main material used for injection moulding, but PP and PET may also be processed in this way. Food applications include cups and tubs for cream, yoghurt, mousses as well as phials and jars for a variety of uses.

9.3.9.4 **Compression Moulding**

Compression moulding is used from thermosetting plastics, such as urea formaldehyde. The plastic powder is held under pressure between heated male and female moulds. It melts and takes up the shape of the mould after which it is cooled, the mould opened and the item ejected. The main application for this method is to produce screw caps [13, 16, 17, 24, 27–29].

9.3.10 Metal Materials and Containers

The metal materials used in food packaging are aluminium, tinplate and electrolytic chromium-coated steel (ECCS). Aluminium is used in the form of foil or rigid metal.

9.3.10.1 Aluminium Foil

Aluminium foil is produced from aluminium ingots by a series of rolling operations down to a thickness in the range 0.15–0.008 mm. Most foil used in packaging contains not less than 99.0% aluminium, with traces of silicon, iron, copper and, in some cases, chromium and zinc. Foil used in semirigid containers also contains up to 1.5% manganese. After rolling, foil is annealed in an oven to control its ductility. This enables foils of different tempers to be produced from fully annealed (dead folding) to hard, rigid material. Foil is a bright, attractive material, tasteless, odourless and inert with respect to most food materials. For contact with acid or salty products, it is coated with nitrocellulose or some polymer material. It is mechanically weak, easily punctured, torn or abraded. Coating or laminating it with polymer materials will increase its resistance to such damage. Relatively thin foil, less than 0.03 mm thick, will contain perforations and will be permeable to vapours and gases. Again, coating or laminating it with polymer material will improve its barrier properties. Foil is stable over a wide temperature range. Relatively thin, fully annealed foil is used for wrapping chocolate and processed cheese portions. Foil is used as a component in laminates, together with polymer materials and, in some cases, paper. These laminates are formed into sachets or pillow packs on FFS equipment (see Section 9.3.6). Examples of foods packaged in this way include dried soups, sauce mixes, salad dressings and jams. Foil is included in laminates used for retortable pouches and rigid plastic containers for ready meals. It is also a component in cartons for UHT milk and fruit juices. Foil in the range 0.040–0.065 mm thick is used for capping glass and rigid plastic containers. Plates, trays, dishes and other relatively shallow containers are made from foil in the thickness range 0.03–0.15 mm and containing up to 1.5% manganese. These are used for frozen pies, ready meals and desserts, which can be heated in the container.

9.3.10.2 Tinplate

Tinplate is the most common metal material used for food cans. It consists of a low-carbon, mild steel sheet or strip, 0.50–0.15 mm thick, coated on both sides with a layer of tin. This coating seldom exceeds 1% of the total thickness of the tinplate. The structure of tinplate is more complex than would appear from this simple description and several detectable layers exist (see Fig. 9.5).

The mechanical strength and fabrication characteristics of tinplate depend on the type of steel and its thickness. The minor constituents of steel are carbon, manganese, phosphorous, silicon, sulphur and copper. At least four types of steel, with different levels of these constituents, are used for food cans. The corrosion resistance and appearance of tinplate depend on the tin coating. The stages in the manufacture of tinplate are shown in Fig. 9.6. These result in two types of tinplate, i.e. single (or cold) reduced electroplate (CR) and double reduced electroplate (DR). DR electroplate is stronger in one direction than CR plate and can be used in thinner gauges than the latter for certain applications.

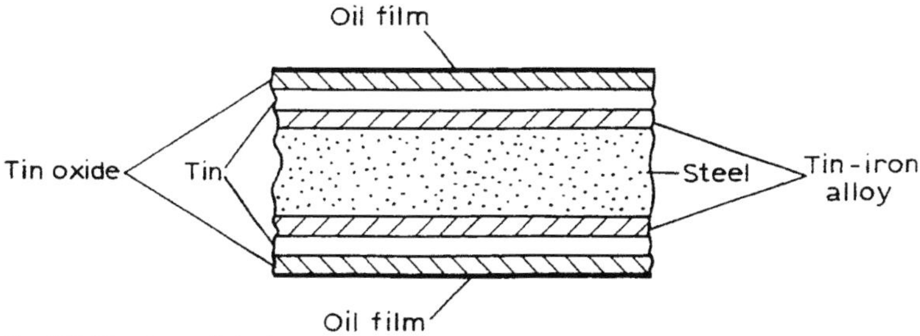

Fig. 9.5 Structure of tinplate.

The thickness of tinplate used for food can manufacture is at the lower end of the range given above. CR plate thickness may be as low as 0.17 mm and DR plate 0.15 mm. The amount of tin coating is now usually expressed as x g m^{-2}. This may be the same on both sides of the plate or a different coating weight may be applied to each side. The latter is known as differentially coated plate. In general, the more corrosive the product the higher the coating weight used. Coating weights range over 11.2–1.1 g m^{-2} (represented as E.11.2/11.2 to E.1.1/1.1) if the same weight is applied to both sides. Differentially coated plate is identified by the letter D followed by the coating weights on each side. For example, D.5.6/2.8 plate has 5.6 g m^{-2} of tin on one side and 2.8 g m^{-2} on the other. Usually, the higher coating weight is applied to the side that will form the inside of the can. Lacquer (enamel)

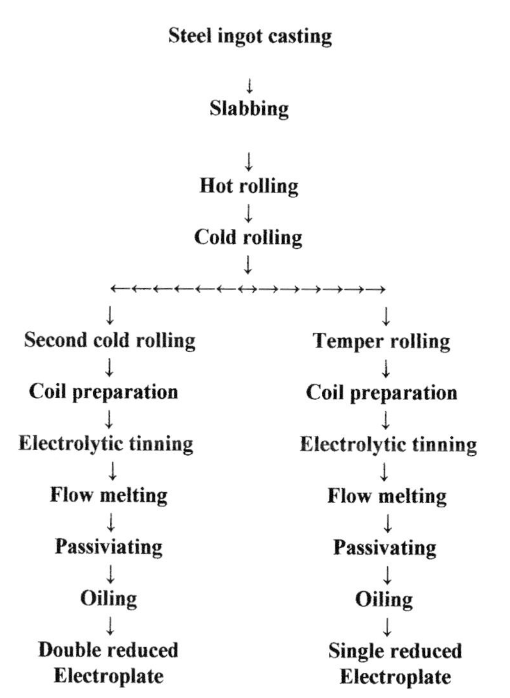

Fig. 9.6 Stages in the manufacture of tinplate.

Table 9.1 Main types of internal can lacquer (enamel); extracted from [32].

General type of resin and compounds blended to produce it	Sulphide stain resistance	Typical uses	Comments
Oleo-resinous	Poor	Acid fruits	Good general purpose natural range at relatively low cost
Sulfur-resistant oleoresinous with added zinc oxide	Good	Vegetables, soups	Not for use with acid products
Phenolic (phenol or relatively low-substituded phenol with vegetables formaldehyde)	Very good	Meat, fish, soups	Good at cost but film thickness restricted by flexibility
Epoxy-phenolic (epoxy resins with phenolic resins)	Poor	Meat, fish, soups vegetables, beer, beverages (top coat)	Wide range of properties may be obtained by modifications
Epoxy-phenolic with zinc oxide	Good	Vegetables, soups (especially can ends)	Not for use with acid products; possible colour change with green vegetables
Aluminized epoxy-phenolic (metallic aluminium powder added)	Very good	Meat products	Clean but dull appearance
Vinyl solution (vinyl chloride-vinyl acetate copolymers)	Not applicable	Spray on can bodies, roller coating on ends, as topcoat for beer and beverages	Free from flavour taints; not usually suitable for direct application to tinplate
Vinyl organosol or plastisol (high MW vinyl resins suspended in a solvent)	Not applicable	Beer and beverage topcoat on ends, drawn cans	As for vinyl solutions, but giving a thicker, tougher layer
Acrylic (acrylic resin usually pigmented white)	Very good when pigmented	Vegetables, soups, prepared foods containing sulphide stainers	Clean appearance
Polybutadiene (hydrocarbon resins)	Very good if zinc oxide added	Beer and beverage first coat, vegetables and soups with ZnO	Costs depend on country

may be applied to tinplate to prevent undesirable interaction between the product and the container. Such interactions arise with: (a) acid foods which may interact with tin dissolving it into the syrup and, in some cases, causing a loss in colour in the product, (b) some strongly coloured products where anthocyanin colour compounds react with the tin, causing a loss of colour, (c) sulphur-containing foods where the sulphur reacts with the tin, causing a blue-black stain on the inside of the can, (d) products sensitive to small traces of tin, such as beer. Lacquers can provide certain functional properties, such as a nonstick surface to facilitate the release of the contents of the can e.g. solid meats packs. A number of such lacquers are available, including natural, oleoresinous materials and synthetic materials. Information on some of these lacquers is presented in Table 9.1. Cans may be made from prelacquered plate or the lacquer may be applied to the made-up can.

9.3.10.3 **Electrolytic Chromium-Coated Steel**

Electrolytic chromium-coated steel (ECCS), sometimes described as tinfree steel, is finding increasing use for food cans. It consists of low-carbon, mild CR or DR steel coated on both sides with a layer of metallic chromium and chromium sesqueoxide, applied electrolytically. It is manufactured by a similar process to that shown in Fig. 9.6, but the flow melting and chemical passivation stages are omitted. A typical coating weight is 0.15 g m^{-2}, much lower than that on tinplate. ECCS is less resistant to corrosion than tinplate and is normally lacquered on both sides. It is more resistant to weak acids and sulphur staining than tinplate. It exhibits good lacquer adhesion and a range of lacquers, suitable for ECCS, is available. The structure of ECCS is shown in Fig. 9.7.

9.3.10.4 **Aluminium Alloy**

Hard-temper aluminium alloy, containing 1.5–5.0% magnesium, is used in food can manufacture. Gauge for gauge, it is lighter but mechanically weaker than tinplate. It is manufactured in a similar manner to aluminium foil. It is less re-

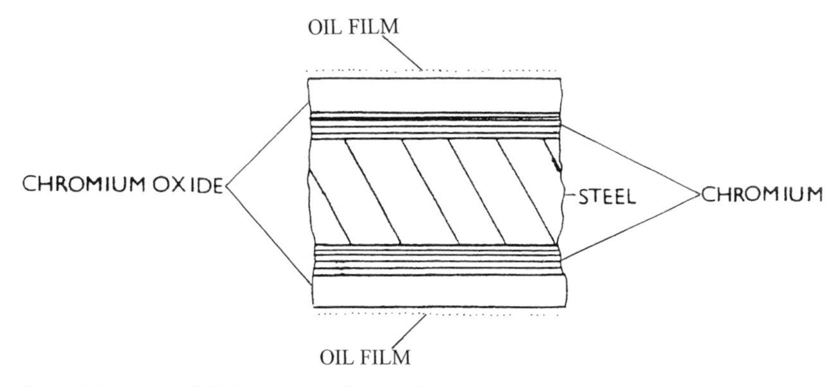

Fig. 9.7 Structure of ECCS plate (tin-free steel).

sistant to corrosion than tinplate and needs to be lacquered for most applications. A range of lacquers suitable for aluminium alloy is available, but the surface of the metal needs to be treated to improve lacquer adhesion.

9.3.10.5 Metal Containers

Metal cans are the most common metal containers used for food packaging. The traditional *three-piece can* (open or sanitary) is still very widely used for heat-processed foods. The cylindrical can body and two ends are made separately. One end is applied to the can body by the can maker, the other (the canners end) by the food processor after the can has been filled with product. The ends are stamped out of sheet metal, the edges curled in and a sealing compound injected into the curl. The body blank is cut from the metal sheet, formed into a cylinder and the lapped, side seam sealed by welding or by polyamide adhesive. Both ends of the cylindrical body are flanged in preparation for the application of the can end (Fig. 9.8a).

The can end is applied to the body by means of double-seaming (Fig. 9.9). The can body and end are clamped tightly between a chuck and a base plate.

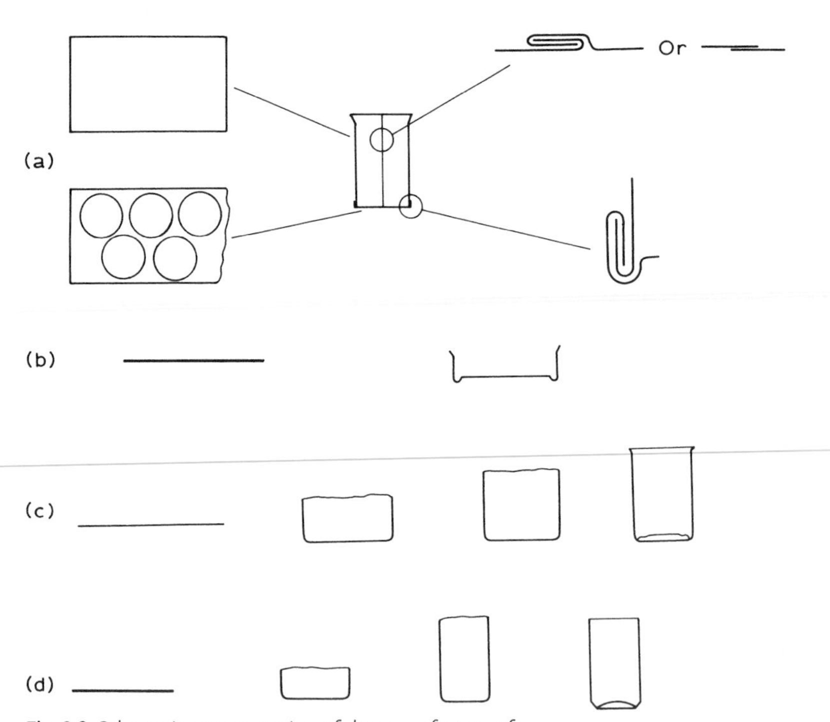

Fig. 9.8 Schematic representation of the manufacture of: (a) three-piece can, (b) drawn can, (c) drawn and redrawn can, (d) drawn and wall-ironed can; from [15] with permission of the authors.

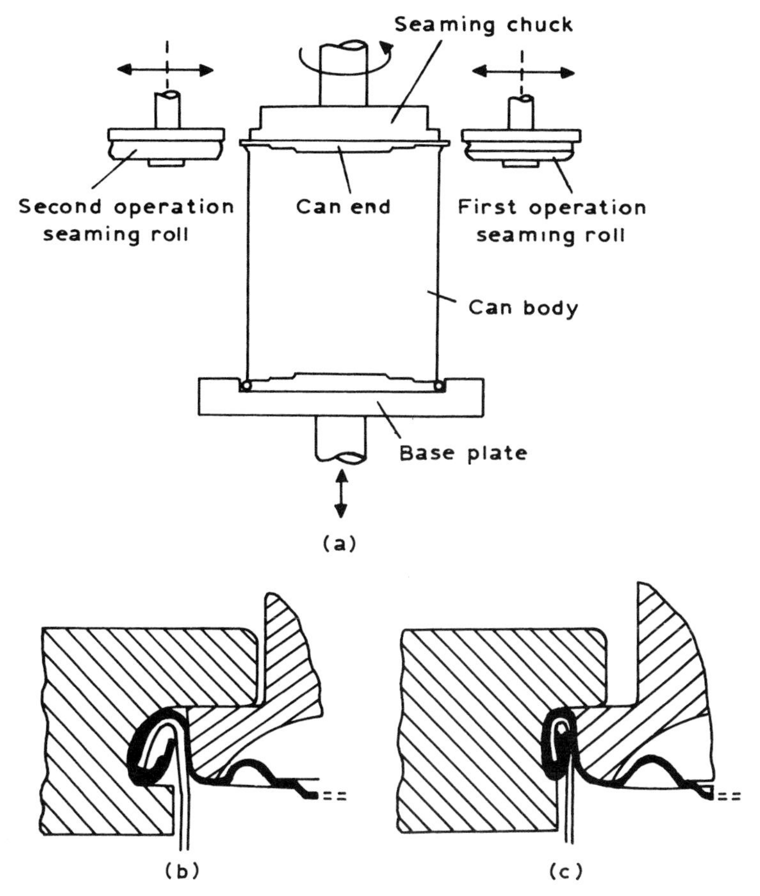

Fig. 9.9 Double-seaming of cans: (a) view of seamer, (b) seam after completion of first operation, (c) seam after completion of second operation; adapted from [15] with permission of the authors.

The chuck is made to rotate rapidly; and the can body and base rotate with the chuck. The first seaming roller moves in and engages with the chuck, forming mating hooks on the can body and end. The first seaming roller moves out and a second roller moves in, tightens the hooks and completes the seam. In some high-speed seaming machines and those used for noncylindrical cans, the can body, end and chuck remain stationary and the seaming rollers rotate on a carriage around them.

The *drawn can* (DR can) is a type of two-piece container. The can body and base are made in one operation from a blank metal sheet by being pressed out with a suitable die. The open end of the body is flanged. The can end, manufactured as described above, is applied to the body by double-seaming after the can is filled with product. Because of the strain on the metal, DR cans are shallow with a maximum height:diameter ratio of 1:2 (Fig. 9.8 b). The *drawn and re-drawn can* (DRD) is another type of two-piece can. It is made by drawing a cup to a smaller diameter in a series of stages to produce a deeper container than the DR can. The can end is applied to the filled can by double-seaming. DRD cans are usually relatively small, cylindrical and have a height:diameter ratio of up to 1.2:1.0 (Fig. 9.8 c). The *drawn and wall-ironed* can (DWI) is made from a disc of metal 0.30–0.42 mm thick. This is drawn into a shallow cup which is forced through a series of ironing rings of reducing internal diameter so that the wall of the cup gets thinner and higher. The top of the body is trimmed, flanged and the end applied by double-seaming after filling the can (Fig. 9.8 d). Because of the very thin body wall, typically 0.10 mm thick, DWI cans are mainly used for packaging carbonated beverages. The internal pressure supports the thin wall.

The dimensions of cylindrical cans are usually specified in diameter and height, in that order. In many countries the units of diameter and height are millimetres. In the UK and USA inches and 16ths of an inch are used. Thus a can specified as 401/411 has a diameter of $4\,^{1}/_{16}$ inches and a height of $4\,^{11}/_{16}$ inches. In the case of rectangular or oval cans, two horizontal dimensions must be given.

Other metal containers used for packaging foods include:

- cylindrical cans with a friction plug closure at the canners end, used for dry powders such as coffee and custard powders or for liquids such as syrups and jams;
- rectangular or cylindrical containers with push-on lids, often sealed with adhesive tape, used for biscuits and sweets;
- rectangular or cylindrical containers, incorporating apertures sealed with screwcaps, used for liquids such as cooking oils and syrups;
- metal drums used for beer and other carbonated drinks [13, 14, 16, 17, 30–36].

9.3.11 Glass and Glass Containers

In spite of the many developments in plastic containers, glass is still widely used for food packaging. Glass is inert with respect to foods, transparent and impermeable to vapours, gases and oils. Because of the smooth internal surface of glass containers, they can be washed and sterilised and used as multitrip containers, e.g. milk and beer bottles. However, glass containers are relatively heavy compared to their metal or plastic counterparts, susceptible to mechanical damage and cannot tolerate rapid changes in temperature (low thermal shock resistance). Broken glass in a food area is an obvious hazard. The composition of a typical UK glass is shown in Table 9.2.

These ingredients, together with up to 30% recycled glass or cullet, are melted in a furnace at temperatures in the range 1350–1600 °C. The viscous mass passes into another chamber which acts as a reservoir for the forming machines. Two forming methods are used, i.e. the blow and blow (B & B) process,

Table 9.2 Composition of a typical British glass.

Silica (from sand)	72.0%
Lime (from limestone)	11.0%
Soda (from synthetic sodium carbonate)	14.0%
Alumina (from aluminium minerals)	1.7%
Potash (as impurity)	0.3%

which is used for narrownecked containers, and the press and blow (P & B) process, which is used for widemouthed containers (see Fig. 9.10).

After forming, the containers are carried on a conveyor belt through a cooling tunnel, known as an annealing lehr. In this lehr, they are heated to just below the softening temperature of the glass, held at that temperature for about 5 min and then cooled in a controlled manner. This is done to remove any stresses in the glass that may have developed during forming and handling. These stresses would weaken the containers and make them less resistant to mechanical damage. As the containers leave the lehr, they are inspected for faults. To produce coloured containers, colouring compounds such as metal oxides, sulphides or selenides are included in the formulation. It is important that the dimensions and capacities of glass containers only vary within specified tolerance limits. Otherwise, breakages and hold ups may occur in the bottling plant and customer complaints may arise. When a delivery of new glass containers is received at the bottling plant, samples should be removed on a statistical basis and their di-

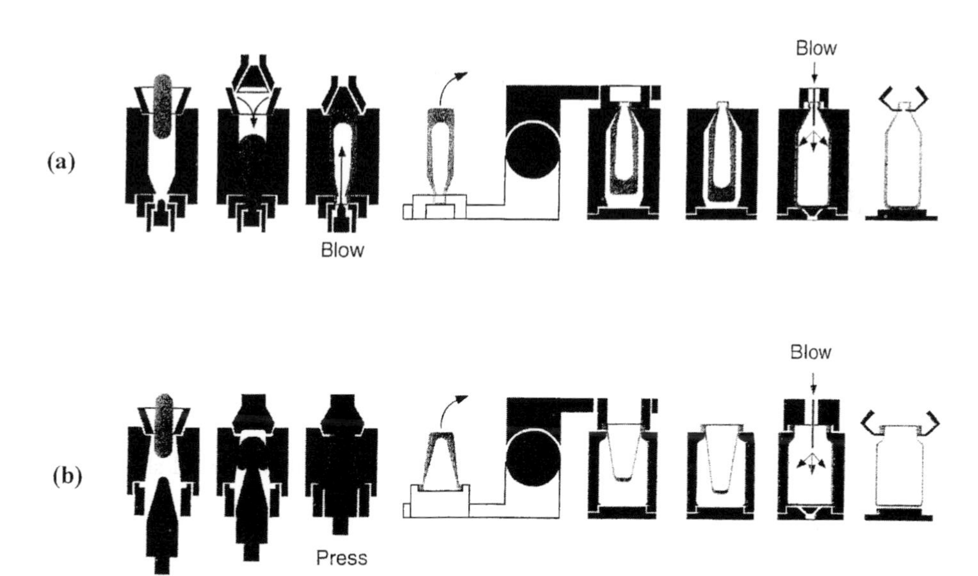

Fig. 9.10 Methods of forming glass containers: (a) blow and blow forming, (b) press and blow forming for wide-mouthed containers; by courtesy of Rockware Glass Ltd.

mensions and capacities measured and checked against specifications. In the case of cylindrical containers, the dimensions that are usually measured are the height, diameter, verticality (how truly vertical the container is) and ovality (how truly cylindrical it is). In the case of noncylindrical containers, other dimensions may be measured. The mechanical strength of glass containers, i.e. their resistance to internal pressure, vertical loads and impacts, increases with increasing thickness of the glass in the bodies and bases. The design of the container also influences its strength. Cylindrical containers are more durable than more complex shapes featuring sharp corners. The greater the radius of curvature of the shoulder, the more resistant the container is to vertical loads (Fig. 9.11). The thickness of the glass in the base is usually greater than that in the body. The

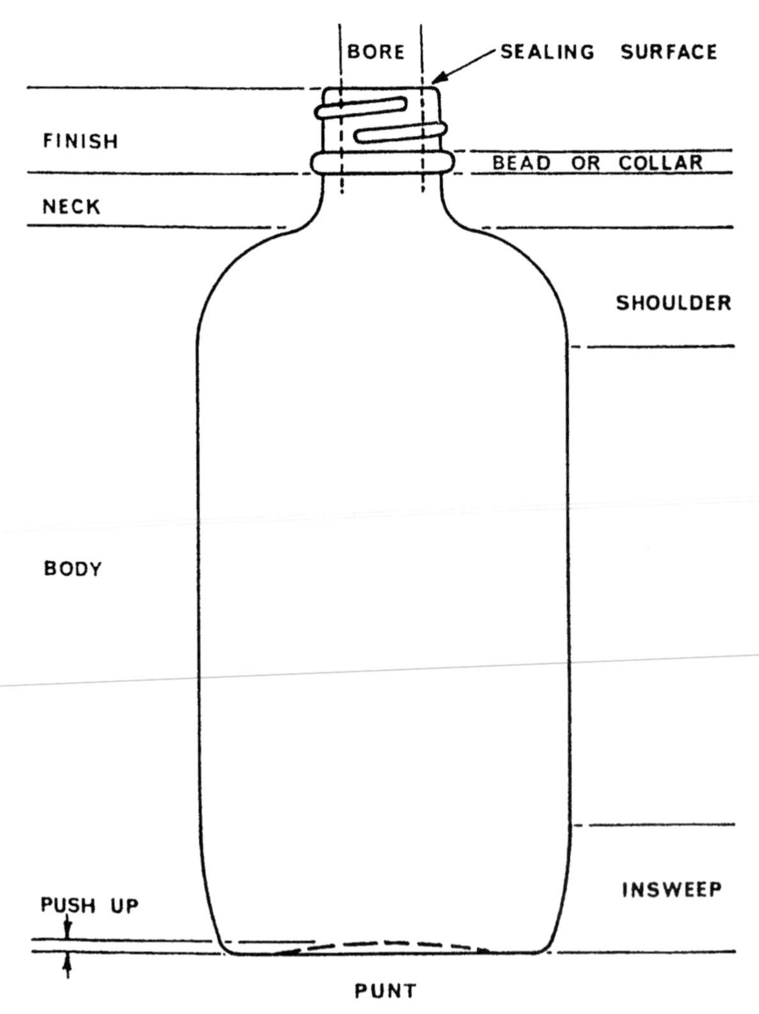

Fig. 9.11 Important features of glass containers.

circle where the body joins the base is weak due to the change in thickness. The insweep (Fig. 9.11) minimises container to container contact in this weak area.

Glass containers become weaker with use, due to abrasion of the outer surface as a result of container to container contact or contact with other surfaces. Treating the surface with compounds of titanium or tin and replacement of the sodium ions at the surface with potassium ions can reduce this problem. The resistance of glass containers to sudden changes in temperature is reduced as the thickness of the glass increases. Thus, when designing glass containers which are to be subjected to heating or cooling, e.g. when the product is to be sterilised or pasteurised in its bottle or jar, or if the container is to be hot-filled with product, a compromise has to be achieved between their mechanical strength and thermal shock resistance. Heating and cooling should be carried out relatively slowly to avoid thermal damage to glass containers.

Glass containers are sealed by compressing a resilient disc, ring or plug against the sealing surface of the container and maintaining it in the compressed condition by means of a retaining cap. The resilient material may be cork, rubber or plastic. The cap is made of metal or plastic. The cap may be screwed on, crimped on or pushed in or onto the finish of the container. Roll-on caps are used as tamper-evident closures. Different closures are effective when: (a) the pressure inside the container is close to atmospheric pressure (normal seal), (b) the pressure inside the container is less than that outside (vacuum seal), (c) the pressure inside the container is higher than that outside (pressure seal). Pressure seals are necessary when packaging carbonated drinks.

Singletrip glass containers are used for liquids such as some beers, soft drinks, wines, sauces, salad dressings and vinegars and for dry foods such as coffee and milk powders. Multitrip containers are used for pasteurised milk, some beers and soft drinks. Products heated in glass containers include sterilised milk, beer, fruit juices and pickled vegetables [13, 16, 17, 37–40].

9.4 Modified Atmosphere Packaging

Modified atmosphere packaging (MAP) is a procedure which involves replacing air inside a package with a predetermined mixture of gases prior to sealing it. Once the package is sealed, no further control is exercised over the composition of the in-package atmosphere. However, this composition may change during storage as a result of respiration of the contents and/or solution of some of the gas in the product.

Vacuum packaging is a procedure in which air is drawn out of the package prior to sealing but no other gases are introduced. This technique has been used for many years for products such as cured meats and cheese. It is not usually regarded as a form of MAP.

In MAP proper, the modified atmosphere is created by one of two methods. In the case of trays, the air is removed by a vacuum pump and the appropriate mixture of gases introduced prior to sealing. In the case of flexible packages, such as pouches, the air is displaced from the package by flushing it through with the gas mixture before sealing. In the case of horticultural products, a modified in-package atmosphere may develop as a result of respiration of the food. The concentration of oxygen inside the package will fall and that of carbon dioxide will rise. The equilibrium composition attained inside the package will largely depend on the rate of respiration of the food and the permeability of the packaging material to gases (see Section 9.2.2).

The gases involved in modified atmosphere packaging, as applied commercially today, are carbon dioxide, nitrogen and oxygen.

Carbon dioxide reacts with water in the product to form carbonic acid which lowers the pH of the food. It also inhibits the growth of certain microorganisms, mainly moulds and some aerobic bacteria. Lactic acid bacteria are resistant to the gas and may replace aerobic spoilage bacteria in modified atmosphere packaged meat. Most yeasts are also resistant to carbon dioxide. Anaerobic bacteria, including food poisoning organisms, are little affected by carbon dioxide. Consequently, there is a potential health hazard in MAP products from these microorganisms. Strict temperature control is essential to ensure the safety of MAP foods. Moulds and some gram negative, aerobic bacteria, such as *Pseudomonas* spp, are inhibited by carbon dioxide concentrations in the range 5–50%. In general, the higher the concentration of the gas, the greater is its inhibitory power. The inhibition of bacteria by carbon dioxide increases as the temperature decreases. Bacteria in the lag phase of growth are most affected by the gas.

Nitrogen has no direct effect on microorganisms or foods, other than to replace oxygen, which can inhibit the oxidation of fats. As its solubility in water is low, it is used as a bulking material to prevent the collapse of MAP packages when the carbon dioxide dissolves in the food. This is also useful in packages of sliced or ground food materials, such as cheese, which may consolidate under vacuum.

Oxygen is included in MAP packages of red meat to maintain the red colour, which is due to the oxygenation of the myoglobin pigments. It is also included in MAP packages of white fish, to reduce the risk of botulism.

Other gases have antimicrobial effects. Carbon monoxide will inhibit the growth of many bacteria, yeasts and moulds, in concentrations as low as 1%. However, due to its toxicity and explosive nature, it is not used commercially. Sulphur dioxide has been used to inhibit the growth of moulds and bacteria in some soft fruits and fruit juices. In recent years, there has been concern that some people may be hypersensitive to sulphur dioxide.

So called noble gases, such as argon, helium, xenon and neon, have also been used in MAP of some foods. However, apart from being relatively inert, it is not clear what particular benefits they bring to this technology.

MAP packages are either thermoformed trays with heat-sealed lids or pouches. With the exception of packages for fresh produce, these trays and pouches need to be made of materials with low permeability to gases (CO_2, N_2, O_2). Laminates are used, made of various combinations of polyester (PET), polyvinylidene chloride (PVdC), polyethylene (PE) and polyamide (PA, Nylons; see Section 9.3.4). The oxygen permeability of these laminates should be less than 15 cm^3 m^{-2} day^{-1} at a pressure of 1 atm (101 kPa). The following are some examples of MAP (see Fig. 9.12) [41–50].

- Meat products: fresh red meat packaged in an atmosphere consisting of 80% oxygen and 20% carbon dioxide or 70% oxygen, 20% carbon dioxide and 10% nitrogen should have a shelf life of 7–12 days at $2 \pm 1\,^{\circ}C$. The meat is usually placed on an absorbent pad, contained in a deep tray with a heat-sealed lid. Poultry can be MA-packaged in a mixture of nitrogen and carbon dioxide. However, this is not widely practised because of cost considerations. Cooked and cured meats may be packaged in a mixture of nitrogen and carbon dioxide.
- Fish: fresh white fish, packaged in a mixture of 30% oxygen, 30% nitrogen and 40% carbon dioxide, should have a shelf life of 10–14 days at a temperature of $0\,^{\circ}C$. Such packages should not be exposed to a temperature above $5\,^{\circ}C$, because of the risk of botulism. Fatty fish are packaged in mixtures of carbon dioxide and nitrogen.
- Fruits and vegetables: respiration in such products leads to a build-up of carbon dioxide and a reduction in the oxygen content (see Section 9.2.2). Some build up of carbon dioxide may reduce the rate of respiration and help to prolong the shelf life of the product. However, if the oxygen level is reduced to 2% or less, anaerobic respiration will set in and the product will spoil. The effect of the build up of carbon dioxide varies from product to product. Some fruits and vegetables can tolerate high levels of this gas while others cannot. Each fruit or vegetable will have an optimum in-package gas composition which will result in a maximum shelf life. Selection of a packaging film with an appropriate permeability to water vapour and gases can lead to the development of this optimum composition. For fruits with very high respiration rates, the package may need to be perforated. A range of microperforated films are available for such applications.
- Cheese: portions of hard cheese may be packaged by flushing with carbon dioxide before sealing. The gas will be absorbed by the cheese, creating a vacuum. Cheese packaged in this way may have a shelf life of up to 60 days. To avoid collapse of the package, some nitrogen may be included with the carbon dioxide. Mould ripened cheese may be packaging in nitrogen.
- Bakery products and snack foods: the shelf life of bread rolls, crumpets and pita bread may be significantly increased by packaging in carbon dioxide or nitrogen/carbon dioxide mixtures. Nuts and potato crisps benefit by being MA-packaged in nitrogen.
- Pasta: Fresh pasta may be MA-packaged in nitrogen or carbon dioxide.
- Other foods: pizza, quiche, lasagne, and many other prepared foods may benefit from MAP. It is very important to take into account the microbiological implications of MA-packaging such products. Maintenance of low temperatures during storage, distribution, in the retail outlet and in the home is essential.

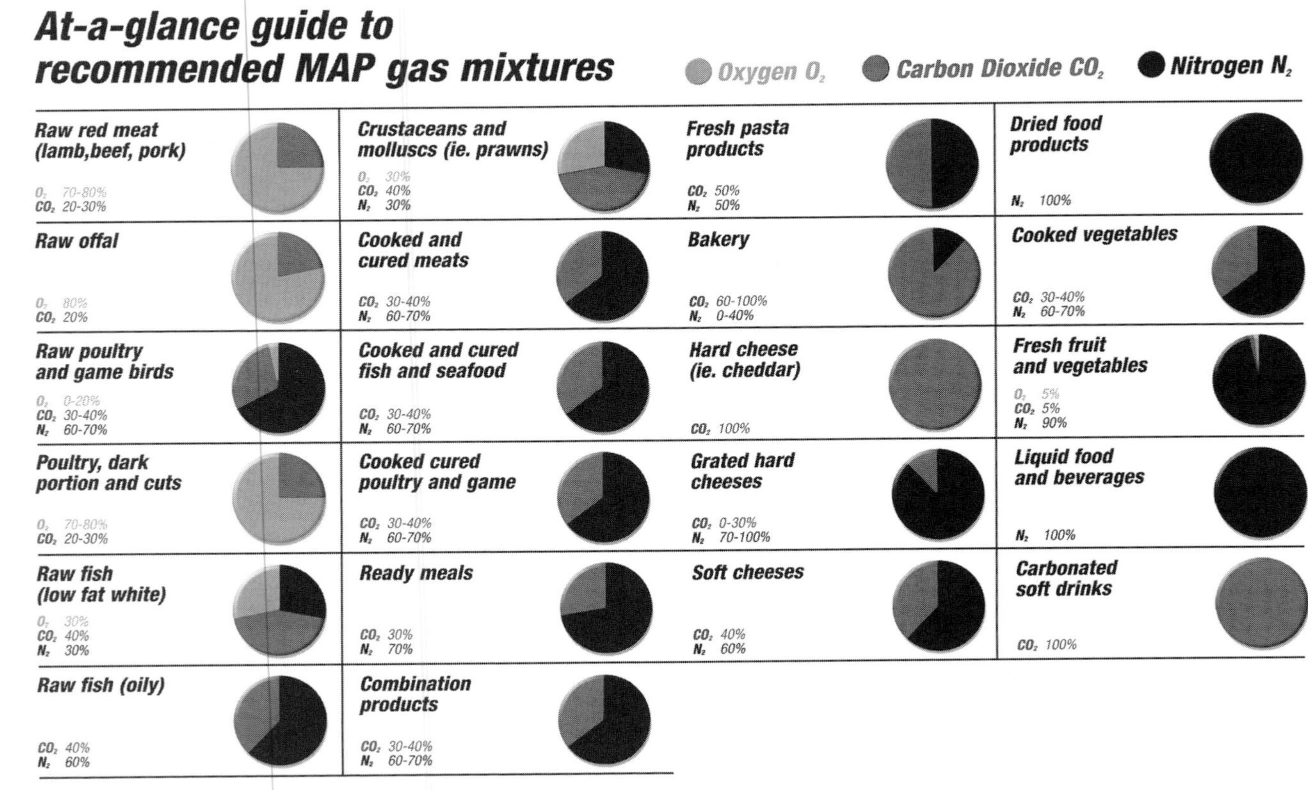

Fig. 9.12 Gas compositions commonly used in MAP food containers; by courtesy of Air Products.

9.5 Aseptic Packaging

Chapter 2 discussed the advantages, in terms of product quality, to be gained by heat processing foods in bulk prior to packaging (UHT treatment) as compared with heat treating the packaged product. UHT-treated products have to be packaged under conditions which prevent microbiological contamination, i.e. aseptically packaged. With some high-acid foods ($pH < 4.5$), it may be sufficient to cool the product after UHT treatment to just below 100 °C, fill it into a clean container, seal the container and hold it at that temperature for some minutes before cooling it. This procedure will inactivate microorganisms that may have been in the container or entered during the filling operation and which might grow in the product. The filled container may need to be inverted for some or all of the holding period. However, in the case of low-acid foods ($pH > 4.5$) this procedure would not be adequate to ensure the sterility of the product. Consequently for such products, aseptic filling must involve sterilising the empty container or the material from which the container is made, filling it with the UHT-treated product and sealing it without it being contaminated with microorganisms.

In the case of rigid metal containers, superheated steam may be used to sterilise the empty containers and maintain a sterile atmosphere during the filling and sealing operations. Empty cans are carried on a stainless steel conveyor through a stainless steel tunnel. Superheated steam, at a temperature of approximately 260 °C, is introduced into the tunnel to sterilise the cans. They then move into an enclosed filling section, maintained sterile by superheated steam. They are sprayed on the outside with cool sterile water before being filled with the cooled UHT product. The filled cans move into an enclosed seaming section, which is also maintained in a sterile condition with superheated steam. The can ends are also sterilised with superheated steam and double-seamed onto the filled cans in the sterile seaming section. The filled and seamed cans then exit from the tunnel. The whole system has to be presterilised and the temperatures adjusted to the appropriate levels before filling commences. This aseptic filling procedure is known as the *Dole process* [51]. Glass containers and some plastic and composite containers may be aseptically filled by this method. Cartons made from a laminate of paper/aluminium foil/polyethylene are widely used for UHT products such as liquid milk and fruit juices. This type of packaging material cannot be sterilised by heat alone. A combination of heat and chemical sterilant is used. Treatments with hydrogen peroxide, peracetic acid, ethylene oxide, ionising radiation, ultraviolet radiation and sterile air have all been investigated. Hydrogen peroxide at a concentration of 35% in water and 90 °C is very effective against heat-resistant, sporeforming microorganisms and is widely used commercially as a sterilant in aseptic packaging in laminates. Form-fill-seal systems are available, an example being the Tetra Brik system, offered by Tetra Pak Ltd. (Fig. 9.13).

The packaging material, a polyethylene/paper/polyethylene/foil/polyethylene laminate, is unwound from a reel and a plastic strip is attached to one edge,

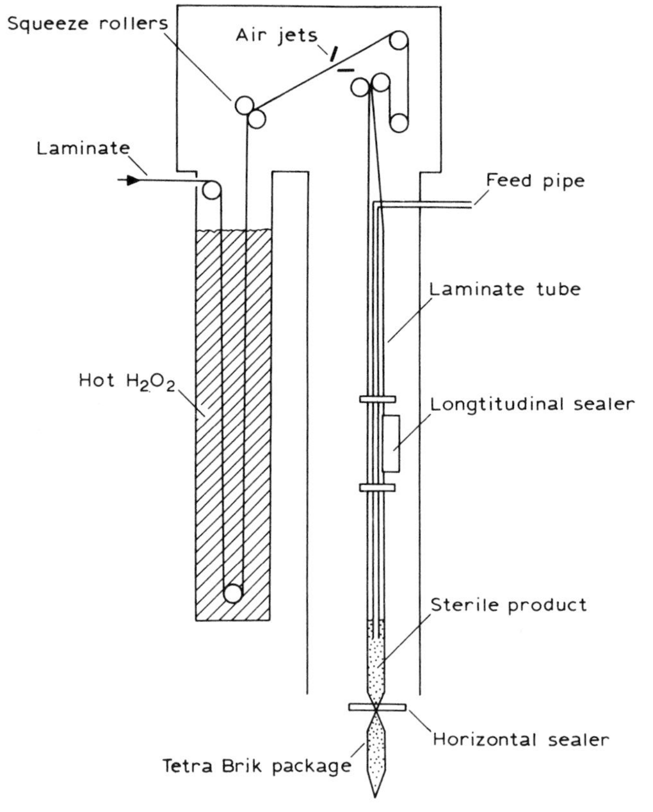

Fig. 9.13 Principle of the Tetra Brik aseptic packaging system; from [15] with permission.

which will eventually overlap the internal longitudinal seal in the carton. It then passes through a deep bath of hot hydrogen peroxide, which wets the laminate. As it emerges from the bath, the laminate passes between squeeze rollers, which express liquid hydrogen peroxide for return to the bath. Next, a high-velocity jet of hot sterile air is directed onto both sides of the laminate to remove residual hydrogen peroxide, as a vapour. The laminate, which is now sterile and dry, is formed into a tube with a longitudinal seal in an enclosed section which is maintained sterile by means of hot, sterile air under pressure. The product filling tube is located down the centre of the laminate tube. The presterilised product is fed into the sterile zone near the bottom of the tube, which is heat-sealed. The air containing the vaporised hydrogen peroxide is collected in a cover and directed to a compressor where it is mixed with water, which washes out the residual hydrogen peroxide. The air is sterilised by heat and returned to the filling zone. In another system, the laminate is in the form of carton blanks which are erected and then sterilised by a downward spray of hydrogen peroxide followed by hot sterile air. This completes the sterilisation and removes residual hydrogen peroxide. The presterilised product is filled into the cartons and the top sealed within a sterile zone. Similar systems are available to aseptically fill into preformed plastic cups. The lidding material is sterilised with hydrogen peroxide or infrared radiation before being heat-sealed onto the cups within a sterile zone. Thermoform filling systems are available to aseptically fill into polymer laminates. The web of laminate passes through a bath of hydrogen peroxide and then is contacted by hot sterile air which completes the sterilisation, removes residual hydrogen peroxide and softens the laminate. The laminate is then thermoformed into cups and filled with presterilised product within a sterile zone. The sterilised lidding material is applied before the cups leave the sterile zone. Thermoforming systems are usually used to fill small containers e.g. for individual portions of milk, cream and whiteners [52–58].

9.6 Active Packaging

9.6.1 Background Information

Active packaging refers to the incorporation of certain additives into packaging film or within packaging containers with the aim of maintaining and extending product shelf life [59]. Packaging may be termed active when it performs some desired role in food preservation other than providing an inert barrier to external conditions [60, 61]. Active packaging includes additives or 'freshness enhancers' that are capable of scavenging oxygen, adsorbing carbon dioxide, moisture, ethylene and/or flavour/odour taints, releasing ethanol, sorbates, antioxidants and/or other preservatives and/or maintaining temperature control. Table 9.3 lists examples of active packaging systems, some of which may offer extended shelf life opportunities for new categories of food products [59].

Active packaging has been used with many food products and is being tested with numerous others. Table 9.3 lists some of the food applications that have benefited from active packaging technology. It should be noted that all food products have a unique deterioration mechanism that must be understood before applying this technology. The shelf life of packaged food is dependent on numerous factors, such as the intrinsic nature of the food (e.g. pH, water activity, nutrient content, occurrence of antimicrobial compounds, redox potential, respiration rate, biological structure) and extrinsic factors (e.g. storage temperature, relative humidity, surrounding gaseous composition). These factors directly influence the chemical, biochemical, physical and microbiological spoilage mechanisms of individual food products and their achievable shelf life. By carefully considering all of these factors, it is possible to evaluate existing and developing active packaging technologies and apply them for maintaining the quality and extending the shelf life of different food products [59].

Table 9.3 Selected examples of active packaging systems.

Systems	Mechanisms	Food applications
Oxygen scavengers	1. Iron-based 2. Metal/acid 3. Metal (e.g. platinum) catalyst 4. Ascorbate/metallic salts 5. Enzyme-based	Bread, cakes, cooked rice, biscuits, pizza, pasta, cheese, cured meats, cured fish, coffee, snack foods, dried foods and beverages
Carbon dioxide scavengers/ emitters	1. Iron oxide/calcium hydroxide 2. Ferrous carbonate/metal halide 3. Calcium oxide/activated charcoal 4. Ascorbate/sodium bicarbonate	Coffee, fresh meats, fresh fish, nuts, other snack food products and sponge cakes
Ethylene scavengers	1. Potassium permanganate 2. Activated carbon 3. Activated clays/zeolites	Fruit, vegetables and other horticultural products
Preservative releasers	1. Organic acids 2. Silver zeolite 3. Spice and herb extracts 4. BHA/BHT antioxidants 5. Vitamin E antioxidant 6. Volatile chlorine dioxide/ sulphur dioxide	Cereals, meats, fish, bread, cheese, snack foods, fruit and vegetables
Ethanol emitters	1. Alcohol spray 2. Encapsulated ethanol	Pizza crusts, cakes, bread, biscuits, fish and bakery products
Moisture absorbers	1. PVA blanket 2. Activated clays and minerals 3. Silica gel	Fish, meats, poultry, snack foods, cereals, dried foods, sandwiches, fruit and vegetables
Flavour/odour adsorbers	1. Cellulose triacetate 2. Acetylated paper 3. Citric acid 4. Ferrous salt/ascorbate 5. Activated carbon/clays/ zeolites	Fruit juices, fried snack foods, fish, cereals, poultry, dairy products and fruit
Temperature control packaging	1. Non-woven plastics 2. Double-walled containers 3. Hydrofluorocarbon gas 4. Lime/water 5. Ammonium nitrate/water	Ready meals, meats, fish, poultry and beverages

Table 9.4 Selected commercial oxygen scavenger systems.

Manufacturer	Country	Trade name	Scavenger mechanism	Packaging form
Mitsubishi Gas Chemical Co. Ltd	Japan	Ageless	Iron-based	Sachets and labels
Toppan Printing Co. Ltd	Japan	Freshilizer	Iron-based	Sachets
Toagosei Chemical Industry Co. Ltd	Japan	Vitalon	Iron-based	Sachets
Nippon Soda Co. Ltd	Japan	Seagul	Iron-based	Sachets
Finetec Co. Ltd	Japan	Sanso-Cut	Iron-based	Sachets
Toyo Seikan Kaisha Ltd	Japan	Oxyguard	Iron-based	Plastic trays
Multisorb Technologies Inc.	USA	FreshMax FreshPax Fresh Pack	Iron-based Iron-based Iron-based	Labels Labels Labels
Ciba Speciality Chemicals	USA	Shelf-plus	PET copolyester	Plastic film
Chevron Chemicals	USA	N/A	Benzyl acrylate	Plastic film
W.R. Grace Co. Ltd	USA	PureSeal	Ascorbate/ metallic salts	Bottle crowns
Food Science Australia	Australia	$ZERO_2$	Photosensitive dye/organic compound	Plastic film
CMB Technologies	France	Oxbar	Cobalt catalyst	Plastic bottles
Standa Industrie	France	ATCO Oxycap	Iron-based Iron-based	Sachets Bottle crowns
EMCO Packaging Systems	UK	ATCO	Iron-based	Labels
Johnson Matthey Plc	UK	N/A	Platinum group metal catalyst	Labels
Bioka Ltd	Finland	Bioka	Enzyme-based	Sachets
Alcoa CSI Europe	UK	O_2-displacer system	Unknown Trade name	Bottle crowns Scavenger mechanism

Active packaging is not synonymous with intelligent or smart packaging, which refers to packaging which senses and informs [62, 63]. Intelligent packaging devices are capable of sensing and providing information about the function and properties of packaged food and can provide assurances of pack integrity, tamper evidence, product safety and quality, as well as being utilised in applications such as product authenticity, anti-theft and product traceability [62, 63]. Intelligent packaging devices include time-temperature indicators, gas-sensing dyes, microbial growth indicators, physical shock indicators and numerous examples of tamper-proof, anti-counterfeiting and anti-theft technologies. Information on intelligent packaging technology can be obtained from other reference sources [62–64].

It is not the intention of this section to extensively review all active packaging technologies but rather to describe the different types of devices, the scientific principles behind them, the principal food applications and the food safety and regulatory issues that need to be considered by potential users. The major focus of this section is on oxygen scavengers but other active packaging technologies are described and some recent developments are highlighted. More detailed information on active packaging can be obtained from the numerous references listed.

9.6.2 Oxygen Scavengers

Oxygen can have considerable detrimental effects on foods. Oxygen scavengers can therefore help maintain food product quality by decreasing food metabolism, reducing oxidative rancidity, inhibiting undesirable oxidation of labile pigments and vitamins, controlling enzymic discoloration and inhibiting the growth of aerobic microorganisms [59, 60, 62].

Oxygen scavengers are by far the most commercially important subcategory of active packaging. The global market per annum for oxygen scavengers was estimated to exceed ten billion units in Japan, several hundred million in the USA and tens of millions in Europe in 1996 [65, 66]. The global value of this market was estimated to exceed $200 million in 1996 and to top $1 billion by 2002, particularly since the recent introduction of oxygen-scavenging PET bottles, bottle caps and crowns for beers and other beverages is now fully commercialised [62, 65, 66]. More recent market information has estimated that the global market per annum for oxygen scavengers was $480 million in 2001, with 12 billion units sold in Japan and 300 million units sold in Europe [67].

The most common oxygen scavengers take the form of small sachets containing various iron-based powders containing an assortment of catalysts. These chemical systems often react with water supplied by the food to produce a reactive hydrated metallic reducing agent that scavenges oxygen within the food package and irreversibly converts it to a stable oxide. The iron powder is separated from the food by keeping it in a small, highly oxygen permeable sachet that is labelled “Do not eat”. The main advantage of using such oxygen scavengers is that they are capable of reducing oxygen levels to less than 0.01%, which is much lower that the typical 0.3–3.0% residual oxygen levels achievable by modified atmosphere packaging (MAP). Oxygen scavengers can be used alone or in combination with MAP. Their use alone eliminates the need for MAP machinery and can increase packaging speeds. However, it is usually more common commercially to remove most of the atmospheric oxygen by MAP and then use a relatively small and inexpensive scavenger to mop up the residual oxygen remaining within the food package.

Nonmetallic oxygen scavengers have also been developed to alleviate the potential for metallic taints being imparted to food products. The problem of inadvertently setting off inline metal detectors is also alleviated even though some modern detectors can now be tuned to phase out the scavenger signal whilst retaining high sensitivity for ferrous and nonferrous metallic contaminants [68]. Nonmetallic scavengers include those that use organic reducing agents such as ascorbic acid, ascorbate salts or catechol. They also include enzymic oxygen scavenger systems using either glucose oxidase or ethanol oxidase which could be incorporated into sachets, adhesive labels or immobilised onto packaging film surfaces [69].

Oxygen scavengers were first marketed in Japan in 1976 by the Mitsubishi Gas Chemical Co. Ltd. under the trade name 'Ageless'. Since then, several other Japanese companies, including Toppan Printing Co. Ltd. and Toyo Seikan Kaisha Ltd., have entered the market but Mitsubishi still dominates the oxygen scavenger business in Japan with a market share of 73% [60]. Oxygen scavenger technology has been successful in Japan for a variety of reasons, including the acceptance by Japanese consumers of innovative packaging and the hot and humid climate in Japan during the summer months, which is conducive to mould spoilage of food products. In contrast to the Japanese market, the acceptance of oxygen scavengers in North America and Europe has been slow, although several manufacturers and distributors of oxygen scavengers are now established in both these continents and sales have been estimated to be growing at a rate of 20% annually [60]. Table 9.4 lists selected manufacturers and trade names of oxygen scavengers, including some which are still under development or have been suspended because of regulatory controls [60, 66, 70, 71].

It should be noted that discrete oxygen-scavenging sachets suffer from the disadvantage of possible accidental ingestion of the contents by the consumer, and this has hampered their commercial success, particularly in North America and Europe. However, in the last few years, the development of oxygen-scavenging adhesive labels that can be adhered to the inside of packages and the incorporation of oxygen-scavenging materials into laminated trays and plastic films have enhanced and help the commercial acceptance of this technology. For example, Marks & Spencer Ltd were the first UK retailer to use oxygen-scavenging adhesive labels for a range of sliced cooked and cured meat and poultry products which are particularly sensitive to deleterious light and oxygen-induced colour changes [62]. Other UK retailers, distributors and caterers are now using these labels for the above food products as well as for coffee, pizzas, speciality

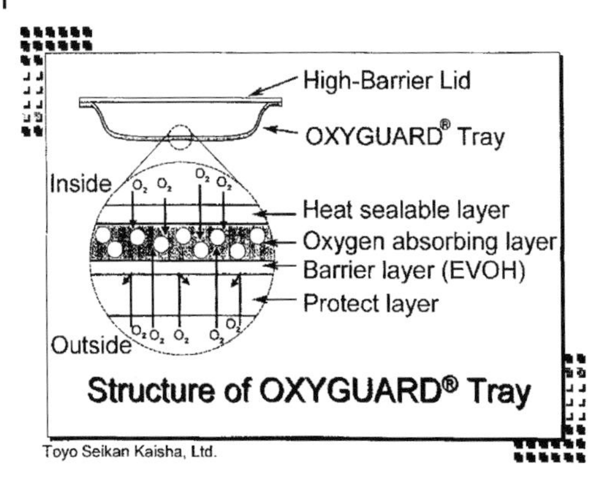

Fig. 9.14 Structure of the Oxyguard tray.

bakery goods and dried food ingredients [72]. Other common food applications for oxygen scavenger labels and sachets include cakes, breads, biscuits, croissants, fresh pastas, cured fish, tea, powdered milk, dried egg, spices, herbs, confectionery and snack food [62]. In Japan, Toyo Seikan Kaisha Ltd. have marketed a laminate containing a ferrous oxygen scavenger, which can be thermoformed into an 'Oxyguard' tray, which has been used commercially for cooked rice, see Fig. 9.14.

The use of oxygen scavengers for beer, wine and other beverages is potentially a huge market that has only recently begun to be exploited. Iron-based label and sachet scavengers cannot be used for beverages or high a_w foods because, when wet, their oxygen-scavenging capability is rapidly lost. Instead, various nonmetallic reagents and organometallic compounds which have an affinity for oxygen have been incorporated into bottle closures, crown and caps or blended into polymer materials so that oxygen is scavenged from the bottle headspace and any ingressing oxygen is also scavenged too. The 'PureSeal' oxygen-scavenging bottle crowns (marketed by W.R. Grace Co. Ltd. and ZapatA Technologies Inc., USA), oxygen-scavenging plastic (PET) beer bottles (manufactured by Continental PET Technologies, Toledo, USA) and light-activated 'ZERO2' oxygen scavenger materials (developed by Food Science Australia, North Ryde, NSW, Australia) are just three of many oxygen scavenger developments aimed at the beverage market but also applicable to other food applications [60, 66, 70, 73]. It should be noted that the speed and capacity of oxygen-scavenging plastic films and laminated trays are considerably lower than iron-based oxygen scavenger sachets or labels [72].

More detailed information on the technical requirements (i.e. for low, medium and high a_w foods and beverages, speed of reaction, storage temperature, oxygen scavenging capacity, necessary packaging criteria) of the different types of oxygen scavengers can be obtained from Rooney [60, 66, 69, 73, 74].

9.6.3
Carbon Dioxide Scavengers/Emitters

There are many commercial sachet and label devices that can be used to either scavenge or emit carbon dioxide. The use of carbon dioxide scavengers is particularly applicable for fresh roasted or ground coffees that produce significant volumes of carbon dioxide. Fresh roasted or ground coffees cannot be left unpackaged since they absorb moisture and oxygen and lose desirable volatile aromas and flavours. However, if coffee is hermetically sealed in packs directly after roasting, the carbon dioxide released builds up within the packs and eventually causes them to burst [50]. To circumvent this problem, two solutions are currently used. The first is to use packaging with patented oneway valves that allow excess carbon dioxide to escape. The second solution is to use a carbon dioxide scavenger or a dual-action oxygen and carbon dioxide scavenger system. A mixture of calcium oxide and activated charcoal has been used in polyethylene coffee pouches to scavenge carbon dioxide but dual-action oxygen and carbon dioxide scavenger sachets and labels are more common and are commercially used for canned and foil pouched coffees in Japan and the USA [59, 60, 75]. These dual-action sachets and labels typically contain iron powder for scavenging oxygen and calcium hydroxide which scavenges carbon dioxide when it is converted to calcium carbonate under sufficiently high humidity conditions [60]. Commercially available dual-action oxygen and carbon dioxide scavengers are available from Japanese manufacturers, e.g. Mitsubishi Gas Chemical Co. Ltd. ('Ageless' type E, 'Fresh Lock') and Toppan Printing Co. Ltd. ('Freshilizer' type CV).

Pack collapse or the development of a partial vacuum can also be a problem for foods packed with an oxygen scavenger. To overcome this problem, dual-action oxygen scavenger/carbon dioxide emitter sachets and labels have been developed which absorb oxygen and generate an equal volume of carbon dioxide. These sachets and labels usually contain ferrous carbonate and a metal halide catalyst although nonferrous variants are available. Commercial manufacturers include Mitsubishi Gas Chemical Co. Ltd. ('Ageless' type G), and Multisorb Technologies Inc. ('Freshpax' type M). The main food applications for these dual-action oxygen scavenger/carbon dioxide emitter sachets and labels have been with snack food products, e.g. nuts, and sponge cakes [60, 76].

9.6.4
Ethylene Scavengers

Ethylene (C_2H_4) is a plant hormone that accelerates the respiration rate and subsequent senescence of horticultural products such as fruit, vegetables and flowers. Many of the effects of ethylene are necessary, e.g. induction of flowering in pineapples and colour development in citrus fruits, bananas and tomatoes, but in most horticultural situations it is desirable to remove ethylene or to suppress its effects. Consequently, much research effort has been undertaken to incorporate ethylene scavengers into fresh produce packaging and storage areas.

Some of this effort has met with commercial success, but much of it has not [60, 77].

Table 9.5 lists selected commercial ethylene scavenger systems. Effective systems utilise potassium permanganate ($KMnO_4$) immobilised on an inert mineral substrate such as alumina or silica gel. $KMnO_4$ oxidises ethylene to acetate and ethanol and in the process changes colour from purple to brown and hence indicates its remaining ethylene-scavenging capacity. $KMnO_4$-based ethylene scavengers are available in sachets to be placed inside produce packages or inside blankets or tubes that can be placed in produce storage warehouses [60, 74, 77].

Activated carbon-based scavengers with various metal catalysts can also effectively remove ethylene. They have been used to scavenge ethylene from produce warehouses or incorporated into sachets for inclusion into produce packs or embedded into paper bags or corrugated board boxes for produce storage. A dual-action ethylene scavenger and moisture absorber has been marketed in Japan by Sekisui Jushi Limited. 'Neupalon' sachets contain activated carbon, a metal catalyst and silica gel and are capable of scavenging ethylene as well as acting as a moisture absorber [60, 77].

In recent years, numerous produce packaging films and bags have appeared on the market place which are based on the putative ability of certain finely ground minerals to adsorb ethylene and to emit antimicrobial far-infrared radiation. However, little direct evidence for these effects has been published in peer-reviewed scientific journals. Typically these activated earth-type minerals include clays, pumice, zeolites, coral, ceramics and even Japanese Oya stone. These

Table 9.5 Selected commercial ethylene scavenger systems.

Manufacturer	Country	Trade name	Scavenger mechanism	Packaging form
Air Repair Products Inc.	USA	N/A	$KMnO_4$	Sachets/blankets
Ethylene Control Inc.	USA	N/A	$KMnO_4$	Sachets/blankets
Extenda Life Systems	USA	N/A	$KMnO_4$	Sachets/blankets
Kes Irrigations Systems	USA	Bio-Kleen	Titanium dioxide catalyst	Not known
Sekisui Jushi Ltd	Japan	Neupalon	Activated carbon	Sachet
Honshu Paper Ltd	Japan	Hatofresh	Activated carbon	Paper/board
Mitsubishi Gas Chemical Co. Ltd	Japan	Sendo-Mate	Activated carbon	Sachets
Cho Yang Heung San Co. Ltd	Korea	Orega	Activated clays/ zeolites	Plastic film
Evert-Fresh Corp.	USA	Evert-Fresh	Activated zeolites	Plastic film
Odja Shoji Co. Ltd	Japan	BO Film	Crysburite ceramic	Plastic film
PEAKfresh Products Ltd	Australia	PEAKfresh	Activated clays/ zeolites	Plastic film

minerals are embedded or blended into polyethylene film bags that are then used to package fresh produce. Manufacturers of such bags claim extended shelf life for fresh produce partly due to the adsorption of ethylene by the minerals dispersed within the bags. The evidence offered in support of this claim is generally based on the extended shelf life of produce and reduction of headspace ethylene in mineral-filled bags in comparison with common polyethylene bags. However, independent research has shown that the gas permeability of mineral-filled polyethylene bags is much greater and consequently ethylene will diffuse out of these bags much faster, as is also the case for commercially available microperforated film bags. In addition, a more favourable equilibrium modified atmosphere is likely to develop within these bags compared with common polyethylene bags, especially if the produce has a high respiration rate. Therefore, these effects can improve produce shelf life and reduce headspace ethylene independently of any ethylene adsorption. In fact, almost any powdered mineral can confer such effects without relying on expensive Oya stone or other speciality minerals [60, 77].

9.6.5 Ethanol Emitters

The use of ethanol as an antimicrobial agent is well documented. It is particularly effective against mould but can also inhibit the growth of yeasts and bacteria. Ethanol can be sprayed directly onto food products just prior to packaging. Several reports have demonstrated that the mould-free shelf life of bakery products can be significantly extended after spraying with 95% ethanol to give concentrations of 0.5–1.5% (w/w) in the products. However, a more practical and safer method of generating ethanol is through the use of ethanol-emitting films and sachets [60].

Primarily Japanese manufacturers have patented many applications of ethanol-emitting films and sachets. These include Ethicap, Antimold 102 and Negamold (Freund Industrial Co. Ltd.), Oitech (Nippon Kayaku Co. Ltd.), ET Pack (Ueno Seiyaku Co. Ltd.) and Ageless type SE (Mitsubishi Gas Chemical Co. Ltd.). All of these films and sachets contain absorbed or encapsulated ethanol in a carrier material that allows the controlled release of ethanol vapour. For example, Ethicap, which is the most commercially popular ethanol emitter in Japan, consists of food grade alcohol (55%) and water (10%) adsorbed onto silicon dioxide powder (35%) and contained in a sachet made of a paper and ethyl vinyl acetate (EVA) copolymer laminate. To mask the odour of alcohol, some sachets contain traces of vanilla or other flavours. The sachets are labelled "Do not eat contents" and include a diagram illustrating this warning. Other ethanol emitters such as Negamould and Ageless type SE are dual-action sachets that scavenge oxygen as well as emitting ethanol vapour [60].

The size and capacity of the ethanol-emitting sachet used depends on the weight of food, the a_w of the food and the shelf life required. When food is packed with an ethanol-emitting sachet, moisture is absorbed by the food and ethanol vapour is released and diffuses into the package headspace. Ethanol emitters are used extensively in Japan to extend the mould-free shelf life of high ratio cakes and other high moisture bakery products by up to 2000% [60, 78]. Research has also shown that such bakery products packed with ethanol-emitting sachets did not get as hard as the controls and results were better than those using an oxygen scavenger alone to inhibit mould growth. Hence, ethanol vapour also appears to exert an anti-staling effect in addition to its anti-mould properties. Ethanol-emitting sachets are also widely used in Japan for extending the shelf life of semi-moist and dry fish products [60].

9.6.6
Preservative Releasers

Recently there has been great interest in the potential use of antimicrobial and antioxidant packaging films that have preservative properties for extending the shelf life of a wide range of food products. As with other categories of active packaging, many patents exist and some antimicrobial and antioxidant films have been marketed but the majority have so far failed to be commercialised because of doubts about their effectiveness, economic factors and/or regulatory constraints [60].

Some commercial antimicrobial films and materials have been introduced, primarily in Japan. For example, one widely reported product is a synthetic silver zeolite that has been directly incorporated into food contact packaging film. The purpose of the zeolite is apparently to allow slow release of antimicrobial silver ions into the surface of food products. Many other synthetic and naturally occurring preservatives have been proposed and/or tested for antimicrobial activity in plastic and edible films [60, 79–81]. These include organic acids, e.g. propionate, benzoate and sorbate, bacteriocins, e.g. nisin,, spice and herb extracts, e.g. from rosemary, cloves, horseradish, mustard, cinnamon and thyme, enzymes, e.g. peroxidase, lysozyme and glucose oxidase, chelating agents, e.g. EDTA, inorganic acids, e.g. sulphur dioxide and chlorine dioxide, and anti-fungal agents, e.g. imazalil and benomyl. The major potential food applications for antimicrobial films include meats, fish, bread, cheese, fruit and vegetables.

An interesting commercial development in the UK is the recent exclusive marketing of food contact approved Microban (Microban International, Huntersville, USA) kitchen products such as chopping boards, dish cloths and bin bags by J. Sainsbury Plc. These Microban products contain triclosan, an antibacterial aromatic chloroorganic compound, which is also used in soaps, shampoos, lotions, toothpaste and mouth washes [82-84]. Another interesting development is the incorporation of methyl salicylate (a synthetic version of wintergreen oil) into RepelKote paperboard boxes by Tenneco Packaging (Lake Forest, Illinois, USA). Methyl salicylate has antimicrobial properties, but RepelKote is primarily being marketed as an insect repellent and its main food applications are dried foods that are very susceptible to insect infestations [85].

Two influences have stimulated interest in the use of antioxidant packaging films. The first of these is the consumer demand for reduced antioxidants and other additives in foods. The second is the interest of plastic manufacturers in using natural approved food antioxidants, e.g. vitamin E, for polymer stabilisation instead of synthetic antioxidants developed specifically for plastics [60]. The potential for evaporative migration of antioxidants into foods from packaging films has been extensively researched and commercialised in some instances. For example, the cereal industry in the USA has used this approach for the release of butylated hydroxytoluene (BHT) and butylated hydroxyanisole (BHA) antioxidants from waxed paper liners into breakfast cereal and snack food products [74]. Recently there has been a lot of interest in the use of α-tocopherol (vitamin E) as a viable alternative to BHT/BHA-impregnated packaging films [86]. The use of packaging films incorporating natural vitamin E can confer benefits to both film manufacturers and the food industry. There have been questions raised regarding the safety of BHT and BHA and hence using vitamin E is a safer alternative. Research has shown vitamin E to be as effective as an antioxidant compared with BHT, BHA or other synthetic polymer antioxidants for inhibiting packaging film degradation during film extrusion or blow moulding. Vitamin E is also a safe and effective antioxidant for low to medium a_w cereal and snack food products where the development of rancid odours and flavours is often the spoilage mechanism limiting shelf life [60, 74, 86].

9.6.7
Moisture Absorbers

Excess moisture is a major cause of food spoilage. Soaking up moisture by using various absorbers or desiccants is very effective at maintaining food quality and extending shelf life by inhibiting microbial growth and moisture-related degradation of texture and flavour. Several companies manufacture moisture absorbers in the form of sachets, pads, sheets or blankets. For packaged dried food applications, desiccants such as silica gel, calcium oxide and activated clays and minerals are typically contained within Tyvek (Dupont Chemicals, Wilmington, Delaware, USA) tear-resistant permeable plastic sachets. For dual-action purposes, these sachets may also contain activated carbon for odour adsorption or iron powder for oxygen scavenging [60, 87]. The use of moisture absorber sachets is commonplace in Japan where popular foods feature a number of dried products which need to be protected from humidity damage. The use of moisture absorber sachets is also quite common in the USA where the major suppliers include Multisorb Technologies Inc. (Buffalo, New York), United Desiccants (Louisville, Kentucky) and Baltimore Chemicals (Baltimore, Maryland). These sachets are not only utilised for dried snack foods and cereals but also for a wide array of pharmaceutical, electrical and electronic goods. In the UK, Marks & Spencer Plc have used silica gel-based moisture absorber sachets for maintaining the crispness of filled ciabatta bread rolls.

In addition to moisture absorber sachets for humidity control in packaged dried foods, several companies manufacture moisture drip absorbent pads, sheets and blankets for liquid water control in high a_w foods such as meats, fish, poultry, fruit and vegetables. Basically they consist of two layers of a microporous nonwoven plastic film, such as polyethylene or polypropylene, between which is placed a superabsorbent polymer that is capable of absorbing up to 500 times its own weight with water. Typical superabsorbent polymers include polyacrylate salts, carboxymethyl cellulose (CMC) and starch copolymers which both have a very strong affinity for water. Moisture drip absorber pads are commonly placed under packaged fresh meats, fish and poultry to absorb unsightly tissue drip exudate. Larger sheets and blankets are used for absorption of melted ice from chilled seafood during airfreight transportation or for controlling transpiration of horticultural produce [60]. Commercial moisture absorber sheets, blankets and trays include Toppan Sheet (Toppan Printing Co. Ltd., Japan), Thermarite (Thermarite Pty Ltd., Australia) and Fresh-R-Pax (Maxwell Chase Inc., Douglasville, Georgia).

Another approach for the control of excess moisture in high a_w foods, is to intercept the moisture in the vapour phase. This approach allows food packers or even householders to decrease the water activity on the surface of foods by reducing in-pack relative humidity. Placing one or more humectants between two layers of water-permeable plastic film can do this. For example, the Japanese company Showa Denko Co. Ltd has developed a Pitchit film which consists of a layer of humectant, carbohydrate and propylene glycol, sandwiched between two layers of polyvinyl alcohol (PVA) plastic film. Pitchit film is marketed for home use in a roll or single sheet form for wrapping fresh meats, fish and poultry. After wrapping in this film, the surface of the food is dehydrated by osmotic pressure, resulting in microbial inhibition and shelf life extension of 3–4 days under chilled storage [60, 74]. Another example of this approach has been applied in the distribution of horticultural produce. In recent years, microporous sachets of desiccant inorganic salts such as sodium chloride have been used for the distribution of tomatoes in the USA [60]. Yet another example is an innovative fibreboard box which functions as a humidity buffer on its own without relying on a desiccant insert. It consists of an integral water vapour barrier on the inner surface of the fibreboard, a paper-like material bonded to the barrier which acts as a wick and an unwettable but highly permeable to water vapour layer next to the fruit or vegetables. This multilayered box is able to take up water in the vapour state when the temperature drops and the relative humidity rises. Conversely, when the temperature rises, the multilayered box can release water vapour back in response to a lowering of the relative humidity [88].

9.6.8 Flavour/Odour Adsorbers

The interaction of packaging with food flavours and aromas has long been recognised, especially through the undesirable flavour scalping of desirable food components. For example, the scalping of a considerable proportion of desirable limonene has been demonstrated after only 2 weeks storage in aseptic packs of orange juice [60]. Commercially, very few active packaging techniques have been used to selectively remove undesirable flavours and taints, but many potential opportunities exist. An example of such an opportunity is the debittering of pasteurised orange juices. Some varieties of orange, such as Navel, are particularly prone to bitter flavours caused by limonine, a tetraterpenoid which is liberated into the juice after orange pressing and subsequent pasteurisation. Processes have been developed for debittering such juices by passing them through columns of cellulose triacetate or Nylon beads [60]. A possible active packaging solution would be to include limonine adsorbers, e.g. cellulose triacetate or acetylated paper, into orange juice packaging material.

Two types of taints amenable to removal by active packaging are amines, which are formed from the breakdown of fish muscle proteins, and aldehydes that are formed from the autooxidation of fats and oils. Volatile amines with an unpleasant smell, such as trimethylamine, associated with fish protein breakdown are alkaline and can be neutralised by various acidic compounds [89]. In Japan, Anico Co. Ltd have marketed Anico bags that are made from film containing a ferrous salt and an organic acid such as citrate or ascorbate. These bags are claimed to oxidise amines when they are adsorbed by the polymer film [60].

Removal of aldehydes, such as hexanal and heptanal, from package headspaces is claimed by Dupont's Odour and Taste Control (OTC) technology [90]. This technology is based upon a molecular sieve with pore sizes of around 5 nm and Dupont claim that their OTC removes or neutralises aldehydes, although evidence for this is lacking. The claimed food applications for this technology are snack foods, cereals, dairy products, fish and poultry [90]. A similar claim of aldehyde removal has been reported recently [91]. Swedish company EKA Noble in cooperation with Dutch company Akzo, have developed a range of synthetic aluminosilicate zeolites which they claim adsorb odorous gases within their highly porous structure. Their BMH powder can be incorporated into packaging materials, especially those that are paper-based, and apparently odorous aldehydes are adsorbed in the pore interstices of the powder [91].

9.6.9 Temperature Control Packaging

Temperature control active packaging includes the use of innovative insulating materials, self-heating and self-cooling cans. For example, to guard against undue temperature abuse during storage and distribution of chilled foods, special insulating materials have been developed. One such material is Thinsulate (3M Company, USA) that is a special nonwoven plastic with many air pore spaces. Another approach for maintaining chilled temperatures is to increase the thermal mass of the food package so that it is capable of withstanding temperature rises. The Adenko Company of Japan has developed and marketed a Cool Bowl that consists of a double-walled PET container in which an insulating gel is deposited in between the walls [74].

Self-heating cans and containers have been commercially available for decades and are particularly popular in Japan. Self-heating aluminium and steel cans and containers for sake, coffee, tea and ready meals are heated by an exothermic reaction when lime and water positioned in the base are mixed. In the UK, Nestlé recently introduced a range of Nescafé coffees in self-heating insulated cans that use the lime and water exothermic reaction. Self-cooling cans have also been marketed in Japan for raw sake. The endothermic dissolution of ammonium nitrate and chloride in water is used to cool the product. Another self-cooling can that has recently been introduced is the Chill Can (The Joseph Company, USA) that relies on a hydrofluorocarbon (HRC) gas refrigerant. The release of HRC gas is triggered by a button set into the can's base and can cool a drink by 10 °C in 2 min. However, concerns about the environmental impact of HRCs are likely to curtail the commercial success of the Chill Can [92].

9.6.10 Food Safety, Consumer Acceptability and Regulatory Issues

At least four types of food safety and regulatory issues related to active packaging of foods need to be addressed. First, any need for food contact approval must be established before any form of active packaging is used. Second, it is important to consider environmental regulations covering active packaging materials. Third, there may be a need for labelling in cases where active packaging may give rise to consumer confusion. Fourth, it is pertinent to consider the effects of active packaging on the microbial ecology and safety of foods [60]. All of these issues are currently being addressed in an EC-funded 'Actipack' project which aims to evaluate the safety, effectiveness, economic and environmental impact and consumer acceptance of active and intelligent packaging [93].

Food contact approval will often be required because active packaging may affect foods in two ways. Active packaging substances may migrate into the food or may be removed from it. Migrants may be intended or unintended. Intended migrants include antioxidants, ethanol and antimicrobial preservatives which would require regulatory approval in terms of their identity, concentration and possible toxicological effects. Unintended migrants include various metal compounds that achieve their active purpose inside packaging materials but do not need to or should not enter foods. Food additive regulations require identification and quantification of any such unintended migration.

Environmental regulations covering reuse, recycling, identification to assist in recycling or the recovery of energy from active packaging materials needs to be addressed on a case by case basis. European Union companies using active packaging as well as other packaging need to meet the requirements of the Packaging Waste Directive (1994) and consider the environmental impact of their packaging operations.

Food labelling is currently required to reduce the risk of consumers ingesting the contents of oxygen scavenger sachets or other in-pack active packaging devices. Some active packages may look different from their passive counterparts. Therefore it may be advisable to use appropriate labelling to explain this difference to the consumer even in the absence of regulations.

Finally, it is very important for food manufacturers using certain type of active packaging to consider the effects this will have on the microbial ecology and safety of foods. For example, removing all the oxygen from within packs of high a_w chilled perishable food products may stimulate the growth of anaerobic pathogenic bacteria such as *Clostridium botulinum*. Specific guidance is available to minimise the microbial safety risks of foods packed under reduced oxygen atmospheres [94]. Regarding the use of antimicrobial films, it is important to consider what spectrum of microorganisms will be inhibited. Antimicrobial films that only inhibit spoilage microorganisms without affecting the growth of pathogenic bacteria will raise food safety concerns.

In the USA, Japan and Australia, active packaging concepts are already being successfully applied. In Europe, the development and application of active packaging is limited because of legislative restrictions, fear of consumer resistance, lack of knowledge about effectiveness, economic and environmental impact of concepts [95]. No specific regulations exist on the use of active packaging in Europe. Active packaging is subjected to traditional packaging legislation, which requires that compounds are registered on positive lists and that the overall and specific migration limits are respected. This is more or less contradictory to the concept of some active packaging systems in which packaging releases substances to extend shelf life or improve quality [93]. The food industry's main concern about introducing active components to packaging seems to be that consumers may consider the components harmful and may not accept them. In Finland a consumer survey conducted in order to determine consumer attitudes towards oxygen scavengers revealed that the new concepts would be accepted if consumers are well informed by using reliable information channels [96]. More information is needed about the chemical, microbiological and physiological effects of various active concepts on the packaged food, i.e. in regard to its quality and safety. So far research has mainly concentrated on development of various methods and their testing in a model system, but not so much on functioning in food preservation with real food products. Furthermore, the benefits of active packaging need to be considered in a holistic approach to environmental impact assessment. The environmental effect of plastics-based active packaging will vary with the nature of the product/package combination and additional additives need to be evaluated for their environmental impact [95].

9.6.11 Conclusions

Active packaging is an emerging and exciting area of food technology that can confer many preservation benefits on a wide range of food products. Active packaging is a technology developing a new trust because of recent advances in packaging, material science, biotechnology and new consumer demands. The objectives of this technology are to maintain sensory quality and extend the shelf life of foods whilst at the same time maintaining nutritional quality and ensuring microbial safety.

Oxygen scavengers are by far the most commercially important subcategory of active packaging and the market has been growing steadily for the last 10 years. It is predicted that the recent introduction of oxygen scavenging films and bottle caps will further help stimulate the market in future years and the unit costs of oxygen scavenging technology will drop. Other active packaging technologies are also predicted to be used more in the future, particularly carbon dioxide scavengers and emitters, moisture absorbers and temperature control packaging. Food safety and regulatory issues in the EU are likely to restrict the use of certain preservative releasers and flavour/odour adsorber active packaging technologies. Nevertheless, the use of active packaging is becoming increasingly popular and many new opportunities in the food and nonfood industries will open up for utilising this technology in the future.

References

1 Levy, G. M. (ed.) **1993**, *Packaging in the Environment*, Blackie Academic & Professional, London.

2 Lauzon, C. and Wood, G. (eds.) **1995**, *Environmentally Responsible Packaging – a Guide to Development, Selection and Design*, Pira International, Leatherhead.

3 Levy, G. M. (ed.) **2000**, *Packaging Policy and the Environment*, Aspen Publishers, Gaithersburgh.

4 McCormack, T. **2000**, Plastics Packaging and the Environment, in *Materials and Development of Plastic Packaging for the Consumer Market*, ed. G. A. Giles and D. R. Bain, Sheffield Academic Press, Sheffield, pp 152–176.

5 Dent, I. S. **2000**, Recycling and Reuse of Plastics Packaging for the Consumer Market, in *Materials and Development of Plastic Packaging for the Consumer Market*, ed. G. A. Giles and D. R. Bain, Sheffield Academic Press, Sheffield, pp 177–202.

6 Ashby, R, Cooper, I, Harvey, S and Tice, P. **1997**, *Food Packaging Migration and Legislation*, Pira International, Leatherhead.

7 Watson, D. H. and Meah, M. N. (eds.) **1994**, *Chemical Migration from Food Packaging* (*Food Science Reviews, vol 2*), Ellis Horwood, London.

8 Crosby, N. T. **1981**, *Food Packaging Materials-Aspects of Analysis and Migration of Contaminants*, Applied Science Publishers, London.

9 Ackermann, P., Jagerstad, M. and Ohlsson, T. (eds.) **1995**, *Food Packaging Materials – Chemical Interactions*, Royal Society of Chemistry, Cambridge.

10 Highland, H. A. **1978**, Insect Resistance of Food Packages – A Review, *J. Food Process. Preserv.* 2, 123–130.

11 Wohlgemoth, R. **1979**, Protection of Stored Foodstuffs Against Insect Infestation by Packaging, *Chem. Ind.* May, 330–334.

12 Paine, F. A. **1989**, *Tamper Evident Packaging – A Literature Review*, Pira, Leatherhead.

13 Paine, F. A. (ed.) **1991**, *The Packaging User's Handbook*, Blackie and Sons, Glasgow.

14 Fellows, P. J. **2000**, *Food Processing Technology*, 2nd edn, Woodhead Publishing, Cambridge.

15 Brennan, J. G., Butters, J. R., Cowell, N. D. and Lilly, A. E. V. **1990**, *Food Engineering Operations*, 3rd edn, Elsevier Applied Science, London.

16 Paine, F. A. and Paine, H. Y. (eds.) **1992**, *A Handbook of Food Packaging*, 2nd edn, Blackie Academic & Professional, London.

17 Robertson, G. L. **1993**, *Food Packaging, Principles and Practice*, Marcel Dekker, New York.

18 Paine, F. **1990**, *Packaging Design and Performance*, Pira, Leatherhead.
19 DeMaria, K. **2000**, *The Packaging Design Process*, Technomic Publishing Co., Lancaster.
20 Kirwan, M. J. **2003**, Paper and Paperboard Packaging, in *Food Packaging Technology*, ed. R. Coles, D. McDowell and M. J. Kirwan, Blackwell Publishing, Oxford, pp 241–281.
21 Anon. **1997**, Paper and Paperboard, in *The Wiley Encyclopedia of Packaging Technology*, 2nd edn, ed. A. L. Brody and K. S. Marsh, John Wiley & Sons, New York, pp 714–723.
22 Foster, G. E. **1997**, Boxes, Corrugated, in *The Wiley Encyclopedia of Packaging Technology*, 2nd edn, ed. A. L. Brody and K. S. Marsh, John Wiley & Sons, New York, pp 100–108.
23 Hernandez, R. J. **1997**, Polymer Properties, in *The Wiley Encyclopedia of Packaging Technology*, 2nd edn, ed. A. L. Brody and K. S. Marsh, John Wiley & Sons, New York, pp 738–764.
24 Kirwan, M. J. and Strawbridge, J. W. **2003**, Plastics in Food Packaging, in *Food Packaging Technology*, ed. R. Coles, D. McDowell and M. J. Kirwan, Blackwell Publishing, Oxford, pp 174–240.
25 Bakish, R. **1997**, Metallizing, Vacuum, in *The Wiley Encyclopedia of Packaging Technology*, 2nd edn, ed. A. L. Brody and K. S. Marsh, John Wiley & Sons, New York, pp 629–638.
26 Anon. **1997**, Sealing, Heat in *The Wiley Encyclopedia of Packaging Technology*, 2nd edn, ed. A. L. Brody and K. S. Marsh, John Wiley & Sons, New York, pp 823–827.
27 Staines, G. **2000**, Injection Moulding in *Development of Plastic Packaging for the Consumer Market*, ed. G. A. Giles and D. H. Bain, Sheffield Academic Press, Sheffield, pp 8–24.
28 Hind, V. **2001**, Extrusion Blow-Moulding in *Technology of Plastics Packaging for the Consumer Market*, ed. D. A. Giles and D. R. Bain, Sheffield Academic Press, Sheffield, pp 25–52.
29 Bain, D. R. **2001**, Thermoforming Technologies for the Manufacture of Rigid Plastics Packaging in *Technology of Plastics Packaging for the Consumer Market*, ed. D. A. Giles and D. R. Bain, Sheffield Academic Press, Sheffield, pp 146–159.
30 Page, B., Edwards, M. and May, N. **2003**, Metal Cans, in *Food Packaging Technology*, ed. R. Coles, D. McDowell and M. J. Kirwan, Blackwell Publishing, Oxford, pp 120–151.
31 Morgan, E. **1985**, *Tinplate and Modern Canmaking Technology*, Pergamon Press, Oxford.
32 Britten, S. C. **1975**, *Tin Versus Corrosion (ITRI Publication No. 510)*, International Tin Research Institute, Middlesex.
33 Good, R. H. **1988**, Recent Advances in Metal Can Interior Coatings, in *Food and Packaging Interactions*, ed. J. H. Horchkiss, American Chemical Society, Washington, D.C., pp 203–219.
34 Turner, T. A. **1998**, *Canmaking – The Technology of Metal Protection and Decoration*, Blackie Academic & Professional, London.
35 Selbereis, J. **1997**, Metal Cans, Fabrication, in *The Wiley Encyclopedia of Packaging Technology*, 2nd edn, ed. A. L. Brody and K. S. Marsh, John Wiley & Sons, New York, pp 616–629.
36 Kraus, F. J. **1997**, Cans, Steel, in *The Wiley Encyclopedia of Packaging Technology*, 2nd edn, ed. A. L. Brody and K. S. Marsh, John Wiley & Sons, New York, pp 144–154.
37 Girling, P. J. **2003**, Packaging of Food in Glass Containers, in *Food Packaging Technology*, ed. R. Coles, D. McDowell and M. J. Kirwan, Blackwell Publishing, Oxford, pp 152–173.
38 Moody, B. E. **1977**, *Packaging in Glass*, Hutchinson and Benham, London.
39 Cavanagh, J. **1997**, Glass Container Manufacturing, in *The Wiley Encyclopedia of Packaging Technology*, 2nd edn, ed. A. L. Brody and K. S. Marsh, John Wiley & Sons, New York, pp 475–484.
40 Tooley, F. V. **1974**, *The Handbook of Glass Manufacture*, Ashlee Publishing, New York.
41 Zagory, D. **1997**, Modified Atmosphere Packaging, in *The Wiley Encyclopedia of Packaging Technology*, 2nd edn, ed. A. L. Brody and K. S. Marsh, John Wiley & Sons, New York, pp 650–656.

42 Mullan, M. and McDowell, D. **2003**, Modified Atmosphere Packaging, in *Food Packaging Technology*, ed. R. Coles, D. McDowell and M.J. Kirwan, Blackwell Publishing, Oxford, pp 303–338.
43 Blakistone, B.A. **1998**, Meats and Poultry, in *Principles and Applications of Modified Atmosphere Packaging of Foods,* 2nd edn, ed. B.A. Blakistone, Blackie Academic and Professional, London, pp 240–290.
44 Gill, C.O. **1995**, MAP and CAP of Fresh Red Meats, Poultry and Offal, in *Principles of Modified-Atmosphere and Sous Vide Product Packaging*, ed. J.M. Fraber and K.L. Dodds, Technomic Publishing Co., Lancaster, pp 105–136.
45 Davis, H.K. **1998**, Fish and Shellfish, in *Principles and Applications of Modified Atmosphere Packaging of Foods,* 2nd ed., B.A. Blakistone, Blackie Academic and Professional, London, pp 194–239.
46 Gibson, D.M. and Davis, H.K. **1995**, Fish and Shellfish Products in Sous Vide and Modified Atmosphere Packs, in *Principles of Modified-Atmosphere and Sous Vide Product Packaging*, ed. J.M. Faber and K.L. Dodds, Technomic Publishing Co., Lancaster, pp 150–174.
47 Garrett, E.H. **1998**, Fresh-Cut Produce, in *Principles and Applications of Modified Atmosphere Packaging of Foods,* 2nd edn, ed. B.A. Blakistone, Blackie Academic and Professional, London, pp 125–134.
48 Zagory, D. **1995**, Principles and Practice of Modified Atmosphere Packaging of Horticultural Commodities, in *Principles of Modified-Atmosphere and Sous Vide Product Packaging*, ed. J.M. Faber and K.L. Dodds, Technomic Publishing Co., Lancaster, pp 175–206.
49 Seiler, D.A.L. **1998**, Bakery Products, in *Principles and Applications of Modified Atmosphere Packaging of Foods,* 2nd edn, ed. B.A. Blackistone, Blackie Academic and Professional, London, pp 135–157.
50 Subramanian, P.J. **1998**, Dairy Foods, Multi-Component Products, Dried Foods and Beverages, in *Principles and Applications of Modified Atmosphere Packaging of Foods,* 2nd edn, ed. B.A. Blackistone, Blackie Academic and Professional, London, pp 158–193.
51 White F.S. **1993**, The Dole Process, in *Aseptic Processing and Packaging of Particulate Foods,* ed. E.M.A. Willhoft, Blackie Academic and Professional, London, pp 148–154.
52 Hersom, A.C. and Hulland, E.D. **1980**, Canned Foods, *Thermal Processing and Microbiology*, 7th edn, Churchill Livingstone, Edinburgh.
53 Burton, H. **1988**, *Ultra-High Temperature Processing of Milk and Milk Products,* Elsevier Applied Science Publishers, London.
54 Holdsworth, S.D. **1992**, *Aseptic Processing and Packaging of Food Products,* Elsevier Applied Science Publishers, London.
55 Buchner N. **1993**, Aseptic Processing and Packaging of Food Particulates, in *Aseptic Processing and Packaging of Particulate Foods,* ed. E.M.A. Willhoft, Blackie Academic and Professional, London, pp 1–22.
56 Joyce D.A. **1993**, Microbiological Aspects of Aseptic Processing and Packaging, in *Aseptic Processing and Packaging of Particulate Foods,* ed. E.M.A. Willhoft, Blackie Academic and Professional, London, pp 155–180.
57 Wakabayashi S. **1993**, Aseptic Packaging of Liquid Foods, in *Aseptic Processing and Packaging of Particulate Foods,* ed. E.M.A. Willhoft, Blackie Academic and Professional, London, pp 181–187.
58 Jairus, R.D., David, R.H., Graves, R.H. and Carlson, V.R. **1996**, *Aseptic Processing and Packaging of Food – A Food Industry Perspective,* CRC Press, London.
59 Day, B.P.F. **1989**, Extension of Shelf life of Chilled Foods, *Eur. Food Drink Rev.* 4, 47–56.
60 Rooney, M.L. **1995**, *Active Food Packaging,* Chapman & Hall, London.
61 Hotchkiss, L. **1994**, Recent Research in MAP and Active Packaging Systems, in *Abstracts of the 27th Annual Convention,* Australian Institute of Food Science and Technology, Canberra.
62 Day, B.P.F. **2001**, Active Packaging – A Fresh Approach, *J. Brand Technol.* 1, 32–41.
63 Summers, L. **1992**, *Intelligent Packaging,* Centre for Exploitation of Science and Technology, London.

64 Day, B. P. F. **1994**, Intelligent Packaging of Foods (*New Technologies Bulletin No. 10*), Campden & Chorleywood Food Research Association, Chipping Campden, pp 1–7.

65 Anon. **1996**, Oxygen Absorbing Packaging Materials Near Market Debuts, *Packaging Strategies Supplement* (31 January edition), Packaging Strategies, West Chester.

66 Rooney, M. L. **1998**, Oxygen Scavenging Plastics for Retention of Food Quality, *Proceedings of a Conference on "Advances in Plastics – Materials and Processing Technology for Packaging"*, Pira International, Leatherhead.

67 Pira International **2002**, Full of gas – the active packaging market is dominated by oxygen and moisture scavengers, *Active Intel. Pack. News* 1, 5.

68 Anon. **1995 a**, Pursuit of Freshness Creates Packaging Opportunities, *Jpn. Pack. News* 12, 14–15.

69 Hume, E. and Ahvenainen, R. **1996**, Active and Smart Packaging of Ready-Made Foods, *Proc. Int. Symp. Minimal Process. Ready Made Foods* 1996, 1–7.

70 Castle, D. **1996**, Polymer Advance Doubles Shelf life, *Pack. Week*, 12, 1.

71 Glaskin, M. **1997 a**, Plastic Bag Keeps Bread Fresh for Three Years, *Financial Times*, 29 June, p. 13.

72 Hirst, J. **1998**, *Personal Communication*, EMCO Packaging Systems, Worth.

73 Rooney, M. L. **2000**, Applications of ZERO2 Oxygen Scavenging Films for Food and Beverage Products, *Proceedings of the "International Conference on Active and Intelligent Packaging"*, Campden & Chorlytwood Food Research Association, Chipping Campden.

74 Labuza, T. P. and Breene, W. M. **1989**, Applications of Active Packaging for Improvement of Shelf life and Nutritional Quality of Fresh and Extended Shelf life Foods, *J. Food Process. Preserv.* 13, 1–69.

75 Anon. **1995 b**, Scavenging Solution, *Pack. News*, December, p. 20.

76 Naito, S., Okada, Y. and Yamaguchi, N. **1991**, Studies on the Behaviour of Microorganisms in Sponge Cake During Anaerobic Storage, *Pack. Technol. Sci.* 4, 4333–4344.

77 Abeles, F. B., Morgan, P. W. and Saltveit, M. E. **1992**, *Ethylene in Plant Biology*, Academic Press, London.

78 Hebeda, R. E. and Zobel, H. F. **1996**, *Baked Goods Freshness – Technology, Evaluation and Inhibition of Staling*, Marcel Dekker, New York.

79 Anon. **1994**, Fresh Produce is Keen as Mustard to Last Longer, *Pack. Week* 10, 6.

80 Weng, Y. M., Chen, M. J. and Chen, W. **1997**, Benzoyl Chloride Modified Ionomer Films as Antimicrobial Packaging Materials, *Int. J. Food Sci. Technol.* 32, 229–234.

81 Gray, P. N. **2000**, Generation of Active Microatmosphere Environments from and in Packages, *Proceedings of the "International Conference on Active and Intelligent Packaging"*, Campden & Chorleywood Food Research Association, Chipping Campden.

82 Goddard, R. **1995 a**, Beating Bacteria, *Pack. Week* 11, 13.

83 Jamieson, D. **1997**, Sainsbury's to set new standards in food hygiene, *Press Release 14 May*, J. Sainsbury, London

84 Rubinstein, W. S. **2000**, Microban Antibacterial Protection for the Food Industry, *Proceedings of the "International Conference on Active and Intelligent Packaging"*, Campden & Chorleywood Food Research Association, Chipping Campden.

85 Barlas, S. **1998**, Packagers Tell Insects: "Stop Bugging Us", *Pack. World* 5, 31.

86 Newcorn, D. **1997**, Not Just for Breakfast Anymore, *Pack. World* 4, 23–24.

87 Rice, J. **1994**, Fighting Moisture – Dessicant Sachets Poised for Expanded Use, *Food Process.* 63, 46–48.

88 Patterson, B. D. and Joyce, D. C. **1993**, *A Package Allowing Cooling and preservation of Horticultural Produce Without Condensation or Dessicants*, International patent application PCT/AU93/00398.

89 Franzetti, L., Martinoli, S, Piergiovanni, L. and Galli, A. **2001**, Influence of Active Packaging on the Shelf life of Minimally Processed Fish Products in a Modified Atmosphere, *Pack. Technol. Sci.* 14, 267–274.

90 Anon. **1996**, Odour Eater, *Pack. News*, August, p. 3.
91 Goddard, R. **1995 b**, Dispersing the Scent, *Pack. Week* 10, 28.
92 Anon. **1997 b**, Things Get Hot for Cool Can of the Year, *Pack. News*, July, p. 1.
93 De Kruijf, N. **2000**, Objectives, Tasks and Results From EC FAIR "Actipak" Research Project, *Proceedings of the "International Conference on Active and Intelligent Packaging"*, Campden & Chorleywood Food Research Association, Chipping Campden.
94 Betts, G. D. **1996**, *Code of Practice for the Manufacture of Vacuum and Modified Atmosphere Packaged Chilled Foods with Particular Regards to the Risks of Botulism* (Guideline No. 11), Campden & Chorleywood Food Research Association, Chipping Campden.
95 Vermeiren, L., Devlieghere, F., van Beest, M, deKruitjf, N and Debevere, L. **1999**, Developments in the active packaging of foods, *Trends Food Sci. Technol.* 10, 77–86.
96 Ahvenainen, R. and Hurme, E. **1997**, Active and Smart Packaging for Meeting Consumer Demands for Quality and Safety, *Food Addit. Contamin.* 14, 753–763.

10
Safety in Food Processing

Carol A. Wallace

10.1
Introduction

It is a fundamental requirement of any food process that the food produced should be safe for consumption. Food safety is a basic need but there is a danger that it may be overlooked in the development of effective and efficient processes.

There are three key elements to ensuring food safety is achieved in food manufacture:
1. safe design of the process, recipe and packaging format;
2. prerequisite programmes or good manufacturing practice to control the manufacturing environment;
3. use of the HACCP system of food safety management.

This chapter will outline these current approaches to effective food safety management and consider how they fit with the design and use of different food processing technologies.

10.2
Safe Design

"When designing a new food product it is important to ask if it is possible to manufacture it safely. Effective HACCP systems (and prerequisite programmes) will manage and control food safety but what they cannot do is make safe a fundamentally unsafe product" [1].

It is important, therefore, to understand the criteria involved in designing and manufacturing a safe product. These include:
- an understanding of the likely food safety hazards that may be presented through the ingredients, processing and handling methods;
- the intrinsic factors involved in developing a safe recipe;
- a thorough knowledge of the chosen food processing and packaging technologies;

Food Processing Handbook. Edited by James G. Brennan

ISBN: 3-527-30719-2

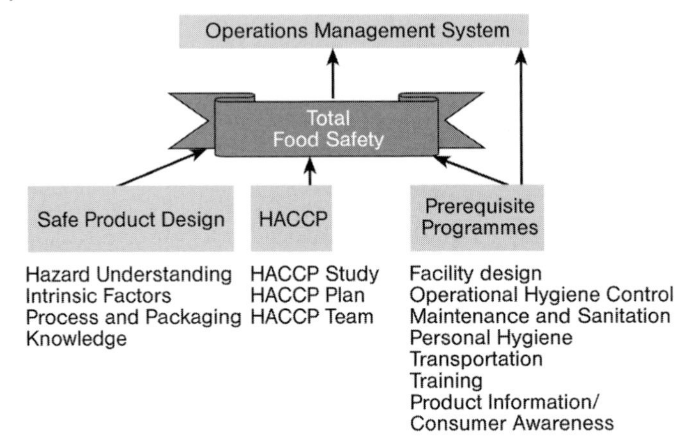

Fig. 10.1 Safe food processing achievement model; adapted from [3].

- manufacturing in a facility operating to prerequisite good manufacturing practice systems;
- management of production within the framework of a validated HACCP programme.

Prerequisite good manufacturing practice programmes and HACCP will be covered in Sections 10.3 and 10.4, respectively. Before further considering the design of safe products, we can look at how these different criteria fit together to ensure safe food processing. Fig. 10.1 shows a model for the achievement of safe food processing. The safety management criteria, i.e. safe product design, HACCP and Prerequisite Good Manufacturing Practice programmes are all managed within the framework of the Operational Management system, which could be a Quality Management system such as ISO 9000:2000 [2].

10.2.1
Food Safety Hazards

Food safety hazards are contaminants that may cause a food product to be unsafe for production. Hazards are defined by Codex 1997 [4] as follows:

"Hazard: a biological, chemical or physical agent in, or condition of, food with the potential to cause an adverse health effect".

Hazards may enter a food product from its ingredients or may contaminate during processing or handling. Table 10.1 shows examples of common hazard types for consideration.

At the product design stage, it is important to understand the likely hazards that might be encountered in the chosen ingredient types, or that might be present in the processing environment. This allows the development team to identify the best ways to control these hazards, either by preventing their entry to the process, destroying them or reducing the contamination to a level where

Table 10.1 Examples of food safety hazards. Note: this table provides examples only and is not intended to be an exhaustive list of food safety hazards.

	Type of hazard		
	Biological	**Chemical**	**Physical**
Considerations	Organisms that can cause harm through infection or intoxication	Chemicals that can cause harm through toxic effects, either immediate or long-term	Items that can cause harm through direct injury or choking
Examples	Pathogenic bacteria, e.g. *Escherichia coli*, *Bacillus cereus*, *Campylobacter jejuni*, *Clostridium botulinum*, *C. botulinum* (non-proteolytic), *C. perfringens*, *Salmonella* spp, *Shigella* spp, *Staphylococcus aureas*, *Vibrio parahaemoliticus*; Viruses, Protozoan parasites, e.g. *Cryptosporidium parvum*, *Giardia intestinalis*, *Cyclospora cayetanensis*	Mycotoxins, e.g. aflatoxins, patulin, vomitoxin, fumonisin; pesticides, allergenic materials, heavy metals, PCBs dioxins, cleaning chemicals	Glass, metal, stones, wood, plastic, pests, intrinsic natural materials, e.g. bone, nut shell

it no longer poses a food safety risk. This information on likely hazards and proposed control options should link with the prerequisite good manufacturing practice programmes and HACCP systems to ensure everyday control is established in the manufacturing operation.

Consideration of likely hazards at an early stage in the development process can also, in some cases, help to design these hazards out of the product, either through careful choice and sourcing of ingredients or through identification of appropriate processing technologies and/or equipment. For example, if there is a concern about physical hazards gaining entry to a product during manufacture due to the use of open vessels, the redesign of the equipment to use enclosed vessels would prevent this hazard from ever occurring at that processing step. Similarly, if there is concern about pathogen contamination in a raw ingredient, e.g. *Salmonella* spp contamination in coconut that is to be used as a topping ingredient after heat processing, it may be possible to replace this ingredient with a preprocessed ingredient, in this example pasteurized coconut.

10.2.2
Intrinsic Factors

Intrinsic factors are the formulation criteria that control the ability of microorganisms to survive and grow in foods. These factors have been used traditionally to prevent problems with spoilage organisms and pathogens in a wide variety of foodstuffs. The most commonly used intrinsic factors in food processing are water activity, pH, organic acids and preservatives.

Water activity (a_w) is a measure of the amount of water available in a foodstuff for microbial growth. Pure water has an a_w of 1.0 and, as solutes such as salt and sugar are added to make a more concentrated solution, the a_w decreases. Table 10.2 shows the a_w limit for growth of a number of key microbial pathogens. There is a characteristic pH range across which microorganisms can grow; and the limiting pH for growth varies widely between species. The use of pH to control the growth of microorganisms is very common in food processing, finding uses in pickled foods such as pickled vegetables and fermented foods such as cheese and yoghurt. The pH limit for growth of a number of key microbial pathogens is also given in Table 10.2.

Organic acids, such as acetic, citric, lactic and sorbic acids, are widely used as preserving factors in food processing. The antimicrobial effect of organic acids is due to undissociated molecules of the acid and, since the dissociation of the molecules is pH-dependent, the effectiveness is related to pH.

Chemical preservatives may be added to food products to prevent the growth of pathogens and spoilage organisms. The use of preservatives is normally controlled by legislation, with different levels of various preservatives allowed for use in different groups of foodstuffs. Further detailed information on the effects of intrinsic factors on a wide range of microorganisms can be found in other publications, such as ICMSF [6], Kyriakides [19].

Table 10.2 Control of key microbiological hazards through intrinsic factors (adapted from [5, 6]).

Organism	Minimum pH for growth	Minimum water activity (a_w) for growth
Bacillus cereus	5.0	0.93
Campylobacter jejuni	4.9	0.99
Clostridium botulinum	4.7	0.94
Clostridium botulinum (non-proteolitic)	5.0	0.97
Clostridium perfringens	5.5	0.93
Escherichia coli	4.4	0.95
Listeria monocytogenes	4.4	0.92
Salmonella spp	3.8	0.94
Shigella spp	4.9	0.97
Staphylococcus aureuas	4.0	0.85
Vibrio parahaemoliticus	4.8	0.94

10.2.3
Food Processing Technologies

A wide variety of food processing technologies is available, as highlighted in the other chapters of this book. It is important for food safety that the chosen food process is thoroughly understood so that any potential food safety hazards can be effectively controlled. Table 10.3 shows the effects of various food processing techniques on food safety hazards.

It can be seen that most types of food processing illustrated in Table 10.3 are designed to control microbiological hazards and involve either destruction, reduction of numbers or prevention from growth of various foodborne pathogens. To a lesser extent, a number food processing techniques, e.g. cleaning and separation, involve the removal of physical hazards. Very few food processing techniques are designed to control chemical hazards in foods; and therefore it is important to source high quality ingredients that are free from chemical hazards.

10.2.4
Food Packaging Issues

The chosen packaging type should also be evaluated as part of the 'Safe Design' process. Food packaging systems have evolved to prevent contamination and ensure achievement of desired shelf life; however there may be hazard considerations if inappropriate to the type of food or proposed storage conditions. Table 10.4 lists a number of considerations for choosing a safe packaging system.

10.3
Prerequisite Good Manufacturing Practice Programmes

Prerequisite programmes or 'Good Manufacturing Practice' (GMP) provide the hygienic foundations for any food operation. The terms 'prerequisite programmes' and 'Good Manufacturing Practice' are used interchangeably in different parts of the world but have the same general meaning. For simplicity, the term prerequisite programmes will be used in this chapter.

Several groups have suggested definitions for the term prerequisites and the most commonly used are reproduced here. Prerequisite programmes are:

- practices and conditions needed prior to and during the implementation of HACCP and which are essential to food safety (World Health Organisation WHO [7]);
- universal steps or procedures that control the operating conditions within a food establishment, allowing for environmental conditions that are favourable for the production of safe food (Canadian Food Inspection Agency [8]);
- procedures, including GMP, that address operational conditions, providing the foundation for the HACCP system (USA National Advisory Committee for Microbiological Criteria for Foods [9]).

Table 10.3 Effects of food processing on food safety hazards.

Processing operation		Intended effect on food safety hazards	Example food types
Cleaning	Dry	Removal of foreign material and dust	Grain crops
	Wet	Reduction in level of microorganisms and foreign material	Raw foods, e.g. vegetables, fruit, dried fruit
Antimicrobial dipping/spraying		Reduction in levels of microorganisms	Fruit and vegetables
Fumigation		Destruction of certain microorganisms and pests	Nuts, dried fruit, cocoa beans
Thermal processing	Pasteurisation/ cooking	Destroys vegetative pathogens, e.g. *Salmonella* spp, *Listeria monocytogenes*	Milk products, meat, fish, ready meals
	Sterilisation: UHT/aseptic	Destroys pathogens and prevents recontamination in packaging system	UHT milk, fruit juices
	Sterilisation: cans/pouches	Destroys pathogens	Canned meats, soups, pet food, etc.
Evaporation/dehydration		Halts growth of pathogenic bacteria at a_w 0.84, all microorganisms at a_w 0.60	Various foodstuffs, e.g. dried fruit, milk powder, cake mixes, etc.
Salt preserving		Halts growth of pathogenic bacteria at a_w 0.84, all microorganisms at a_w 0.60; growth of many microorganisms halted at ca. 10% salt	Fish, meats, vegetables
Sugar preserving		Halts growth of pathogenic bacteria at a_w 0.84, all microorganisms at a_w 0.60	Jam, fruits, syrups, jellies, confectionery
Chilling (<5 °C)		Slows or prevents growth of most pathogens	Cooked meats, dairy products, fruit juices
Freezing (at least −10 °C)		Prevents growth of all microorganisms. Destroys some parasites	Many foodstuffs, e.g. fruit, vegetables, meat, fish, ice cream, etc.
Irradiation		Destroys microorganisms	Can be used for various products, e.g. fruit, shellfish, however consumer pressure has limited its application
High pressure processing		Destroys/inactivates microorganisms; affects functional and organoleptic properties	Fruit juice, guacomole, yoghurt, oysters

Table 10.3 (continued)

Processing operation	Intended effect on food safety hazards	Example food types
Pulsed electric field processing	Destroys microorganisms, inactivates some enzymes	Potential applications include fruit juice, milk
Fermentation/acidification	Halts growth of pathogens; destroys some organisms, depending on pH/acid used	Cheese, yogurt, vegetables, fruit, etc.
Separation (e.g. filtration)	Removes physical hazards and/or pathogens (depending on filter pore size), adjusts chemical concentration (e.g. reverse osmosis)	Various foodstuffs, e.g. sugar, grains, water, etc.

A number of groups have published helpful material on prerequisite programmes; however the internationally accepted requirements for prerequisites are defined in the Codex general principles of food hygiene [10]. Box 10.1 shows the section headings from this Codex document.

10.3.1 Prerequisite Programmes – The Essentials

Using the headings given in [10], the following notes describe the general requirements for prerequisite programmes in each area. Further, more detailed, information can be found in other publications such as the Codex document itself [10], CFIA [8], IFST [11], Sprenger [12], Engel [18], Mortimore [20], and Wallace [21].

Establishment: Design and Facilities The location of food premises is important and care should be taken to identify and consider the risks of potential sources of contamination in the surrounding environment. Suitable controls to prevent contamination should be developed and implemented.

The design and layout of the premises and rooms should permit good hygiene and protect the products from cross-contamination during operation. Internal structures and equipment should be built of materials able to be easily cleaned/disinfected and maintained. Surfaces should be smooth, impervious and able to withstand the normal conditions of the operation, e.g. moisture and temperature ranges.

Facilities should be provided to include adequate potable water supplies, suitable drainage and waste disposal, appropriate cleaning facilities, storage areas, lighting, ventilation and temperature control. Suitable facilities should also be provided to promote personal hygiene for the workforce, including adequate changing areas, lavatories and hand washing and drying facilities.

Table 10.4 Food packaging considerations.

Packaging type	Considerations
Retortable containers: cans	Hygienic container suitable for a wide range of foods. Suitable for retort sterilisation/pasteurisation and ambient storage, dependant on formulation suitability (pasteurisation). Careful handling required after sealing and retorting. Type of can and inner laquer needs to be matched to food type to prevent degradation and leaching of metal into the product.
Retortable containers: pouches	Hygienic container suitable for a wide range of foods. Suitable for retort sterilisation/pasteurisation and ambient storage, dependant on formulation suitability (pasteurisation). Careful handling required after sealing and retorting. Need to check that film constituents, e.g. plasticisers and additives, cannot transfer to food during packaging use.
Glass	Hygienic container suitable for a wide range of foods. Suitable for hot and cold fill. For ambient storage, need to ensure that the recipe intrinsic factors keep the product safe over the shelf-life. High quality glass required and container design for strength necessary. Careful handling required to prevent breakage and glass hazards. Glass breakage procedures needed.
Gas-flushed containers	Intended to extend life of product by preventing growth of spoilage organisms. Need to ensure that any pathogens present, e.g. anaerobic spore formers, cannot grow in the chosen gas mix.
Vacuum packaging	Intended to extend life of product by preventing growth of spoilage organisms. Anaerobic conditions provided can allow growth of some pathogens, e.g. *Clostridium botulinum.* May need to use vacuum packaging in conjunction with additional control measures such as chilling.
Gas-permeable packaging	Is it possible for other materials, e.g. moisture, to pass through into the product and cause contamination?
Product contact films and plastics	Need to check that constituents, e.g. plasticisers and additives, cannot transfer to food during packaging use.
Paper/cardboard	Most suitable for secondary/tertiary packaging. Need to ensure that inks and adhesives cannot transfer to foodstuff.
Wood	May introduce hazards, e.g. splinters. In most cases wood is best kept for secondary/tertiary packaging rather than direct product contact.

Box 10.1 Prerequisite programme topics for manufacturing facilities (adapted from [10], which also includes recommended general principles of food hygiene for primary production facilities).

Establishment: design and facilities	**Control of operation**
Location	Control of food hazards
Premises and rooms	Key aspects of hygiene control systems
Equipment	Incoming material requirements
Facilities	Packaging
	Water
	Management and supervision
	Documentation and records
	Recall procedures
Establishment: maintenance and sanitation	**Establishment: personal hygiene**
Maintenance and cleaning	Health status
Cleaning programmes	Illness and injuries
Pest control systems	Personal cleanliness
Waste management	Personal behaviour
Monitoring effectiveness	Visitors
Transportation	**Product information and consumer awareness**
General	Lot identification
Requirements	Product information
Use and maintenance	Labelling
	Consumer education
Training	
Awareness and responsibilities	
Training programmes	
Instruction and supervision	
Refresher training	

Control of Operation The rationale for operational control listed in [10] is "to reduce the risk of unsafe food by taking preventive measures to assure the safety and suitability of food at an appropriate stage in the operation by controlling food hazards". This includes the need to control potential food hazards by using a system such as HACCP.

Codex also describes key aspects of hygiene control systems, including:
- time and temperature control;
- microbiological and other specifications;
- microbiological cross-contamination risks;
- physical and chemical contamination.

Incoming material requirements and systems to ensure the safety of materials and ingredients at the start of processing are necessary, along with a suitable packaging design (see also Section 10.2.4).

Codex [10] also lists the importance of hygienic control of water, ice and steam, appropriate management and supervision, the need to keep adequate documentation and records and the need to develop and test suitable recall procedures so that product can be effectively withdrawn and recalled in the event of a food safety problem.

Establishment: Maintenance and Sanitation Maintenance and cleaning are important both to keep the processing environment, facilities and equipment in a good state of repair where they function as intended and to prevent cross-contamination with food residues and microorganisms that might otherwise build up. Facilities should operate preventative maintenance programmes as well as attending to breakdowns and faults without delay.

Cleaning programmes should be developed to encompass all equipment and facilities as well as general environmental cleaning. Cleaning methods need to be developed that are suitable for the item to be cleaned, including the use of appropriate chemical cleaning agents, disinfectants, hot/cold water and cleaning tools, e.g. brushes, scrapers, cloths, etc. Methods should describe how the item is to be cleaned and personnel should be trained to apply the methods correctly. A cleaning schedule should also be developed to identify the frequency of cleaning needed in each case and records of cleaning and monitoring should be kept.

Cleaning in place (CIP) solutions may be used in certain types of equipment, e.g. tanks and lines. Here it is important that the CIP programme is properly designed for the equipment to be cleaned, taking into account the flow rates, coverage and the need for rinsing and disinfection cycles.

Pest control systems are important to prevent the access of pests that might cause contamination to the product. Pest management is often contracted out to a professional pest control contractor. Buildings need to be made pestproof and regularly inspected for potential ingress points. Interior and exterior areas need to be kept clean and tidy to minimise potential food and harbourage sources. Suitable interior traps and monitoring devices should also be considered and any pest infestations need to be dealt with promptly, without adversely affecting food safety.

Waste management should ensure that waste materials can be removed and stored safely so that they do not provide a cross-contamination risk or become a food or harbourage source for pests.

All maintenance and sanitation systems should be monitored for effectiveness, verified and reviewed, with changes made to reflect operational changes.

Establishment: Personal Hygiene The objectives for personal hygiene stated in [10] are: "To ensure that those who come directly or indirectly into contact with food are not likely to contaminate food by:
- maintaining an appropriate degree of personal cleanliness;
- behaving and operating in an appropriate manner."

Food companies should, therefore, have standards and procedures in place to define the requirements for personal hygiene and staff responsibility; and staff should be appropriately trained. This should include the establishment of health status where individuals may be carrying disease that can be transmitted through food, a consideration of illness and injuries where affected staff members may need to be excluded or wear appropriate dressings, the need for good personal cleanliness and effective hand washing, the wearing of adequate protective clothing and the prevention of inappropriate behaviour such as smoking, eating or chewing in food handling areas. Visitors to processing and product handling areas should be adequately supervised and required to follow the same standards of personal hygiene as employees.

Transportation To ensure continuation of food safety throughout transportation, transport facilities need to be designed and managed to protect food products from potential contamination and damage and to prevent the growth of pathogens. This includes the need for cleaning and maintenance of vehicles and containers and the use of temperature control devices where appropriate.

Product Information and Consumer Awareness It is important that sufficient information is easily identifiable on the products so that the lot or batch can be identified for recall purposes and that the product can be handled correctly, e.g. stored at $<5\,^{\circ}C$. Product information and labelling should be clear such that it facilitates consumer choice and correct storage/use.

Codex [10] also highlights the importance of consumer education, particularly the importance of following handling instructions and the link between time/temperature and foodborne illness.

Training Food hygiene training is essential to make personnel aware of their roles and responsibilities for food control. Companies should develop and implement appropriate training programmes and should include adequate supervision and monitoring of food hygiene behaviour. Training should be evaluated and reviewed with refresher or update training implemented as necessary.

10.3.2 Validation and Verification of Prerequisite Programmes

Prerequisite programmes are the basic standards for the food facility, in which the safely designed product can be manufactured. They form the hygiene foundations on which the HACCP System is built to control food safety every day of operation. As such, it is essential that prerequisite programmes are working effectively at all times and it is therefore necessary that each prerequisite element is validated to establish that it will be effective and that an ongoing programme of monitoring and verification is developed and implemented.

10.4
HACCP, the Hazard Analysis and Critical Control Point System

The acronym HACCP stands for the 'hazard analysis and critical control point' system, a method of food control based on the prevention of food safety problems. The HACCP story began in the early 1960s, when the Pillsbury Company was working with NASA and the US Army Natick laboratories to provide food for the American manned space programme. Up until this time, most food safety control systems had been based on end product testing but it was realised that this would not give enough assurance of food safety for such an important mission. Taking the failure mode and effect analysis (FMEA) approach as a starting point, the team adapted this into the basis of the HACCP system that we know today: a system that looks at what can go wrong at each step in the process and builds in control to prevent the problem from occurring.

The HACCP system has become the internationally accepted approach to food safety management [4, 9]. It is based on the application of seven principles (Box 10.2) that show how to develop, implement and maintain a HACCP system.

The use of HACCP is promoted by the WHO and is increasingly being seen by government groups worldwide as a cornerstone of food safety legislation.

HACCP systems can be linear, where the principles are applied to the whole operation from ingredients to end product, or modular, where the operation is split into process stages or modules and HACCP plans [1)] are developed for each module. Modular systems are common in complex manufacturing operations and are practical to develop; however a key point is to ensure that the modules add up to the entire operation and that no process stages are missed out. Fig. 10.2 shows the linear and modular approaches to HACCP.

10.4.1
Developing a HACCP System

In order to develop a HACCP system, a food company applies the Codex HACCP principles to its operations. This is most easily achieved using the following logic sequence (Box 10.3), also proposed by Codex [4].

Step 1. Assemble HACCP Team HACCP is normally applied by a multidisciplinary team, so that the system is the output of a group with the necessary combined experience and knowledge to take decisions about product safety. This approach works well in manufacturing operations and normally includes, as a minimum, the following disciplines:

- manufacturing or operations personnel who understand the process operations on site;

1) HACCP plan: A document prepared in accordance with the principles of HACCP to ensure control of hazards that are significant for food safety in the segment of the food chain under consideration [4]. The HACCP plan is simply the documentation produced that shows how significant hazards will be controlled.

Box 10.2 The HACCP principles (from [4]).

Principle 1
Conduct a hazard analysis.

Principle 2
Determine the critical control points (CCPs).

Principle 3
Establish critical limit(s).

Principle 4
Establish a system to monitor control of the CCP.

Principle 5
Establish the corrective action to be taken when monitoring indicates that a particular CCP is not under control.

Principle 6
Establish procedures for verification to confirm that the HACCP system is working effectively.

Principle 7
Establish documentation concerning all procedures and records appropriate to these principles and their application.

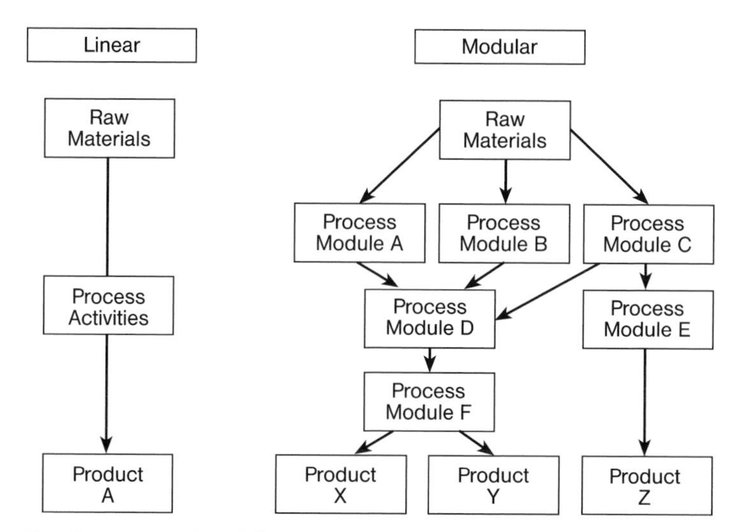

Fig. 10.2 Linear and modular HACCP system example layouts.

- quality or technical personnel who understand the product's technical characteristics regarding hazard control and have up to date information on likely hazards in that sector of the food industry;
- engineering personnel who have knowledge and experience of the equipment and process operations in use on site.

Box 10.3 Logic sequence for application of the Codex HACCP principles (adapted from [4]).

Logic sequence for application of HACCP

Step 1 Assemble HACCP team
Step 2 Describe product
Step 3 Identify intended use
Step 4 Construct flow diagram
Step 5 On-Site confirmation of flow digram
Step 6 List all potenzial hazards, conduct a hazard analysis and consider control measures
Step 7 Determine CCPs
Step 8 Establish critical limits for each CCP
Step 9 Establish a monitoring system for each CCP
Step 10 Establish corrective actions
Step 11 Establish verification procedures
Step 12 Establish documentation and record keeping

In addition to the above disciplines, it can be helpful to include personnel from the following areas; however the total size of a HACCP team is best kept to 4–6 personnel for ease of management:

- microbiology;
- supplier/vendor assurance;
- storage and distribution;
- product development.

Step 2. Describe Product It is important for all members of the HACCP team to understand the background to the product/process that they are about to study. This is achieved by constructing a product description (also known as a process description). The product description is not simply a specification for the product, but rather contains information important to making safety judgments. The following criteria are normally included:

- hazard types to be considered;
- main ingredient groups to be used in the product/process line;
- main processing technologies;
- key control measures;
- intrinsic (recipe) factors;
- packaging system;
- start and end points of the study.

The task of constructing a product description helps to familiarize all HACCP team members with the product/process under study. It is normal practice to document the product description and include it with the HACCP plan paperwork. The document is also useful at later stages as a familiarization tool for HACCP system auditors or any personnel who need to gain an understanding of the HACCP plan.

Step 3. Identify Intended Use It is necessary to identify the intended use of the product, including the intended consumer target group, because different uses may involve different hazard considerations and different consumer groups may have varying susceptibilities to the potential hazards. This information is usually included as part of the product description (Step 2).

Step 4. Construct Flow Diagram A process flow diagram, outlining all the process activities in the operation being studied, needs to be constructed. This should list all the individual activities in a stepwise manner and should show the interactions of the different activities. The purpose of the process flow diagram is to document the process and provide a foundation for the hazard analysis (Step 5). A simple example of a process flow diagram is shown in Fig. 10.3. This shows a process module taken from a modular HACCP system at a milk processing plant.

Notice that the steps are shown as activities. A common error in HACCP is to list the names of the process equipment rather than the process activity (Fig. 10.4). This error can cause difficulties, particularly where more than one process activity takes place in the same piece of equipment, since different haz-

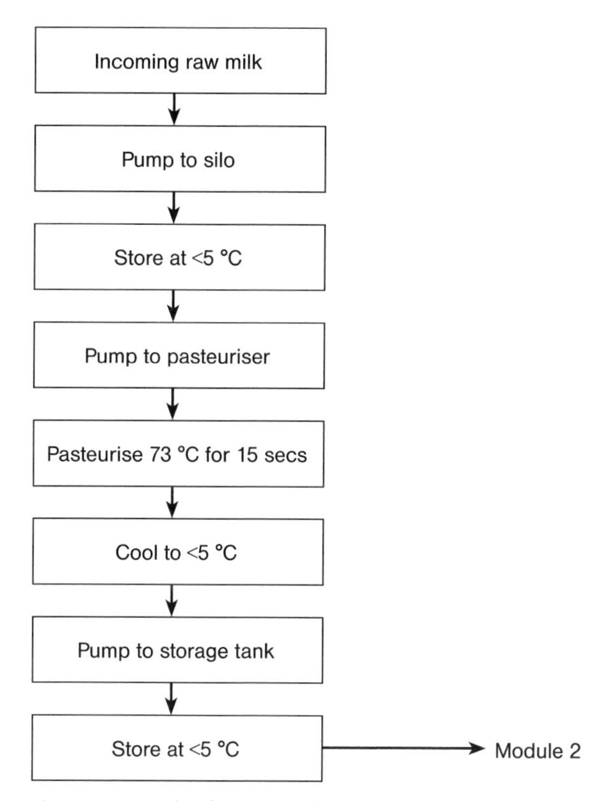

Fig. 10.3 Example of a process flow diagram.

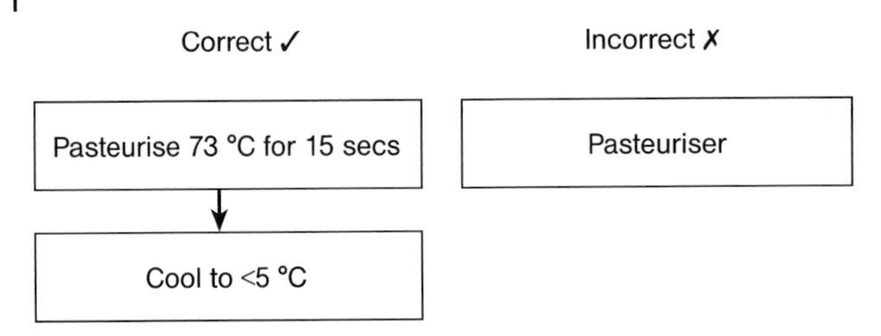

Fig. 10.4 Process flow diagrams – process activities vs equipment names.

ards can apply. The example shown in Fig. 10.4 is for the pasteurisation process where two different steps, the pasteurisation heat treatment and the cooling process, take place in the same piece of equipment: the pasteuriser. These steps have different potential hazards, the former being survival of vegetative pathogens if the heat process is not effective and the latter being potential cross-contamination with pathogens from raw milk during cooling due to inadequate pressure differential in the pasteuriser.

Step 5. On Site Confirmation of Flow Diagram Since the process flow diagram is used as a tool to structure the hazard analysis, it is important to check and confirm that it is correct. This is done by walking the line and comparing the documented diagram with the actual process activities, noting any changes necessary. This exercise is normally done by members of the HACCP team but could also be done by process line operators. The completed process flow diagram should be signed off as valid by a responsible member of staff, e.g. the HACCP team leader.

Step 6. List all Potential Hazards, Conduct a Hazard Analysis and Consider Control Measures Using the process flow diagram, the HACCP team now needs to consider each step in turn and list any potential hazards that might occur. They should then carry out an analysis to identify the significant hazards and identify suitable control measures. These terms are defined by Codex [4] as follows:

Hazard: a biological, chemical or physical agent in, or condition of, food with the potential to cause an adverse health effect;

Hazard Analysis: the process of collecting and evaluating information on hazards and conditions leading to their presence to decide which are significant for food safety and therefore should be addressed in the HACCP plan;

Control Measure: an action or activity that can be used to prevent, eliminate or reduce a hazard to an acceptable level.

An example of hazard analysis for two steps from the milk process flow diagram (see Fig. 10.3) is given in Table 10.5. Note, only one potential hazard has been detailed for each process step – there may be others.

The process of hazard analysis requires the team to transcribe each process activity to a table such as the example given, consider any potential hazards

Table 10.5 Example of hazard analysis process.

Process step	Hazard and source/cause	Significant hazard? (yes or no)	Control measure
Incoming raw milk	Presence of vegetative pathogens, e.g. *Salmonella*, due to contamination from animal	Yes	Control by pasteurisation step in process
Pasteurisation	Survival of vegetative pathogens, e.g. *Salmonella*, due to incorrect heat process	Yes	Effective heat process (correct time/temperature combination)

along with their sources or causes and then evaluate their significance. To identify the significant hazards, it is necessary to consider the likelihood of occurrence of the hazard in the type of operation being studied as well as the severity of the potential adverse effect. This may be done using judgement and experience or using a structured 'risk assessment' method, where different degrees of likelihood and severity are weighted to help with the significance decision. Effective control measures then need to be identified for each significant hazard.

Step 7. Determine CCPs Critical control points (CCPs) are the points in the process where the hazards must be controlled in order to ensure product safety. They are defined by Codex [4] as follows:

Critical control point (CCP): a step at which control can be applied and is essential to prevent or eliminate a food safety hazard or reduce it to an acceptable level.

It is important to identify the correct points as CCPs so that resource can be focused on their management during processing. CCPs can be identified using HACCP team knowledge and experience or by using tools such as the Codex CCP decision tree (see Fig. 10.5). More detailed explanations on the identification of CCPs and use of decision trees can be found in other publications, e.g. [1, 13].

Step 8. Establish Critical Limits for each CCP Critical limits are the safety limits that must be achieved for each CCP to ensure that the products are safe. As long as the process operates within the critical limits, the products will be safe but if it goes beyond the critical limits then the products made will be potentially unsafe. Critical limits are defined by Codex [4] as follows:

Critical limit: a criterion that separates acceptability from unacceptability.

Critical limits are expressed as absolute values (never a range) and often involve criteria such as temperature and time, pH and acidity, moisture, etc.

Step 9. Establish a Monitoring System for each CCP Monitoring is necessary to demonstrate that the CCPs are being controlled within the appropriate critical limits. Monitoring requirements are specified by the HACCP team during the

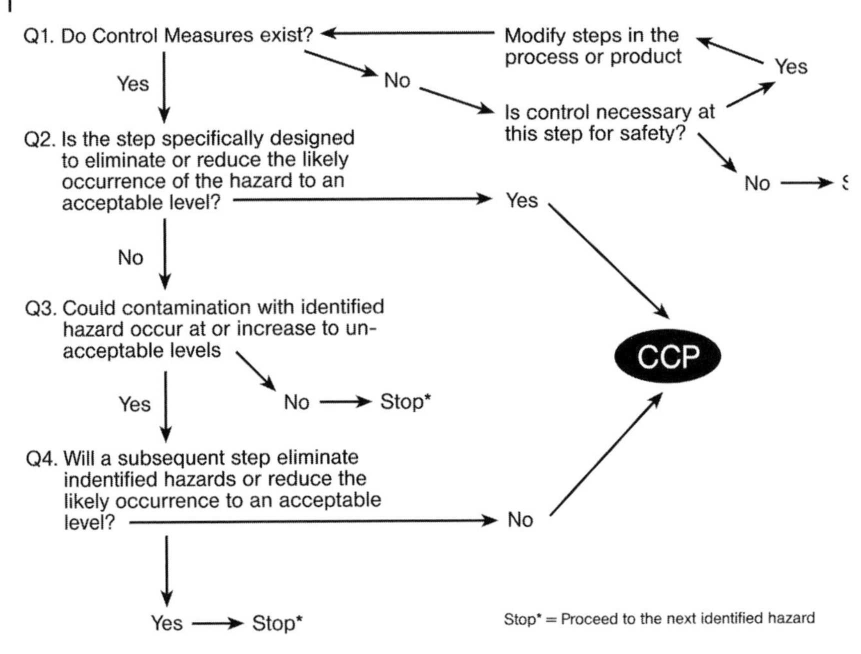

Fig. 10.5 CCP decision tree; from [4].

HACCP study but will usually be done by the process operators when the HACCP plan is implemented in the operation.

Monitoring: the act of conducting a planned sequence of observations or measurements of control parameters to assess whether a CCP is under control; Codex [4].

Monitoring should be defined in terms of the monitoring activity itself, along with the frequency and responsibility for doing the task.

Step 10. Establish Corrective Actions Corrective action needs to be taken where monitoring shows that there is a deviation from a defined critical limit. Corrective actions will deal with the material produced while the process is out of control and will also bring the process back under control.

Corrective action: any action to be taken when the results of monitoring at the CCP indicate a loss of control; Codex [4].

As for monitoring, the corrective action procedures and responsibility need to be identified by the HACCP team during the HACCP study, but will be implemented by the appropriate operations personnel if deviation occurs.

A completed table demonstrating control of CCPs using critical limits, monitoring and corrective action is shown as Table 10.6.

Step 11. Establish Verification Procedures The HACCP team needs to consider how to determine if the HACCP system is valid and working effectively over

Table 10.6 Example of CCP control.

Process step	Hazard	Control measure	Critical Limit	Monitoring			Corrective Action	
				Procedure	Frequency	Responsibility	Activity	Responsibility
Pasteurisation	Survival of vegetative pathogens, e.g. *Salmonella*	Correct temperature and time regime: effective heat process	71.7 °C for 15 s	Chart recorder: visual check and sign off	Each batch	Pasteuriser operator	Report to supervisor; contact QA and discuss; ensure divert working correctly; if not, dump/ re-process	Pasteuriser operator, production supervisor, QA manager, plant engineer
				Check auto-divert function	Daily at start up and shutdown	Pasteuriser operator	Hold product until correct heat process verified; dump/ reprocess if not	Pasteuriser operator, production supervisor, QA manager, plant engineer

time. Verification procedures are the methods that will be used to demonstrate compliance and verification is defined by Codex [4] as:

Verification; the application of methods, procedures, tests and other evaluations, in addition to monitoring, to determine compliance with the HACCP plan.

Commonly used verification procedures include:

- HACCP audits;
- review of CCP monitoring records;
- validity assessment of HACCP plan elements;
- product testing – microbiological and chemical;
- review of deviations, including product disposition and customer complaints.

Step 12. Establish Documentation and Record Keeping It is important to document the HACCP system and to keep adequate records. The HACCP plan will form a key part of the documentation, outlining the CCPs and their management procedures (critical limits, monitoring, corrective action). It is also necessary to keep documentation describing how the HACCP plan was developed, i.e. the hazard analysis, CCP determination and critical limit identification processes.

When the HACCP plan is implemented in the operation, records will be kept on an ongoing basis. Essential records include:

- CCP monitoring records;
- records of corrective actions associated with critical limit deviation;
- records of verification activities;
- records of modifications to processes and the HACCP plans.

10.4.2 Implementing and Maintaining a HACCP System

The twelve steps of the HACCP logic sequence outlined above describe how to develop HACCP plans and their associated verification and documentation requirements. However they do not describe how to implement the HACCP plans into everyday practice. Implementing HACCP requires careful preparation and training of the workforce and is, perhaps, best managed as a change management process. Depending on the maturity of the operation, this may be a straightforward implementation of the HACCP requirements or may require a culture change.

The implementation stage is where the HACCP plans are handed over from the HACCP team(s) that worked on the development process to the operations personnel who will manage the CCPs on a day to day basis. Training for the personnel who will monitor CCPs and take corrective action is essential and HACCP awareness training for the operations workforce is advisable. HACCP monitoring personnel need to understand the monitoring procedures and frequency, as well as how to record results and when corrective action must be taken.

After implementation, the HACCP verification procedures identified in Step 11 of the HACCP logic sequence need to commence. Results of verification should be reviewed regularly and actions should be taken where necessary to strengthen the HACCP system.

10.4.3 Ongoing Control of Food Safety in Processing

In order to ensure ongoing control of food safety, the prerequisite programmes, HACCP and safe design processes need to work together as a cohesive system. The keys points to ongoing control of food safety are:

- verification of food safety system elements effectiveness;
- review of system elements and their suitability for food safety;
- change control procedures that require safety assessment and approval for all proposed changes to ingredients, process activities and products;
- ongoing management and update of system elements;
- training of staff.

As shown at the start of this chapter (see Fig. 10.1), the management of food safety system elements is often done using an overall operations management system, e.g. the quality management framework ISO 9001:2000 [2]. At the time of writing, a new ISO standard for food safety management, ISO 22000 [14], is at the draft stage. This document includes HACCP, quality management and prerequisite programme requirements. It remains to be seen what the take up of this ISO Standard will be by food processing companies. Other external Standards also require food companies to manage food safety through prerequisite programmes, management practices and HACCP. These include retail-driven Standards such as the BRC Global Technical Standard – Food [15], manufacturing Standards such as the American NFPA-Safe Program [16] and national expert Standards such as the Netherlands National Board of HACCP Experts HACCP Code [17].

External standards for food safety management can be helpful in giving an external perspective as well as keeping the requirements for food safety in the forefront of people's minds. These schemes are now a requirement for doing business in many areas, required by manufacturers and retailers alike.

The essential requirement for any food processor is that they can manage their facility, ingredients, processes and products to ensure that only safe products reach the customer. The food safety system elements – safe design, prerequisite programmes and HACCP – described in this chapter will allow the requirement for safe food to be achieved. The use of an external audit standard to assess the operation of system elements may be a business requirement for some companies or may be regarded as an optional extra by others. Either way, it is important to assess regularly whether the systems are working and therefore that there is ongoing control of food safety.

References

1 Mortimore, S. E., Wallace, C. A. **1998**, *HACCP: A Practical Approach*, 2nd edn, Aspen Publishers, Gaithersburgh.

2 International Organisation for Standardisation **2000**, *ISO 9001:2000, Quality Managment Systems – Requirements*, International Organisation for Standardisation, London.

3 Mortimore, S. E. **2001**, How to Make HACCP Work in Practice, *Food Control* 12, 209–215.

4 Codex Committee on Food Hygiene **1997**, HACCP System and Guidelines for its Application, in *Food Hygiene Basic Texts*, ed. Food and Agriculture Organisation of the United Nations, World Health Organisation, Rome, pp 33–45.

5 Kyriakides, A. L. **2005**, The Principles of Food Safety, in *Training and Education for Food Safety*, ed. C. A. Wallace, Blackwell Science, Oxford.

6 International Commission on Microbiological Specifications of Foods **1996**, *Microbiological Specifications of Food Pathogens* (*Microorganisms in Foods* 5), Blackie Academic & Professional, London.

7 World Health Organisation **1999**, *Strategies for implementing HACCP in small and/or less developed businesses* (Report of a WHO consultation, WHO/SDE/PHE/FOS/99.7), WHO, Geneva.

8 Canadian Food Inspection Agency **2000**, Prerequisite Programs, in *Guidelines and Principles for the Development of HACCP Generic Models*, 2nd edn (*Food Safety Enhancement Program Implementation Manual*, vol. 2), available at: www.inspection.gc.ca/english/fssa/polstrat/haccp/manu/vol2/volii/pdf.

9 NACMCF **1997**, *Hazard Analysis and Critical Control Point Principles and Appli-*

cations Guidelines, www.seafood_ucdavis.edu/Guidelines/nacmcfl.htm.

10 Codex Committee on Food Hygiene **1999**, *Recommended International Code of Practice – General Principles of Food Hygiene* (CAC/RCP 1-1969, rev. 3-1997, amended 1999), Food and Agriculture Organisation of the United Nations, Rome.

11 Institute of Food Science and Technology **1998**, *Food & Drink – Good Manufacturing Practice: A Guide to its Responsible Management,* 4th edn, Institute of Food Science and Technology, London.

12 Sprenger, R. **2003**, *Hygiene for Management,* Highfield Publications, Doncaster.

13 Mortimore, S. E., Wallace, C. A. **2001**, *HACCP – Food Industry Briefing,* Blackwell Science, Oxford.

14 International Organisation for Standardisation **2003**, *Food Safety Management Systems – Requirements (ISO/CD 22000, draft),* ISO/TC 34 N60, International Organisation for Standardisation, London.

15 British Retail Consortium **2003**, *BRC Global Standard – Food* (Issue 3), The Stationery Office, London.

16 National Food Processors Association **2004**, *NPFA – Safe Program Policies and Procedures Manual,* National Food Processors Association, Washington D.C., available at: www.nfpa-safe.org.

17 Netherlands National Board of HACCP Experts **2002**, *Dutch HACCP Code (Requirements for a HACCP Based on Food Safety System,* Ver. 3), Netherlands National Board of HACCP Experts, The Hague, available at: www.foodsafetymanagement.info.

18 Engel, D., MacDonald, D. **2001**, *Managing Food Safety,* Chadwick House Group, London.

19 Kyriakides, A., Bell, C. **1999–2005**, *The Practical Food Microbiology Series (Listeria 2005, E. coli 1999, Clostridium botulinum 2000, Salmonella 2001*), Blackwell Publications, Oxford.

20 Mortimore, S. E., Mayes, T. **2001**, *Making the Most of HACCP – Learning from Others' Experience,* Woodhead Publishing, Cambridge.

21 Wallace, C. A., Williams, A. P. **2001**, Pre-Requisites – a Help or a Hindrance to HACCP, *Food Control* 12, 235–240.

11
Process Control In Food Processing

Keshavan Niranjan, Araya Ahromrit and Ahok S. Khare

11.1
Introduction

Process control is an integral part of modern processing industries; and the food processing industry is no exception. The fundamental justification for adopting process control is to improve the economics of the process by achieving, amongst others, the following objectives: (1) reduce variation in the product quality, achieve more consistent production and maximise yield, (2) ensure process and product safety, (3) reduce manpower and enhance operator productivity, (4) reduce waste and (5) optimise energy efficiency [1, 2].

Processes are operated under either steady state, i.e. process conditions do not change, or unsteady state conditions, process conditions depend on time. The latter occurs in most real situations and requires control action in order to keep the product within specifications. Although there are many types of control actions and many different reasons for controlling a process, the following two steps form the basis of any control action:

1. accurate measurement of process parameters;
2. manipulation of one or more process parameters using control systems in order to alter or correct the process behaviour.

It is essential to note that a well designed process ought to be easy to control. More importantly, it is best to consider the controllability of a process at the very outset, rather than attempt to design a control system after the process plant has been developed [1].

11.2
Measurement of Process Parameters

As mentioned earlier, accurate measurement of the process parameters is absolutely critical for controlling any process. There are three main classes of sensors used for the measurement of key processing parameters, such as tempera-

Food Processing Handbook. Edited by James G. Brennan

ISBN: 3-527-30719-2

ture, pressure, mass, material level in containers, flow rate, density, viscosity, moisture, fat content, protein content, pH, size, colour, turbidity, etc. [3]:

- Penetrating sensors: these sensors penetrate inside the processing equipment and come into contact with the material being processed.
- Sampling sensors: these sensors operate on samples which are continuously withdrawn from the processing equipment.
- Nonpenetrating sensors: these sensors do not penetrate into the processing equipment and, as a consequence, do not come into contact with the materials being processed.

Sensors can also be characterised in relation to their application for process control as follows [3]:

- Inline sensors: these form an integral part of the processing equipment, and the values measured by them are used directly for process control.
- Online sensors: these too form an integral part of the processing equipment, but the measured values can only be used for process control after an operator has entered these values into the control system.
- Offline sensors: these sensors are not part of the processing equipment, nor can the measured values be used directly for process control. An operator has to measure the variable and enter the values into a control system to achieve process control.

Regardless of the type of the sensor selected, the following basic characteristics have to be evaluated before using it for measurement and control: (1) response time, gain, sensitivity, ease and speed of calibration, (2) accuracy, stability and reliability, (3) material of construction and robustness and (4) availability, purchase cost and ease of maintenance.

Detailed information on sensors, instrumentation and automatic control for the food industry can be obtained from references [4, 5].

11.3 Control Systems

Control systems can be of two types: manual control and automatic control.

11.3.1 Manual Control

In manual control, an operator periodically reads the process parameter which requires to be controlled and, when its value changes from the set value, initiates the control action necessary to drive the parameter towards the set value.

Fig. 11.1 shows a simple example of manual control where a steam valve is adjusted to regulate the temperature of water flowing through the pipe [3].

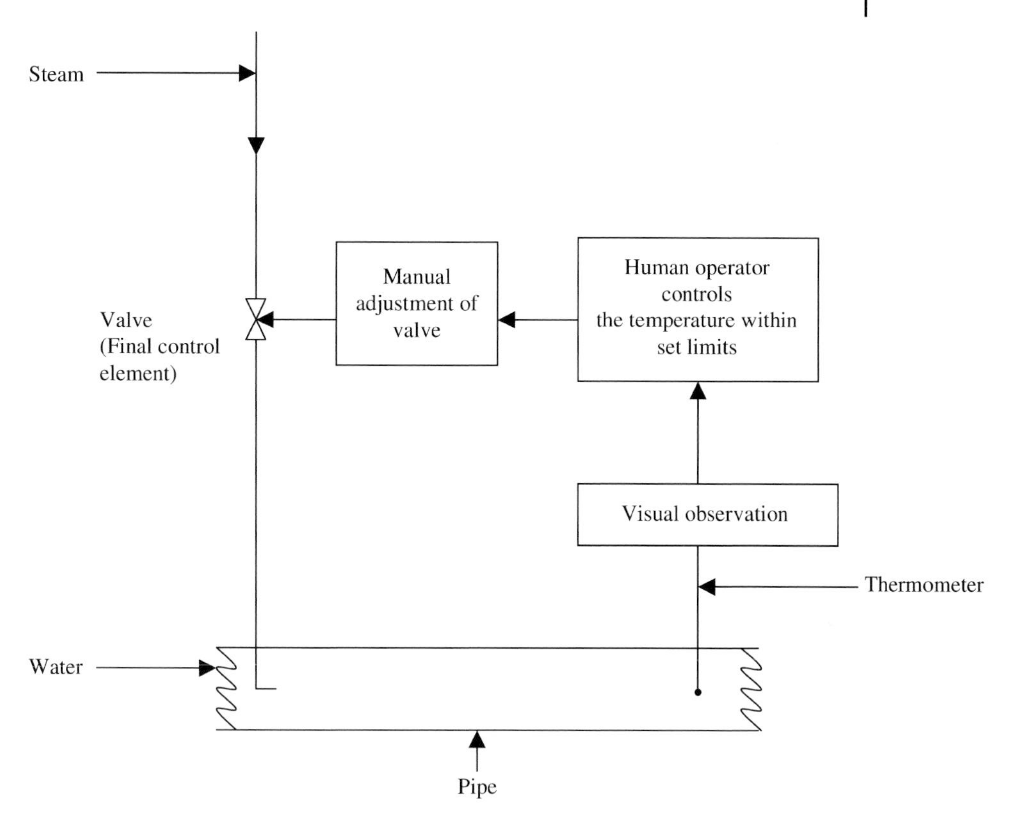

Fig. 11.1 A simple example of manual control; from [3].

An operator is constantly monitoring the temperature in the pipe. As the temperature changes from the set point on the thermometer, the operator adjusts the steam valve, i.e. either increases or decreases its flow, in order to get the temperature in the pipe back to the set point. Further action may be required if the temperature does not return to the set point within a reasonable time.

It is clear that the success of manual control operation depends on the skill of individual operators in knowing when and how much adjustment to make. Therefore, manual control may be used in those applications where changes in the manipulated parameter cause the process to change slowly and by a small amount. This is possible in plants where there are few processing steps with infrequent process upsets and the operator has sufficient time to correct before the process parameter overshoots acceptable tolerance. Otherwise, this approach can prove to be very costly in terms of labour, product inconsistencies and product loss.

11.3.2 Automatic Control

In automatic control, the process parameters measured by various sensors and instrumentation may be controlled by using control loops. A typical control loop consists of three basic components [3]:

- Sensor: the sensor senses or measures process parameters and generates a *measurement signal* acceptable to the controller.
- Controller: the controller compares the measurement signal with the set value and produces a *control signal* to counteract any difference between the two signals.
- Final control element: the final control element receives the control signal produced by the controller and adjusts or alters the process by bringing the measured process property to return to the set point, e.g. liquid flow can be controlled by changing the valve setting or the pump speed.

An automatic control system can be classified into four main types:

- on/off (two position) controller
- proportional controller (P-controller)
- proportional integral controller (PI controller)
- proportional integral derivative controller (PID controller).

11.3.2.1 On/Off (Two Position) Controller

This is the simplest automatic controller for which the final control element, e.g. valve, is either completely open or at maximum, or completely closed or at minimum. There are no intermediate values or positions for the final control element. Thus, final control elements often experience significant wear, as they are continually and rapidly switched from open to closed positions and back again. To protect the final control element from such wear, on/off controllers are provided with a *dead band*. The dead band is a zone bounded by an upper and a lower set point. As long as the measured process parameter remains between these set points, no changes in the control action are made. On/off controllers with a dead band are found in many instances in our daily lives: home heating system, oven, refrigerator and air conditioner. All these appliances oscillate periodically between an upper and lower limit around a set point. Fig. 11.2 illustrates the action of the control system.

It is interesting to note from the figure that the use of dead band reduces the wear and tear on the final control element, but amplifies the oscillations in the measured process parameter. Such controllers have three main advantages: (1) low cost, (2) instant response and (3) ease of operation. However, it is important to ascertain that the upper and lower limit values are acceptable for a specific process. The main disadvantages of this type of control action are: (1) it is not suitable for controlling any process parameter likely to suffer large sudden deviations from the set point and (2) the quality of control is inferior to the continuous controller.

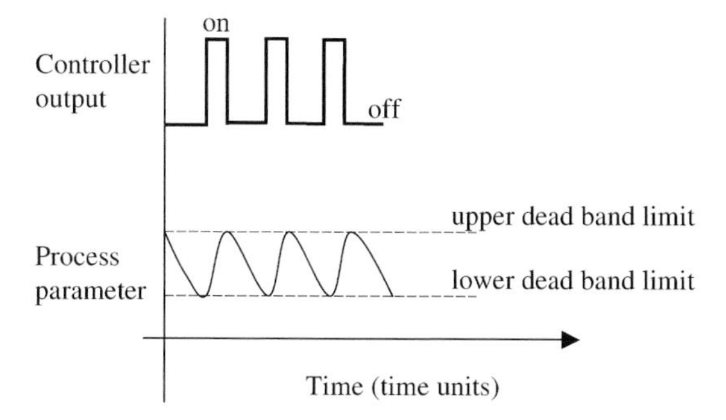

Fig. 11.2 Action of an on/off controller with dead bands; from [6].

11.3.2.2 **Proportional Controller**

The P-controller is one of the most commonly used controllers; and it produces an output signal to the final control element that is proportional to the difference between the set point and the value of the measured process parameter given by the sensor (this difference is also known as controller error or offset). Mathematically, it can be expressed as:

$$COS_{(t)} = COS_{(NE)} + K_C E_t \tag{11.1}$$

where $COS_{(t)}$ is the controller output signal at any time t, $COS_{(NE)}$ is the controller output signal when there is no error, K_C is known as the *controller gain* or *sensitivity* (controller tuning parameter) and E_t is the controller error or offset. The proportional controller gain or sensitivity (K_C) can also be expressed as:

$$K_C = \frac{100}{PB} \tag{11.2}$$

where PB is known as the proportional band, which expresses the value necessary for 100% controller output. The proportional controller gain (K_C), through the value of PB, describes how aggressively the P-controller output will move in response to changes in offset or controller error (E_t). When PB is very small, K_C is high and the amount added to $COS_{(NE)}$ in Eq. (11.1) is large. The P-controller will therefore respond aggressively like any simple on/off controller, with no offset, but a high degree of oscillations. In contrast, when PB is very high, K_C is small and the controller will respond sluggishly, with reduced oscillations, but increased offset. Thus, K_C through PB can be adjusted for each process to make the P-controller more or less active by achieving a compromise between degree of oscillations and offset. The main disadvantage of the P-controller is the occurrence of the offset, while its key advantage is that there is only one controller tuning parameter: K_C. Hence, it is relatively easy to achieve a best final tuning. Fig. 11.3 shows a step set point change under a P-controller for two different K_C values [6].

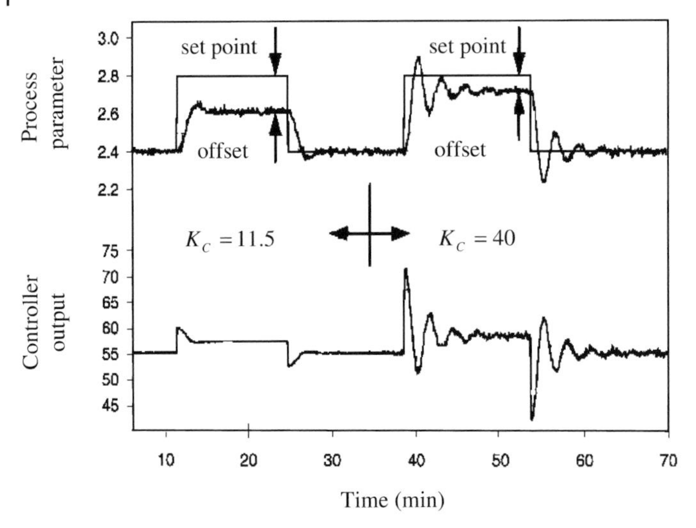

Fig. 11.3 P-controller with two different K_c values; from [6].

11.3.2.3 **Proportional Integral Controller**

The PI controller produces an output signal to the final control element which can be mathematically expressed as:

$$COS_{(t)} = COS_{NE} + K_C E_t + \frac{K_C}{\tau_I} \int E_t dt \tag{11.3}$$

where τ_I is a tuning parameter called the *reset time*; and the remaining notations are explained under Eq. (11.1). The integral term continually sums the controller error and its history over time to reflect how long and how far the measured process parameter has deviated from the set point. Thus, even if a small error persists over a long duration of time, the effects will add up. However, according to Eq. (11.3), the contribution of this integral term depends on the values of the tuning parameters K_C and τ_I. It is evident from Eq. (11.3) that higher values of K_C and lower values of τ_I will increase the contribution of the integral term.

Fig. 11.4 shows a typical PI controller response. It is clear from the case considered in the figure that, from 80 min onwards, E_t is constant at zero, yet the integral of the complete transient has a final residual value [obtained by subtracting A_2 from the sum (A_1+A_3)]. This residual value, when added to $COS_{(NE)}$, effectively creates a new overall $COS_{(NE)}$ value, which corresponds to the new set point value.

The consequence of this is that integral action enables the PI controller to eliminate the offset, which is the key advantage of the PI controller over a P-controller. However, it is important to note that in a PI controller, two tuning parameters interact and it is difficult to find the 'best' tuning values once the controller is placed in automatic mode. Moreover, a PI controller increases the oscillatory behaviour, as shown in Fig. 11.4.

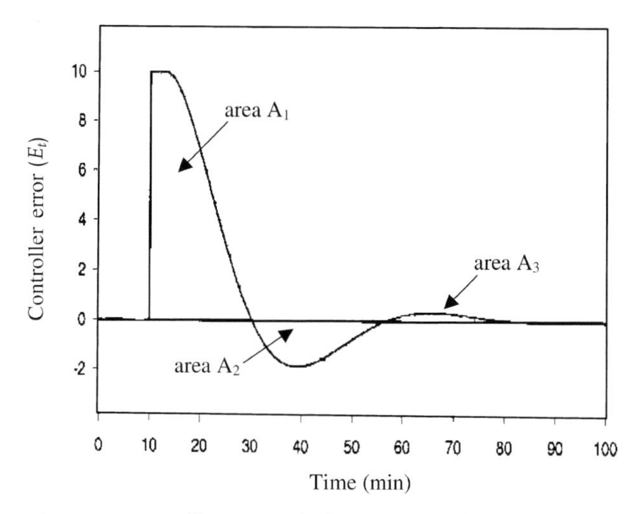

Fig. 11.4 PI controller – integral of error continually increases and decreases with time; from [6].

11.3.2.4 **Proportional Integral Derivative Controller**

In line with the PI controller, the PID controller produces an output signal to the final control element which can be expressed as:

$$COS_{(t)} = COS_{(NE)} + K_C E_t + \frac{K_C}{\tau_I} \int E_t dt + K_C \tau_D \frac{dE_t}{dt} \tag{11.4}$$

where τ_D is a new tuning parameter called the *derivative time*; and the remaining notations are already explained above. Higher values of τ_D provide a higher weighting to the fourth, i.e. derivative, term which determines the rate of change of the controller error (E_t), regardless of whether the measurement is heading towards or away from the set point, i.e. whether E_t is positive or negative. This implies that an error which is changing rapidly will yield a larger derivative. This will cause the derivative term to dominate in determining $COS_{(t)}$, provided K_C is positive. Fig. 11.5 shows a situation where the derivative values are positive, negative and zero (when the derivative term momentarily makes no contribution to the control action; note that the proportional and integral terms definitely influence at that point in time).

The major advantage of the PID controller, i.e. the introduction of the derivative term, is that it modifies the drawback of the PI controller: it works to decrease the oscillating behaviour of the measured process parameter. A properly tuned PID controller action can achieve a rapid response to error (proportional term), offset elimination (integral term) and minimise oscillations (derivative term). The key disadvantage of the PID controller is that it has three tuning parameters, which interact and must be balanced to achieve the desired controller performance. Just as in the case of the PI controller, the tuning of a PID controller can be quite chal-

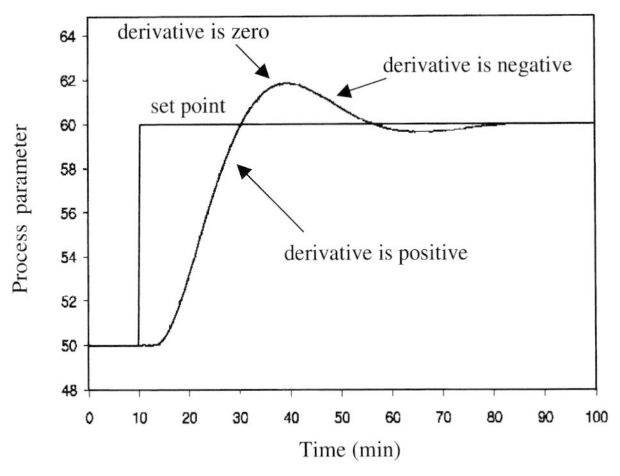

Fig. 11.5 PID controller – positive, negative and zero derivative values; from [6].

lenging, as it is often hard to determine which of the three tuning parameters is dominantly responsible for an undesirable performance.

To summarise, in all the above control actions, there are one or more parameters to be set when the controller is installed. However, it is most likely that the process information will be insufficient to give best values for these variables. Most control loops are therefore capable of becoming unstable and potentially result in serious consequences. Formal procedures are therefore necessary to arrive at the right controller settings. A number of procedures and techniques, each with their own advantages and disadvantages, have been developed over the years for *single* and *multivariable* control [7, 8].

A range of mathematical concepts have been applied in order to seek improvements in control quality. These are generally known as *advanced control* and include parameter estimation, fuzzy logic [9] and neural networks [10]. Without going into any further details, we will now consider how process control is implemented in modern food processes.

11.4 Process Control in Modern Food Processing

Control applications in food processing, according to McFarlane [1], can be discussed in the context of three categories of products: (1) bulk commodity processing, e.g. grain milling, milk, edible oil, sugar and starch production, where control is arguably most advanced, (2) manufactured products, e.g. pasta, cheese, in-container and aseptically processed products, and (3) products which have been subjected to processing methods essentially designed to retain their original structure, e.g. meat, fish, fruits and vegetables. Regardless of the nature of the products, process control in food processing has moved on from just attempting to control single variables, e.g. level, temperature, flow, etc., to systems which ensure smooth plant operation with timely signalling of alarms. The systems are also geared to provide vital data at shop floor level right through to vertically structured systems which encompass supervisory control and data acquisition (SCADA), manufacturing execution systems (MES) and interfacing with complex enterprise resource planning systems (ERP), which may be connected across multiple production sites.

11.4.1 Programmable Logic Controller

The programmable logic controller (PLC) is the most common choice in modern control [11]. It is a microprocessor-based system which can communicate with other process control components through data links. PLCs commonly use the so called *ladder logic* which was originally developed for electrical controls using relay switches. Programs can be written using a variety of languages. Once the program sequence has been entered into the PLC, the keyboard may be locked or removed altogether for security. When a process is controlled by a PLC, it uses inputs from sensors to take decisions and update outputs to drive actuators, as shown in Fig. 11.6.

Thus, a control loop is a continuous cycle of the PLC reading inputs, solving the ladder logic and then changing the outputs. A real process will inevitably change over time and the actuators will drive the system to new states (or modes of operation). This implies that the control performance relies on the sensors available and its performance is limited by their accuracy.

11.4.2 Supervisory Control and Data Acquisition

The supervisory control and data acquisition (SCADA) system is not a full control system, but is a software package that is positioned at a supervisory level on

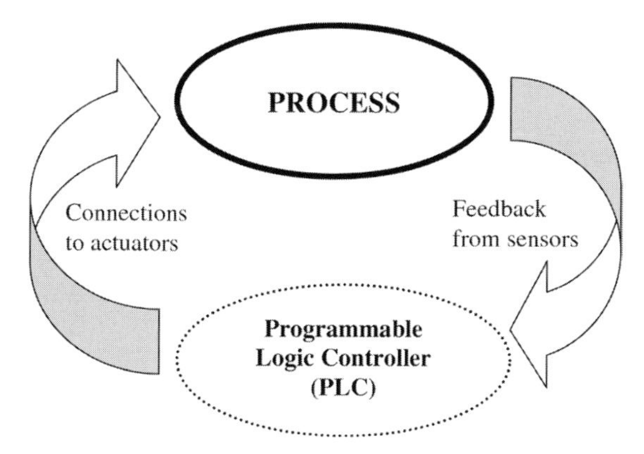

Fig. 11.6 Schematic illustration of the working of a programmable logic controller (PLC).

top of hardware to which it is interfaced, generally via PLCs, or other hardware modules [12]. SCADA systems are designed to run on common operating systems. Two basic layers can be distinguished in a SCADA system: (1) the *client layer* which serves as the human-machine interface and (2) the *data server layer* which handles process data and control activities, by communicating with devices such as PLCs and other data servers. Such communication may be established by using common computing networks. Modern data servers and client stations often run on Windows NT or Linux platforms. SCADA-based control systems also lend themselves to being scalable by adding more process variables, more specialised servers, e.g. for alarm handling, or more clients. This is normally done by providing multiple data servers connected to multiple controllers. Each data server has its own configuration database and realtime database (RTDB) and is responsible for handling a section of the process variables, e.g. data acquisition, alarm handling or archiving. Reports can also be produced when needed, or be automatically generated, printed and archived. SCADA systems are generally reliable and robust and, more often than not, technical support and maintenance are provided by the vendor.

11.4.3 Manufacturing Execution Systems

Manufacturing execution systems (MES) [13] are software packages which have been used for a number of years in process industries to support key operations and management functions ranging from data acquisition to maintenance management, quality control and performance analysis. However, it is only in recent years that there has been a concerted attempt to integrate factory floor information with *enterprise resource planning systems* (ERP). Modern MES include supply chain management, combine it with information from the factory floor and deliver results to plant managers in real time, thus integrating supply chain and production systems with the rest of the enterprise. This gives a holistic view of the business that is needed to support a 'manufacture to order' model. MES can manage production orders and can track material use and material status information. The software collects data and puts it into context, so that the data can be used for both realtime decision making and performance measurement and for historical analysis.

A typical control system currently available for dairy plants is shown in Fig. 11.7 [14].

The system combines supervision and control into a single concept, but its architecture is essentially open and modular, thereby delivering the operational flexibility needed. *Operator control* is the link between the plant operator and the process control modules, which enables actions such as routing and storage tank selection to be initiated at the click of a mouse. Using a variety of software tools, it is possible to incorporate a graphical presentation of the plant and other written information. Using the zoom facility, an operator can quickly access more detail on any given section of the plant. This feature can also be used to

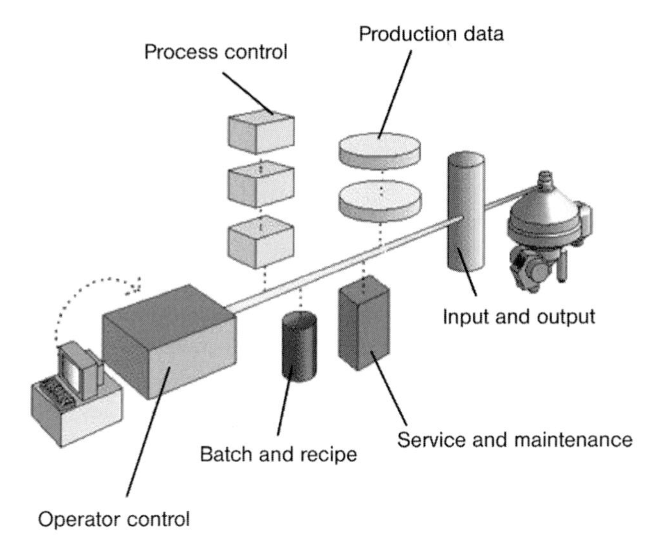

Fig. 11.7 A typical control system for dairy plant: FOOCOS system from ABB Automation [14]; reproduced with permission.

train new operators. Process data are also stored in this section, which can be easily accessed, for instance via 'pop up' control windows. The *process control* model contains information on the process and the parameters which have to be controlled. It can also include flow routing and control, storage information and sequences of the cleaning cycles. These modules can provide realtime control capability. The *batch and recipe* section contain information on recipe, ingredients and product specifications. This ensures use of ingredients in the right proportion and consistent product quality. All critical parameters can be monitored in realtime graphs and displayed on screen as well as logged for reports. The *production data* modules form a logbook of the activities of each processing unit – giving the origin of the product, how it has been treated and its final destination. The logbook can also be used for process evaluation and establishing traceability. The *service and maintenance* modules log information on equipment run times, number of valve strokes, alarm limits on equipment, etc., which identify poor functioning or worn units, well ahead of a possible breakdown. This enables the implementation of a preventative maintenance programme. Finally, the *input and output modules* handle the actual connections from the process control and supervision to the physical parts (like valves, pumps, etc.) to give a complete and detailed inventory.

11.5
Concluding Remarks

Most food businesses are currently facing constraints on capital budgets. There is a perpetual need for improving efficiency and seeking out valuable incremental improvements in manufacture. Proper implementation of process control is one way to release latent potential in existing processing facilities. Savings to be made by various sectors of the food and drink industry by implementing appropriate control of the manufacturing process are given in 'Energy Wizard', which is an interactive energy efficiency guide aiming to provide companies with free, independent and authoritative advice on many aspects of efficient energy use [15].

References

1 McFarlane, I. **1995**, *Automatic Control of Food Manufacturing Process*, 2nd edn, Chapman & Hall, London.
2 Pyle, D.L., Zaror, C.A. **1997**, Process control, in *Chemical Engineering for the Food Industry*, ed. P.J. Fryer, D.L. Pyle, C.D. Rielly, Blackie Academic and Professional, London, pp 250–294.
3 Stanbury, P.F., Whitaker, A., Hall, S.J. **1995**, *Principles of Fermentation Technology*, 2nd edn, Elsevier, Oxford, pp 215–241.
4 Kress-Rogers, E., Brimelow, C.J.B. **2001**, *Instrumentation and Sensors for the Food Industry*, 2nd edn, Woodhead, Cambridge.
5 Moreira, R.G. **2001**, *Automatic Control for Food Processing Systems*, Aspen, New-York.
6 Cooper, D. **2002**, *Practical Process Control using Control Station*, Course Notes, Control Station. www.controlstation.com.
7 Wellstead, P.E., Zarrop, M.B. **1995**, *Self-Tuning Systems: Control and Signal Processing*, John Wiley & Sons, Chichester.
8 Albertos, P., Antonio, S. **2004**, *Multivariable Control Systems: an Engineering Approach*, Springer, New York.
9 Zhang, Q., Litchfield, J.B. **1991**, Fuzzy mathematics for product development, *Food Technol.* 45, 108–115.
10 Eerikainen, T., Linko, P., Linko, S., Siimes, T., Zhu Y.-H. **1993**, Fuzzy logic and neural network applications in food science and technology, *Trends Food Sci. Technol.* 4, 237–242.
11 http://claymore.engineer.gvsu.edu/~jackh/books/plcs/chapters/plc_intro.pdf, accessed 3 July 2005.
12 Daneels, A., Salter, W. **1999**, Selection and evaluation of commercial SCADA systems for the controls of the CERN LHC experiments, *Proc. Int. Conf. Accelerator Large Exp. Phys. Control Syst.*, p. 353, available at: http://ref.web.cern.ch/ref/CERN/CNL/2000/003/scada/.
13 Fraser, J. **2003**, MES as enterprise application, not just plant system, available at: http://www.entegreat.com/eg_news_msifraser_20031110.htm.
14 http://www.abb.com/global/abbzh/abbzh251, accessed 3 July 2005.
15 http://www.actionenergy.org.uk, accessed 3 July 2005.

12
Environmental Aspects of Food Processing

Niharika Mishra, Ali Abd El-Aal Bakr and Keshavan Niranjan

12.1
Introduction

Waste is inevitably produced in all human endeavours; and its volume is proportional to the resources consumed. Waste is generally thought of as something that is no longer needed by the original user and is subsequently discarded. It is defined in UK legislation as: "any substance which constitutes a scrap material or an effluent or other unwanted surplus substance arising from the application of any process" [1]. It is further defined as: "any substance or article which requires to be disposed of as being broken, worn out, contaminated or otherwise spoiled" [1], or "that the holder discards, intends to or is required to discard" [2] (based on the definition of waste in EC Directive 91/156/EEC) [3].

The increased culture of consumerism within our societies has escalated the problem of waste because of the use of disposable goods. Processed food wastes constitute one of the largest fractions of municipal waste these days. Manufacturing processes operate under strict quality control and retailing has stringent 'sell by' date regulations, which has resulted in the generation of large volumes of food and packaging waste. The food industry is facing increasing pressure to reduce its environmental impact, both from consumers and regulators. Initial results from the study of Environment Agency indicate that the food, drink and tobacco sector contributes $8–11 \times 10^6$ t year^{-1} to the industrial/commercial total of $70–100 \times 10^6$ t year^{-1}. This partly reflects the importance and size of the food and drink industry within the UK [4].

Transferring food from the field to the plate involves a sophisticated production and supply chain, but for the purposes of waste production this can be simplified into three main steps: agriculture, food processors/manufacturers and the retail/commercial sector. Each of the sectors generates waste and wash water. Given the complexity of the food chain, environmental impacts can occur at various points in the chain, even for a single food product. It is therefore necessary to take a holistic systems-based approach to tackle the problem. This, however, demands that the entire food chain be considered in the context of

Food Processing Handbook. Edited by James G. Brennan

ISBN: 3-527-30719-2

dealing with environmental issues. Since such an approach would become too unwieldy in the context of this book, this chapter merely aims to identify key environmental issues relating to food processing and manufacture; and it discusses food waste characteristics, the relation between processing operations and the types of waste generated, waste processing options, energy issues in food manufacture, the environmental impact of refrigerants and packaging wastes.

12.2 Waste Characteristics

The quality and quantity of wastes produced depend on the type of food being processed. There are big differences from sector to sector, and even site to site: generalisation is not only difficult, but could also be misleading. Food wastages levels are often inferred from mass balances. It is estimated that about 21% of food product at the farm gate is lost, much due to spoilage, and only about 7%, on an average, is lost during processing [5]. From the data cited in [5] (see Table 12.1), it can be inferred that, although the percentage loss during food processing is low, wastage mass or volumes are very high. The wastes produced in any food industry depend mainly on the type of food being processed.

Food processing operations produce many varied types of wastes that can be categorised into solid, liquid and gaseous wastes.

Table 12.1 Solid wastes generated in selected processes [5].

Processed food waste	Total solids ($g\ kg^{-1}$)	Liquid volume ($m^3\ kg^{-1}$)
Vegetables		
Kale	16	0.004
Spinach	20	–
Mustard greens	16	–
Turnip greens	15	–
Potatoes	66	0.012
Peppers (caustic peeling)	65	0.020
Tomatoes (caustic peeling)	14	0.010
Dairy		
Cheese whey	–	9.00
Skim milk	–	0.07
Ice cream	–	0.08
Meat		
Red meat	0.440	25.00
Poultry	0.270	50.00
Eggs	0.111	–

12.2.1 Solid Wastes

Solid wastes emanating from food processing plants may include: the unnecessary leftover from the preliminary processing operations, residues generated as an integral part of processing, wastes resulting from processing inefficiencies, sludge produced from the treatment of wastewater, containers for the raw materials and finished products. Table 12.1 summaries typical solid wastes generated from a selection of food processes [7, 8]. In general, solid wastes are poorly characterised, both in terms of quality and quantity; and estimates of solid wastes are usually inferred from mass balances [7, 8].

12.2.2 Liquid Wastes

Wastewater from the processing industry is the main stream that is produced. It includes: wastewater resulting from using water as a coolant, water produced by different processing operations like washing, trimming, blanching and pasteurising and a large amount of wastewater produced from cleaning equipment [8].

12.2.3 Gaseous Wastes

The gaseous emissions from the food processing industry are mainly manifested in terms of emanating odors and, to a lesser extent, in terms of dust pollution. Other emissions include solvent vapors commonly described as volatile organic emissions and gases discharged by combustion of fuels.

Even though the characteristics of food wastes can be discussed in terms of their physical states, it is necessary to note that solid wastes contain a substantial proportion of water, just as liquid wastes may contain a significant proportion of solids. It is therefore absolutely critical to note that food wastes are not only multicomponent but also multiphase in nature [8].

12.3 Wastewater Processing Technology

Treatment of the wastes produced from food industries is an important concern from the environmental point of view. As discussed earlier, the waste products from food processing facilities include bulky solids, wastewater and airborne pollutants. All of these cause potentially severe pollution problems and are subject to increasing environmental regulations in most countries. Generally, wastewater is most common, because food processing operations involve a number of unit operations, such as washing, evaporation, extraction and filtration. The wastewaters resulting from these operations normally contain high concentrations of suspended solids and soluble organics, such as carbohydrates, proteins and lipids, which cause disposal problems. To remove these contaminants from water, different technologies are adopted in the food industry, which are described in detail in Chapter 13 and in [9, 10].

12.4 Resource Recovery From Food Processing Wastes

The wastes from food industry, after recovery and further processing, can be used for different purposes: the recovered materials can either be recycled, or be used to recover energy by incineration or anaerobic digestion. Recycling not only reduces the environmental impact of the material, but also helps to satisfy the increasing demands for raw materials. In addition, it also reduces disposal costs, a key driver of recycling technologies. For instance, fruit, vegetables and meat processors generate large quantities of solid wastes. Table 12.2 lists examples of useful materials which can be recovered from fruit and vegetable wastes.

Recovered materials can be used in various ways. Solid food wastes can be used as animal feed after reducing their water content. A good example of this practice is soybean meal, a byproduct of soybean oil extraction, which was simply discarded previously but is now used as animal feed on account of its high nutritive value [11]. Solid wastes can also be upgraded by fermentation. A number of fermented foods are produced this way. Composting and ensilaging are also examples of solid waste fermentation process [6]. Solid wastes rich in carbohydrate can also be converted to sugars by enzyme-assisted hydrolysis: an example is the enzymatic hydrolysis of lactose and galactose sugar using β-galactosidase [12]. Solid wastes rich in sugar can be fermented to produce carbon dioxide and ethanol. The latter a valuable product, and has also been earmarked as an alternative fuel for the future [13].

As mentioned above, solid wastes can also be utilised as fuel directly or converted to methane by anaerobic digestion in a bioreactor. Biological hydrogen is produced by fermentation of both glucose and sucrose in food processing wastes under slightly acidic conditions in the absence of oxygen. This can be achieved by using a variety of bacteria through the actions of well studied anae-

Table 12.2 Some examples of products which can be recovered or made from fruit and vegetable wastes [8].

Source of waste	Product
Apple pomace	Pectins
Apple skin	Aromatics
Tomato pomace	Pectins, tomato seed oil, colour from skin
Stalk of paparika and pumpkin seeds	Natural colouring agents
Green pea pods	Leaf proteins, chlorophyll
Stones from stoned fruits	Active carbon, kernels (after debittering)

robic metabolic pathways and hydrogenase enzymes. Hydrogen has 2.4 times the energy content of methane, i.e. on a mass basis; and its reaction with oxygen in fuel cells produces only water, a harmless byproduct. Hydrogen gas has valuable potential for producing clean and economical energy in the near future [14].

12.5 Environmental Impact of Packaging Wastes

Packaging is acknowledged to perform a number of useful functions. It acts as a physical barrier between a product and the external environment, thereby protecting it from external contamination and maintaining hygienic conditions, it protects and preserves the product during handling and transportation, it serves to attract the attention of consumers thereby giving the product a good market value and it also serves to provide information on the product and instructions on how to use it (see also Chapter 9). Despite these advantages, the environmental impact of packaging wastes is considerably high and, in many cases, outweighs their benefits. Recent studies have shown that, in Europe, packaging forms ca. 16% of municipal solid waste (MSW) and 2% of nonMSW [15].

The key environmental issues related to packaging are:

- the use of packaging materials like plastics and steel which are either nonrecyclable or uneconomic to recycle (a large amount of such wastes invariably end up in landfills);
- the use of material intensive packaging, which requires an energy-intensive process to manufacture;
- the use of substances in the packages having high chemical and biological oxygen demand (some even hazardous and toxic to the environment) which cannot be discharged safely into natural water streams.

In most countries, regulations are in place for reducing the impact of packaging and packaging wastes on the environment. This is mostly done by limiting the production of packaging wastes, enforcing the recycle of packaging material and reuse of packages where possible and encouraging the use of minimal packaging at source.

12.5.1 Packaging Minimisation

The foremost strategy in packaging waste management is to reduce the use of packaging to a bare minimum level at all stages of production, marketing and distribution. This can be achieved by: (a) decreasing the weight of material used in each pack (known as lightweighting or downgauging), (b) decreasing the size or volume of the package or using less material in the first place, e.g. reducing the thickness of the packaging material, (c) using consumable or edible package and (d) modifying the product design, e.g. avoiding unnecessary multiple wrapping of a product with different materials [16].

12.5.2
Packaging Materials Recycling

The purpose of recycling is to use a material as raw material for the production of a new product after it has already been used successfully. If recycling is done properly and in conjunction with good design, many materials can be recovered after their first useful life is over. The two major objectives should be to conserve limited natural resources and to reduce and rationalise the problems of managing municipal solid waste disposal [15].

Recycling is defined as the reprocessing of the waste material in a production process either for the original purpose or for other purposes. The EU definition [17] also includes organic recycling, i.e. aerobic or anaerobic treatment of the biodegradable part of the packaging waste to produce stabilised organic residues or methane. In general, recycling involves physical and/or chemical processes which convert collected and sorted packaging, or scrap, into secondary raw materials or products. Secondary raw material is defined as the material recovered as a raw material from used products and from production scrap.

Before sending packaging materials for recycling, they should be properly sorted (i.e. separated from other packaging materials) and cleaned (i.e. free from any contamination). Sorting and cleaning are two important operations before processing, since they affect the quality of the input stream which finally determines the quality and value of the secondary materials. The materials commonly used for food packaging are: paper and board, plastic, glass, aluminum and steel.

Given the widespread use of paper and board as packaging material, their recycling is critical from the environmental point of view as well as resource recovery. Recycled paper is a major source of raw material for the paper industry. About 44.7×10^6 t of waste paper were recycled in Europe in 2003, which is substantially higher than 10 years ago, when only ca. 26×10^6 t were recycled. This represents 53.2% of the total paper used in Europe [18]. Packaging is the largest sector; and it uses almost two-thirds of the recycled paper in Europe to manufacture case materials, corrugated board, wrapping, etc.

The total consumption of plastics in Europe was about 36.8×10^6 t in 2000, of which 13.7×10^6 t (37.3%) were used for manufacturing packaging materials. Plastics account for 17% of the total packaging usage in Western Europe [1]. The most widely use packaging plastics include low and high density polyethylene (LDPE, HDPE), linear low density polyethylene (LLDPE), polypropylene (PP), polystyrene (PS), polyethylene tetraphthalate (PET) and polyvinyl chloride (PVC). After collection, the material is sorted, to separate the plastics from other materials like paper, steel, aluminum, etc. The sorting step also includes the separation of plastics by their resin type (like PET, HDPE, etc.). The sorted plastics are then recycled by different technologies, such as mechanical, feedstock and chemical recycling [19].

Mechanical recycling involves processes like extrusion, coextrusion, injection, blow moulding, etc. (see also Chapter 9). Feedstock recycling includes pyrolysis, in which plastics are subjected to high temperature in the absence of oxygen which enables the hydrocarbon content of the polymer to be recycled. Pyrolytic processes have been studied extensively for the last two decades. However, most of this research has been undertaken using pure and clean plastics, or mixtures of pure plastics. There is a strong need to develop processes capable of dealing with wastes that have plastics attached to other contaminants, such as paper, metals or bioproducts. Microwave-induced pyrolysis of plastics is a novel process in which microwave energy is applied to carbon mixed with plastic waste [20].

Chemical recycling involves depolymerisation of PET, resulting in the monomers terepthalic acid and ethylene glycol which, after purification, can be reused to produce new polymers.

Another method, called the 'super clean recycling process', uses mechanical and nonmechanical procedures to recycle high quality postconsumer material, producing polymers suitable for use in monolayer application, i.e. use in direct contact with food. The processes are proprietary, but they are believed to involve a combination of standard mechanical recycling processes with nonmechanical procedures such as high-temperature washing, high-temperature and pressure treatments, use of pressure/catalysts and filtration to remove polymer-entrained contaminants [21].

Recycled plastics have been used in food contact applications since 1990 in various countries around the world. To date, there have been no reported issues concerning health or off taste resulting from the use of recycled plastics in food contact applications. This is due to the fact that the criteria that have been established regarding safety and processing are based on extremely high standards that render the finished recycled material equivalent in virtually all aspects to virgin polymers [22].

Various food contact materials and constituents can be used, provided they do not pose health concerns to consumers, which may occur when some substances from the food packaging migrate into the food. To ensure the safety of such materials, food packaging regulations in Europe require that the packaging materials must not cause mass transfer (migration) of harmful substances to the food, by imposing restrictions on substances from the materials itself that could migrate into the food. Consequently, food packaging materials must comply with many chemical criteria and prescribed migration limits. The migration of substances from the materials into the foodstuffs is a possible interaction that must be minimised or even avoided, since it may affect the food or pose longer-term health concerns to the consumer [23] (see also Chapter 9).

With regard to recycled PET, there is strong need to have relevant analytical data on the nature and the concentration of the contaminants that can be found in the recycled material, in order to ascertain the safety of reusing PET for food purposes. Knowledge of the contaminants and information on practical and effective test methods would help in the formulation of future legislation [24].

With a view to make packaging from sustainable materials, a number of biodegradable alternatives have been developed. Traditionally, biobased packaging materials have been divided into three types, which illustrate their historical development. First generation materials consist of synthetic polymers and 5–20% starch fillers. These materials do not biodegrade after use, but will biofragment, i.e. they break into smaller molecules. Second generation materials consist of a mixture of synthetic polymers and 40–75% starch. Some of these materials are fully biobased and biodegradable [25]. The market value of biobased food packaging materials is expected to incorporate niche products, where the unique properties of the biobased materials match the food product concept [26]. Packaging of high-quality products such as organic products, where extra material costs can be justified, may form the starting point. Biopolymer-based materials are not expected to replace conventional materials on a short-term basis. However, due to their renewable origin, they are indeed the materials of the future [27].

According to [4], targets for recovery and recycling have been set by EU as follows: 50–65% by weight of packaging waste to be recovered, 25–45% to be recycled, and 15% to be recovered by materials. These targets refer to packaging composed of plastics, paper, glass, wood, aluminum and steel. Further, the combined content of lead, mercury, cadmium and chromium (VI) has been limited to 100 ppm. The law also ensures that packaging materials are introduced in the marketplace only if they meet 'essential requirements', i.e. characteristics that include minimisation of weight and volume, and suitability for material recycling.

12.6 Refrigerents

Refrigeration systems are essential for the production, storage and distribution of chilled foods. The commonly used refrigerants in these systems are chlorofluorocarbons (CFC) and hydrofluorocarbons (HCFC). Although highly efficient, these refrigerants have been shown to be responsible for severe environmental threats like global warming and depletion of the ozone layer.

CFCs are organic compounds containing chlorine, fluorine and carbon atoms and having ideal thermodynamic properties for use as refrigerants. But their chlorine content is mainly responsible for the depletion of the ozone layer in our environment. When CFCs are released into atmosphere, they dissociate in the presence of ultraviolet (UV) light to give free chlorine. This free chlorine atom decomposes ozone to oxygen and regenerates itself by interacting with a free oxygen atom, as follows [28].

$$CF_2Cl_2 \rightarrow CF_2Cl + Cl$$

$$Cl + O_3 \rightarrow ClO + O_2$$

$$Cl + O \rightarrow Cl + O_2$$

The regeneration of chlorine sustains the process and depletes the ozone layer. This layer is known to protect life on earth from UV radiation, by absorbing a large portion of it and allowing only a small fraction to reach the earth. But its depletion will expose us, causing skin cancer, damage to eyes, damage to crops, global warming, climate change, etc. [28]. Besides this effect, such refrigerants are also known to contribute to global warming, along with CO_2 and other gases such as methane, nitrous oxides, chlorofluorocarbons, and halocarbons.

The extent to which a substance can destroy the ozone layer is measured in terms of a parameter called ozone-depleting potential (ODP). The ODPs of CFCs are significantly greater than ODPs of hydrofluorocholorcarbons (HCFC) and hydroflurocarbons (HFC). Hence, CFCs are gradually being replaced by these other two. It may be noted that HFCs have zero ODP, since they do not contain any chlorine atoms. However, the F-C bonds in CFCs, HCFCs and HFCs are very strong in absorbing infrared radiations escaping from the earth's surface. Their absorption capacity is much more than CO_2 [29]. To measure the contribution of different gases to global warming, a scale called the global warming potential (GWP) has been set up. Table 12.3 lists the ODP and GWP of different refrigerants. It is quite obvious from the table that CFCs have high ODP and GWP compared to HCFCs and HFCs [30].

To control the production and consumption of substances which cause ozone depletion, the 'Montreal protocol on substances that deplete the ozone layer' was signed in 1987 and has been effective since 1989 [31]. The purpose of this agreement was to phase out CFCs by the year 2000 and to regularly review the use of transitional ozone-safe alternative refrigerants, which are scheduled to be replaced by 2040. Similarly, though HCFCs are used as replacements for CFCs,

Table 12.3 Refrigerant characteristics [30].

Refrigerant	Ozone-depleting potential	Global warming potential (CO_2=1.0)	Stratospheric lifetime
CFC 11	1.0	4100	55.0
CFC 12	1.0	7400	116.0
CFC 113	1.07	4700	110.0
CFC 114	0.8	6700	220.0
CFC 115	0.5	6200	550.0
HCFC 22	0.055	2600	15.2
HCFC 123	0.02	150	1.6
HCFC 124	0.022	760	6.6
HCFC 141b	0.11	980	7.8
HCFC 142b	0.065	2800	19.1
HFC 125	0.0	4500	28.0
HFC 134a	0.0	1900	15.5
HFC 143a	0.0	4500	41.0
HFC 152a	0.0	250	1.7

they are still responsible for ozone depletion and need to be phased out by 2020 as specified by the amended Montreal protocol [15]. HCFCs are expected to be replaced by HFCs.

12.7 Energy Issues Related to Environment

The energy consumed by the food and drink industry, in most countries, is a significant proportion of the total energy used in manufacturing industries. For instance, this proportion within UK is around one-tenth [32]. Energy is consumed by the food industry to keep food fresh and safe for consumption. This is achieved by different processing operations (boiling, evaporation, pasteurisation, cooking, baking, frying, etc.), safe and convenient packaging (aseptic packaging) and storage (freezing, chilling). The energy required for these processes is obtained from either electricity or burning fossil fuel. When the cost of energy consumption is considered, it has received very low priority in many organisations because it accounts for only 2–3% of the total production cost [32]. But considering the other side of the coin, i.e. the environmental effects, the energy consumption cannot be ignored. The food industry will be affected by all international measures aimed at reducing industrial energy consumption. The background to some of the international measures is discussed below.

The burning of fossil fuel results in emission of large amounts of CO_2, the most important greenhouse gas, which is responsible for about two-thirds of potential global warming. CO_2 produced from burning fossil fuel is responsible for 80% of the world's annual anthropogenic emissions of CO_2. Methane (CH_4), the second most important greenhouse gas, is responsible for about 15% of the build up, and nitrous oxide (N_2O), which also has a high stratospheric lifetime, is responsible for 3% of the build up [33]. Other greenhouse gases, e.g. CFCs, HCFCs, Perfluorinated Carbons (PFCs), sulfur hexafluoride (SF_6), etc., are produced from various sources, which include the refrigeration systems used in food processing. The average temperature rise experienced by the planet on account of greenhouse emissions has been estimated to be approximately 0.5 °C over the past 100 years. But sophisticated computer models solely based on CO_2 emissions are predicting a temperature rise of 5 °C over the next 200 years [34]. The average rate of warming due to emission of these gases would probably be greater than ever seen in the last 10 000 years. This increasing temperature may cause many catastrophic events, like melting of the polar ice cap, rising of global sea levels and unbearably hot climates all over the world. The global sea level has risen by 10–25 cm in the last 100 years and it is expected to increase in between 13 cm and 94 cm by the year 2100, which might cause widespread flooding [34]. Burning of fossil fuels also gives rise to SO_2, which is converted to sulphate in the atmosphere, known as sulphate aerosols. These aerosol particles absorb and scatter solar radiation back into space and hence tend to cool the earth. But, due to their shorter lifespans, it is difficult to assess the impact of aerosols on the global climate. However, it has been concluded that the increase in sulphate aerosols has had a cooling effect since 1850 [34].

To minimise the chances of catastrophic events occurring in the future, we must slow down the emission of greenhouse gases. This can be achieved by limiting the combustion of fossil fuels, which ultimately leads to reduced energy demand by increasing the drive for energy efficiency and improving its use.

With limited use of electricity and fuel, the energy efficiency can be achieved by the use of combined heat and power (CHP) or renewable energy. CHP is a fuel-efficient energy technology in which a major part of the heat that is being wasted to the environment is recovered and used in other heating systems. CHP can increase the overall efficiency of fuel use to more than 75%, compared with around 40% from conventional electricity generation. CHP plays an important role in the UK Government's new energy policy, whose ambition is to achieve a 60% reduction in CO_2 emissions by 2050 [35]. Following good process design practices can also make a difference [36]: for instance, insulating valves, flanges, autoclaves, heated vessels, pipes, etc. during steam production can prevent leakage of steam and hence reduce heat loss; also, using the optimum air-fuel ratio prevents unnecessary burning of fuel, etc. Renewable sources of energy like solar radiations, wind, sea waves and tides, biomass, etc. and the use of fuel containing low or no carbon (e.g. hydrogen) can reduce the emission of greenhouse gases to a significant extent. Another option is to capture the CO_2 emitted by a burning fuel and then utilise it or store it for later use [37]. CO_2 could be captured by various methods like adsorption onto molecular sieves, absorption into chemically reacting solvents (e.g. ethanolamines), membrane separation methods, etc. After separation, it can be used as a feedstock for the manufacture of chemicals which enhance the production of crude oil in the growth of plants or algae which could be used as a biofuel. Several methods of storing the CO_2 have been proposed, such as storing it inside ocean beds, in deep saline reservoirs, in depleted oil and gas reservoirs, etc. [38]. All these options may not be economically viable at this stage, but technology may need to be improved so that these options could be exercised more easily.

It is evident from the above discussion that environmental problems resulting from energy consumption cannot be resolved by nations unilaterally. A number of international treaties and agreements have been formulated to protect the environment from the hazards of greenhouse gases, such as the Kyoto protocol. During the 1992 'Framework convention on climate change' (FCCC), the first formal international statement of concern and agreement was formulated to take a concerted action for stabilising atmospheric CO_2 concentrations. In this context, the 1997 Kyoto protocol was negotiated (which includes several decisions such as reducing greenhouse gas emissions, based on 1990 levels, by 5.2% in the period 2008 to 2012) by the industrialised countries [39]. The UK voluntarily committed to reducing emissions by 12.5% by 2010. Other measures include enhancement of energy efficiency in different sectors, increased use of new and renewable forms of energy, advanced innovative technology for CO_2

separation and the protection and enhancement of sinks and reservoirs of greenhouse gases. In addition, there was a commitment to reduce fiscal incentives, tax and duty exemptions and subsidies in all greenhouse gas-emitting sectors that ran counter to the objective of the Convention [40]. The European Union aimed to control the environmental impacts of industrial activities by formulating an 'integrated pollution prevention and control' (IPPC) directive (Directive 96/61/EC of 24 September 1996), which sets out measures to ensure the sensible management of natural resources. These provisions enable a move towards a sustainable balance between human activity and the environment's resources and regenerative capacity [4]. The 'climate change levy' (CCL) was introduced as a tax on fuels or energy sources used by industry on 1 April 2001. The levy package aims to reduce CO_2 emissions of at least 2.5×10^6 t year^{-1} by 2010. The levy does not apply to waste used as fuel. To encourage the reduction of fuel consumption, the UK government has also announced that a discount of 80% from the levy will be given to companies who agree to reduce the CO_2 emission by reducing their energy consumption [41]. Food processing industries will be expected to work within the above parameters and there is no doubt that manufacturing practices will continue to change for the foreseeable future to comply with national and international regulations formulated to protect our environment.

12.8 Life Cycle Assessment

The life cycle assessment (LCA) is a tool standardised by the International Standardisation Organisation (ISO) to evaluate the environmental risks associated with a product from 'cradle' to 'grave'. It takes into account the environmental impact associated with its production starting from the raw materials and energy needed to produce it, to its disposal, along with processing, transportation, handling, distribution, etc. in between [42]. LCA studies have been carried out for a variety of products, including food. The first LCA studies on food products were undertaken at the beginning of the 1990s [43].

LCA identifies the material, energy and waste flows associated with a product during the different stages of life cycle and the resulting environmental impact. For example, if we consider the production of orange juice, the LCA analysis will involve: the weight of oranges and energy associated with transporting raw materials, the amount of wastes produced (both processing and packaging wastes), the energy or power consumed during processing, the mass and energies associated with the use of utilities like cleaning water, steam and air, the emissions released into air, water and land from the processing site and other relevant factors depending on the operating technology and regional location of the processing facility.

References

1 Association of Plastics Manufacturers in Europe **2002**, *An Analysis of Plastics Consumption and Recovery in Western Europe 2000*, APME, London.
2 Balch, W. E., Fox, G. E., Magrm, L. J., Woese, C. R., Wolfe, R. S. **1979**, Methanogens: Reevaluation of a Unique Biological Group, *Microbiol. Rev.* 43, 260–269.
3 European Commission **1991**, Council Directive 91/156/EEC of 18 March 1991 Amending Directive 75/442/EEC on Waste, *Eur. Comm. Off. J.* L 78, 32–27.
4 Cybulska, G. **2000**, *Waste Management in the Food Industry, an Overview*, Campden & Chorleywood Food Research Association Group, Chipping Campden.
5 Gorsuch, T. T. (ed.) **1986**, *Food Processing Consultative Committee Report*, Ministry of Agriculture, Fisheries and Foods, London.
6 Litchfield, J. H. **1987**, Microbiological and Enzymatic Treatments for Utilizing Agricultural and Food Processing Wastes, *Food Biotechnol.*, **1**, 27–29.
7 Mardikar, S. H., Niranjan, K. **1995**, Food Processing and the Environment, *Environ. Manage. Health*, 6, 23–26.
8 Niranjan, K. **1994**, An Assessment of the Characteristics of Food Processing Wastes, in *Environmentally Responsible Food Processing* (*Symposium Series No. 300*), ed. K. Niranjan, M. R. Okos, M. Rankowitz, American Institute of Chemical Engineers, Washington D.C., pp 1–7.
9 Hansen, C. L., Hwang, S. **2003**, Waste Treatment, in *Environmentally Friendly Food Processing*, ed. B. Mattsson, U. Sonesson, Woodhead Publishing, Cambridge, pp 218–240.
10 Eckenfelder, W. W. J. R. **1961**, *Biological Treatment*, Pergamon Press, Oxford.
11 Lancaster Farming, available at: http://www.lancasterfarming.com/18.html.
12 Kirsop, B. H. **1986**, Food Wastes, *Prog. Ind. Microbiol.*, 23, 285–306.
13 Gong, C. S. **2001**, Ethanol production from renewable resources, *Fuel Energy Abstr.*, 42, 10.
14 Yang, S. T. **2002**, Effect of pH on hydrogen production from glucose by a mixed culture, *Bioresour. Technol.*, **82**, 87–93.
15 Sturges, M. **2002**, Packaging and Environment – Arguments For and Against Packaging and Packaging Waste Legislation, Pira International, Leatherhead.
16 De Leo, F. **2003**, The Environmental Management of Packaging: an Overview, in *Environmentally-Friendly Food Processing*, ed. B. Mattsson, U. Sonesson, Woodhead Publishing, Cambridge, pp 130–153.
17 European Communities **1994**, Council Directive 94/62/EC, *Eur. Comm. Off. J.*, 31.
18 Confederation of European Paper Industry **2004**, *Special Recycling 2003 Statistics*, available at: http://www.forestindustries.se/pdf/SpecRec2003-092022A.pdf.
19 Dainelli, D. **2003**, Recycling of Packaging Materials, in *Environmentally Friendly Food Processing*, ed. B. Mattsson, U. Sonesson, Woodhead Publishing, Cambridge, pp 154–179.
20 Ludlow-Palafox, C. **2002**, Microwave Induced Pyrolysis of Plastic Wastes. *PhD thesis*, University of Cambridge, Cambridge.
21 Recoup **2002**, Fact Sheet: Use of Recycled Plastics in Food Grade Applications, available at: www.recoup.org.
22 Bayer, A. L. **1997**, The Threshold of Regulation and its Application to Indirect Food Additive Contaminants in Recycled Plastics, *Food Addit. Contam.*, 14, 661–670.
23 Simoneau, C., Raffael, B., Franz, R. **2003**, Assessing the Safety and Quality of Recycled Packaging Materials, in *Environmentally-Friendly Food Processing*, ed. B. Mattsson, U. Sonesson, Woodhead Publishing, Cambridge, pp 241–265.
24 Baner, A. L., Franz, R., Piringer, O. **1994**, Alternative Fatty Food Stimulants for Polymer Migration Testing, in *Food Packaging and Preservation*, ed. M. Mathlouthi, Chapman & Hall, Glasgow, pp 23–47.
25 Gontard, N., Gulbert, S. **1994**, Bio-Packaging: Technology and Properties of Edible and/or Biodegradable Materials of Agricultural Origin, in *Food Packaging and Preservation*, ed. M. Mathlouthi,

Blackie Academic & Professional, Glasgow, pp 159–181.

26 Weber, C. J., Haagard, V., Festersen, R., Bertelsen, G. **2002**, Production and Applications of Biobased Packaging Materials for the Food Industry, *Food Addit. Contam.*, 19 (*Suppl.*), 172–177.

27 Weber, C. J. (ed.) **2000**, *Biobased Packaging Materials for the Food Industry, Status and Perspectives* (EU Concerted Action Project Report, Contract PL98 4045), available at: http://www.nf-2000.org/publications/f4046fin.pdf.

28 http://www.infoplease.com/ce6/sci/A0812001.html.

29 http://chemcases.com/fluoro/fluoro17.htm.

30 Dellino, C. V. J., Hazle, G. **1994**, Cooling and Temperature Controlled Storage and Distribution Systems, in *Food Industry and the Environment*, ed. J. M. Dalzell, Blackie Academic & Professional, Glasgow, pp 259–282.

31 http://www.afeas.org/montreal_protocol.html.

32 Walshe, N. M. A. **1994**, Energy Conservation and the Cost Benefits to the Food Industry, in *Food Industry and the Environment*, ed. J. M. Dalzell, Blackie Academic & Professional, Glasgow, pp 76–105.

33 http://www.iclei.org/EFACTS/GREENGAS.HTM.

34 http://www.ieagreen.org.uk/ghgs.htm.

35 http://www.defra.gov.uk/environment/chp/index.htm.

36 http://cleanerproduction.curtin.edu.au/industry/foods/energy_efficiency-foods.pdf.

37 http://www.ieagreen.org.uk/doc3a.htm.

38 http://www.ieagreen.org.uk/removal.htm.g

39 http://www.uic.com.au/nip24.htm.

40 http://www.defra.gov.uk/environment/chp/index.htm

41 http://www.defra.gov.uk/environment/ccl.

42 Berlin, J. **2000**, Life Cycle Assessment – an Introduction, in *Environmentally-Friendly Food Processing*, ed. B. Mattsson, U. Sonesson, Woodhead Publishing, Cambridge, pp 5–15.

43 Mattsson, B., Olsson, P. **2001**, Environmental Audits and Life Cycle Assessment, in *Auditing in the Food Industry*, ed. M. Dillon, C. Griffith, Woodhead Publishing, Cambridge.

13
Water and Waste Treatment

R. Andrew Wilbey

13.1
Introduction

Most food manufacturers use much more water than the ingredients or raw materials that they are processing. While some water may be used as an ingredient, the greater use will be for cleaning of raw materials, plant cleaning, cooling water and boiler water feed, each use potentially requiring water to a different specification.

This, in turn, creates similar quantities of used water containing variable concentrations of food components, cleaning chemicals, biocides and boiler treatment chemicals. Small enterprises may find it preferable to discharge their waste to the municipal sewage system, though larger process plants normally need to carry out either partial or complete treatment of their trade effluent.

13.2
Fresh Water

Water quality requirements in food processing will vary from product to product. Extreme cases of product-led specifications are to be found with beer and whisky production. For English beer production, a very hard water is needed and product waters are 'Burtonised', i.e. hardened by the addition of salts to approximate the composition of ground water from Burton on Trent. At the other extreme, water used in Scotch malt whisky production is very soft and may contain soluble organic compounds from the peaty highland soils.

Thus the water quality in an area, which is largely determined by its geology, can be a historical determinant of the development of specific sectors within the food industry. Modern water treatment can overcome this constraint by the introduction of a range of physical and chemical treatments, which will be adjusted to the source and end use of the water. These treatments must cope with suspended matter, from trees in floodwater at one extreme to grit and microor-

Food Processing Handbook. Edited by James G. Brennan

ISBN: 3-527-30719-2

ganisms at the other, plus dissolved minerals, gases and organic compounds that may give rise to colour, taste and odour problems in the final product.

While most food processors draw their water from the municipal supply and need to carry out very little treatment themselves, some will be required to provide part or all of their water from an untreated source.

13.2.1
Primary Treatment

Surface waters, whether drawn from rivers or lakes, are assumed to contain large suspended matter so intakes must be of robust construction and located away from direct flow so that collision damage may be avoided. The intakes would typically be faced with 15–25 mm vertical mild steel bars, gap width 25–75 mm. Flow through the intake should be $<0.6 \text{ m s}^{-1}$, ideally $<0.15 \text{ m s}^{-1}$, which would minimise drawing in silt [1]. Where ice formation is expected, the intakes must be in sufficiently deep water to permit adequate flow in cold weather despite the surface being frozen over.

Incoming water should then be passed through an intermediate filter, typically with an aperture of 5–10 mm (sometimes down to 1 mm), to remove the smaller debris. Drum or travelling band screens are often used as frequent back washing is needed to prevent blockage. Flow through the screens should be at $<0.15 \text{ m s}^{-1}$. With ground water sources, there should be little suspended matter and only light-duty intermediate screening is needed. The screened water should then be pumped to the treatment plant, the velocity in the pipeline being $\geq 1 \text{ m s}^{-1}$ to avoid deposition in the pipe.

Sedimentation may be employed if there are significant quantities of suspended matter in the water, the process being described by the Stokes Law equation:

$$v = \frac{d^2 g(\rho_s - \rho)}{18\mu} \qquad (13.1)$$

where v is the velocity of the particle, d is the equivalent diameter of the particle, g is the acceleration due to gravity, ρ_s is the density of the suspended particle, ρ is the density of the water and μ is the viscosity of the water at the prevailing temperature.

Some water treatment systems use a simple, upflow sedimentation basin for pretreatment. In this case the throughput or surface overflow rate, Q, is given by:

$$Q = uA \qquad (13.2)$$

Where u is the upward velocity of the water and A is the cross-sectional area.

Providing that $u < v$, then sedimentation will occur. Brownian motion will prevent very small particles (≤ 1 µm diam.) from separating.

Water that has undergone this primary treatment is adequate for cooling refrigeration plant, for example ammonia compressors, where there is no risk of contact with foodstuffs. For other uses, further treatment is required.

13.2.2
Aeration

Some ground waters may contain gases and volatile organic compounds that could give rise to taints, off flavours and other problems. For instance, while hydrogen sulphide may be regarded as a curiosity in spa waters, it is considered objectionable in potable water. Fresh water may be treated by aeration, mainly using either waterfall aerators or diffusion/bubble aerators. The transfer of a volatile substance to or from water is dependent upon:
(a) characteristics of that compound
(b) temperature
(c) gas transfer resistance
(d) partial pressures of the gases in the aeration atmosphere
(e) turbulence in the gas and liquid phases
(f) time of exposure.

Henry's Law for sparingly soluble gases states that the weight of a gas dissolved by a definite volume of liquid at constant temperature is directly proportional to the pressure. However, there is seldom time for equilibrium to be achieved, so the extent of the interchange will depend on the gas transfer that has occurred in the interfacial film, the diffusion process being described by Fick's First Law. Reducing the bubble size will considerably increase aeration efficiency. With very high surface:volume ratios, such as with a spray nozzle, an exposure time of 2 s may be adequate, while an air bubble in a basin aerator may need to have a contact time of at least 10 s [2]. All aeration systems must be well ventilated, not only to maximise efficiency but also to avoid safety risks, e.g. explosion with methane, asphyxia with carbon dioxide and poisoning with hydrogen sulphide. Hydrogen sulphide can be difficult to remove from water as it ionises, the anions being nonvolatile. However, removal may be accelerated by enriching the atmosphere to over 10% carbon dioxide in order to lower the pH and thus reduce ionisation. In aerating such waters, there is the risk that oxygen could also oxidise the hydrogen sulphide, giving colloidal sulphur which is difficult to remove.

13.2.3
Coagulation, Flocculation and Clarification

Particles of about 1 μm in size, including microbes, are maintained in suspension by Brownian motion. Many particles are also stabilised by their net negative charge within the normal pH range of water. In soft waters, colloidal dispersions of humic and fulvic acids may give rise to an undesirable 'peat stain'. This suspended matter may be destabilised by addition of salts, sometimes combined with alkali to raise the pH of the water. Trivalent are more effective than divalent cations, which are much more effective than monovalent cations.

Coagulant dosing may be preceded by injection with ozone or hypochlorite to oxidise organic compounds as well as to reduce the microbial loading.

Aluminium sulphate is effective as a coagulant over a pH range of 5.5–7.5, but its popularity in the UK has declined since an accident at Camelford, Cornwall, in 1988. Ferric chloride or ferric sulphate, with an effective pH range of 5.0–8.5, are now more commonly used in the UK. The salts may be used in conjunction with a cationic polymer. Where necessary, lime may be added prior to coagulant addition, allowing about 10 s for mixing, though the pH correction and coagulant dosing may be carried out at the same time if high energy mixing is employed. A wide range of mixing systems has been used, including air injection which, although aiding aeration of the water, can cause scum problems. The principle is that rapid mixing of the chemicals enables a homogeneous mixing so that aggregation of the colloidal particles to form flocs can then progress under low-shear conditions, followed by settlement [3].

Inorganic cations, for example Pb, As, Se, may also be removed from the water, the extent depending on the pH, coagulant and oxidation state of the cation.

Clarification of the water is achieved by allowing the flocs to settle out in settling basins, which may be rectangular or circular and be of downflow or upflow contact design. Figure 13.1 illustrates the principles of a circular clarifier basin.

The pH correcting and coagulating agents may be dosed into the feed line to the clarifier or at the discharge into the basin. In the latter case, higher shear mixing would be needed. Coagulent dosing may be accompanied by the addition of water-softening agents, such as calcium hydroxide and sodium carbonate

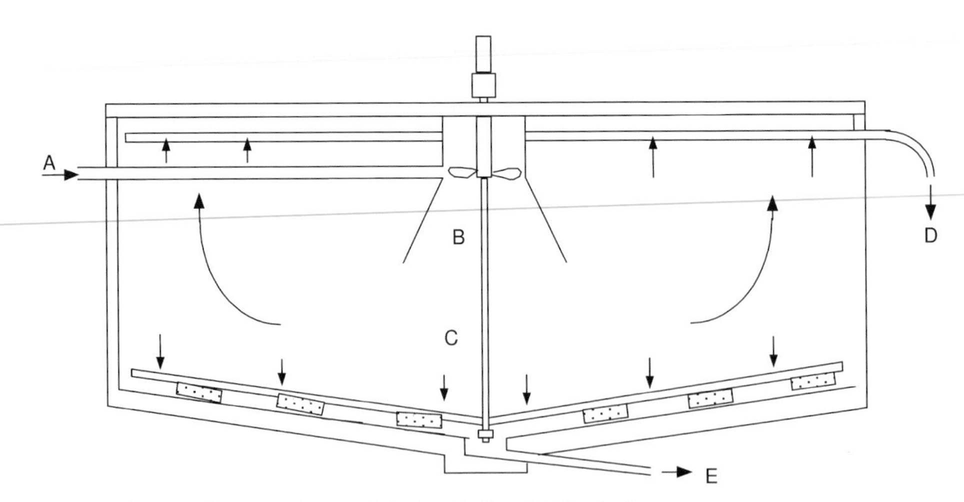

Fig. 13.1 Schematic layout of circular clarifier: A, inlet; B, flocculation zone; C, sedimentation; D, overflow of clarified water; E, sludge removal.

at 100–200 $g m^{-3}$ [4, 5]. Flocculation is encouraged by gentle mixing in the central portion of the basin (shown as B in Fig. 13.1) as the influent water moves downwards within the central ducting. Sedimentation then takes place as the water slowly moves outwards and then up to the radial collecting trays at the surface. The sediment forms a sludge, which is directed down the sloping base of the basin towards the centre by the slowly rotating sweep arms and is then periodically pumped away [6]. This sludge is of no further use for fresh water treatment but may be used to aid flocculation of effluent in waste water treatment.

13.2.4 Filtration

For large-scale fresh water filtration, for example with water utility companies, and where extensive land use can be justified, slow sand filters are very effective. The efficiency is due to a combination of physical separation and biological activity. Each tank is approximately 3 m deep, with gravel over a porous base to allow the filtered water to drain away. Fine silica sand, particle size typically 0.2–0.4 mm, is deposited evenly to a depth of up to 1 m, with the influent water being maintained at 1–2 m above the sand. Influent water must be distributed evenly across the filter to avoid disturbing the bed. The top 20–30 mm of the sand rapidly develops a complex biofilm, commonly referred to as zoogleal slime or Schmutzdecke. This biofilm is made up from polysaccharide secreted by bacteria such as *Zoogloea ramigera*, which is also colonised by protozoa. Bacteria and fine particles become trapped in the slime and are ingested by the protozoa. The resulting increase in the efficiency of filtration can give a two order (99%) drop in the microbial population of the water, as well as reducing the levels of dissolved organic and nitrogeneous matter [5, 7]. The removal of organic matter may be further improved by including a layer of granular activated carbon in the sand bed and by ozone injection prior to filtration. Throughput can be up to $0.7\, l\, m^{-2}\, s^{-1}$ ($\approx 60\, m^3\, m^{-2}\, day^{-1}$). The build up of biofilm and retained debris plus algal growth slowly reduces the throughput of the filter, so that the surface layer needs to be removed and cleaned at intervals of 1–6 months. Complete replacement of the sand is needed at longer intervals, an expensive operation.

In most industrial sites space is at a premium, so either high-rate or pressure filters are employed (see Fig. 13.2). High-rate filters are also commonly used by water utilities, either instead of or in conjunction with slow filters. High-rate filters may be upflow or downflow, the flow rate for the former being limited so that fluidisation of the bed does not occur. These filters are normally built as a series of modules with the pipework so arranged that one module can be taken out of service at a time for regeneration by backflushing, which is usually carried out on a daily basis [8]. Units are normally up to $\approx 200\, m^2$ in surface area.

High-rate filters use a coarse silica sand, e.g. particle size 0.4–0.7 mm, in a thinner layer (0.4–0.7 m deep) than for the trickling filters. Sometimes dual media beds are used, which can almost double the throughput to $\approx 24\, l\, m^{-2}\, s^{-1}$

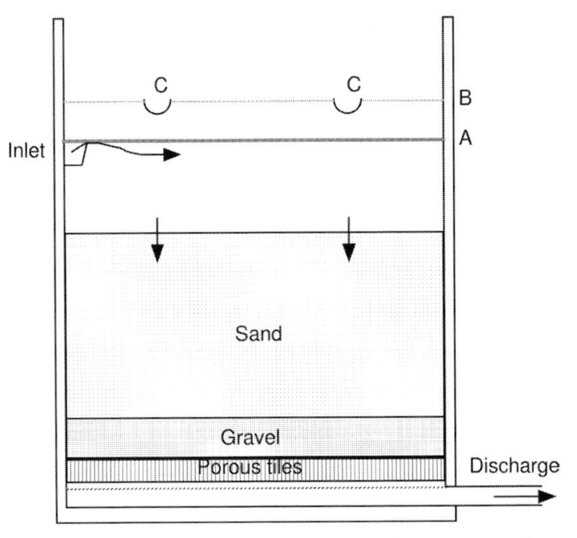

Fig. 13.2 Schematic of high rate down-flow filter: A, normal water level; B, backflushing level; C, washwater collection troughs.

($2000\ m^3\ m^{-2}\ day^{-1}$). The faster throughput and more frequent cleaning prevent the build up of a biofilm and bacterial removal is typically less than 80% (less than one order). The sand must be periodically topped up as some is lost during backflushing when the bed is fluidised.

Pressure filters are suitable for smaller-scale operations where space is limited (see Fig. 13.3). These units are normally supplied as prefabricated mild steel pressure vessels, up to 3 m in diameter for ease of transport. The main axis may be horizontal or vertical to suit the site. Operating pressures are higher, with drops of up to 80 kPa across the filter medium. Filter media may range from sand or anthracite to diatomaceous earth on stainless support mesh or plastic formers wound with monofilament. Pressure filters should be installed in a duplex arrangement to allow one to be cleaned while the other remains in operation.

Cartridge filters have been used extensively for medium and small-scale water filtration, using a range of filter media from stainless meshes and sintered materials to plastics and paper filters. While paper filters are single-use, others are more robust and can be cleaned by backflushing. Ceramic filters have been used for small-scale and portable filtration equipment. These can operate down to the micron level and be used for removal of microorganisms, that is for microfiltration, which has been used commercially to provide water with a high microbiological quality. In small-scale personal filters, the ceramic filter may also contain silver to add a bactericidal stage.

Occasionally, where ground water has to be used and is contaminated by heavy metals, the water may be treated by nanofiltration (NF) or reverse osmo-

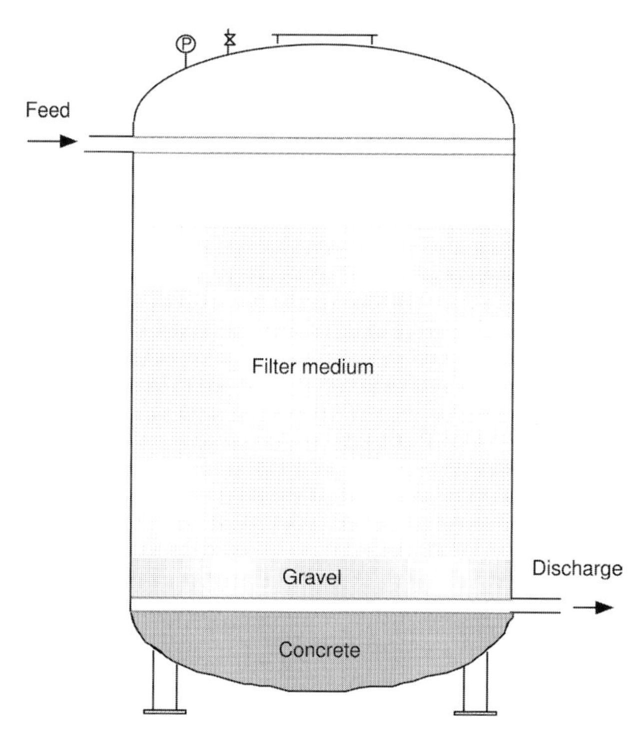

Fig. 13.3 Pressurised vertical filter unit.

sis (RO), the latter also being used for treatment of saline waters (see Chapter 14) [9–11]. In these treatments, the pore size is so small that the filtration is carried out at the molecular level, RO being regarded as a diffusion-based process. With NF there will be appreciable leakage of small ions, e.g. Na^+, through the membrane, but larger ions and molecules will be retained. The retention of small molecules and ions leads to an increase in the osmotic pressure of the retentate, requiring a higher driving force (>2 MPa) than for other filtration processes [11]. This energy requirement is less than would be required for distillation processes and RO provides an economic alternative for desalination. Ion exchange processes may also be employed for removal of heavy metals [10, 12].

The application of membrane processes for water treatment has been covered in detail by Duranceau [13]. NF has been applied to the treatment of contaminated river water to provide drinking water. At Méry sur Oise, 80% of the water being treated was put through an additional treatment, with MF followed by NF with 230 Da cutoff at a maximum throughput of 180000 m^3 day^{-1}. The system achieved a 90% reduction in total organic compounds from a maximum intake value of 3.5 mg l^{-1} to typical output values of 0.1–0.25 mg l^{-1} [14, 15].

13.2.5 Disinfection

It is assumed that any water used within a food processing plant must, at the very least, be of potable quality. This means freedom from taints, chemical contaminants and pathogenic organisms. Potable water often contains low levels of organisms capable of causing product spoilage problems. The microbiology of water is discussed in detail by McFeters [16]. Disinfection is any process whereby pathogenic organisms are removed or inactivated so that there is no risk of infection from consuming the treated water. Both chemical and physical methods may be used. Chemical methods are the most common and are normally based on chlorine or ozone, while physical methods may include microfiltration, irradiation with ultraviolet (UV) and heat. Since the disinfected water is only safe until it is recontaminated, disinfection should be regarded as the terminal treatment in the process. It is imperative that disinfected water be protected from recontamination. Both pasteurisation and UV treatments have been used for preparing washwaters and additive water where the addition of trace compounds is not desired, e.g. washwater for cottage cheese curd preparation and the dilution of orange concentrates.

Chemical disinfection using chlorine or chlorine derivatives is the most commonly applied method for large-scale water treatment, followed by the use of ozone, the latter particularly in Canada, France and Germany.

The Chick-Watson theory remains the principal theory to explain the kinetics of disinfection, where the lethality of the process may be described by the following equation [17]:

$$\ln \frac{N}{N_0} = kC^n t \qquad (13.3)$$

where N is the number of pathogens surviving, N_0 is the number of pathogens at t_0, C is the concentration of disinfectant, t is time, k is the coefficient of specific lethality and n is the dilution coefficient.

Specific lethalities vary considerably between disinfectants and with the target microorganisms. These lethalities also change at different rates with pH and temperature; and thus Table 13.1 should be treated as a semiquantitative indication of the ranking of disinfectants and their effectiveness against groups of organisms.

13.2.5.1 Chlorination

Liquid chlorine was the most commonly used agent for large-scale chlorination. Smaller-capacity plant has also used chlorine dioxide, sodium or calcium hypochlorites.

On dissolution of chlorine in water, chloride ions and hypochlorous acid are generated:

Table 13.1 Specific lethalities of common disinfectants ($mg\ l^{-1}\ min^{-1}$, $n=1$). Source: [1], by courtesy of McGraw-Hill.

Disinfectant	Enteric bacteria	Viruses	Spores	Amoebic cysts
Ozone (O_3)	500.0	5.0	2.0	0.5
Chlorine dioxide (ClO_2)	10.0	1.5	0.6	0.1
Hypochlorous acid (HOCl)	20.0	>1.0	0.05	0.05
Hypochlorite ion (OCl^-)	0.2	0.02	0.0005	0.0005
Monochloramine (NH_2Cl)	0.1	0.005	0.001	0.002

$$Cl_2 + 2H_2O = H_3O^+ + Cl^- + HOCl \tag{13.4}$$

The equilibrium is temperature-sensitive, $pK_H=3.64$ at 10 °C, 3.42 at 25 °C. Ionisation of the hypochlorous acid is pH-sensitive, $pK_I \approx 7.7$ at 10 °C, 7.54 at 25 °C [1].

$$HOCl + H_2O = H_3O^+ + OCl^- \tag{13.5}$$

Thus, below pH 7, the un-ionised form predominates and at pH 6–9, the proportion increases rapidly with pH. Chlorine existing in solution as chlorine, hypochlorous acid or hypochlorite ions is known as free available chlorine.

Safety concerns have led to the replacement of chlorine and chlorine dioxide by sodium hypochlorite solutions. Concentrated hypochlorite solutions are both corrosive and powerful oxidising agents, so it is now generated as a dilute solution by electrolysis of brine in all but the smallest water treatment plants [4]. Care must be taken to avoid accumulation of hydrogen.

$$NaCl + H_2O \rightarrow NaOCl + H_2 \tag{13.6}$$

Hypochlorous acid reacts with ammonia, ammonium compounds and ions to form a range of compounds, e.g.:

$$NH_4^+ + HOCl \rightarrow NH_2Cl + H_2O + H^+ \tag{13.7}$$

$$NH_2Cl + HOCl \rightarrow NHCl_2 + H_2O \tag{13.8}$$

$$NHCl_2 + HOCl \rightarrow NCl_3 + H_2O \tag{13.9}$$

$$2NH_4^+ + 3HOCl \rightarrow N_2 + 3Cl^- + 3H_2O + 5H^+ \tag{13.10}$$

$$NH_4^+ + 4HOCl \rightarrow NO_3^- + H_2O + 6H^+ + 4Cl^- \tag{13.11}$$

Up to 8.4 mg of chlorine could be needed to react with 1 mg of ammonia. Doses of 9–10 mg chlorine per 1 mg ammonia are recommended in order to guarantee that there would be free residual chlorine in the treated water. The addition level at which the added chlorine has oxidised the ammonia is referred to as the break point. Addition of chlorine beyond the break point enables the disinfection process to continue after treatment and confers some resistance to postprocess contamination. This can be critical in maintaining the safety of some food processes, for example the cooling of canned products.

Superchlorination may be used when there are potential problems, e.g. due to a polluted source or breakdown. The water is chlorinated well beyond the break point to ensure rapid disinfection. If the water is left in this state it will be unpalatable, thus the water may be partially dechlorinated by adding sulphur dioxide, sodium bisulphite, ammonia or by adsorption onto activated carbon. Superchlorinated water should be used in the food industry for cooling cans and other retorted products to avoid postprocess contamination due to leakage during the cooling stage.

In contrast to superchlorination, the ammonia-chlorine or combined residual chlorination methods make use of chloramines or 'bound chlorine' left by incomplete oxidation. Chloramine is 40–80 times less effective as a disinfecting agent than hypochlorous acid but is more persistent and causes less problems with off flavours and odours than chlorine or dichloramine ($\approx$200 times less effective). Thus, any residual chloramine may have a bactericidal effect during distribution. The most common way to apply this principle is to first chlorinate and then add ammonia to the disinfected water [17].

13.2.5.2 **Ozone**

Ozone treatment has been used widely, both for general supplies and for smaller-scale treatment of water for breweries, mineral water plants and swimming pools. The advantage with ozone is its rapid bactericidal effect, good colour removal, taste improvement and avoidance of problems with chlorophenols. Chlorophenol production can be critical in the reconstitution of orange juice from concentrates; and chlorine-free water is essential if this taint is to be avoided. There is no cost advantage over chlorine.

Ozone is more soluble than oxygen in water and is a more powerful oxidising agent than chlorine. As such, it is frequently used in the pretreatment of water to break down pesticide residues and other organic compounds in the raw water. This has a secondary benefit of inactivating protozoan contaminants such as *Cryptosporidium*, providing a higher quality intermediate for filtration and final treatment. The effectiveness of ozone is related to oxygen, superoxide and hydroperoxyl radicals formed on its autodecomposition [18].

$$O_3 \rightarrow O_2 + O^\bullet \text{ (air)} \tag{13.12}$$

$$O_3 + OH^- \rightarrow 2HO_2 + O_2^- \text{ (initiation in water by hydroxide ion)} \tag{13.13}$$

$$HO_2 \rightarrow H^+ + O_2^- \tag{13.14}$$

$$O_2^- + O_3 \rightarrow O_2 + O_3^- \quad (13.15)$$

$$H^+ + O_3^- \rightarrow HO_3 \quad (13.16)$$

$$HO_3 \rightarrow HO^\bullet + O_2 \quad (13.17)$$

$$HO^\bullet + O_3 \rightarrow HO_2 + O_2 \quad (13.18)$$

Autodecomposition rates increase with pH, radicals, UV and hydrogen peroxide, but are reduced by high concentrations of carbonate or bicarbonate ions. Ignoring the effect of carbonate, the autodecomposition rate was described [1] as:

$$\frac{[O_3]_t}{[O_3]_0} = 10^{-At} \quad (13.19)$$

where $[O_3]_t$ is the concentration at time t, $[O_3]_0$ is the concentration at time 0, t is timeand the value of A is $10^{(0.636\ \mathrm{pH}\ -6.97)}$.

Thus at pH 8, the halflife of ozone is about 23 min, depending upon water quality.

13.2.6 Boiler Waters

Boiler waters must be soft so that total solids build up only slowly in the boiler, but the water should not be corrosive. Simple treatments rely on lime and ferric chloride for softening and/or carbonate reduction with sodium hydroxide addition to give a final pH of 8.5–10.0. Residual carbon dioxide should be removed. Further softening can be achieved by phosphate addition to sequester calcium, by ion exchange or by NF. Dissolved oxygen must be removed, either by deaeration or by dosing in a scavenger such as sodium sulphite. Dispersants such as tannins or polyacrylates should be added to aid dispersal of the accumulating suspended matter in the boiler, while antifoaming agents can reduce the carryover of water droplets in the steam [19]. Boiler water demand should be reduced and run time extended by returning condensate to the boiler feed [20]. Where relatively unpolluted evaporator condensate is available, this can be treated to provide either boiler feed or soft cleaning water.

While all boiler water additives must be compatible with food production, any steam being generated for direct injection into food must be of exceptional quality and should be supplied from a dedicated generator.

13.2.7
Refrigerant Waters

Any water used in a heat exchanger for food should be of good microbiological quality, on the assumption that, despite precautions, leakage may occur. The risk of leakage may be reduced by operating the refrigerant at a lower pressure than the product. Any refrigerant must be checked regularly to ensure that it has not been contaminated by leakage of the food product.

For refrigerant waters at 3–8 °C, potable water should be sufficient. Below 3 °C, antifreeze additives may be necessary, becoming essential for operating temperatures below 1 °C unless an extensive ice bank system is used. Such additives are usually covered by national legislation. Glycol solutions are often used and calcium chloride ($CaCl_2$) solutions have been widely used throughout the food industry, both as a refrigerant and for freezing, e.g. in ice lolly baths.

13.3
Waste Water

All processes will create waste and byproducts to a greater or lesser degree, as illustrated in Table 13.2. These represent not only a loss of ingredients and hence reduced profit from their conversion but an increased fresh water cost plus the additional cost of disposal of the waste created. It is essential that any manufacturing process should be designed and managed so as to minimise both the amount of fresh water used and the quantity of waste produced.

Waste minimisation must start at raw material delivery. Bulk tankers must be adequately drained before cleaning and a burst rinse should minimise waste before cleaning [20]. As in the factory, cleaning solutions should be collected and reused. Wherever possible, solid waste should be collected for separate disposal

Table 13.2 Examples of effluent loads from food processing.

	COD (mg l^{-1})	SS (mg l^{-1})	Water:product ratio (W : P)	Source
Beet sugar refining	1600	1015	–	[21]
Bread, biscuits, confectionery	5100 (275–9500)	3144	–	[22]
Brewing	2105 (1500–3500)	441	4	[22]
Cocoa, chocolate confectionery	9500 (up to 30 000)	500	4	[22]
Fruit and vegetables	3500 (1600–11 100)	500	15–20	[22]
Meat, meat products, poultry	2500 (500–8600)	712	10	[22]
Milk and milk products	4500 (80–9500)	820	1.5	[22]
Milk: liquid processing	ca. 700	–	12	[23, 25]
Milk: cheesemaking	≥2000	–	3	[23, 25]
Potato products	2300	656	–	[22]
Starch, cereals	1900	390	–	[22]

or recycling and certainly not flushed into the drains to add to effluent disposal problems.

The level of contamination of waste water is normally measured in terms of the biochemical oxygen demand over 5 days (BOD_5). This is defined as the weight of oxygen (mg l^{-1}, or g m^{-3}) which is absorbed by the liquid on incubation for 5 days at 20 °C. The method is time-consuming and may underestimate the potential for pollution if the sample is deficient in microflora capable of degrading the materials present. Measurement of the chemical oxygen demand (COD), based on the oxidation by potassium dichromate in boiling sulphuric acid, provides a more rapid measure of the capacity for oxygen uptake. Sometimes the milder oxidation by potassium permanganate may be used to yield a permanganate value. These values are typically lower than the total oxygen demand based on incineration and CO_2 measurement [21]. There is no direct link between BOD_5 and COD, since the relationship depends on the biodegradability of the components in the waste stream. For readily biodegradable wastes there is an empirical relationship: 1 BOD_5 unit $\approx$ 1.6 COD units, rising for less-biodegradable materials. The variability in these values may be less important if citing the difference between influent and effluent values for a treatment step or process, as these differences are based on changes in readily metabolised components.

It can be argued that the role of a food factory is to produce food and not to become involved with a nonproductive issue such as effluent disposal, which should be left to specialists. This could certainly be true of small enterprises but for larger factories, it may prove more cost-effective to undertake either partial or complete treatment of its trade effluent. In the UK, the charge for treatment of effluent is calculated by the Mogden formula, based primarily on the volume, COD and total suspended solids (TSS) [22].

Where an enterprise is to treat its own trade waste, the treatment plant should be located as far away from the production plant as possible, downwind and yet not being a nuisance to neighbours.

In most cases where effluent treatment is undertaken, this is kept separate from sewage, which poses greater public health problems and is normally dealt with on a community basis. Occasionally a plant may be built as a joint operation to handle sewage as well as trade waste, in which case more rigorous isolation from the food plant is essential.

13.3.1 Types of Waste from Food Processing Operations

The types of waste water produced by food processing operations reflect the wide variety of ingredients and processes carried out. Washing of root vegetables, including sugar beet, can give rise to high TSS levels in the effluent. Further processing of vegetables, involving peeling and/or dicing, increases the dissolved solids, as is also the case with fruit processing where sugars are likely to be the major dissolved component. Cereals processing and brewing create a carbohydrate-rich effluent, while effluent from processing legumes contains a higher level of protein. The processing of oilseeds results in some loss of fats, usually as suspended matter. Milk processing creates an effluent with varying proportions of dissolved lactose and protein plus suspended fat. Meat and poultry processing gives rise to effluents rich in both protein and fat.

In most of these examples, there is particulate waste, i.e. particles greater than 1 mm in size, in addition to the fine suspended matter. These should be removed by screening prior to disposal of the plant effluent into the drain and strainers should be fitted into each drain to collect those particles that bypass the screens. Both screens and strainers must be cleaned daily.

Material recovered from screens within the process plant may be suitable for further processing. If screens are not fitted into the process plant, then particles collected on screens at the entry to effluent treatment cannot be reclaimed within the processing plant and must go to solid waste disposal.

13.3.2 Physical Treatment

Sedimentation and/or flotation usually form the first stage of effluent treatment, depending on the particular effluent. Both processes are applications of Stoke's Law (see Section 13.2.1). Where the fat may be recovered and recycled within the process, the flotation must be carried out within the production area. Centrifugation provides a rapid and hygienic technique.

The simplest flotation technique is to use a long tank. Waste water enters at one end over a distribution weir. Flows at $\geq 0.3\ \mathrm{m\ s^{-1}}$ prevent sedimentation of fine suspended matter, but a residence time of about 1 h can be needed. Flotation can be hastened by aeration, fine gas bubbles being introduced at the base of the tank by air or oxygen injection, or by electrolysis. Fat globules associate with the gas bubbles, forming larger, less-dense particles that rise more rapidly to the surface. Bubbles 0.2–2.0 mm in diameter rise at $0.02\text{–}0.2\ \mathrm{m\ s^{-1}}$ in water. The presence of free, that is unemulsified, liquid fat acts as an antifoam and prevents excessive foaming, as the fat agglomerates into a surface layer. This can be scraped off, dewatered and, for instance, sold off for fatty acid or soap production.

Production processes starting with dirty raw materials such as root vegetables produce an effluent with high TSS, some of which may get past the primary screening within the plant. In this instance a sedimentation or grit tank is needed. A rectangular tank is often used, with flows $\geq 0.3\ \mathrm{m\ s^{-1}}$ to allow grit and mineral particles to sediment without loss of suspended organic material. The grit may be removed from the tank by a jog conveyor and dumped into a skip for disposal, either back onto the farmland, if relatively uncontaminated, or by landfill.

Following either of these pretreatments, the effluent should be collected into balance tanks. These serve to even out the fluctuations in pH, temperature and concentration throughout the day. Some form of mixing is desirable, both to aid standardisation and to maintain an aerobic environment, thus reducing off odour generation. This minimises the use of acids and alkali to standardise the pH of the waste water and render it suitable for subsequent treatments. Lime (calcium hydroxide) or sodium hydroxide has been used to raise pH, while hydrochloric acid is a common acidulant. With dairy wastes, the use of sodium hydroxide as the principal cleaning agent normally results in an alkaline effluent, while plants handling citrus products could expect an acid waste stream.

13.3.3 Chemical Treatment

Most of the organic contaminants remaining in the waste water are either in solution or in colloidal dispersion. At around neutral pH, these colloidal particles usually have a net negative charge, so the addition of polyvalent cations, for instance aluminium sulphate at pH 5.5–7.5, or ferric chloride (or sulphate) at pH 5.0–8.5, promotes the formation of denser agglomerates that can be sedimented and recovered as sludge. Chemical addition is usually by dosing a solution into the waste stream followed by rapid mixing to ensure even distribution. The treated waste water is then allowed to stand, to permit formation of the flocs and their sedimentation. Sedimentation may be carried out in rectangular or circular basins. This process is similar to that employed for fresh water treatment (see Section 13.2.3), but the quantities of sludge settling out are much higher. Effluent leaves the settling tank via an overflow weir, which should be protected in order to prevent any surface fat and scum from overflowing too. Such fat should be scraped off periodically.

Sludge from settlement vessels typically contains about 4% solids and must be pumped over to an additional settlement tank where about half of the volume can be removed as supernatant and returned to the beginning of the treatment process.

The effluent from the settlement tank may then either be discharged to the sewer as partially treated effluent, incurring a much lower disposal charge, or else taken on to biological treatment.

13.3.4 Biological Treatments

Biological treatments may be divided into aerobic and anaerobic processes. In aerobic processes, oxygen acts as the electron acceptor so the primary products are water and carbon dioxide. In anaerobic treatment, the primary products are methane and carbon dioxide, with sulphur being reduced to hydrogen sulphide.

While properly run aerobic treatments produce the less polluting effluents, anaerobic treatment has great potential for large-scale treatment of sludge and highly polluted waste waters. Relative advantages are summarised in Table 13.3.

In general, smaller plants opt for aerobic treatment while the larger plants may use a combination of aerobic and anaerobic methods.

Table 13.3 Comparison of aerobic and anaerobic processes.

Factor	Aerobic	Anaerobic
Capital cost	(Lower)	Higher
Energy cost	Medium-high	Net output
Influent quality	Flexible	Demanding
Sludge retention	High	Low
Effluent quality	Potentially good	Poor

13.3.4.1 Aerobic Treatment – Attached Films

The trickling or percolating filter provides a simple and flexible means of oxidising dilute effluents. It takes the form of a circular (typically 7–15 m diam.) or rectangular concrete containment wall, 2–3 m high, on a reinforced concrete base which includes effluent collecting channels (see Fig. 13.4). The infill is preferably a light, porous material (about 50% voidage) with a high surface area (up to 100 m^2 m^{-3}), for example coke or slag. Solid rock may also be used. The particle size varies over 30–50 mm, sometimes up to 75–125 mm for pretreatment prior to discharge to a sewer. The use of less dense, synthetic filter media allows deeper beds to be constructed.

The influent of clarified waste water is spread over the top surface of the filter by nozzles, mounted either on rotating arms for circular filters or on reciprocating bars for the rectangular beds. The surface must be evenly wetted, the liquid then trickling down through the filter. Filamentous algae often grow on the surface while, within the bed, the medium provides a physical support for a complex ecosystem. This biofilm contains bacteria, protozoa, fungi, rotifers, worms, and insect larvae. *Zoogloea ramigera* is the predominant colonising bacterial species, producing an exopolysaccharide-based support medium, the thickness of which increases with the richness of the nutrients in the influent liquor. The slime may be less than 1 mm thick with lean wastes but reach several millimetres with concentrated wastes, with a higher proportion of fungal mycelium in the latter. The biofilm also contains a range of other heterotrophic species, including *Pseudomonas, Flavobacterium* and *Alcaligenes* spp, which absorb soluble nutrients from the influent [21, 24]. Fungi and algae are also present. Fine suspended matter is trapped by the slime and ingested by protozoa, while the bur-

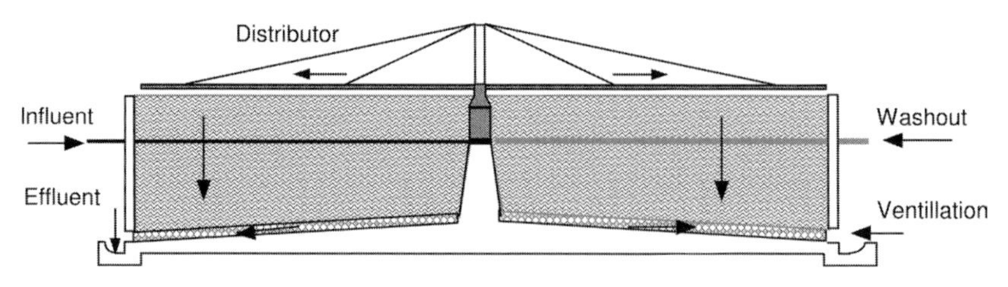

Fig. 13.4 Section through a percolating filter.

rowing activity of the larvae helps maintain the flow through the biomass on the filter. Most of the BOD_5 reduction occurs in upper layers, while the oxidation of ammonia to nitrate take place in the lower portion of the bed.

With such a complex ecosystem, trickling filters take time to adapt to changes of feedstock. Initiation time can be reduced by seeding the filter with material from another filter running on similar effluent. Care must be taken to avoid feeding inhibitory materials or making sudden, major changes to the feedstock.

Careful management of the filter is essential. Since the biofilm builds up when fed rich effluents, blockage of the channels can occur with ponding of feedstock on the top of the filter. Algae can also grow on the surface and sometimes weeds can grow too, requiring the surface layers to be periodically turned over with a fork. The biofilm can be most easily managed by using two filters in series, alternating their position at intervals of 10–20 days, as illustrated in Fig. 13.5. The filter with a rich biofilm is then supplied with a much leaner nutrient stream and loses biofilm, which is sloughed off into the effluent at a higher rate than before. This system is referred to as alternating double filtration (ADF). In some small effluent plants a pseudoADF system is used, where the effluent from the trickling filter is collected and then passed back through the filter a second time, while the primary effluent is held back. The pseudo-ADF system could be run on a daily cycle basis.

The hydraulic loading on the filter should ideally be less than 1 $m^3\ m^{-3}\ day^{-1}$ (or doubled with an ADF system), with a BOD loading of less than 300 $g\ m^{-3}\ day^{-1}$, normally $\approx$ 60 $g\ m^{-3}\ day^{-1}$. Filter effluent contains flocs of biofilm, which

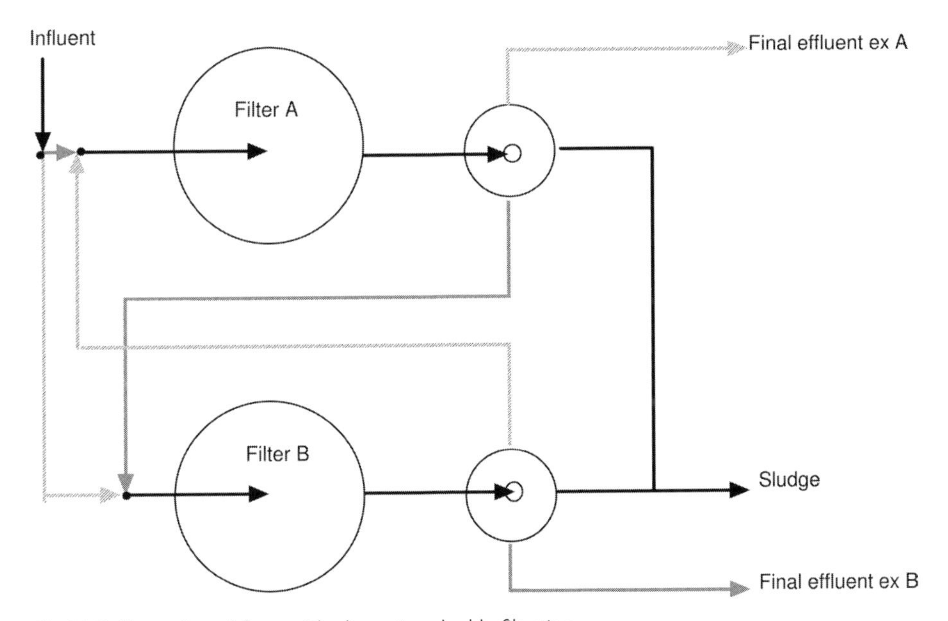

Fig. 13.5 Illustration of flows with alternating double filtration.

must be removed by sedimentation. Clarification is achieved by passing the effluent through a settling tank, for example as shown in Fig. 13.6. The sludge from the settling tank may be combined with that from earlier settling processes. The settled effluent BOD from the primary filter should be less than 30 mg l^{-1}, while that from the final filter should be less than 15 mg l^{-1}, typically $\approx$ 4 mg l^{-1}.

While trickling filters are relatively tolerant of varying effluent quality, their oxidative capacity is fairly low. In all but the smallest effluent plants, there can be an advantage in using higher rate aerobic systems, if only for pretreatment.

High-rate aerobic filters use a very low-density medium, e.g. plastic tube section, with high viodage (up to 90%) and specific surface areas up to 300 m^2 m^{-3}. Being very light, high-rate filters can be mounted above ground level, sometimes being used as modifications mounted above preexisting treatment plants. The standardised, pretreated effluent is pumped over the filter at a relatively high rate. The hydraulic load is typically 5–10 m^3 m^{-2} day^{-1} with high recirculation rates, to give more than 50% removal of BOD_5 [24].

Various types of disc and other rotating contactors have been used for effluent treatment, although initially these were more successful for general sewage than for food industry wastes. Rotating contactors use slowly rotating surfaces, which are less than half immersed in the effluent. As the device rotates, the damp film is taken up into the air and oxygenated, enabling the surface biofilm to metabolise the effluent. The build-up of biomass must be removed by peri-

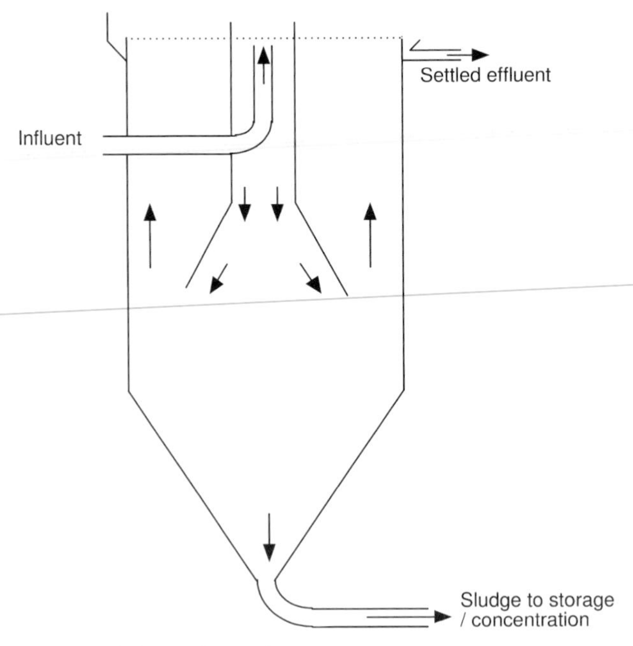

Fig. 13.6 Sludge settling tank.

Table 13.4 Examples of aerobic treatments used for food effluents. Source: [19]

Sector	Organisation	Location	Reactor	Capacity (t day^{-1} BOD$_5$)	COD[a)]
Brewery	San Miguel	Philippines	Trickling filter + activated sludge	14.0	22.4
Dairy	Entrement	Malestroit, France	Extended aeration	3.0	4.8
Vegetables	Findus	Beauvais, France	Activated sludge	7.5	12.0

a) COD converted from BOD, assuming 1.6 COD $\approx$ 1 BOD$_5$, to aid comparison with data in Table 13.5

odic flushing. Greater success has been achieved with submerged filters, operating with forced aeration in either upflow or downflow modes. The method has found use in smaller plants where activated sludge treatment may be difficult to use.

Some examples of aerobic treatments are given in Table 13.4.

13.3.4.2 Aerobic Treatment – Suspended Biomass

Activated sludge processes have been adapted successfully to the treatment of food wastes, and are attractive for larger plants (>8 t COD day^{-1}) despite the higher energy costs, as they take up less land than trickling filters [25].

The biomass, consisting primarily of bacteria, is suspended in the medium as flocs. Protozoa are present in the flocs as well as free-swimming ciliated species [21]. Oxygen is introduced via compressed air injection, oxygen injection or by rigorous stirring using surface impellers. Surface mixing is less efficient but simpler and is used in smaller plants. The suspended biomass requires a constant influent composition for optimal operation. The efficiency in removing BOD depends on the rate of oxygenation, which can be reduced as the medium flows through the tank. Where phosphate reduction is also desired, part of the tank is run anaerobically to encourage additional phosphate uptake into the biomass once aerobic conditions are reintroduced. Partial denitrification is achieved by recycling biomass, a 1:1 recycling ratio being associated with 50% denitrification as a result of anoxic conditions being set up in the first stage of the digestion. Figure 13.7 illustrates the basic principles of an activated sludge plant. Where air or oxygen injection is employed, a tall cylindrical shape may also be used.

Though activated sludge plants are effective in reducing BOD$_5$, energy costs and sludge production are relatively high. A typical BOD$_5$ reduction of 95% can be achieved with hydraulic retention times of 10–20 h and biomass retention times of 5–10 days [25]. Higher BOD$_5$ removal can be achieved by reducing the substrate concentration and throughput, the longer residence time also resulting in lower sludge production. Increasing the substrate loading and/or throughput increases sludge production while giving a lower % BOD$_5$ removal.

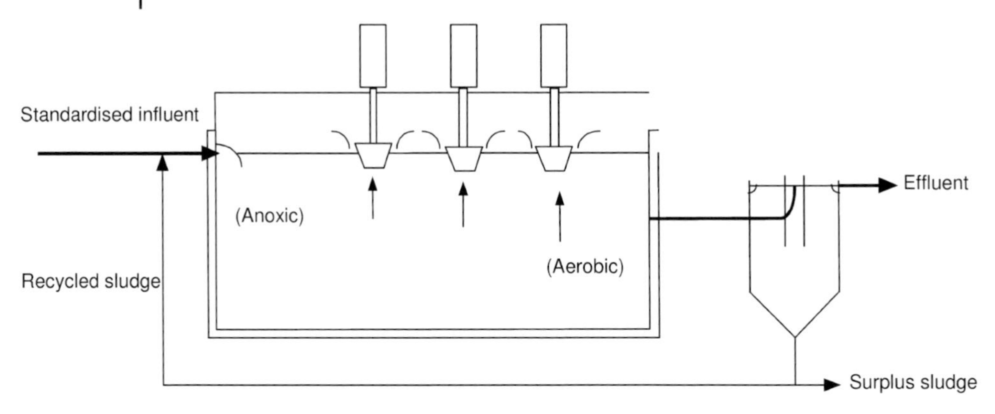

Fig. 13.7 Activated sludge fermentation plant.

With dilute effluents, similar results can be achieved using an oxidation ditch where, again, the biomass is largely suspended. Waste water is circulated under turbulent conditions ($\geq 0.3\ \mathrm{m\ s^{-1}}$) around a channel, 1–2 m deep. The propulsion and aeration is carried out by a series of rotating brushes, as illustrated in Fig. 13.8.

Waste water is constantly fed to the ditch, the overflow passing through a settling tank. Retention times vary over 1–4 days, with sludge retention times of 20–30 days. Part of the sludge can be fed back to the ditch, giving a relatively low, net sludge production. While the oxidation ditch is simpler to operate than the activated sludge system, more land is required.

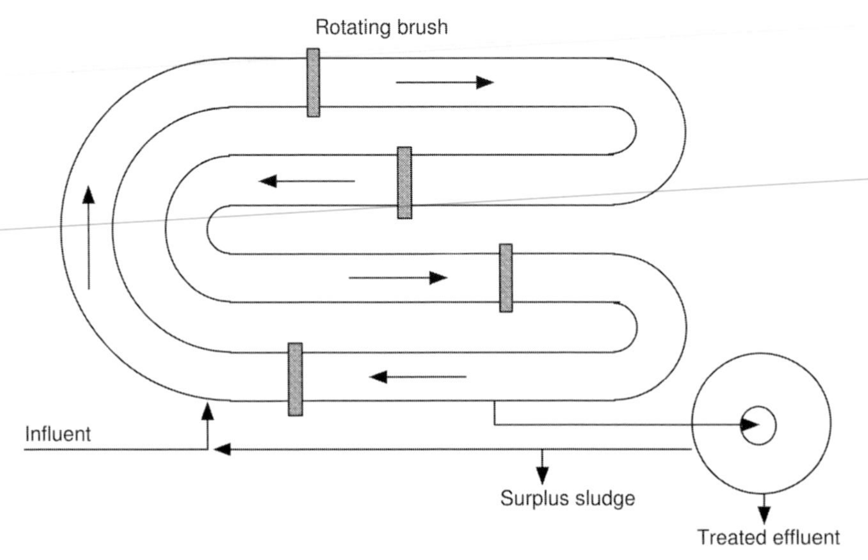

Fig. 13.8 Plan view of an oxidation ditch.

13.3.4.3 **Aerobic Treatment – Low Technology**

In some areas, where land is inexpensive and rainfall moderate or low, it may be possible to simply use a shallow lagoon, typically 0.9–1.2 m deep, for slow oxidation and settlement of effluents prior to irrigation [8]. Some lagoons may be up to 2 m deep, in which case anaerobic conditions will occur in the lower levels [26]. The lagoon should be lined with clay or other impermeable material to minimise the loss of polluted water into the surrounding soil. This approach has been used for effluent from seasonal canning operations and some dairy wastes. The ecosystem in such lagoons is extremely complex, with bacteria, protozoa and invertebrates plus algae and aquatic plants. Though these systems tolerate high organic shock loadings, care must be taken to site such a lagoon well away from the factory, to avoid overloading its oxidation capacity and to prevent leakage into any watercourse. Such partially treated effluent can be used for irrigation, using low pressure sprays to minimise drift. Up to 25 mm ($1\ l\ m^{-2}\ day^{-1}$, including rainfall) have been used per 25-day irrigation cycle on grassland.

The lagoon approach can be improved by providing a number of ponds operating in series, providing a residence time of up to 3 weeks for stabilisation of BOD_5 levels [21].

Treatment of effluents has been achieved by trickling through beds of reeds and/or other semiaquatic plants, where the root structure supports a complex aerobic ecosystem, similar to that found in lagoons. Soil based wetland has been found to be more stable than gravelbased systems, which can block up. Findlater et al. [27] reported 70–80% BOD_5 removal at loadings of $4–20\ g\ m^2\ day^{-1}$, while Halberl & Perfler [28] reported 80–90% BOD_5 removal at slightly lower loadings of $1.5–15\ g\ m^2\ day^{-1}$. Extending the loading to $2–25\ g\ m^2\ day^{-1}$ gave a still wider range of BOD_5 removal (56–93%), the higher loading giving a reduced percent removal [29]. A reduced reduction in BOD_5 was also noted when feeding the reed beds with treated waste water where the BOD_5 was $\approx 50\ mg\ l^{-1}$. In experiments with meat processing effluent in gravelbed wetland trenches ($18\ m^2$) and soilbased surfaceflow beds ($250\ m^2$), van Oostrom & Cooper [30] found that COD removal rates increased with loadings up to $\approx 20\ g\ m^{-2}\ day^{-1}$. BOD_5 reduction was 79% with untreated effluent from balance tanks at a loading of $117\ g\ m^{-2}\ day^{-1}$, rising to 84% for partially treated, anaerobic effluent fed at $24\ g\ m^{-2}\ day^{-1}$. These data suggest a bed requirement of $50\ m^2\ kg^{-1}\ BOD_5\ day^{-1}$ with feed rates of $\approx 50\ l\ m^{-2}\ day^{-1}$, depending upon influent strength. A complex of beds in both parallel and series would be desirable to consistently produce a low BOD_5 effluent, with an annual harvest of the aboveground biomass.

13.3.4.4 **Anaerobic Treatments**

Anaerobic treatments have been applied to both the sludges from aerobic treatment and to treat highly polluted waste streams from food plants. Aerobic treatments can produce 0.5–1.5 kg of biomass per 1 kg of BOD_5 removed, so potentially large quantities of sludge may need to be treated. Much of the sludge is

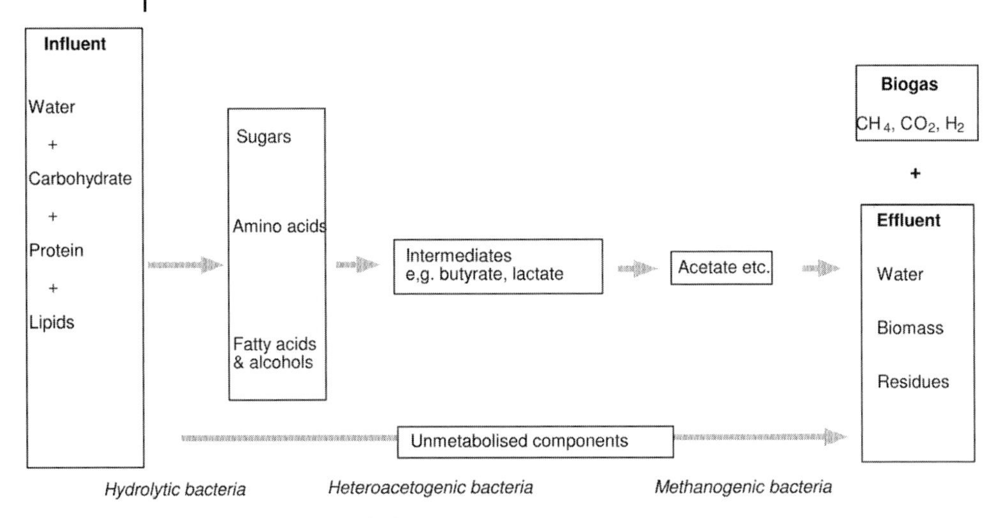

Fig. 13.9 Anaerobic catabolism.

low in solids and should first be concentrated, by settlement or centrifugation, to at least 8–10% dry matter.

Most anaerobic reactors are run at $35 \pm 5\,^{\circ}C$, to ensure methanogenesis. The catabolism of the food components is summarised in Fig. 13.9. While the main gaseous products are methane and carbon dioxide, there are also small quantities of hydrogen sulphide and other noxious compounds. The gases are normally collected and used on site, e.g. to drive a combined heat and power (CHP) plant, with scrubbing of the waste gasses where necessary.

There are five main types of anaerobic reactor:

- stirred tank reactor
- upflow sludge blanket
- upflow filter
- downflow filter
- fluid bed.

The stirred tank reactor is similar to the sludge fermenters used for domestic sewage, but with part of the effluent sludge recycled to the influent (see Fig. 13.10). Sludge from domestic sewage treatment is increasingly being heat-treated before fermentation to remove pathogens; and similar treatment may be required for some food wastes such as slaughterhouse effluent. Mixing may also be achieved by returning biogas from the gas separator. Heat from the CHP plant or another low-grade source is used to preheat the influent and to maintain the temperature at $35\,^{\circ}C$ in the reactor. Biogas may be recovered from both the reactor and separator.

The upflow sludge anaerobic blanket (USAB) reactor does not use mixers but relies on the evenly distributed upflow of the influent (see Fig. 13.11), plus the bubbles from gas generation during fermentation. The bulk of the biomass

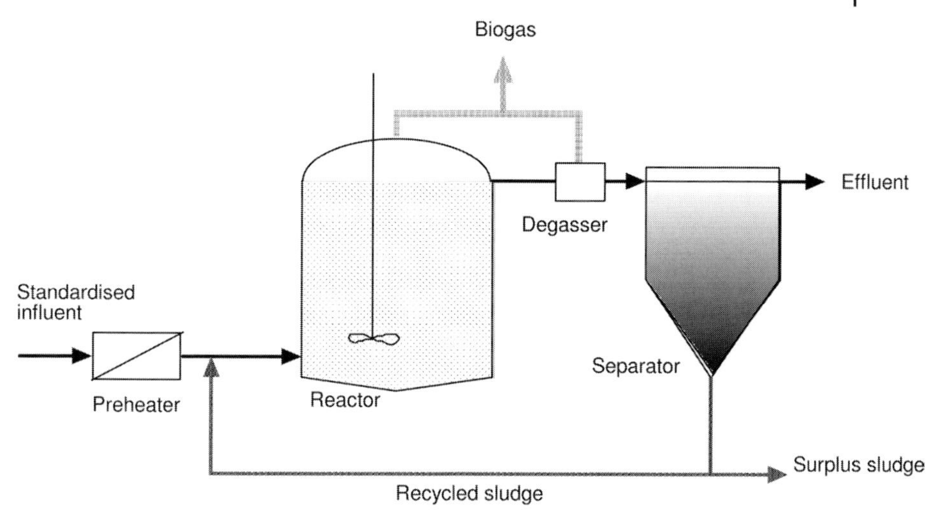

Fig. 13.10 Stirred tank reactor. Mixing may also be carried out by recycling biogas.

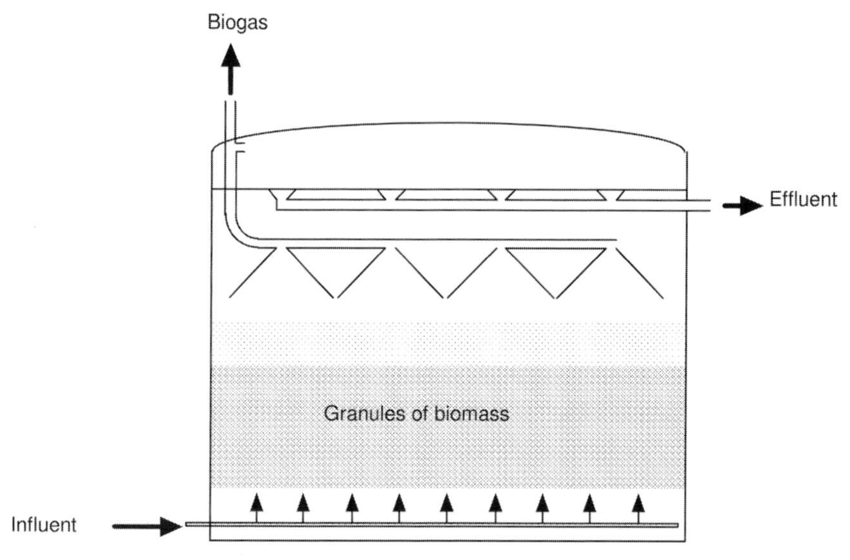

Fig. 13.11 Upflow stirred anaerobic blanket (USAB) reactor.

forms a granular floc in the lower layer of the reactor, encouraged by a high proportion of short chain fatty acids (which may be produced by prefermentation), the presence of Ca^{2+} and pH >5.5, preferably pH ≈ 7.5. For many influents, dosing with $Ca(OH)_2$ is needed. A lighter floc of biomass also covers the granular floc. The influent quality demands have restricted the application of this type of reactor to wastes from yeast, sugar beet and potato processing

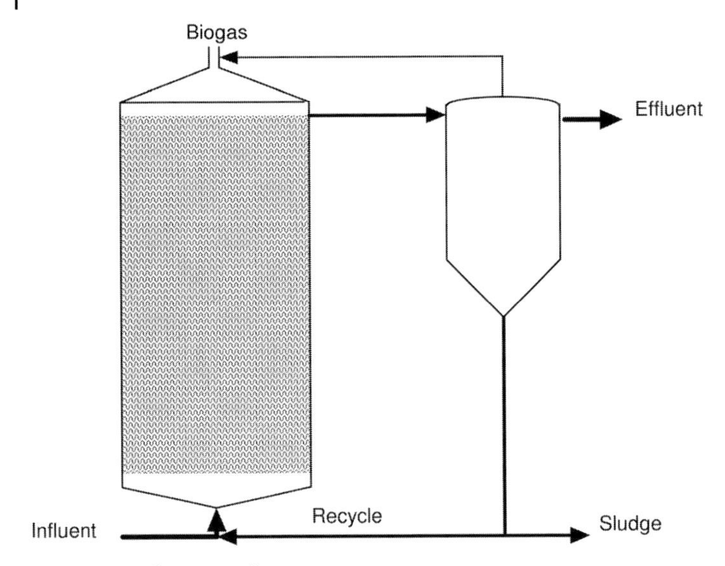

Fig. 13.12 Upflow anaerobic reactor.

wastes. Problems have been encountered with abattoir, dairy, distillery and maize processing wastes [31].

Anaerobic upflow filter reactors (see Fig. 13.12), contain a media fill. Crushed rock is the lowest-cost fill, similar to that with the aerobic trickling filters, with 25–65 mm rock giving approximately 50% void. About half of the biomass is attached to the medium, the rest being in suspension in the voids. More expensive media can give up to 96% void but with less biomass attached to the medium. The high proportion of suspended biomass limits the throughput as excessive biomass can be lost from the reactor. Similarly, the risk of excessive biomass growth also limits the influent concentration.

Anaerobic downflow reactors tend to use random packed high-void media. Though the bulk of the biomass is in suspension, the downward flow of the influent is opposed by the upward movement of the gas bubbles (see Fig. 13.13). This upward movement both buoys up the biomass and promotes mixing within the reactor.

Fluid bed reactors provide an improvement on the USAB and upflow filter reactors, combining some of their properties. Fine particulate material, typically sand (particle size ≤ 1 mm) is fluidised by the upflow of the influent liquid plus the gas evolved, expanding the bed volume by 20–25%. Upflow velocity is critical, typically being $3–8 \times 10^{-3}$ m s^{-1}. The sand and attached biomass are retained within the reactor by reducing the upflow velocity in the wider section at the top of the reactor, suspended biomass being returned to the primary vessel, as shown in Fig. 13.14.

The various types of reactor have been increasingly adopted for large-scale processing of food processing wastes, particularly where these wastes are rela-

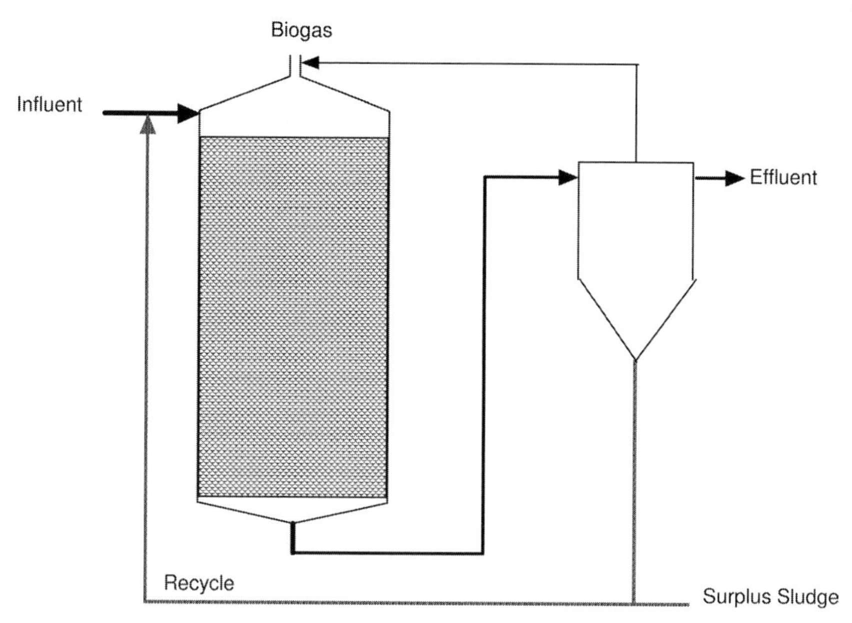

Fig. 13.13 Downflow anaerobic reactor.

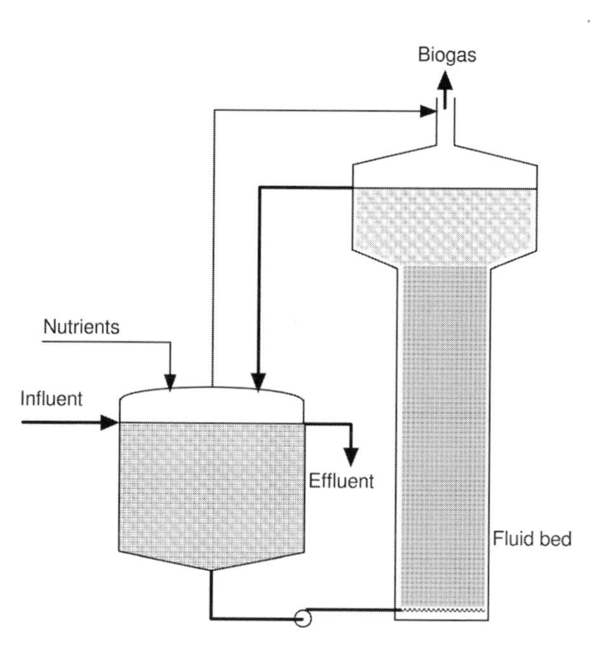

Fig. 13.14 Fluidised bed.

Table 13.5 Examples of anaerobic treatments used for food effluents. Source: [18]

Sector	Organisation	Location	Reactor	Capacity (t day^{-1} COD)
Brewing	El Aguila	San Sebastian de Los Reyes, Spain	Fluidised bed	50.0
Brewing	Sébastien-Artois	Armentières, France	Sludge blanket	10.0
Canning	Bonduelle	Renescue, France	Digester	18.0
Dairy	St Hubert	Magnières, France	Upflow filter	2.8
Distillery	APAL	Paraguay	Digester	54.0
Distillery	DAA	Ahausen, Germany	Upflow filter	12.0
Sugar	Julich	Julich, Germany	Digester + activated sludge	30.0
Sugar	Südzucker	Platting, Germany	Digester	30.0–38.0
Winery	Canet C.C.	Canet, France	Fluidised bed	4.2

Table 13.6 Comparison of anaerobic reactor designs. Sources: [19, 25, 32]

Type	Loading (kg COD m^{-3} day^{-1})	Feed (g l^{-1})	Retention time (h)	COD removal (%)
Stirred tank	0.2–2.5	5–10	24–120	80–90
Filter	2–15	1	10–50	70–80
USAB	2–15	10	10–50	70–90
Fluid bed	2–60	2.5	0.5–24.0	70–80

tively concentrated (Table 13.5). A comparison between types is given in Table 13.6. Typically, 10–30% of the influent COD remains in the effluent so, irrespective of the type of reactor, further processing of the effluent is required. This entails separation of the sludge and final aerobic treatment of the water.

13.3.4.5 Biogas Utilisation

Gas collected from anaerobic treatments contains primarily methane and carbon dioxide in ratios varying from 1:1 to 3:1, with traces of hydrogen, hydrogen sulphide and other volatiles, depending on the substrate and operating conditions [32]. It is normally collected in a floating-dome gas holder at relatively low pressure, 1.0±0.5 kPa. The capacity of the gas holder depends on the output from the digester and whether the gas is used constantly or only during part of the day. Gas may be used in boilers to raise steam or in a CHP engine. The power produced by the CHP should exceed that used in the anaerobic process, the heat being used to maintain the fermentation temperature in the reactor(s). A flare may be used to automatically burn off surplus gas in the event of a breakdown.

13.4
Sludge Disposal

Sludge production is a major problem with both aerobic and anaerobic processes, in terms of both their immediate offensive nature and the potential for pollution. Sludge disposal at sea is no longer permitted, so the choice is between disposal on land or incineration. Sludges, particularly those from primary settlement, can be highly putrescible but have an advantage over municipal sewage sludge in that their levels of heavy metals and organic contaminants are likely to be low.

If agricultural land is nearby, it may be economic to dispose of the unconcentrated sludge direct to the land, preferably by injection below the surface to minimise nuisance and avoid runoff [33]. In many cases, it is necessary first to concentrate or thicken the sludge. This may be carried out by gravity settlement for 2–5 days, sometimes aided by the addition of polyelectrolytes. The supernatant must be fed back to the effluent treatment plant. Disposal of sludge from food processing by landfill is uncommon in the UK but has been used widely in other parts of the EU.

Sludge has been used to aid bioremediation of contaminated land, where its nutrients and humus help raise the activity of the soil bacteria. It has also been useful in raising the productivity of the poor soils often used for forestry. With further dehydration, such as by belt drying, there is potential for mixing the sludge with straw and composting. The composting process is an aerobic batch process, during which the temperature can rise to 70 °C so that further dehydration occurs. The resulting compost is suitable for horticultural use as a soil conditioner.

Exceptionally, dehydration by belt press could be carried out to give solids in excess of 30%, for instance with fibrous sludges from vegetable processing. These high-solids sludges could be disposed of by high-temperature incineration. As with other waste incineration, the waste gases must be brought up to 800–900 °C to destroy any volatile organic compounds, particularly dioxins, that may be formed at lower temperatures in the initial stages of combustion. Within the EU, Germany is the greatest user of incineration for sludge disposal [25]. Incineration still yields a solid waste, the ash which makes up to 30% of the original solids normally being sent for landfill as a hazardous waste.

13.5
Final Disposal of Waste Water

In most cases, final disposal of treated waste water is into a water course where it will be diluted by the existing flow. General requirements are covered by regulations, in the EU based on the Urban wastewater directive (91/271/EC), usually complemented by consent limits based on avoidance of pollution. The EU approach is now complemented by a move to integrated pollution prevention

Table 13.7 Simplified example of a river classification. BOD_5 (mg ml^{-1}) to the 90th percentile, DO (min. % oxygen saturation) to the tenth percentile, NH_3 (mg N l^{-1}) to the 90th percentile. Source: adapted from [35].

Class	Description	BOD_5	DO	NH_3	Biology
A	Very good	2.5	80	0.25	No problems
B	Good	4.0	70	0.6	No significant problems
C	Fairly good	6.0	60	1.3	Some restriction to fish species
D	Fair	8.0	50	2.5	Extraction for potable water after advanced treatment
E	Poor	15.0	20	9.0	Only low-grade abstraction, eg cooling water
F	Bad	>15.0	<20	>9.0	Very polluted, severely restricted eco-system and potential nuisance

and control [34] under Directive 96/61/EC. While it may be desirable to recycle water within a factory [35], with food processing such recycling would be constrained by aesthetic as well as cost considerations.

Measurement of river quality is complex, as the river is effectively an aerobic fermenter, with the flow rate and hence oxygenation playing a vital part in the natural bioremediation processes [36]. An illustration of a simplified river classification is given in Table 13.7.

Discharge licences may include maxima for flow, temperature, suspended solids, dissolved solids, BOD_5, nitrogen, phosphorous and turbidity. One processor's waste stream may subsequently (certainly eventually) be another's water source.

References

1 American Water Works Association, American Society of Civil Engineers **1990**, *Water Treatment Plant Design*, 2nd edn, McGraw-Hill, New York.

2 Anon. **1998**, Aeration, in *Water Treatment Plant Design*, 3rd edn, ed. AWWA/ASCE, McGraw-Hill, New York.

3 Wesner G. M. **1998**, Mixing Coagulation and Flocculation, in *Water Treatment Plant Design*, 3rd edn, ed. AWWA/ASCE, McGraw-Hill, New York.

4 Twort A. C., Law F. M., Crowley F. W., Ratnayake D. D. **1994**, *Water Supply*, 4th edn, Edward Arnold, London.

5 Benefield L. D., Morgan J. M. **1998**, Chemical Precipitants, in *Water Treatment Plant Design*, 3rd edn, ed. AWWA/ASCE, McGraw-Hill, New York.

6 Willis J. R. **1998**, Clarification, in *Water Treatment Plant Design*, 3rd edn, ed. AWWA/ASCE, McGraw-Hill, New York.

7 Choreser M., Broder M. Y. **1998**, Slow Sand and Diatomaceous Earth Filtration, in *Water Treatment Plant Design*, 3rd edn, ed. AWWA/ASCE, McGraw-Hill, New York.

8 Degrément **1991**, *Water Treatment Handbook*, vol. 1, 6th edn, Lavoisier Publishing, Paris.

9 Taylor J. S., Wiesner M., Membranes, in *Water Quality and Treatment*, 5th edn, ed.

R. D. Letterman, McGraw-Hill, New York.

10 Bergman R. A. **1998**, Membrane Processes, in *Water Treatment Plant Design*, 3rd edn, ed. AWWA/ASCE, McGraw-Hill, New York.

11 Lewis M. J. **1996**, Pressure-Activated Membrane Processes, in *Separation Processes in the Food and Biotechnology Industries*, ed. A. S. Grandison, M. J. Lewis, Woodhead Publishing, Cambridge.

12 Gottlieb M. C., Meyer P. **1998**, Ion Exchange Processes, in *Water Treatment Plant Design*, 3rd edn, ed. AWWA/ASCE, McGraw-Hill, New York.

13 Duranceau S. J. (ed.) **2001**, *Membrane Practices for Water Treatment*, American Water Works Association, Denver.

14 Peltier S. J., Benezet M., Gatel D., Cavard J. **2001**, What are the Expected Improvements of a Distributed System by Nanofiltered Water?, in *Membrane Practices for Water Treatment*, ed. S. J. Duranceau, American Water Works Association, Denver.

15 Ventresque C, Gisclon V., Bablon G., Chagneau G. **2001**, First-Year Operation of the Méry-sur-Oise Membrane Facility, in *Membrane Practices for Water Treatment*, ed. S. J. Duranceau, American Water Works Association, Denver.

16 McFeters G. A. (ed.) **1990**, Drinking Water Microbiology, Springer, New York.

17 Haas C. N. **1999**, Disinfection, in *Water Quality and Treatment*, 5th edn, ed. R. D. Letterman (AWWA), McGraw-Hill, New York.

18 Singer P. C., Reckhow D. A. **1999**, Chemical Oxidation, in *Water Quality and Treatment*, 5th edn, ed. R. D. Letterman (AWWA), McGraw-Hill, New York.

19 Anon. **1991**, *Water Treatment Handbook*, vol. II, 6th edn, Degrément (Lavoisier Publishing), Paris.

20 Hills J. S. **1995**, *Cutting Water and Effluent Costs*, 2nd edn, Institution of Chemical Engineers, Rugby.

21 Horan N. J. **1990**, *Biological Wastewater Treatment Systems: Theory and Operation*, John Wiley & Sons, Chichester.

22 Anon. **1986**, *Food Processing Research Consultative Committee Report to the Priorities Board*, Ministry of Agriculture, Fisheries and Food, London.

23 Walker S. **2000**, Water Charges: the Mogden Formula Explained? *Int. J. Dairy Technol.* 53, 37–40.

24 Gray N. F. **1989**, Biology of Wastewater Treatment, Oxford University Press, Oxford.

25 Wheatley A. S. **2000**, Food and Wastewater, in *Food Industry and the Environment in the European Union: Practical Issues and Cost Implications*, 2nd edn, ed. J. M. Dalzell, Aspen Publishers, Gaithersburg.

26 UNEP International Environmental Technology Centre **2002**, *Environmentally Sound Technologies for Wastewater and Stormwater Management: An International Source Book*, IWA Publishing, London.

27 Findlater B. C., Hobson J. A., Cooper P. F. **1990**, Reed Bed Treatment Systems: Performance Evaluation, in *Constructed Wetlands in Water Pollution Control*, ed. P. F. Cooper, B. C. Findlater, Pergamon, Oxford.

28 Halberl R., Perfler R. **1990**, Seven Years of Research Work and Experience with Wastewater Treatment by a Reed Bed System, in *Constructed Wetlands in Water Pollution Control*, ed. P. F Cooper, B. C. Findlater, Pergamon, Oxford.

29 Coombes C. **1990**, Reed Bed Treatment Systems in Anglian Water, in *Constructed Wetlands in Water Pollution Control*, ed. P. F. Cooper, B. C. Findlater, Pergamon, Oxford.

30 Van Oostram A. J., Cooper R. N. **1990**, Meat Processing Effluent Treatment in Surface-Flow and Gravel Bed Constructed Wastewater Wetlands, in *Constructed Wetlands in Water Pollution Control*, ed. P. F Cooper, B. C. Findlater, Pergamon, Oxford.

31 Stronach S. M., Rudd T., Lester J. N. **1986**, *Anaerobic Digestion Process in Industrial Waste Treatment*, Springer, Berlin.

32 Barnes D., Fitzgerald P. A. **1987**, Anaerobic Wastewater Treatment Processes, in *Environmental Biotechnology*, ed. C. F. Forster, D. A. J. Wase, Horwood, Chichester.

33 Department of the Environment **1989**, *Code of Practice for the Agricultural Use of Sewage Sludge*, HMSO, London.

34 DEFRA **2002**, *Integrated Pollution Prevention and Control*, 2nd edn, Department for Environment, Food and Rural Affairs, London.

35 Environment Agency **2003**, *GQA Methodologies for the Classification of River and Estuary Quality*, available at: http://www.environment-agency.gov.uk/science/219121/monitoring/184353.

36 Lens P., Pol L. H., Wilderer P., Asano T. (eds.) **2002**, *Water Recycling and Resource Recovery in Industry: Analysis, Technologies and Implementation*, IWA Publishing, London.

14
Separations in Food Processing

James G. Brennan, Alistair S. Grandison and Michael J. Lewis

14.1
Introduction

Alistair S. Grandison

Separations are vital to all areas of the food processing industry. Separations usually aim to remove specific components in order to increase the added value of the products, which may be the extracted component, the residue or both. Purposes include cleaning, sorting and grading operations (see Chapter 1), extraction and purification of fractions such as sugar solutions or vegetable oils, recovery of valuable components such as enzymes or flavour compounds, or removal of undesirable components such as microorganisms, agricultural residues or radionuclides. Operations range from separation of large food units, such as fruits and vegetables measuring many centimetres, down to separation of molecules or ions measured in nanometres.

Separation processes always make use of some physical or chemical difference between the separated fractions; examples are size, shape, colour, density, solubility, electrical charge and volatility.

The separation rate is dependent on the magnitude of the driving force and may be governed by a number of physical principles involving concepts of mass transfer and heat transfer. Rates of chemical reaction and physical processes are virtually always temperature-dependent, such that separation rate will increase with temperature. However, high temperatures give rise to degradation reactions in foods, producing changes in colour, flavour and texture, loss of nutritional quality, protein degradation, etc. Thus a balance must be struck between rate of separation and quality of the product.

Separations may be classified according to the nature of the materials being separated, and a brief overview is given below.

Food Processing Handbook. Edited by James G. Brennan

ISBN: 3-527-30719-2

14.1.1 Separations from Solids

Solid foods include fruits, vegetables, cereals, legumes, animal products (carcasses, joints, minced meat, fish fillets, shellfish) and various powders and granules. Their separation has been reviewed by Lewis [1] and can be subdivided as follows.

14.1.1.1 Solid-Solid Separations

Particle size may be exploited to separate powders or larger units using sieves or other screen designs, examples of which are given in Chapter 1.

Air classification can be achieved using differences in aerodynamic properties to clean or fractionate particulate materials in the dry state. Controlled air streams will cause some particles to be fluidised in an air stream depending on the terminal velocity, which in turn is related primarily to size, but also to shape and density. Also in the dry state, particles can be separated on the basis of photometric (colour), magnetic or electrostatic properties.

By suspending particles in a liquid, particles may be separated by settlement on the basis of a combination of size and density differences. Buoyancy differences can be exploited to separate products from heavy materials such as stones or rotten fruit in flotation washing, while surface properties can be used to separate peas from weed seeds in froth flotation.

14.1.1.2 Separation From a Solid Matrix

Plant materials often contain valuable components within their structure. In the case of oils or juices, these may be separated from the bulk structure by expression, which involves the application of pressure. Alternatively, components may be removed from solids by extraction (see Section 14.4), which utilises the differential solubilities of extracted components in a second medium. Water may be used to extract sugar, coffee, fruit and vegetable juices, etc. Organic solvents are necessary in some cases, e.g. hexane for oil extraction. Supercritical CO_2 may be used to extract volatile materials such as in the decaffeination of coffee. A combination of expression and extraction is used to remove 99% of the oil from oilseeds.

Water removal from solids plays an important role in food processing (see Chapter 3).

14.1.2 Separations From Liquids

Liquid foods include aqueous or oil based materials, and frequently contain solids either in true solution or dispersed as colloids or emulsions.

14.1.2.1 Liquid-Solid Separations

Discrete solids may be removed from liquids using a number of principles. Conventional filtration (see Section 14.2) is the removal of suspended particles on the basis of particle size using a porous membrane or septum, composed of wire mesh, ceramics or textiles. A variety of pore sizes and geometric shapes are available and the driving force can be gravity, upstream pressure (pumping), downstream pressure (vacuum) or centrifugal force. Using smaller pore sizes, microfiltration, ultrafiltration and related membrane processes can be used to fractionate solids in true solution (see Section 14.7).

Density and particle size determine the rate of settlement of dispersed solids in a liquid, according to Stokes Law. Settlement due to gravity is very slow, but is widely used in water and effluent treatment. Centrifugation subjects the dispersed particles to forces greatly exceeding gravity which dramatically increases the rate of separation and is widely used for clarifying liquid food products. A range of geometries for batch and continuous processing are available (see Section 14.3).

14.1.2.2 Immiscible Liquids

Centrifugation is again used to separate immiscible liquids of different densities. The major applications are cream separation and the dewatering of oils during refining.

14.1.2.3 General Liquid Separations

Differences in solubility can be exploited by contacting a liquid with a solvent which preferentially extracts the component(s) of interest from a mixture. For example, organic solvents could be used to extract oil soluble components, such as flavour compounds, from an aqueous medium.

An alternative approach is to induce a phase change within the liquid, such that components are separated on the basis of their freezing or boiling points. Crystallisation is the conversion of a liquid into a solid plus liquid state by cooling or evaporation (see Section 14.6). The desired fraction, solid or liquid, can then be collected by filtration or centrifugation. Alternatively, evaporation (see Chapter 3) is used to remove solvent or other volatile materials by vapourisation. In heat-sensitive foods, this is usually carried out at reduced operating pressures and hence reduced temperature, frequently in the range 40–90 °C. Reverse osmosis (see Section 14.7) is an alternative to evaporation in which pressure rather than heat is the driving force.

Ion exchange and electrodialysis (see Sections 14.8 and 14.9) are used to separate dissolved components in liquids, depending on their electrostatic charge.

14.1.3
Separations From Gases and Vapours

These separations are not common in food processing. Removal of solids suspended in gases is required in spray drying and pneumatic conveying and is achieved by filter cloths, bag filters or cyclones. Another possibility is wet scrubbing to remove suspended solids on the basis of solubility in a solvent (see Chapter 3).

14.2
Solid-Liquid Filtration

James G. Brennan

14.2.1
General Principles

In this method of separation the insoluble solid component of a solid-liquid suspension is separated from the liquid component by causing the latter to flow through a porous membrane, known as the *filter medium*, which retains the solid particles within its structure and/or as a layer on its upstream face. If a layer of solid particles does form on the upstream face of the medium it is known as the *filter cake*. The clear liquid passing through the medium is known as the *filtrate*. The flow of the liquid through the medium and cake may be brought about by means of gravity alone (*gravity filtration*), by pumping it through under pressure (*pressure filtration*), by creating a partial vacuum downstream of the medium (*vacuum filtration*) or by centrifugal force (*centrifugal filtration*). Once the filtration stage is complete, it is common practice to wash the cake free of filtrate. This is done to recover valuable filtrate and/or to obtain a cake of adequate purity. When filtering oil, the cake may be blown free of filtrate by means of steam. After washing, the cake may be dried with heated air.

In the early stages of a filtration cycle, solid particles in the feed become enmeshed in the filter medium. As filtration proceeds, a layer of solids begins to build up on the upstream face of the medium. The thickness of this layer and so the resistance to the flow of filtrate increases with time. The pressure drop, $-\Delta p_c$, across the cake at any point in time may be expressed as:

$$-\Delta p_c = \frac{\alpha \eta w V}{A^2}\left(\frac{dV}{dt}\right) \tag{14.1}$$

Where η is the viscosity of the filtrate, w is the mass of solids deposited on the medium per unit volume of filtrate, V is the volume of filtrate delivered in time t, A is the filter area normal to the direction of flow of the filtrate and α is the *specific cake resistance*. a characterises the resistance to flow offered by the cake and physically represents the pressure drop necessary to give unit superficial velocity of filtrate of unit viscosity through a cake containing unit mass of solid per unit filter area.

If a cake is composed of rigid nondeformable solid particles, then a is independent of the pressure drop across the cake and is constant throughout the depth of the cake. Such a cake is known as *incompressible*. In the case of incompressible cakes, it is possible to calculate the value of a. In contrast, a *compressible* cake is made up of nonrigid deformable solid particles or agglomerates of particles. In such cakes, the value of a increases with increase in pressure and also varies throughout the depth of the cake, being highest near the filter medium. The relationship between a and $-\Delta p_c$ is often expressed as:

$$a = a_0(-\Delta p_c)^s \tag{14.2}$$

Where a_0 and s are empirical constants. s is known as the *compressibility coefficient* of the cake and is zero for an incompressible cake rising towards 1.0 as the compressibility increases. In the case of compressible cakes, values of a must be obtained by experiment.

The filter medium also offers resistance to the flow of the filtrate and a pressure drop, $-\Delta p_m$, develops across it. This pressure drop may be expressed as:

$$-\Delta p_m = \frac{R_m \eta}{A}\left(\frac{dV}{dt}\right)\cdots \tag{14.3}$$

Where R_m is known as the *filter medium resistance*. Values of R_m are determined experimentally. It is usual to assume that R_m is constant throughout any filtration cycle.

The total pressure drop across the cake and medium, $-\Delta p$, is obtained by adding the two pressure drops together, thus:

$$-\Delta p = -\Delta p_c - \Delta p_m = \frac{\eta}{A}\left(\frac{dV}{dt}\right)\left(\frac{awV}{A} + R_m\right) \tag{14.4}$$

or:

$$\frac{dV}{dt} = \frac{A(-\Delta p)}{\eta\left(\frac{awV}{A} + R_m\right)} \tag{14.5}$$

A filter cycle may be carried out by maintaining a constant total pressure drop across the cake and medium. This is known as *constant pressure filtration*. As the cake builds up during the cycle, the rate of flow of filtrate decreases. Alternatively, the rate of flow of filtrate may be maintained constant throughout the cycle, in which case the pressure increases as the cake builds up. This is known as *constant rate filtration*. A combination of constant rate and constant pressure filtration may also be employed by building up the pressure in the early stages of the cycle and maintaining it constant throughout the remainder of the cycle. Equation (14.5) may be applied to both constant pressure and constant rate filtration. In the case of a compressible cake, a relationship such as Eq. (14.2) needs to be used to account for the change in α with increase in pressure during constant rate filtration [2–10].

14.2.2 Filter Media

The main functions of the filter medium are to promote the formation of the filter cake and to support the cake once it is formed. Once the filter cake is formed, it becomes the primary filter medium. The medium must be strong enough to support the cake under the pressure and temperature conditions that prevail during the filtering cycle. It must be nontoxic and chemically inert with respect to the material being filtered. Filter media may be flexible or rigid. The most common type of flexible medium is a woven cloth, which may be made of cotton, wool, silk or synthetic material. The synthetic materials used include Nylon, polyester, polyacrylonitrile, polyvinylchloride, polyvinylidenechloride, polyethylene and polytetrafluoroethylene. Such woven materials are available with different mesh counts, mesh openings, thread sizes and weaves. Woven glass fibre and flexible metal meshes are also used as filter media. Nonwoven, flexible media are fabricated in the form of belts, sheets or pads of various shapes. These tend to be used for filtering liquids with relatively low solids content. Most of the solids remain enmeshed within the depth of the media rather than forming a cake on the surface. Rigid media may be fixed or loose. Fixed rigid media are made in the forms of disks, pads and cartridges. They consist of rigid particles set in permanent contact with one another. They include ceramic and diatomaceous materials and foamed plastics made from polyvinylchloride, polyethylene, polypropylene and other polymer materials. Perforated metal plates and rigid wire meshes are used for filtering relatively large particles. Loose rigid media consist of rigid particles that are merely in contact with each other but remain in bulk, loose form. They include sand, gravel, charcoal and diatomaceous material arranged in the form of beds. All types of media are available with different pore sizes to suit particular filtration duties. Practical trials are the most reliable methods for selecting media for particular tasks [7–11].

14.2.3 Filter Aids

Filter aids are employed to improve the filtration characteristics of highly compressible filter cakes or when small amounts of finely divided solids are being filtered. They consist of hard, strong, inert incompressible particles of irregular shape. They form a porous, permeable rigid lattice structure which allows liquid to pass through but retains solid particles. They are usually applied in small amounts in the range 0.01–4.00% of the weight of the suspension. They may be applied in one of two ways. The filter medium may be precoated with a layer of filter aid prior to introducing the suspension to be filtered. This precoat, which is usually 1.5–3.0 mm thick, prevents the suspension particles from becoming enmeshed in the filter medium and reducing the flow of liquid. It may also facilitate the removal of the cake when filtration and washing are complete. Alternatively, the filter aid may be added to the suspension before it is introduced into the filter unit. It increases the porosity of the cake and reduces its compressibility. Sometimes a combination of precoating and premixing is used. The materials most commonly used as filter aids include: diatomaceous material (which is made from the siliceous remains of tiny marine plants known as diatoms, known as diatomite and kieselguhr), expanded perlite (made from volcanic rock), charcoal, cellulose fibres and paper pulp. These materials are available in a range of grades. Experimental methods are used to select the correct grade, amount and method of application for a particular duty [7–10].

14.2.4 Filtration Equipment

Gravity filtration is not widely applied to food slurries, but is used in the treatment of water and waste disposal. These applications are covered in Chapter 13.

14.2.4.1 Pressure Filters

In pressure filters the feed is pumped through the cake and medium and the filtrate exits at atmospheric pressure. The following are examples of pressure filters used in processing of foods.

Plate-and-Frame Filter Press In this type of press grooved plates, covered on both sides with filter medium, alternate with hollow frames in a rack (see Fig. 14.1). The assembly of plates and frames is squeezed tightly together to form a liquid-tight unit. The feed is pumped into the hollow frames through openings in one corner of the frames (see Fig. 14.2). The cake builds up in the frames and the filtrate passes through the filter medium onto the grooved surface of the plates, from where it exits via an outlet channel in each plate. When filtering is complete, wash liquid may be pumped through the press following the same path as the filtrate. Some presses are equipped with special wash plates (see Fig. 14.2). Every second plate in the frame is a wash plate. During filtration, these act as filter plates. During washing, the outlets from the wash plates are closed and the wash liquid is pumped onto their surfaces via an inlet channel (see Fig. 14.2). The wash liquid then passes through the full thickness of the cake and two layers of filter medium before exiting from the filter plates. This is said to achieve more effective washing than that attainable without the wash plates. After washing, the press is opened, the cake is removed from the frames, the filter medium is cleaned and the press is reassembled ready for the

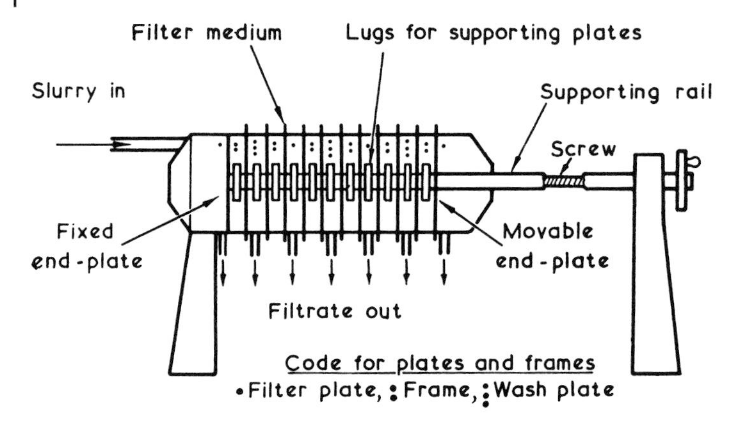

Fig. 14.1 Schematic drawing of assembled plate-and-frame filter press; from [2] with permission of the authors.

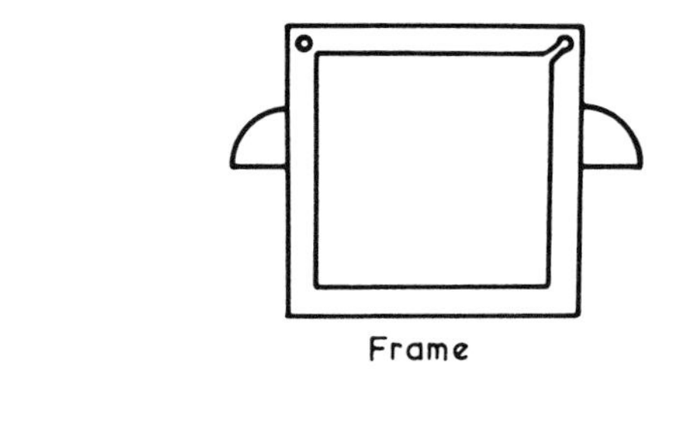

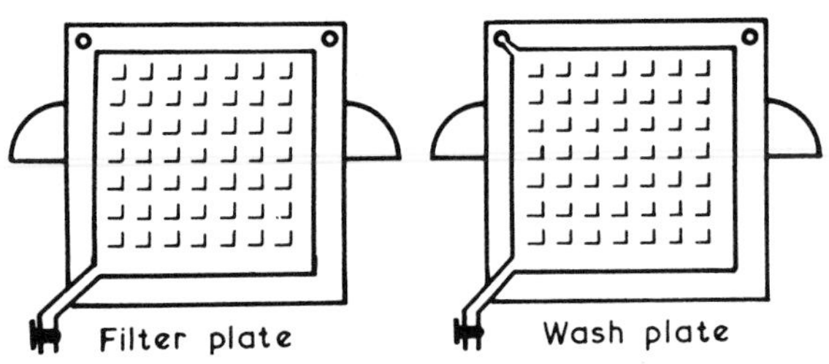

Fig. 14.2 Schematic drawings of plates and frames; from [2] with permission of the authors.

next run. This and other types of vertical plate filters are compact, flexible and have a relatively low capital cost. However, labour costs and filter cloth consumption can be high.

Horizontal Plate Filter In this type of filter, the medium is supported on top of horizontal drainage plates which are stacked inside a pressure vessel. The feed is pumped in through a central duct, entering above the filter medium. The filtrate passes down through the medium onto the drainage plates and exits from them through an annular outlet. The cake builds up on top of the filter medium. After filtration, the feed is replaced by wash liquid which is pumped through the filter. After washing, the assembly of plates is lifted out of the pressure vessel and the cake removed manually. This type of filter is compact. The units are readily cleaned and can be sterilised if required. Labour costs can be high. They are used mainly for removing small quantities of solids and are known as polishing filters.

Shell-and-Leaf Filters A filter leaf consists of a wire mesh screen or grooved plate over which the filter medium is stretched. Leaves may be rectangular or circular in shape. They are located inside a pressure vessel or shell. They are either supported from the bottom or centre or suspended from the top, inside the shell. The supporting member is usually hollow and acts as a takeaway for the filtrate. In horizontal shell-and-leaf filters, the leaves are mounted vertically inside horizontal pressure vessels (see Fig. 14.3).

As the feed slurry is pumped through the vessel, the cake builds up on the filter medium covering the leaves while the filtrate passes through the medium into the hollow leaf and then out through the leaf supports. Leaves may be stationary or they may rotate about a horizontal axis. When filtering is stopped, washing is carried out by pumping wash liquid through the cake and leaves. The cake may be removed by withdrawing the leaf assembly from the shell and cleaning the leaves manually. In some designs, the bottom half of the shell may be opened and the cake sluiced down with water jets. In vertical shell-and-leaf filters, rectangular leaves are mounted vertically inside a vertical pressure vessel. Shell-and-leaf filters are generally not as labour intensive as plate-and-frame presses but have higher capital costs. They are mainly used for relatively long filtration runs with slurries of low or moderate solids content.

Edge Filters In this type of filter a number of stacks of rings or discs, known as filter piles or packs, are fixed to a header plate inside a vertical pressure ves-

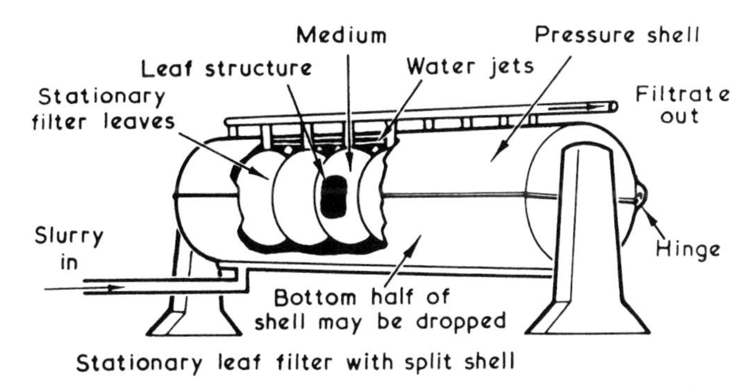

Fig. 14.3 Schematic drawing of a shell-and-leaf filter; adapted from [2] with permission of the authors.

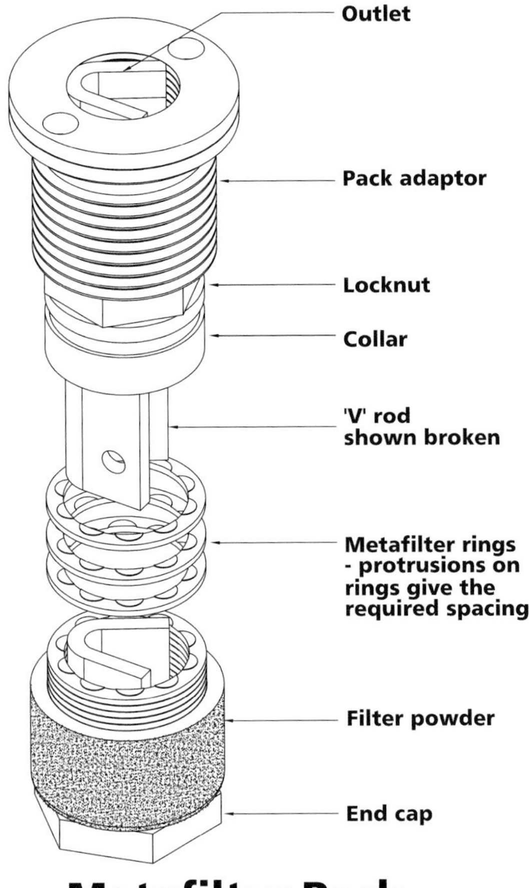

Fig. 14.4 Filter pile; by courtesy of Stella-Meta Ltd.

sel. Each pile consists of a number of discs mounted one above the other on a fluted vertical rod and held together between a boss and nut (see Fig. 14.4).

The clearance between the discs is in the range 25–250 μm. Before filtration commences, a precoat of filter aid is applied to the edges of the discs. When the feed slurry is pumped into the pressure vessel, the cake builds up on top of the precoat of filter aid, while the filtrate passes between the discs and exits via the grooves on the supporting rod. Additional filter aid may be mixed with the feed. When filtering and washing are complete, the cake is removed by back flushing with liquid through the filtrate outlet and removing the cake in the form of a sludge through an outlet in the bottom of the pressure vessel. The discs may be made of metal or plastic. Edge filters have a relatively low labour requirement and use no filter cloth. They are used mainly for removing small quantities of fine solids from liquids.

14.2.4.2 Vacuum Filters

In vacuum filters a partial vacuum is created downstream of the medium and atmospheric pressure is maintained upstream. Most vacuum filters are operated continuously, as it is relatively easy to arrange continuous cake discharge under atmospheric pressure.

Rotary Drum Vacuum Filters There are a number of different designs of this type of filter, one of which is depicted in Fig. 14.5.

A cylindrical drum rotates about a horizontal axis partially immersed in a tank of the feed slurry. The surface of the drum is divided into a number of shallow compartments by means of wooden or metal strips running the length of the drum. Filter medium is stretched over the drum surface, supported on perforated plates or wire mesh. A pipeline runs from each compartment to a rotary valve located centrally at one end of the drum. Consider one of the compartments on the surface of the drum (shown shaded in Fig. 14.5). As the drum rotates, this compartment becomes submerged in the slurry. A vacuum is applied to the compartment through the rotary valve. Filtrate is drawn through the medium and flows through the pipe to the rotary valve, from where it is directed to a filtrate receiver. The solids form a layer of cake on the outer surface of the medium. The cake increases in thickness as long as the compartment remains submerged in the slurry. As it emerges from the slurry, residual filtrate is sucked from the cake. Next the compartment passes beneath sprays of wash

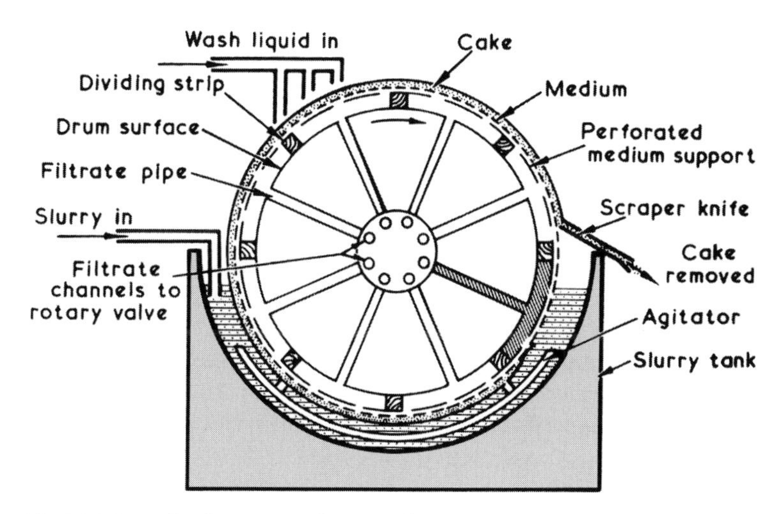

Fig. 14.5 Principle of operation of a rotary drum vacuum filter; from [2] with permission of the authors.

liquid. The washings are directed to a different receiver by means of the rotary valve. As the compartment passes from beneath the sprays, residual wash liquid is sucked from the cake. Next, by means of the rotary valve, the compartment is disconnected from the vacuum source and compressed air introduced beneath the medium for a short period of time. This loosens the cake from the surface of the medium and facilitates its removal by means of a scraper knife.

Many other designs of rotary drum filters are available featuring different methods of feeding the slurry onto the drum surface removing the cake from the medium.

Rotary drum vacuum filters incur relatively low labour costs and have large capacities for the space occupied. However, capital costs are high and they can only handle relatively free-draining solids. In common with all vacuum filters, they are not used to process hot and/or volatile liquids.

For removing small quantities of fine solids from a liquid, a relatively thick layer (up to 7.5 cm) of filter aid may be precoated onto the medium. A thin layer of this precoat is removed together with the cake by the scraper knife.

Rotary Vacuum Disc Filters In a disc filter, instead of a drum, a number of circular filter leaves, mounted on a horizontal shaft, rotate partially submerged in a tank of slurry. Each disc is divided into sections. Each section is covered with filter medium and is connected to a rotary valve, which controls the application of vacuum and compressed air to the section. Scraper knives remove the cake from each disc. Such disc filters have a larger filtering surface per unit floor area, compared to drum filters. However, cake removal can be difficult and damage to filter cloth excessive.

Other designs of continuous vacuum filters are available featuring moving belts, rotating tables and other supports for the filter medium. These are used mainly for waste treatment rather than in direct food applications [2, 5–7, 9, 10].

14.2.4.3 Centrifugal Filters (Filtering Centrifugals, Basket Centrifuges)

In this type of filter, the flow of filtrate through the cake and medium is induced by centrifugal force. The slurry is fed into a rotating cylindrical bowl with a perforated wall. The bowl wall is lined on the inside with a suitable filter medium. Under the action of centrifugal force, the solids are thrown to the bowl wall where they form a filter cake on the medium. The filtrate passes through the cake and medium and leaves the bowl through the perforations in the wall.

Batch Centrifugal Filters The principle of this type of filter is shown in Fig. 14.6. The cylindrical metal bowl in suspended from the end of a vertical shaft within a stationary casing. With the bowl rotating at moderate speed, slurry is fed into the bowl. A cake forms on the medium lining the inside of the perforated bowl wall and the filtrate passes through the perforations into the casing and out through a liquid outlet. The speed of the bowl is increased to recover most of the filtrate. Wash liquid may be sprayed onto the cake and spun off at high speed. The bowl

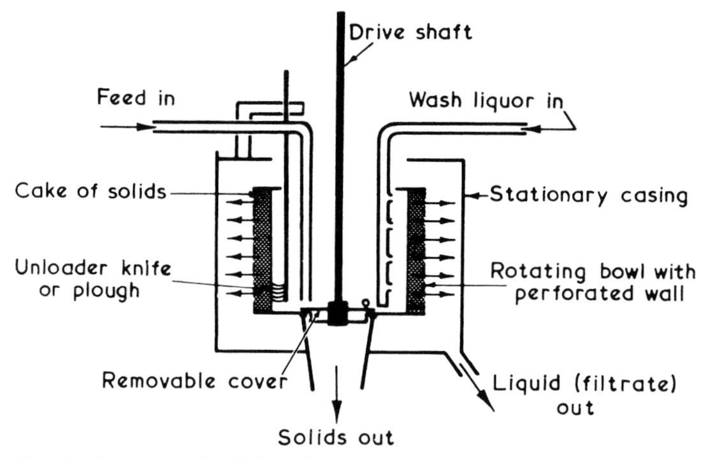

Fig. 14.6 Batch centrifugal filter; from [2] with permission of the authors.

is then slowed down, the cake cut out with an unloader knife or plough and removed through an opening in the bottom of the bowl. Cycle times vary over 3–30 min.

Fully automated versions of these batch filters operate at a constant speed, about horizontal axes, throughout a shorter cycle of 0.5–1.5 min. The feed and wash liquid are introduced automatically and the cake is cut out by a hydraulically operated knife.

Continuous Centrifugal Filters The principle of one type of continuous centrifugal filter is shown in Fig. 14.7.

A conical perforated bowl (basket) rotates about a vertical axis inside a stationary casing. The incline of the bowl causes the separation force to be split be-

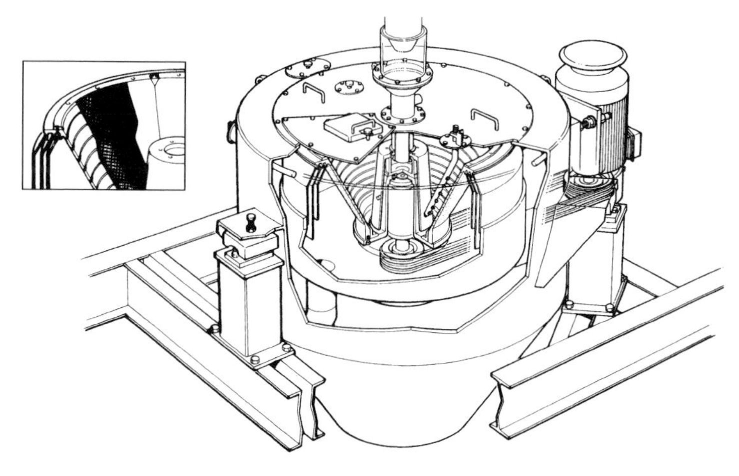

Fig. 14.7 A continuous centrifugal filter; by courtesy of Broadbent Customer Services Ltd.

tween vertical and horizontal elements resulting in the product moving upwards. The vertical force pushes the product up over the basket lip into the casing from where it is discharged. The horizontal element ensures that purging of the liquid phase takes place. This type of centrifuge is used for separating sugar crystals from syrup. Sliding of the product upwards and its discharge form the lip of the bowl, usually at high speed, is a relatively violent process and may damage the product, i.e. fracture of crystals. Washing the solid phase while it is moving may limit its effectiveness.

Other types of continuous centrifugal filters feature reciprocating pushing devices, screw conveyors or vibrating mechanisms to facilitate removal of the cake [2, 5, 6, 11].

14.2.5 Applications of Filtration in Food Processing

14.2.5.1 Edible Oil Refining

Filtration is applied at a number of stages in refining of edible oils. After extraction or expression, crude oil may be filtered to remove insoluble impurities such as fragments of seeds, nuts, cell tissue, etc. For large-scale applications rotary filters are used. Plate-and-frame filters are used for smaller operations. Bleaching earths used in decolourising oils are filtered off using rotary or plate filters. The catalysts used in hydrogenating fats and oils are recovered by filtration. Since hydrogenated fats have relatively high melting points, heated plate filters may be used. During winterisation and fractionation of fats, after cooling, the higher melting point fractions are filtered off using plate-and-frame or belt filters [12–15].

14.2.5.2 Sugar Refining

The juice produced by extraction from sugar cane or sugar beet contains insoluble impurities. The juice is treated with lime to form a flocculent precipitate which settles to the bottom of the vessel. The supernatant liquid is filtered to produce a clear juice for further processing. Plate-and-frame presses, shell-and-leaf and rotary drum vacuum filters are used. The settled 'mud' is also filtered to recover more juice. Plate-and-frame presses or rotary drum vacuum filters are used for this duty. Filtration is also used at a later stage in the refining process to further clarify sugar juice. In the production of granulated sugar, purified sugar juice is concentrated up to 50–60% solids content by vacuum evaporation and seeded with finely ground sugar crystals to initiate crystallisation. When the crystals have grown to the appropriate size, they are separated from the juice in batch or continuous centrifugal filters (see Section 3.1.4.2) [16, 17].

14.2.5.3 **Beer Production**

During maturation of beer, a deposit of yeast and trub forms on the bottom of the maturation tank. Beer may be recovered from this by filtration using plate-and-frame presses, shell-and-leaf or rotary drum vacuum filters. The beer is clarified by treatment with isinglass finings, centrifugation or filtration. If filtration is used, the beer is first chilled and then filtered through plate-and-frame, horizontal plate or edge filters. In the case of plate filters, the filter medium consists of sheets of cellulose, aluminium oxide or zirconium oxide fibres, with added kieselguhr. Insoluble polyvinyl pyrrolidone may also be incorporated into the medium to absorb phenolic materials associated with beer haze. Edge filters are precoated with filter aid and more filter aid is usually added to the beer prior to filtration. Yeasts and bacteria may also be removed from beer by filtration. Although the pore sizes in the media are much larger than the microorganisms, the fibres hold the negatively charged microorganisms electrostatically. The pressure drop across these filters needs to be limited to avoid the microorganisms being forced off the media fibres. When a sterile product is desired, the sealed filter must be presterilised before use [18, 19].

14.2.5.4 **Wine Making**

Wine is filtered at different stages of production: after racking, after decolouring and finally just before bottling. Plate-and-frame presses, shell-and-leaf filters, edge filters and precoated rotary drum vacuum filters have been used. Filter media are mainly sheets made of cellulose incorporating filter aid material (mainly diatomaceous earth) which is bound into the cellulose sheets with bitumen. With edge and precoated drum filters, loose filter aid material is used. Sterile wine may be produced by filtration in presterilised equipment [20–22].

There are many other applications for filtration in the food industry, including the filtration of starch and gluten suspensions and the clarification of brines, sugar syrups, fruit juices, yeast and meat extracts.

14.3 Centrifugation

James G. Brennan

14.3.1 General Principles

Centrifugation involves the application of centrifugal force to bring about the separation of materials. It may be applied to the separation of immiscible liquids and the separation of insoluble solids from liquids.

14.3.1.1 Separation of Immiscible Liquids

If two immiscible liquids, A and B, with different densities, are introduced into a cylindrical bowl rotating about a vertical axis, under the influence of centrifugal force, the more dense liquid A moves towards the wall of the bowl where it forms an annular ring (see Fig. 14.8). The less dense liquid B is displaced towards the centre of the bowl where it forms an inner annular ring.

If the feed is introduced continuously into the bottom of the bowl through a vertical feed pipe, the liquids may be removed separately from each layer by a weir system, as shown in Fig. 14.9. The more dense liquid A flows out over a circular weir of radius R_A and the less dense liquid B over a weir of radius R_B.

The interface between the two layers is known as the neutral zone. The position of this interface can influence the performance of the centrifuge. In the outer zone (A), light liquid is effectively stripped from a mass of dense liquid while, in the inner zone (B), dense liquid is more effectively stripped from a mass of light liquid. Thus, if the centrifuge is being used to strip a mass of dense liquid free of light liquid so that the dense phase leaves in as pure a state

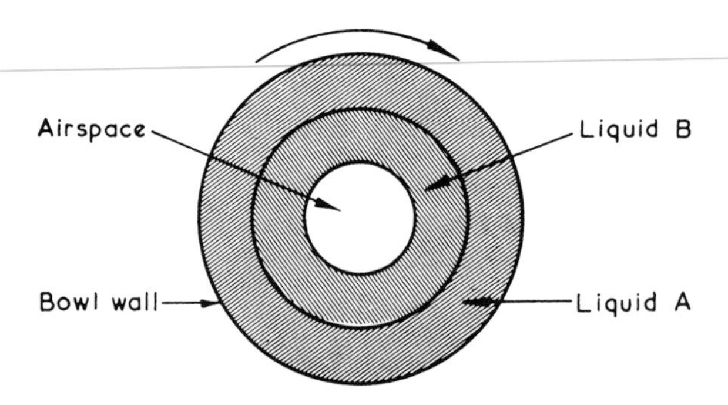

Fig. 14.8 Separation of immiscible liquids in a cylindrical bowl (plan view); from [2] with permission of the authors.

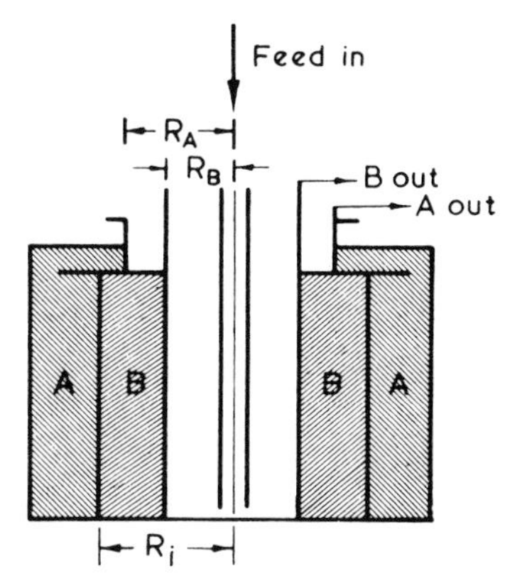

Fig. 14.9 Separation of immiscible liquids in a cylindrical bowl with submerged weir (sectional view); from [2] with permission of the authors.

as possible, then the dwell time in zone A should be greater than that in zone B. For such a duty, the interface is best moved towards the centre of rotation so that the volume of zone A exceeds that of zone B. In this situation, the light component is exposed to a relatively small centrifugal force for a short time, while the heavy component is exposed to a large force for a longer time. An example of such a duty is the separation of cream from milk, where the objective is to produce skim milk with as little fat in it as possible. In contrast, if the duty is to strip a mass of light liquid free of dense liquid, i.e. to produce a pure light phase, the interface is best moved out towards the bowl wall so that the volume occupied by zone B exceeds that of zone A. An example of such a duty would be the removal of small amounts of water from an oil. The actual change in position of the interface is quite small, in the order of 25–50 μm. However, it does affect the performance of the separator. The position of the interface can be changed by altering the radii of the liquid outlets. For example, if the radius of the light liquid outlet R_B is fixed, decreasing the radius of the dense liquid outlet R_A will move the interface towards the centre of rotation. In practice, the radius of either liquid outlet and hence the position of the interface, is determined by fitting a ring with an appropriate internal diameter to the outlet. Such rings are known as *ring dams* or *gravity discs*. It has also been established that the best separation is achieved by introducing the feed to the bowl at a point near the interface. The density difference between the liquids needs to be 3% or more for successful separation. Other factors which influence the performance of liquid-liquid centrifugal separators are bowl speed and the rate of flow of the liquids through them. In general, the higher the speed of the bowl, the better the separation. However, at very high speeds the viscosity of the oil phase may impede its flow through the centrifuge. The higher the rate of flow of the liquids through the bowl, the shorter the dwell time in the action zone; and so the less effective the separation is likely to be. For each duty a compromise needs to be struck between throughput and efficiency of separation.

14.3.1.2 Separation of Insoluble Solids from Liquids

If a liquid containing insoluble solid particles is fed into the bottom of a cylindrical bowl rotating about a vertical axis, under the influence of centrifugal force, the solid particles move towards to the bowl wall. If a particular solid particle reaches the bowl wall before being swept out by the liquid leaving through a central outlet in the top of the bowl (see Fig. 14.10), it remains in the bowl and thus is separated from the liquid. If it does not reach the bowl wall, it is carried out by the liquid. The fraction of the solid particles remaining in the bowl and the fraction passing out in the liquid depend on the rate of feed, i.e. the dwell time in the bowl.

The following expression relates the throughput of liquid through a cylindrical bowl centrifuge to the characteristics of the feed and the dimensions and speed of the bowl:

$$q = 2\left[\frac{g(\rho_s - \rho_l)D_p^2}{18\eta}\right]\left[\frac{\omega^2 V}{2g\ln\left(\frac{R_2}{[(R_1^2+R_2^2)/2]^{1/2}}\right)}\right] \tag{14.6}$$

where q is the volumetric flow rate of liquid through the bowl, g is acceleration due to gravity, ρ_s is the density of the solid, ρ_l is the density of the liquid, D_p is the minimum diameter of a particle that will be removed from the liquid, η is the viscosity of the liquid, ω is the angular velocity of the bowl, V is the volume of liquid held in the bowl at any time, R_1 is the radius of the liquid outlet and R_2 is the inner radius of the bowl. Note the quantities contained within the first set of square brackets relate to the feed material, while those within the second set refer to the centrifuge. This expression can be used to calculate the throughput of a specified feed material through a cylindrical bowl centrifuge of known dimensions and speed. It can also be used for scaling-up calculations. Alternative expressions for different types of bowl (see Section 14.3.2) can be found in the literature [2, 5, 6, 23].

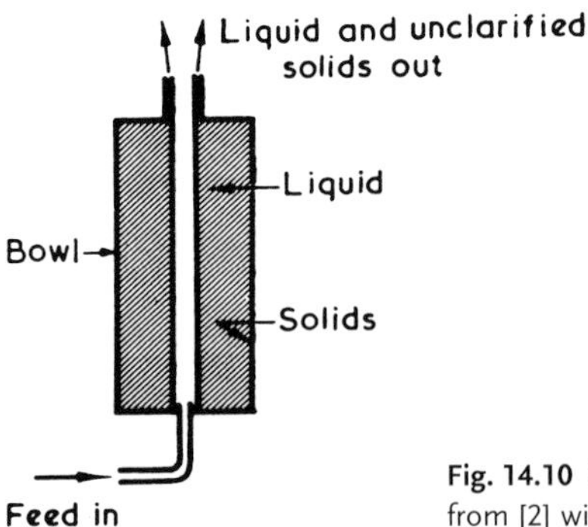

Fig. 14.10 Principle of simple cylindrical centrifugal clarifier; from [2] with permission of the authors.

14.3.2
Centrifugal Equipment

14.3.2.1 Liquid-Liquid Centrifugal Separators

Tubular Bowl Centrifuge This type of centrifuge consists of a tall, narrow bowl rotating about a vertical axis inside a stationary casing. Bowl diameters range from 10 cm to 15 cm with length:diameter ratios of 4–8. The feed enters into the bottom of the bowl through a stationary pipe and is accelerated to bowl speed by vanes or baffles. The light and dense phases leave via a weir system at the top of the bowl and flow into stationary discharge covers. Depending on the duty it has to perform, a gravity disc of appropriate size is fitted to the dense phase outlet, as explained in Section 14.3.1.1. Bowl speeds range from 15 000 rpm (large) to 50 000 rpm (small).

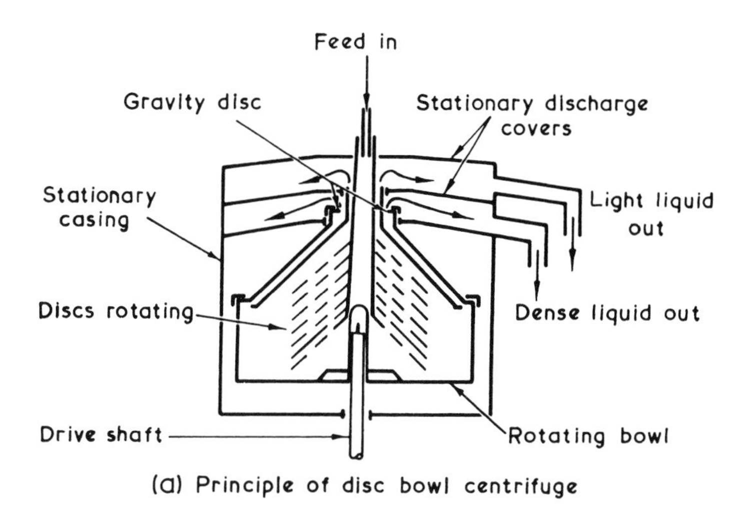

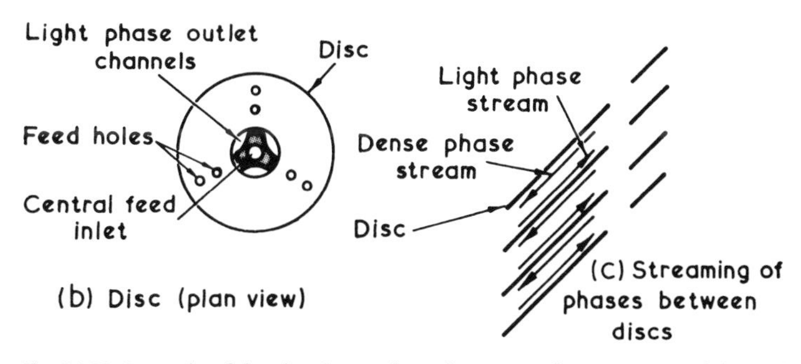

Fig. 14.11 Principle of disc bowl centrifuge; from [2] with permission of the authors.

Disc Bowl Centrifuge In this type of centrifuge a relatively shallow, wide bowl rotates within a stationary casing. The bowl usually has a cylindrical body, with a diameter in the range 20–100 cm and a conical top (see Fig. 14.11). The bowl contains a stack of truncated metal cones, known as discs, which rotate with the bowl. The clearance between the discs is of the order of 50–130 μm. The discs have one or more sets of matching holes which form vertical channels in the stack. The feed is introduced to the bottom of the bowl, flows through these channels and enters the spaces between the discs. Under the influence of centrifugal force, the dense phase travels in a thin layer down the underside of the discs towards the bowl wall while the light phase, displaced towards the centre of rotation, flows over the top of the discs. Thus the space between each pair of discs is a minicentrifuge. The distance any drop of one liquid must travel to get into the appropriate stream is small compared to that in a tubular bowl machine or indeed an empty bowl of any design. In addition, in a disc machine there is considerable shearing at the interface between the two countercurrent streams of liquid which contributes to the breakdown of emulsions. The phases leave the bowl through a weir system fitted with an appropriate gravity disc and flow into stationary discharge covers.

14.3.2.2 Solid-Liquid Centrifugal Separators

Both tubular bowl and disc bowl centrifuges can be used for solid-liquid separation, within certain limitations. For this type of duty, the dense phase outlet is closed off and the clear liquid exits the bowl through the central, light phase outlet. The solid particles which are separated from the liquid remain in the bowl, building up as a deposit on the wall of the bowl. Consequently, the centrifuges are operated on a batch principle and have to be stopped and cleaned at intervals. The tubular bowl machines have a relatively small solids capacity and are only suitable for handling feeds with low solids content, less than 0.5%. However, because of the high speeds they operate at, they are particularly suited to removing very fine solids. Disc bowl machines have up to five times the capacity for solids, compared to tubular bowl centrifuges. However, to avoid frequent cleaning, they are also used mainly with feeds containing relatively low solids content, less than 1.0%.

Solid Bowl Centrifuge (Clarifier) For separating solid particles which settle relatively easily, a bowl similar in shape to that shown in Fig. 14.7 may be used. However, the bowl wall is not perforated and no filter medium is used. The solids build up on the inside of the bowl wall and the clear liquid spills out over the top rim of the bowl into the outer casing. At intervals, the feed is stopped and the solids removed by means of a knife or plough and discharged through an opening in the bottom of the bowl. This type of clarifier can handle feeds with up to 2.0% solids content.

Nozzle-discharge Centrifuge In this type of centrifuge there is provision for the continuous discharge of solids, in the form of a sludge, as well as the clear liquid. There are many different designs available. One design consists of a disc-bowl machine with two to twenty four nozzles spaced around the bowl. The size of the nozzles is in the range 0.75 to 2.00 mm, depending on the size of the solid particles in the feed. From 5 to 50% of the feed is continuously discharged in the form of a slurry through these nozzles. The slurry may contain up to 25% v/v solids. By recycling some of the slurry the solids content may be increased. Up to 75% of the slurry may be recycled, depending on its flowability, and the solids content increased up to 40%.

Self-opening Centrifuge In this type of centrifuge the ports discharging the slurry open at intervals and the solids are discharged under a pressure of up to 3500 kN m^{-2}. The opening of the ports may be controlled by timers. Self-triggering ports are also available. The build-up of solids in the bowl is monitored and a signal is generated which triggers the opening of the ports. An example of one such centrifuge, used in the brewing industry, is shown in Fig. 14.12. The slurry discharged from these self-opening centrifuges usually has a higher solids content compared to that continuously discharged through open nozzles.

Decanting Centrifuge Nozzle and valve discharge centrifuges can only handle feeds containing a few percent or less of solids. For feeds containing a higher percent of solids, decanting or conveyor bowl centrifuges may be used. The principle of operation of one such centrifuge is shown in Fig. 14.13. A solid

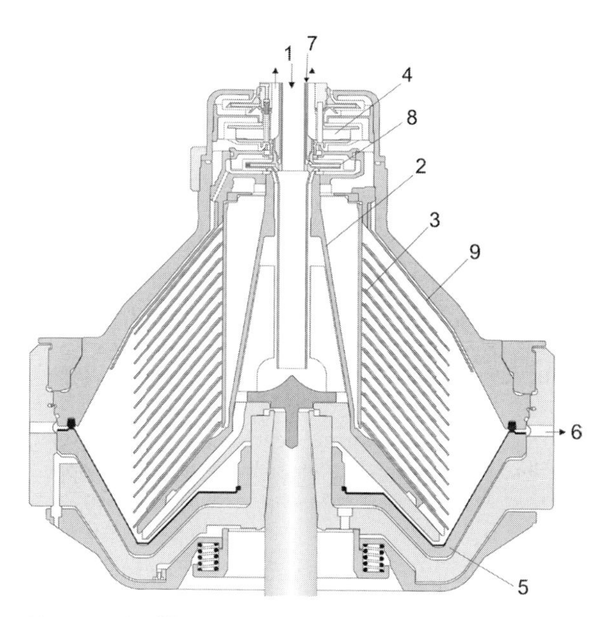

Fig. 14.12 A self-triggering solids-ejecting clarifier centrifuge; by courtesy of Alfa Laval Ltd.

bowl containing a screw conveyor rotates about a horizontal axis. The bowl and conveyor rotate in the same direction but at different speeds. The feed enters the bowl through the conveyor axis. The solids are thrown to the bowl wall and are conveyed to one end of the bowl, up a conical section, from where they are discharged. The clear liquid is discharged through an adjustable weir at the other end of the bowl. Such machines can handle feeds containing up to 90% (v/v) of relatively large solid particles. Particles 2 μm or less in diameter are normally not removed from the liquid. Where necessary, the liquid discharged from decanting centrifuges may be further clarified in tubular or disc bowl centrifuges [2, 5, 6, 10].

14.3.3 Applications for Centrifugation in Food Processing

14.3.3.1 Milk Products

Centrifugation is used in the separation of milk to produce cream and/or skim milk. Disc bowl centrifuges are generally used for this duty. They may be hermetically sealed and fitted with centripetal pumps. Milk is usually heated to between 40 °C and 50 °C prior to separation, to reduce its viscosity and optimise the density difference between the fat and aqueous phases. The fat content of the skim milk may be reduced to less than 0.05%. Although the process is continuous, insoluble solids present in the milk (dirt particles, casein micelles, microorganisms) build up as sludge in the centrifuge bowl. The bowl has to be cleaned out at intervals. Alternatively, nozzle or self-opening centrifuges may be used, but with outlets for the cream and skim milk as well as the sludge.

Fat may be recovered from whey and buttermilk by centrifugation [24].

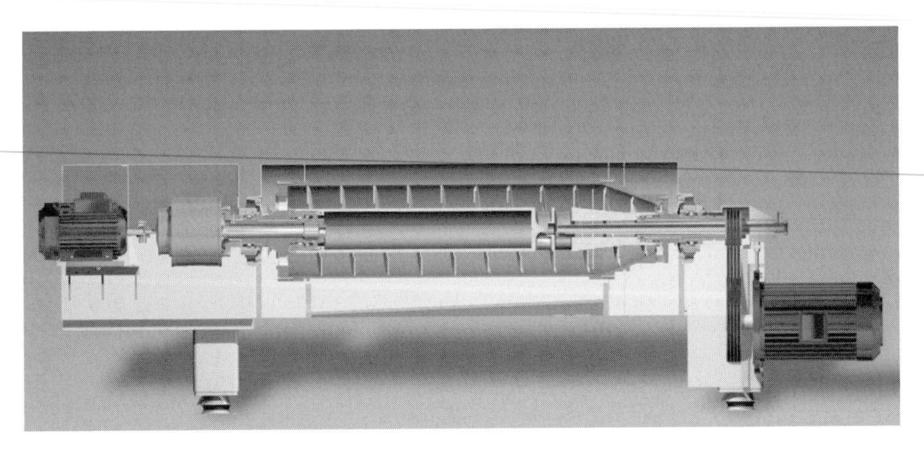

Fig. 14.13 Cut away view – Alfa Leval Decanter, by courtesy of Alfa Laval Ltd.

14.3.3.2 **Edible Oil Refining**

In the early stages of oil refining, the crude oil is treated with water, dilute acid or alkali to remove phosphatides and mucilaginous material. This process is known as degumming. Nozzle or self-opening centrifuges are used to remove the gums after these treatments. In the case of acid-degumming, the degummed oil may be washed with hot water and the washings removed by centrifugation. The next step in oil refining is neutralisation. The free fatty acids, phosphatides and some of the pigments are treated with caustic soda to form soapstock which is then separated from the oil by centrifugation, using nozzle or self-opening centrifuges. The oil is then washed with hot water and the washings removed by centrifugation [13].

14.3.3.3 **Beer Production**

Centrifugation may be used as an alternative to filtration at various stages in the production of beer. Nozzle discharge centrifuges may be used for clarifying rough beer from fermenting vessels and racking tanks. Self-opening centrifuges may be used for wort and beer clarification. Centrifuges used for the treatment of beer may be hermetically sealed to prevent the loss of carbon dioxide and the take-up of oxygen by the beer. Self-opening centrifuges may also be used for the recovery of beer from fermenters and tank bottoms. Decanting centrifuges may be used for clarifying worts and beers containing relatively high contents of yeast or trub. They may also be used as an alternative to self-opening machines to recover beer from fermenters and tank bottoms [18].

14.3.3.4 **Wine Making**

Centrifugation may be used instead of or in combination with filtration at various stages in the production of wine. Nozzle or self-opening centrifuges are generally used. Applications include: the clarification of must after pressing, provided that the solids content is relatively low, the clarification of wine during fermentation to stabilise it by gradual elimination of yeast, the clarification of new wines after fermentation and before filtration, the clarification of new red wines before filling into barrels and the facilitation of tartrate precipitation for the removal of tartrate crystals [20, 21, 22].

14.3.3.5 **Fruit Juice Processing**

Centrifugation may be used for a variety of tasks in fruit juice processing. Self-opening centrifuges are used to remove pulp and control the level of pulp remaining in pineapple and citrus juices. Centrifuged apple juice is cloudy but free from visible pulp particles. Tubular bowl centrifuges were originally used to clarify apple juice but more recently nozzle and self-opening machine are used. The use of hermetically sealed centrifuges prevents excessive aeration of the juice. In the production of oils from citrus fruits centrifugation is applied in two stages. The product from the extractor contains an emulsion of 0.5–3.0% oil. This is concentrated up to 50–70% oil in a nozzle or self-opening centrifuge. The concentrated emulsion is then separated in a second centrifuge to produce the citrus oil [25, 26].

There are many other applications for centrifugation in food processing, e.g. tubular bowl machines for clarifying cider and sugar syrups and separating animal blood into plasma and haemoglobin, nozzle and self-opening machines for dewatering starches and decanting centrifuges for recovering animal and vegetable protein, separating fat from comminuted meat and separating coffee and tea slurries.

14.4 Solid-Liquid Extraction (Leaching)

James G. Brennan

14.4.1 General Principles

This is a separation operation in which the desired component, the *solute*, in a solid phase is separated by contacting the solid with a liquid, the *solvent*, in which the desired component is soluble. The desired component leaches from the solid into the solvent. Thus the compositions of both the solid and liquid phases change. The solid and liquid phases are subsequently separated and the desired component recovered from the liquid phase.

Solid-liquid extraction is carried out in single or multiple stages. A stage is an item of equipment in which the solid and liquid phases are brought into contact, maintained in contact for a period of time and then physically separated from each other. During the period of contact, mass transfer of components between the phases takes place and they approach a state of equilibrium. In an *equilibrium* or *theoretical* stage, complete thermodynamic equilibrium is attained between the phases before they are separated. In such a stage, the compositional changes in both phases are the maximum which are theoretically possible under the operating conditions. In practice, complete equilibrium is not reached and the compositional changes in a *real* stage is less than that attainable in an equilibrium stage. The *efficiency* of a real stage may be defined as the ratio of the compositional change attained in the real stage to that which would have been reached in an equilibrium stage under the same operating conditions. When estimating the number of stages required to carry out a particular task in a multistage system, the number of equilibrium stages is first estimated and the number of real stages calculated by dividing the number of equilibrium stages by the stage efficiency. Graphical and numerical methods are used to estimate the number of equilibrium stages required for a particular duty [2, 4–6].

After the period of contact, the solid-liquid mixture is separated into two streams: a 'clear' liquid stream or *overflow* consisting of a solution of the solute in the solvent and a 'residue' stream or *underflow* consisting of the insoluble solid component with some solution adhering to it. In an equilibrium stage, the composition of the overflow is the same as that of the solution leaving with the insoluble solid in the underflow. In a real stage, the concentration of solute in the overflow is less than that in the solution leaving with the insoluble solid in the underflow.

The extraction of the solute from a particle of solid takes place in three stages. The solute dissolves in the solvent. The solute in solution then diffuses to the surface of the particle. Finally, the solute transfers from the surface of the particle into the bulk of the solution. One or more of these steps can limit the rate of extraction. If the correct choice of solvent has been made, the solution of the solute in solvent is rapid and is unlikely to influence the overall rate of extraction. The rate of movement of the solute to the surface of the solid particle depends on the size, shape and internal structure of the particle and is difficult to quantify. The rate of transfer of the solute from the surface of the solid particle to the bulk of the solution may be represented by the expression:

$$\frac{dw}{dt} = KA(C_s - C) \tag{14.7}$$

Here, $\frac{dw}{dt}$ is the rate of mass transfer of the solute, A is the area of the solid-liquid interface, C_s and C are the concentration of the solute at the surface of the solid particle and in the bulk of the solution, respectively, and K is the mass transfer coefficient.

In a single stage extraction unit where V is the total volume of the solution and is constant, then:

$$\frac{dw}{dt} = VdC$$

and so:

$$\frac{dC}{dt} = \frac{KA(C_s - C)}{V} \tag{14.8}$$

The main factors which influence the rate of extraction include:

1. The solid-liquid interface area. The rate of mass transfer from the surface of the particle to the bulk of the solution increases with increase in this area. Reducing the size of the solid particles increases this area and so increases the rate of mass transfer. In addition, the smaller the particle the shorter the distance the solute has to travel to reach the surface. This is likely to further speed up the extraction. However, very small particles may impede the flow of solvent through the bed of solid in an extractor and some particles may

not come in contact with the solvent. In the case of cellular material, such as sugar beet (see Section 14.4.3.2), the cell wall acts as a semipermeable membrane releasing sugar but retaining larger nonsugar molecules. Therefore, the beet is sliced rather than comminuted to increase the surface area for extraction but to limit cell wall damage.

2. Concentration gradient. To ensure as complete extraction as possible, a gradient must be maintained between the concentration of solute at the surface of the solid particles and that in the bulk of the solution. In a single stage extractor, as the phases approach equilibrium, this gradient decreases and so does the rate of extraction until it ceases. When this occurs, the solid may still contain a significant amount of solute and the solution may be relatively dilute, depending on the equilibrium conditions. This solution could be drained off and replaced with fresh solvent resulting in further extraction of the solute. This could be repeated until the solute content of the solid reached a suitably low level. However, this would result in the production of a large volume of relatively dilute solution. The cost of recovering the solute from this solution increases as its solute content decreases. For example, the lower the concentration of sugar in the solution obtained after extraction of sugar beet, the more water has to be evaporated off before crystallisation occurs. Multistage countercurrent extraction systems enable a concentration gradient to be maintained even when the concentration of solute in the solid is low (see Fig. 14.14b). This results in more complete extraction as compared with that attainable in single-stage or multistage concurrent systems (see Fig. 14.14a).
3. Mass transfer coefficient. An increase in temperature increases the rate of solution of the solute in the solvent and also the rate of diffusion of solute through the solution. This is reflected in a higher value of K in Eq. (14.7) and Eq. (14.8). Thus, the solvent is usually heated prior to and/or during extraction. The upper limit in temperature depends on the nature of the solids. For example, in the extraction of sugar from beet, too high a temperature can result in peptisation of the beet cells and the release of nonsugar compounds into the solution (see Section 14.4.3.2). In the case of the extraction of solubles from ground roasted coffee beans, too high a temperature can result in the dried coffee powder having an undesirable flavour (see Section 14.4.3.3).

Increasing the velocity and turbulence of the liquid as it flows over the solid particles can result in an increase in the value of K in Eq. (14.7) and Eq. (14.8) and hence an increase in the rate of extraction. In some industries, when fine particles are being extracted, they are mechanically stirred. However, in most food applications, this is not the case as agitation of the solid can result in undesirable breakdown of the particles. In most food applications, the solvent is made to flow through a static bed of solids under the influence of gravity or with the aid of a pump. Alternatively, the solids are conveyed slowly, usually countercurrent to the flow of solvent [2, 4–6, 27].

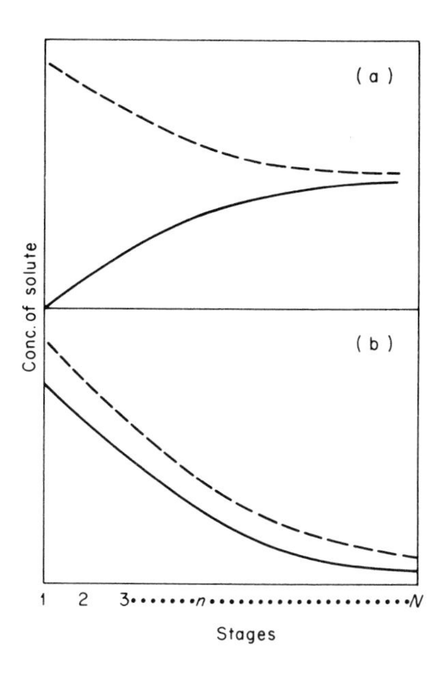

Fig. 14.14 Graphical representation of concurrent and countercurrent extraction systems. (a) Concurrent system, (b) countercurrent system. Solid lines indicates solution, dashed lines indicate solid matter; from [2] with permission of the authors.

14.4.2 Extraction Equipment

14.4.2.1 Single-Stage Extractors

A simple extraction cell consists of a tank fitted with a false bottom which supports a bed of the solids to be extracted. The tank may be open or closed. If extraction is to be carried under pressure, as in the case with extraction of ground roasted coffee (see Section 14.4.3.3), or if volatile solvents are used, as in the case of edible oil extraction (see Section 14.4.3.1), the tank is enclosed (see Fig. 14.15). The solvent is sprayed over the top surface of the bed of solids, percolates down through the bed and exits via an outlet beneath the false bottom. The tank may be jacketed and/or a heater incorporated into the solvent feed line to maintain the temperature of the solution at the optimum level. Usually a pump is provided for recirculating the solution. The spent solid is removed manually or dumped through an opening in the bottom of the tank. In large cells, additional supports may be provided for the bed of solids to prevent consolidation at the bottom of the cell.

Single extraction cells are used for laboratory trials and for small-scale industrial applications. As discussed in Section 14.4.1, the bulked solution from such units is relatively dilute. If a volatile solvent is used, the overflow from the cell may be heated to vapourise the solvent, which is then condensed and recycled through the cell. In this way, a more concentrated solution of the solute may be obtained.

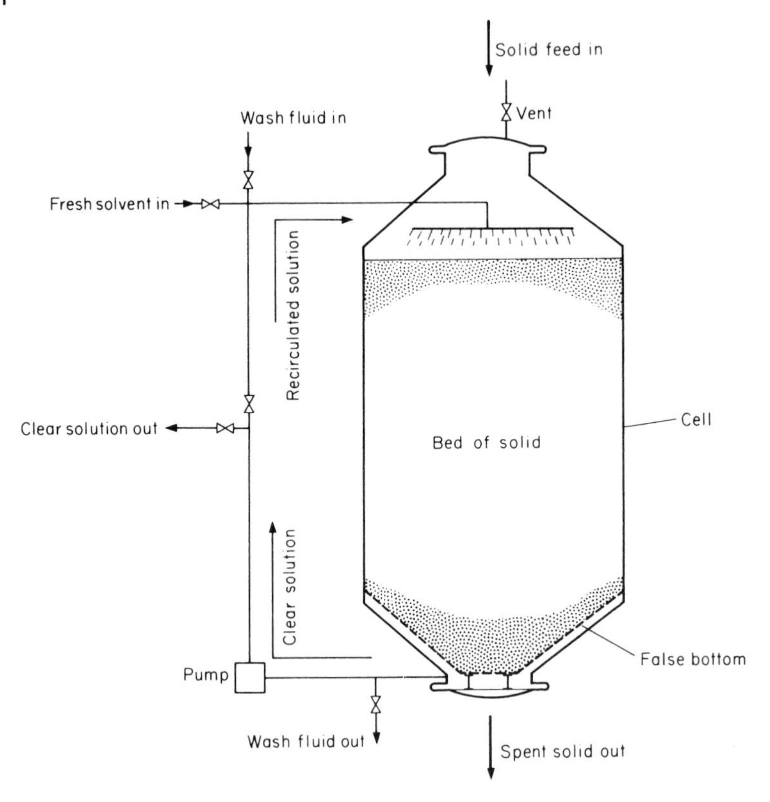

Fig. 14.15 Single-stage, enclosed extraction cell; from [2] with permission of the authors.

14.4.2.2 Multistage Static Bed Extractors

One method of applying multistage countercurrent extraction is to use a number of single cells arranged in a circuit. Each cell contains a charge of solids. The solution from the preceding cell is sprayed over the surface of the bed of solids and percolates down through the bed, becoming more concentrated as it does so. The solution leaving from the bottom of the cell is introduced into the top of the next cell. A typical battery, as used for the extraction of sugar beet, contains 14 cells, as shown in Fig. 14.16. At the time depicted in this figure, three of the cells are excluded from the circuit. Cells 10, 11 and 12 are being filled, washed and emptied, respectively. The fresh water enters cell 13 and the concentrated sugar solution, or overflow, leaves from cell 9. When the beet in cell 9 is fully extracted, this cell is taken out of the circuit and cell 10 brought in to take its place. Fresh water then enters cell 14 and the concentrated sugar solution leaves from cell 10. By isolating cells in turn around the circuit, the principle of countercurrent extraction may be achieved without physically moving the beet from one cell to the next. The number of cells in such a circuit may vary from three to 14.

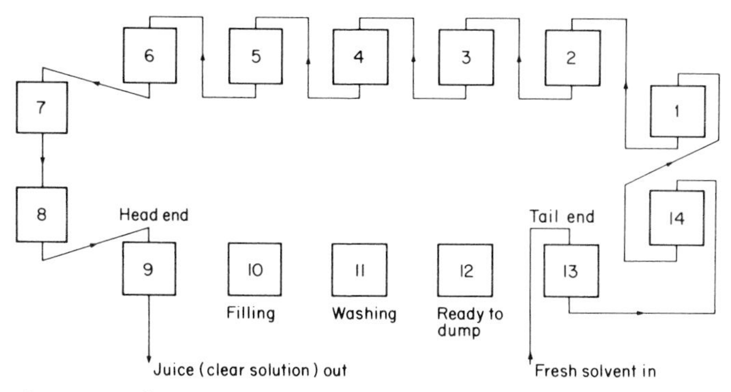

Fig. 14.16 Mulistage, countercurrent extraction battery showing flow of solution at one particular point in time; from [2] with permission of the authors.

14.4.2.3 **Multistage Moving Bed Extractors**

There are many different designs of moving bed extractors available. They usually involve moving the solid gently from one stage to the next, countercurrent to the flow of the solution. One type of continuous extractor consists of a trough set at a small angle to the horizontal containing two screw conveyors with intermeshing flights. The solvent is introduced at the elevated end of the trough. The solid is fed in at the other end and is carried up the slope by conveyors countercurrent to the flow of the solution. The trough is enclosed and capable of withstanding high pressure. Extractors of this type are used for sugar beet and ground roasted coffee. Another type, known as the *Bonotto* extractor is shown in Fig. 14.17. It consists of a vertical tower divided into sections by horizontal plates. Each plate has an opening through which the solid can pass downwards from plate to plate; and each plate is fitted with a wiper blade which moves the solid to the opening. The holes are positioned 180° from each other in successive plates. The solid is fed onto the top plate. The wiper blade moves it to the opening and it falls onto the plate below and so on down the tower from plate to plate. Fresh solvent is introduced at the bottom of the tower and is pumped upwards countercurrent to the solid. The rich solution leaves at the top of the tower and the spent solid is discharged from the bottom. This type of extractor is used for oil extraction from nuts and seeds.

Many other designs of continuous moving bed extractors are in use in industry. One design features moving perforated baskets which carry the solid through a stream of the solvent. In another design, the solid is conveyed by screw conveyors, with perforated blades, through vertical towers, countercurrent to the flow of solvent [2, 4–6, 27].

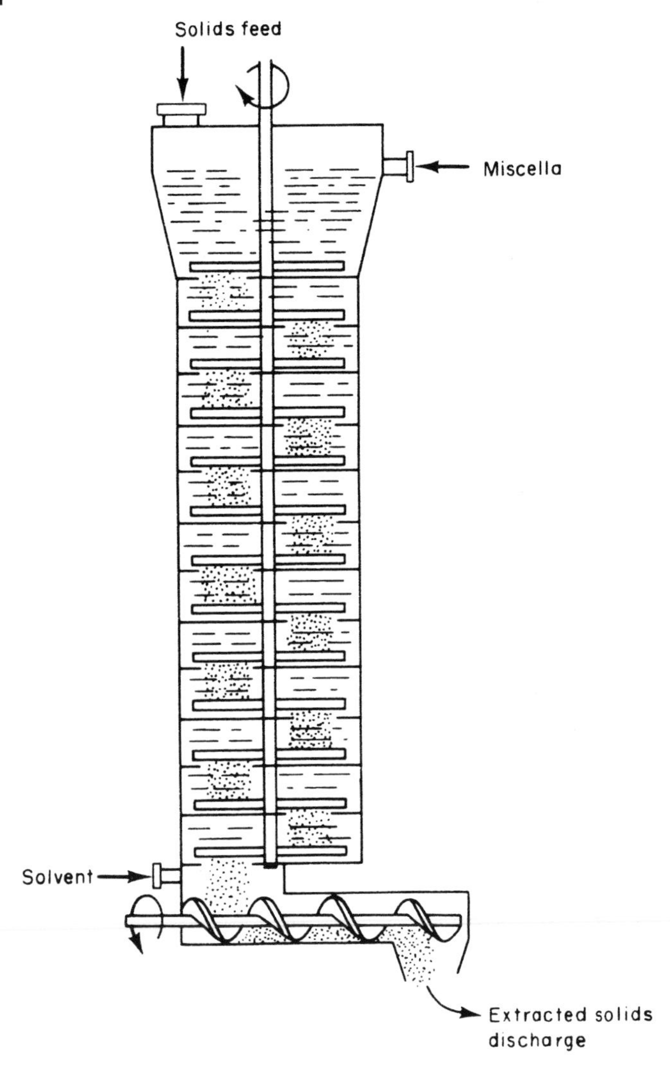

Fig. 14.17 Bonotto extractor.

14.4.3
Applications for Solid-Liquid Extraction in Food Processing

14.4.3.1 Edible Oil Extraction

Solvent extraction may be used as an alternative to or in combination with expression to obtain oil from nuts, seeds and beans. The most commonly used solvent is hexane. This is a clear hydrocarbon, derived from petroleum that boils at 68.9 °C. It is miscible with oil, immiscible with water and does not impart any objectionable odour or taste to the oil or spent solid. Hexane is highly flammable and so the plant must be vapour-tight and care must be taken to avoid the generation of sparks that might ignite the solvent. Other solvents have been investigated, including heptane and cyclohexane. Nonflammable solvents, such as trichloroethylene and carbon disulphide, have been studied but are toxic and difficult to handle. Various alcohols and supercritical carbon dioxide (see Section 14.4.5) have also been investigated. Various designs of moving bed extractors are used in large scale-oil extraction, including those described in Section 14.4.3. After extraction, the solution of oil in solvent is filtered, the solvent removed by vacuum evaporation followed by distillation and the solvent reused. Residual solvent may be removed from the spent solid by direct or indirect heating with steam and the resulting meal used for animal feed. Batch extractors featuring solvent recovery and reuse can be used for small-scale operations. Cottonseed, linseed, rapeseed, sesame, sunflower, peanuts, soybean and corn germ may be solvent-extracted [2, 28].

14.4.3.2 Extraction of Sugar from Sugar Beet

Sugar is extracted from sugar beet using heated water as solvent. The beets are washed and cut into slices, known as cossettes. This increases the surface area for extraction and limits cell wall damage (see Section 14.4.1). Water temperature ranges from 55 °C in the early stages of extraction to 85 °C towards the end. Higher temperatures can cause peptisation of the beet cells and release nonsugar compounds into the extract. Multistage static bed batteries, as depicted in Fig. 14.17, are widely used. So also are various designs of moving bed extractors, including those described in Section 14.4.2.3. The solution leaving the extractor contains about 15% of dissolved solids. This is clarified by settling and filtration, concentrated by vacuum evaporation, seeded and cooled to crystallise the sugar. The crystals are separated from the syrup by centrifugation, washed and air dried (see Section 3.1.5.2) [2, 29].

14.4.3.3 Manufacture of Instant Coffee

A blend of coffee beans is roasted to the required degree, ground to the appropriate particle size range and extracted with heated water. Extraction may be carried out in a multistage, countercurrent static bed system consisting of 5–8 cells. Each cell consists of a tall cylindrical pressure vessel as temperatures above 100 °C are used. Heat exchangers are located between the cells. Water at about 100 °C is introduced into the cell containing the beans that are almost fully extracted and then passes through the other cells, until the rich solution exits from the cell containing the freshly ground beans. The temperature of the solution increases up to a maximum of 180 °C as it passes through the battery of cells. In the later stages of extraction, some hydrolysis of insoluble carbohydrate material occurs, resulting in an increase in the yield of soluble solids. Higher temperatures may impart an undesirable flavour to the product due to excessive hydrolysis. Continuous, countercurrent extractors featuring screw conveyors within pressurised chambers (see Section 14.4.2.3) may be used instead of the static bed system. The rich solution leaving the extractor usually contains 15–28% solids. This may be fed directly to a spray drier. Alternatively, the solution may be concentrated up to 60% solids by vacuum evaporation (see Section 3.1.5.2). The volatiles may be stripped from the extract before or during evaporation and added back to the concentrated extract prior to dehydration either by spray drying or freeze drying [2, 30, 31].

14.4.3.4 Manufacture of Instant Tea

Dried, blended tea leaves may be extracted with heated water in a static bed system consisting of 3–5 cells. Water temperature ranges from 70 °C in the early stages of extraction to 90 °C in the later stages. The cells may be evacuated after filling with the dry leaves and the pressure brought back to atmospheric level by introducing gaseous carbon dioxide. This facilitates the flow of the water through the cells. Continuous tower or other moving-bed extractors are also used to extract tea leaves. The rich solution coming from the extractor usually contains 2.5–5.0% solids. This is concentrated by vacuum evaporation to 25–50% solids. The volatile aroma compounds are stripped from the extract prior to or during evaporation and added back before dehydration by spray drying, vacuum drying or freeze drying.

14.4.3.5 Fruit and Vegetable Juice Extraction

In recent years there has been considerable interest in using extraction instead of expression for recovering juices from fruits and vegetables. Countercurrent screw extractors, some operated intermittently, have been used to extract juice with water. In some cases this results in higher yields of good quality compared to that obtained by expression [32].

14.4.4 The Use of Supercritical Carbon Dioxide as a Solvent

The critical pressure and temperature for carbon dioxide are 73.8 kPa and 31.06 °C, respectively. At pressures and temperatures above these values carbon dioxide exists in the form of a supercritical fluid (supercritical carbon dioxide;

$SC\text{-}CO_2$). In this state it has the characteristics of both a gas and a liquid. It has the density of a liquid and can be used as a liquid solvent, but it diffuses easily like a gas. It is highly volatile, has a low viscosity, a high diffusivity, is nontoxic and nonflammable. These properties make it a very useful solvent for extraction. However, the fact that $SC\text{-}CO_2$ has to be used at high pressure means that relatively expensive pressure resistant equipment is required and running costs are also high. The solvent power of $SC\text{-}CO_2$ increases with increase in temperature and pressure. For the extraction of highly soluble compounds or for deodorisation, pressures and temperatures close to the critical values may be used. When a single component is to be extracted from an insoluble matrix, so called *simple extraction*, the highest pressure and temperature possible for each application should be used. The upper limit on temperature will depend on the heat-sensitivity of the material. The limit on pressure will be determined by the cost of the operation. When all soluble matter is to be extracted, so called *total extraction*, high pressures and temperatures are again necessary.

The following are examples of the industrial application of $SC\text{-}CO_2$ extraction.

Hop Extract A good quality hop extract, for use in brewing, may be obtained by extraction with $SC\text{-}CO_2$. A multistage, countercurrent, static bed system, consisting of four extraction cells, is normally used. The $SC\text{-}CO_2$ percolates down through the hop pellets in each cell in turn. The solution of the extract in the $SC\text{-}CO_2$ leaving the battery is heated and the carbon dioxide evaporates, precipitating out the extract. The carbon dioxide is recompressed and cooled and condenses back to $SC\text{-}CO_2$ which is chilled to 7 °C and recycled through the extraction battery.

Decaffeination of Coffee Beans $SC\text{-}CO_2$ may be used as an alternative to water or methylene chloride for the extraction of caffeine from coffee beans. The beans are moistened before being loaded into the extractor. $SC\text{-}CO_2$ is circulated through the bed of beans extracting the caffeine. The caffeine-laden $SC\text{-}CO_2$ passes to a scrubbing vessel where the caffeine is washed out with water. Alternatively, the caffeine may be removed by passing the caffeine-laden $SC\text{-}CO_2$ through a bed of activated charcoal.

Removal of Cholesterol from Dairy Fats $SC\text{-}CO_2$ at 40 °C and 175 kPa has been used to remove cholesterol from butter oil in a packed column extractor. The addition of methanol as an entrainer increases the solubility of cholesterol in the fluid phase. The methanol is introduced with the oil into the column.

Many other potential applications for $SC\text{-}CO_2$ extraction have been investigated including: extraction of oils from nuts and seeds, extraction of essential oils from roots, flowers, herbs and leaves, extraction of flavour compounds from spices and concentration of flavour compounds in citrus oils [2, 28, 33–37].

14.5 Distillation

James G. Brennan

14.5.1 General Principles

Distillation is a method of separation which depends on there being a difference in composition between a liquid mixture and the vapour formed from it. This difference in composition develops if the different components of the mixture have different vapour pressures or *volatilities*. In *batch distillation*, a given volume of liquid is heated and the vapours formed are separated and condensed to form a product. In batch distillation, the compositions of the liquid remaining in the still and the vapour collected change with time. Batch distillation is still used in some whisky distilleries. However, continuous distillation columns are used in most industrial applications of distillation.

Consider a liquid mixture consisting of two components with different volatilities. If such a mixture is heated under constant pressure conditions, it does not boil at a single temperature. The more volatile component starts to vaporise first. The temperature at which this commences is known as the *bubble point*. If a vapour consisting of two components with different volatilities is cooled, the less volatile component starts to condense first. The temperature at which this commences is known as the *dew point*. A diagram of the temperature composition for liquid- vapour equilibrium of a two-component mixture is presented in Fig. 14.18.

The bottom 'L' line in the phase envelope represents liquid at its bubble point and the top 'V' line represents vapour at its dew point. The 0–1.0 composition axis refers to the more volatile component 'a', x_a and y_a are the mole fractions of 'a' in the liquid and vapour phases, respectively, and x_b and y_b are the mole fractions of component 'b', the less volatile component, in the liquid and vapour phases, respectively. The region below the bubble point curve represents subcooled liquid. The region above the dew point curve represents superheated vapour. Within the envelope, between the L and V lines, two phases exist. Saturated liquid and saturated vapour exist in equilibrium with each other.

If a liquid mixture at temperature θ_1 is heated until it reaches temperature θ_2, its bubble point, it will start to vaporise. The vapour produced at this temperature will contain mole fraction y_2 of component 'a'. Note that the vapour is richer in 'a' than the liquid. As a result of the evaporation the liquid becomes less rich in component 'a' and richer in 'b' so its temperature rises further. At temperature θ_3 the liquid phase contains mole fraction x_3 of 'a' and vapour phase y_3 of 'a'. Note that, at this temperature, the vapour is less rich in 'a' than it was at its bubble point. When temperature θ_4 is reached, all of the liquid is evaporated and the composition of the vapour is the same as the original liquid, $y_4 = x_1$. A similar sequence of events occurs if we start with a superheated va-

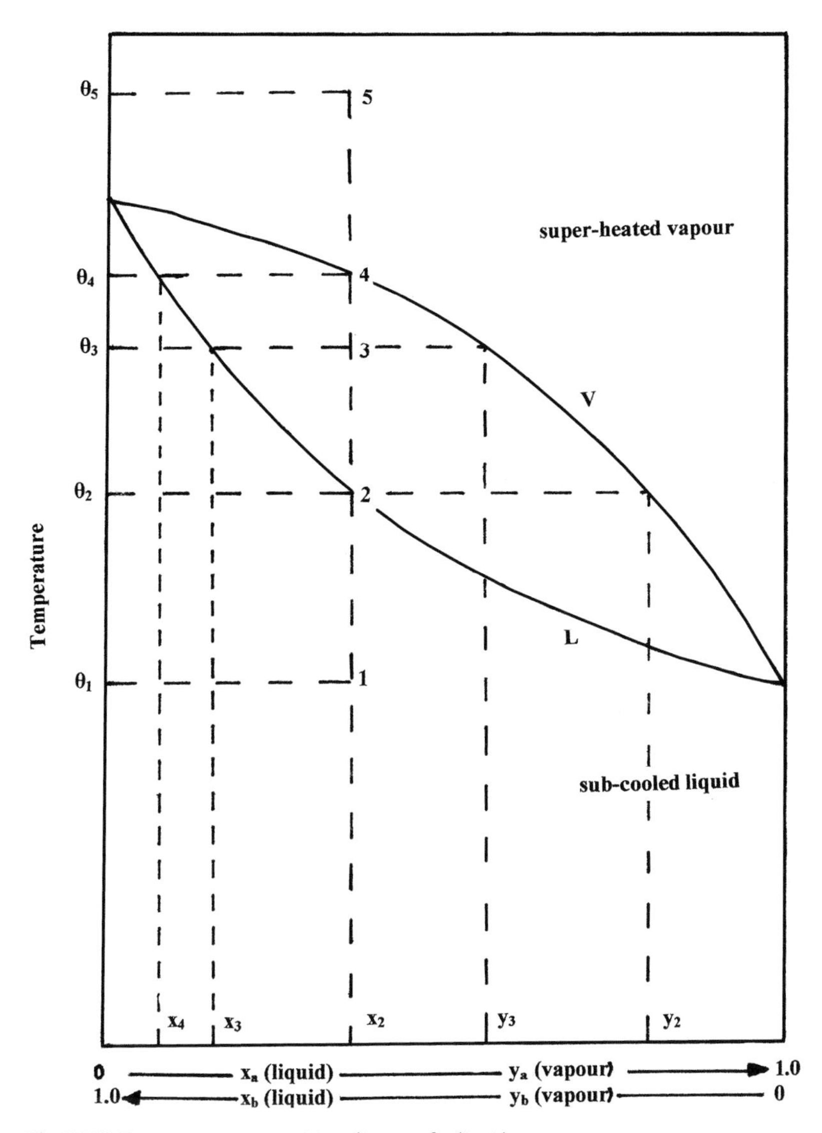

Fig. 14.18 Temperature-composition diagram for liquid-vapour equilibrium for a two-component mixture.

pour at temperature θ_5 and cool it. When it reaches its dew point θ_4, it begins to condense and the liquid contains mole fraction x_4 of 'a'. If cooled further to θ_3, the liquid and vapour contain mole fractions x_3 and y_3 of 'a', respectively. Thus both partial vaporisation and partial condensation bring about an increase of the more volatile component in the vapour phase. A distillation column consists of a series of stages, or *plates*, on which partial vaporisation and condensation takes place simultaneously.

The principle of a continuous distillation column, also known as a *fractionation* or *fractionating column*, is shown in Fig. 14.19.

The column contains a number of plates which are perforated to allow vapour rising from below to pass through them. Each plate is equipped with a weir over which the liquid flows and then through a *downtake* onto the plate below. The liquid contained in the *reboiler* at the bottom of the column is heated. When the liquid reaches its bubble point temperature, vapour is formed and this vapour bubbles through the liquid on the bottom plate. The vapour from the reboiler has a composition richer in the more volatile components than the liquid remaining in the reboiler. This vapour is at a higher temperature than the liquid on the bottom plate. Some of that vapour condenses and causes some of the liquid on the bottom plate to evaporate. This new vapour is richer in the more volatile components than the liquid on the bottom plate. This vapour in turn bubbles through the liquid on the plate above the bottom plate, causing some of it to evaporate and so on up the column. Thus, partial condensation and partial vaporisation takes place on each plate. The vapour rising up the column becomes increasingly rich in the more volatile components while the liquid flowing down from plate to plate becomes richer in the less volatile components. If all the vapour leaving the top of the column is condensed and removed, then the liquid in the column becomes progressively less rich in the volatile components as does the vapour being removed at the top of the column. This is the equivalent of batch distillation. However, if some of the condensed vapour is returned to the column and allowed to flow down from plate to plate, the concentration of the volatile components in the column is maintained at a higher level. The condensed vapour returned to the column is known as *reflux*. If feed material is introduced continuously into the column, then a product rich in volatile components can be withdrawn continuously from the top of the column and one rich in the less-volatile components from the reboiler at the bottom of the column. The feed is usually introduced onto a plate partway up the column.

Each plate represents a stage in the separation process. As is the case in solid-liquid extraction (see Section 14.4.1), the terms *equilibrium* or *theoretical* plate and *plate efficiency* may be applied to distillation. Graphical and numerical methods are used to estimate the number of equilibrium plates required to perform a particular duty and the number of real plates is calculated by dividing the number of equilibrium plates by the plate efficiency.

Steam distillation is applicable to mixtures which have relatively high boiling points and which are immiscible with water. If steam is bubbled through the liquid in the still, some of it will condense and heat the liquid to boiling point. Two liquid layers will form in the still. The vapour will consist of steam and the volatile vapour, each exerting its own vapour pressure. The mixture will boil when the sum of these pressures equals atmospheric pressure. Thus, the distillation temperature will always be lower than 100 °C at atmospheric pressure. By reducing the operating pressure, the distillation temperature will be reduced further and less steam will be used. Steam distillation may be used to separate

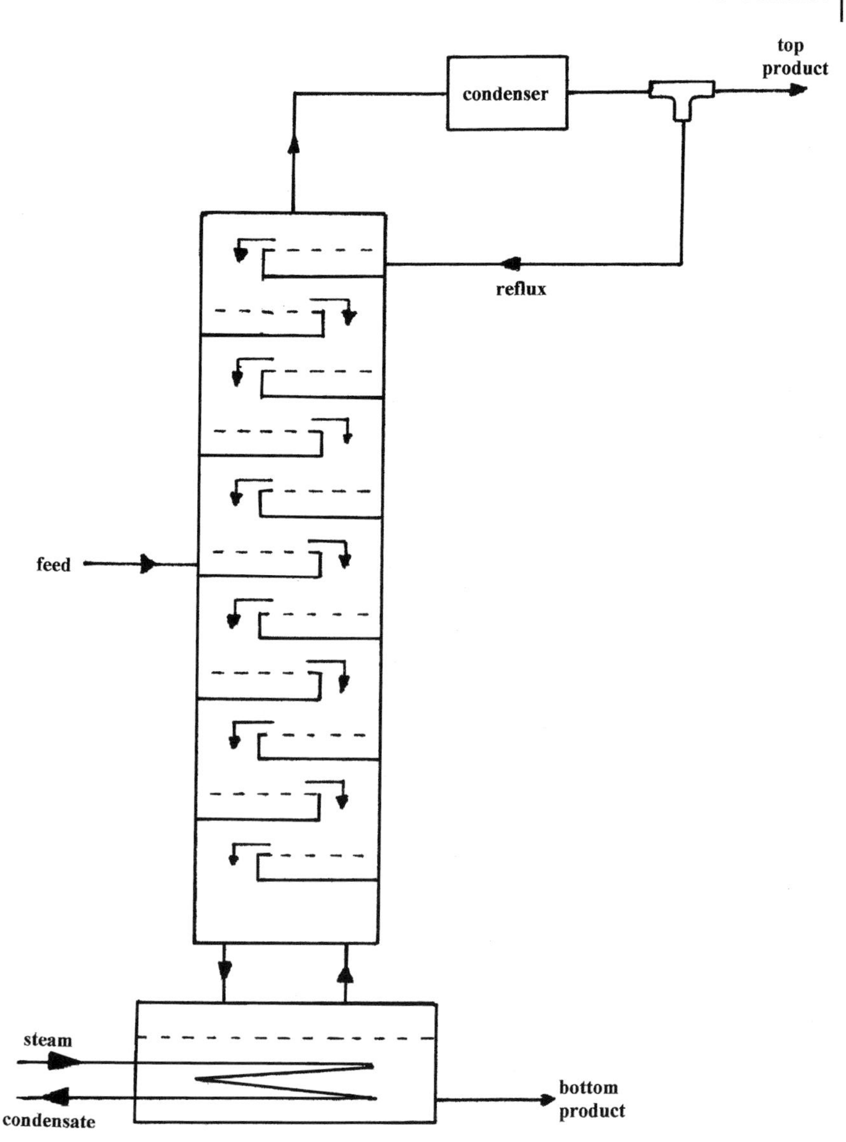

Fig. 14.19 Principle of a continuous distillation (fractionating) column.

temperature-sensitive, high boiling point materials from volatile impurities, or for removing volatile impurities with high boiling points from even less volatile compounds. The separation of essential oils from nonvolatile compounds dissolved or dispersed in water is an example of its application [5, 6, 38, 39].

14.5.2
Distillation Equipment

14.5.2.1 Pot Stills

Pot stills, used in the manufacture of good quality whiskey, are usually made from arsenic-free copper. Apparently the copper has an influence on the flavour of the product. The still consists of a pot which holds the liquor to be distilled and which is fitted with a swan neck. Because of their shape, pot stills are sometimes known as onion stills. A lyne arm, a continuation of the swan neck, tapers towards the condenser. The condenser is usually a shell and tube heat exchanger also made of copper. Heat is applied to the pot by means of steam passing through coils or a jacket. There is very limited use of direct heating by means of a solid fuel furnace beneath the pot. Such stills operate on a batch principle and two or three may operate in series (see Section 14.5.3.1).

14.5.2.2 Continuous Distillation (Fractionating) Columns

A distillation column consists of a tall cylindrical shell fitted with a number of plates or trays. The shell may be made of stainless steel, monel metal or titanium. As described in Section 14.5.1, the vapour passes upwards through the plates while the refluxed liquid flows across each plate, over a weir and onto the plate below via a downtake. There are many different designs of plate including the following examples.

Sieve plates consist of perforated plates with apertures of the order of 5 mm diameter, spaced at about 10 mm centres. The vapour moving up through the perforations prevents the liquid from draining through the holes. Each plate is fitted with a weir and downtake for the liquid.

Bubble cap plates are also perforated but each hole is fitted with a riser or 'chimney' through which the vapour from the plate below passes. Each riser is covered by a bell-shaped cap, which is fastened to the riser by means of a spider or other suitable mounting. There is sufficient space between the top of the riser and the cap to permit the passage of the vapour. The skirt of the cap may be slotted or the edge of the cap may be serrated. The vapour rises through the chimney, is diverted downwards by the cap and discharged as small bubbles through the slots or from the serrated edge of the cap beneath the liquid. The liquid level is maintained at some 5–6 cm above the top of the slots in the cap by means of the weir. The bubbles of vapour pass through the layer of liquid, heat and mass transfer occur and vapour, now richer in the more volatile components, leaves the surface of the liquid and passes to the plate above.

Valve plates are also perforated but the perforations are covered by liftable caps or valves. The caps are lifted as the vapour flows upwards through the perforations, but they fall and seal the holes when the vapour flow rate decreases. Liquid is prevented from falling down through the perforations when the vapour flow rate drops. The caps direct the vapour horizontally into the liquid thus promoting good mixing.

Packed columns may be used instead of plate columns. The cylindrical column is packed with an inert material. The liquid flows down the column in the form of a thin film over the surface of the packing material providing a large area of contact with the vapour rising up the column. The packing may consist of hollow cylindrical rings or half rings, which may be fitted with internal cross pieces or baffles. These rings may be made of metal, various plastics or ceramic materials. The packing is supported on perforated plates or grids. Alternatively, the column may be packed with metal mesh. The liquid flows through the packing in a zigzag pattern providing a large area of contact with the vapour [5, 38–40].

The spinning cone column consists of a vertical cylinder with a rotating shaft at its centre. A set of inverted cones are fixed to the shaft and rotate with it. Alternately between the rotating cones is a set of stationary cones fixed to the internal wall of the cylinder. The feed material, which may be in the form of a liquid, puree or slurry, is introduced into the top of the column. It flows by gravity down the upper surface of a fixed cone and drops onto the next rotating cone. Under the influence of centrifugal force, it is spun into a thin film which moves outwards to the rim of the spinning cone and onto the next stationary cone below. The liquid, puree or slurry moves from cone to cone to the bottom of the column. The stripping gas, usually nitrogen or steam, is introduced into the bottom of the column and flows upwards, countercurrent to the feed material. The thin film provides a large area of contact and the volatiles are stripped from the feed material by the gas or steam. Fins on the underside of the rotating cones create a high degree of turbulence in the rising gas or steam which improves mass transfer. They also provide a pumping action which reduces the pressure drop across the column. The gas or steam flows out of the top of the column and passes through a condensing system, where the volatile aroma compounds are condensed and collected in a concentrated form. This equipment is used to recover aroma compounds from fruits, vegetables and their by-products, tea, coffee, meat extracts and some dairy products [41].

14.5.3 Applications of Distillation in Food Processing

14.5.3.1 Manufacture of Whisky

Whisky is a spirit produced by the distillation of a mash of cereals, which may include barley, corn, rye and wheat, and is matured in wooden casks. There are three types of Scotch and Irish whisky: malt whisky produced from 100% malted (germinated) barley, grain whisky produced from unmalted cereal grains and blended whisky which contains 60–70% grain whisky and 30-40% malt whisky.

Malting of barley is carried out by steeping the grain in water for 2–3 days and allowing it to germinate. The purpose of malting is the production of amylases which later convert grain starch to sugar. Malting is stopped by drying the grain down to a moisture content of 5% in a kiln. In traditional Scotch whisky production, the grain is dried over a peat fire which contributes to the characteristic flavour of the end product. Alternatively, the kiln may be heated indirectly by gas or oil or directly by natural gas. The dried grain is milled by means of corrugated roller mills, hammer mills or attrition mills (see Section 15.3.2), in order to break open the bran layer without creating much fines. The milled grain is mashed. In the case of malted barley, the grain is mixed with water at 63–68 °C for 0.5–1.5 h before filtering off the liquid, known as wort. In the production of grain whisky, the milled grain is cooked for 1.5 h at 120 °C to gelatinise the starch in the nonbarley cereals. It is cooled to 60–65 °C and 10–15% freshly malted barley is added to provide a source of enzymes for the conversion of the starch to the sugar maltose. The wort is filtered, transferred to a fermentation vessel, cooled to 20–25 °C, inoculated with one or more strains of the yeast *Saccharomyces cerevisiae* and allowed to ferment for 48–72 h. This produces a wash containing about 7% ethanol and many flavour compounds. The fermented wash is then distilled. In the batch process for the production of Scotch whisky, two pot stills are used. In the first, known as the *wash still*, the fermented mash is boiled for 5–6 h to produce a distillate, known as *low wines*, containing 20–25% (v/v) ethanol. This is condensed and transferred to a smaller pot still, known as the *low wines still*, and distilled to produce a spirit containing about 70% (v/v) ethanol. A crude fractionation is carried out in this second still. The distillate first produced, known as the *foreshots*, contains aldehydes, furfurols and many other compounds which are not used directly in the product. The foreshots are returned to the second still. The distillation continues and the distillate collected as product until a specified distillate strength is reached. Distillation continues beyond that point but the distillate, known as *feints*, is returned to the second still. A similar process, but incorporating three pot stills, is used in the production of Irish whiskey.

The continuous distillation of fermented mash to produce whisky is usually carried out using two distillation columns. The fermented wash is fed towards the top of the first column, known as the *beer column*. Alcohol-free stillage is withdrawn from the bottom of this column. The vapour from the top of the first column is introduced at the base of the second column, known as the *rectifier column*. The vapour from the top of this column is condensed and collected as product. The bottom product is returned as reflux to top of the first column. The rectifier column contains mainly sieve plates but with some bubble cap plates near the top. If the columns are fabricated from stainless steel, a flat disc of copper mesh may be fitted near the top of the rectifier column to improve the flavour of the product.

Whiskies produced by batch or continuous distillation are matured in wooden barrels, usually oak, for periods ranging from 1 year to more than 18 years, depending on the type and quality of the whisky. The type of barrel used has a pronounced effect on the flavour of the matured product. The matured products are usually blended before bottling.

The following are examples of other distilled beverage spirits. *Brandy* is a distillate from the fermented juice, mash or wine of fruit, *rum* is a distillate from the fermented juice of sugar cane, sugar cane syrup, sugar cane molasses or other sugar cane products and *gin* is obtained by distillation from mash or by redistillation of distilled spirits, or by mixing neutral spirits, with or over juniper berries and other aromatics, or extracts from such materials. It derives its main characteristic flavour from juniper berries. *Tequila* is a distillate produced in Mexico from the fermented juice of the heads of *Agave tequilana* Weber, with or without other fermentable substances [42–45].

14.5.3.2 **Manufacture of Neutral Spirits**

A multicolumn distillation plant is used for producing neutral spirits from fermented mash. A typical system would be comprised of five columns: a whisky-separating column, an aldehyde column, a product-concentrating column, an aldehyde-concentrating column and a fusel oils concentrating column (see Fig. 14.20). The whisky-separating column is fitted with sieve plates, with some bubble cap plates near the top of the column. The other four columns are fitted with bubble cap plates. The fermented mash containing 7% (v/v) of alcohol is fed to near the top of the whisky-separating column. The overhead distillate from this column is fed to the aldehyde column. The bottom product from this column is pumped to the middle of the product-concentrating column. The end product, neutral spirit, is withdrawn from near the top of this column. The top product from the aldehyde column is rich in aldehydes and esters and is fed to the aldehyde-concentrating column. The top product from this column is rich in aldehydes and is removed while the bottom product is recycled to the aldehyde column. Fusel oils concentrate near the bottom of the aldehyde column and from there are fed to the fusel oil-concentrating column. The bottom product from this column is rich in fusel oils and is removed. The top product is recycled to the aldehyde column. The product from the very top of the product-concentrating column is condensed and returned as reflux to the aldehyde column [42, 45].

There are many other applications for distillation including the following examples.

Recovery of solvents from oil after extraction Most of the solvent can be recovered by evaporation using a film evaporator (see Section 3.1.2.3). However, when the solution becomes very concentrated, its temperature rises and the oil may be heat-damaged. The last traces of solvent in the oil may be removed by steam distillation or stripping with nitrogen.

Concentration of Aroma Compounds from Juices and Extracts By evaporating 10–30% of the juice in a vacuum evaporator, most of the volatile aroma compounds leave in the vapour. This vapour can be fed to a distillation column. The bottom product from the column is almost pure water and the aroma concentrate leaves from the top of the column. This is condensed and may be added back to the juice or extract prior to drying. Fruit juices and extract of coffee may be treated in this way (see Sections 3.1.5.2, 3.1.5.3). A spinning cone evaporator (see Section 14.5.2.2) may be used for this duty.

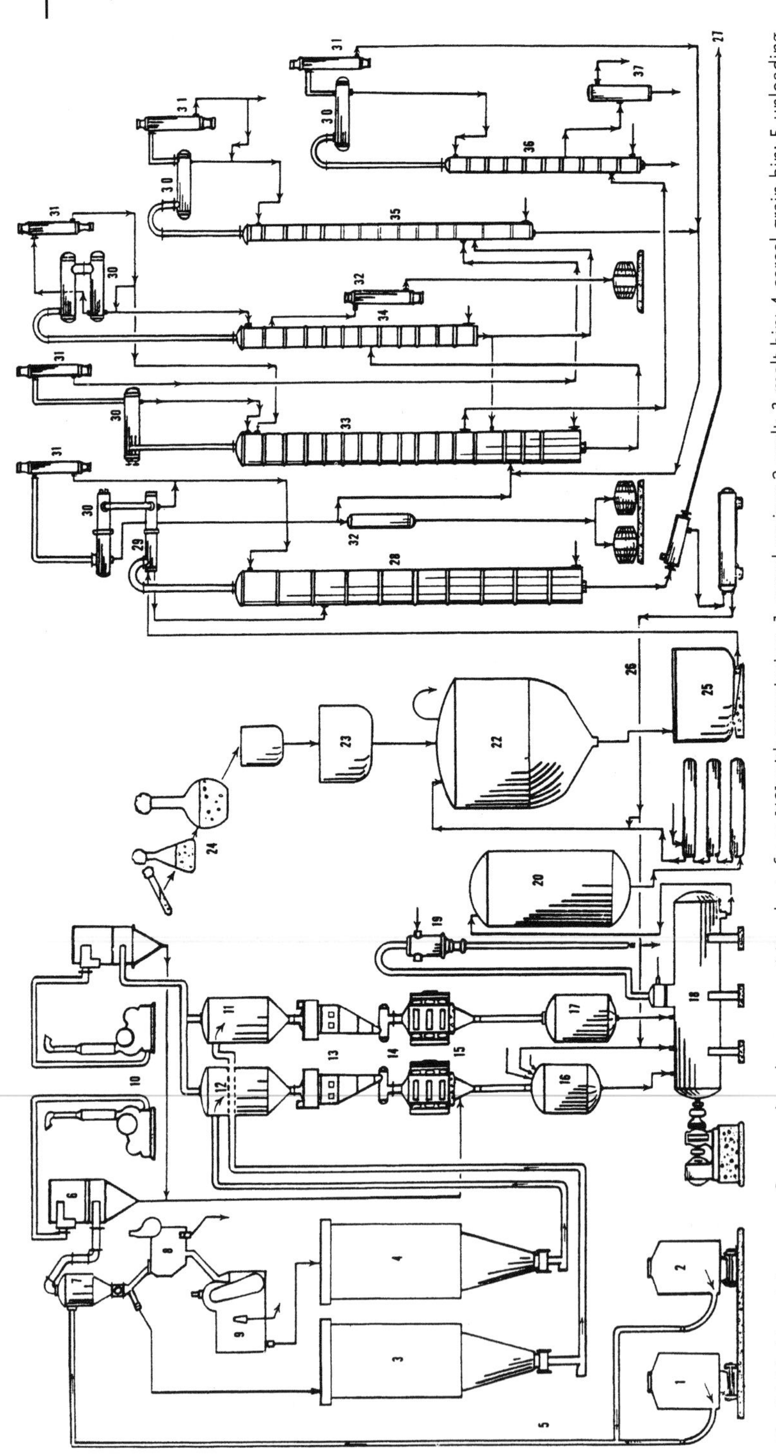

Fig. 14.20 Material process flow, modern beverage spirits plant; from [42] with permission. 1 cereal grains; 2 malt; 3 malt bin; 4 cereal grain bin; 5 unloading elevator; 6 dust filter; 7 collector; 8 scalperator; 9 millerator; 10 reclaiming exhauster; 11 malt receiver; 12 cereal grain receiver; 13 automatic scale; 14 mill feeder; 15 roller mills; 16 precooker; 17 malt infusion; 18 cooker; 19 barometric condenser; 20 converter; 21 mash coolers; 22 fermenter; 23 final yeast propagator; 24 yeast culture and intermediate yeast propagator; 25 fermented-mash holding vessel; 26 stillage return system; 27 stillage flow to recovery system; 28 whisky separating column; 29 heat exchanger; 30 dephlegamtor; 31 vent condenser; 32 product cooler; 33 selective distillation column; 34 product concentrating column; 35 aldehyde concentrating column; 36 fusel oil concentrating column; 37 fusel oil decanter.

Extraction of Essential Oils from Leaves, Seeds, etc. This may be achieved by steam distillation. The material in a suitable state of subdivision is placed on a grid or perforated plate above heated water. In some cases the material is in direct contact with the water or superheated steam may be used. If the oil is very heat sensitive distillation may be carried out under vacuum [6].

14.6 Crystallisation

Alistair S. Grandison

14.6.1 General Principles

Many foods and food ingredients consist of, or contain crystals. Crystallisation has two types of purpose in food processing: (a) the separation of solid material from a liquid in order to obtain either a pure solid, e.g. salt or sugar, or a purified liquid, e.g. winterised salad oil and (b) the production of crystals within a food such as in butter, chocolate or ice cream.

In either case, it is desirable to control the process such that the optimum yield of crystals of the required purity, size and shape is obtained. It is also important to understand crystallisation when considering frozen food (see Chapter 4) or where undesirable crystals are produced, e.g. lactose crystals in dairy products, or precipitated fat crystals in salad oils.

14.6.1.1 Crystal Structure

Crystals are solids with a three-dimensional periodic arrangement of units into a spatial lattice. They differ from amorphous solids in having highly organised structures of flat faces and corners. A limited number of elementary crystal cells are possible with accurately defined angles; and any crystalline material has one of 14 possible lattice structures. Some examples of these structures are shown in Fig. 14.21.

It is important to note that the final macroscopic shape of a crystal is usually not the same as its elementary lattice structure, as growth conditions can change the final 'habit'. The units involved in lattice structures may be metallic nuclei or atoms, but most food crystals are formed of molecular units bonded by van der Waals forces, or in a limited number of cases, ions bonded by ionic bonds. Detailed information on crystal structure and the crystallisation process may be found elsewhere [46–48].

14.6.1.2 The Crystallisation Process

Crystallisation is the conversion of one or more substances from an amorphous solid or the gaseous or liquid state to the crystalline state. In practice, we are

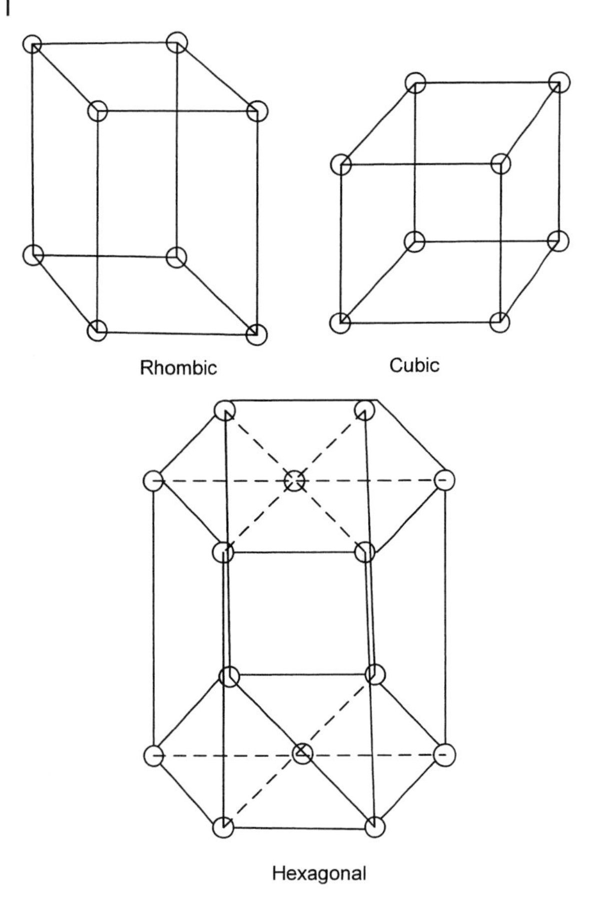

Fig. 14.21 Elementary cells of some crystal systems.

only concerned with conversion from the liquid state. The process involves three stages: supersaturation of the liquid, followed by nucleation (formation of new crystal structures) and crystal growth. A distinction that is commonly used is 'crystallisation from a solution', which is the case where a purified substance is produced from a less pure mixture, as opposed to 'melt crystallisation' in which both or all components of a mixture crystallise into a single solid phase.

Saturation of a solution is the equilibrium concentration which would be achieved by a solution in contact with solute after a long period of time. In most solute/solvent systems, the saturation concentration increases with temperature, although this is not always the case, e.g. with some calcium salts. Supersaturation occurs when the concentration of solute exceeds the saturation point ($S=1$). The saturation coefficient (S) at any temperature is defined as:

$$S = \text{(Concentration of solute in solution)}/\text{(Concentration of solute in saturated solution)}$$

Solutes differ in their ability to withstand supersaturation without crystallisation – for example sucrose can remain in solution when $S = 1.5–2.0$, whereas sodium chloride solutions crystallise with only a very small degree of supersaturation. Crystals cannot form or grow in a solution at or below its saturation concentration ($S \leq 1$) at any given temperature; and thus supersaturation must be achieved in two main ways: cooling or evaporation. Cooling moves the system along the solubility curve such that a saturated solution becomes supersaturated at the same solute concentration, while evaporation increases the concentration into the supersaturated zone. It is also possible to produce supersaturation by chemical reaction or by addition of a third substance which reduces solubility, e.g. addition of ethanol to an aqueous solution, but these are not commercially important in food processing.

Supersaturation does not necessarily result in spontaneous crystallisation because, although the crystalline state is more thermodynamically stable than the supersaturated solution and there is a net gain in free energy on crystallisation, the activation energy required to form a surface in a bulk solution may be quite high. In other words, the probability of aligning the units correctly to form a viable crystal nucleus is low and is dependent on a number of factors. Viscosity is one such factor; and crystallisation occurs less readily as viscosity increases. This can readily be seen in sugar confectionery which frequently consists of supercooled viscous liquids. Miers' theory (Fig. 14.22) defines three regions for a solute/solvent mixture. Below the saturation curve (in the undersaturated zone) there is no nucleation or crystal growth; and in fact crystals dissolve. In the metastable zone, crystal growth occurs, but nucleation does not occur spontaneously. In the labile zone, both nucleation and crystal growth occur and, the higher the degree of supersaturation, the more rapidly these occur. While the solubility curve is fixed, the supersolubility curve is not only a property of the

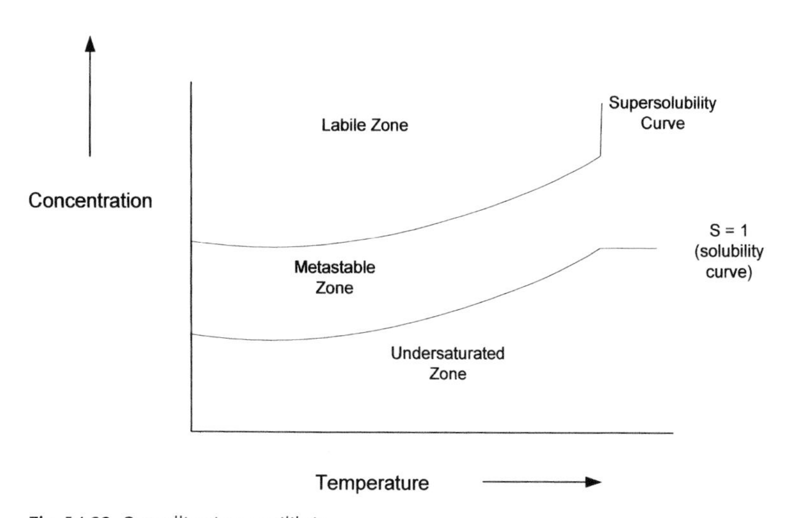

Fig. 14.22 Crystallisation equilibria.

system, but also depends on other factors such as presence of impurities, cooling rate or agitation of the system, in other words factors which affect the activation energy of the system as mentioned above.

In practice, crystals can be produced by several methods:

1. Homogeneous nucleation is the spontaneous production of crystals in the labile zone.
2. Heterogeneous nucleation occurs in the presence of other surfaces, such as foreign particles, gas bubbles, stirrers, which form sites for crystallisation in the metastable zone.
3. Secondary nucleation requires the presence of crystals of the crystallising species itself and also occurs in the upper regions of the metastable zone. The reason for this phenomenon is not clear but may be due to fragments breaking off existing crystals by agitation or viscous drag and forming new nuclei.
4. Nucleation is also stimulated by outside effects such as agitation or ultrasound.

Alternatively, crystalline materials can be produced by 'seeding' solutions in the metastable zone. In this case, very finely divided crystals are added to a supersaturated solution and allowed to grow to the finished size without further nucleation.

Control of crystal growth following nucleation or seeding is essential to obtain the correct size and shape of crystals. The rate of growth depends on the rate of transport of material to the surface and the mechanism of deposition. The rate of deposition is approximately proportional to S, while diffusion to the crystal surface can be accelerated by stirring. Impurities generally reduce the rate.

The final shapes or 'habits' of crystals are distinguished into classes such as platelike, prismatic, dendritic and acicular. Habits are determined by conditions of growth. Invariant crystals maintain the same shape during growth, as deposition in all directions is the same (see Fig. 14.23 a). Much more commonly, the shape changes and overlapping of smaller faces occurs (see Fig. 14.23 b).

Crystals grown rapidly from highly supersaturated solutions tend to develop extreme habits, such as needle-shaped or dendritic, with a high specific surface, due to the need to dissipate heat rapidly. The final shape is also affected by impurities which act as habit modifiers, interacting with growing crystals and causing selective growth of some surfaces, or dislocations in the structure.

Some substances can crystallise out into chemically identical, but structurally different forms. This is known as polymorphism; and perhaps the best recognised examples are diamond and graphite. Polymorphism is particularly important in fats, which can often crystallise out into different polymorphs with different melting points [48, 49]. In some cases, polymorphs can be converted to other, higher melting point crystal forms, which is the basis of tempering some fat-based foods.

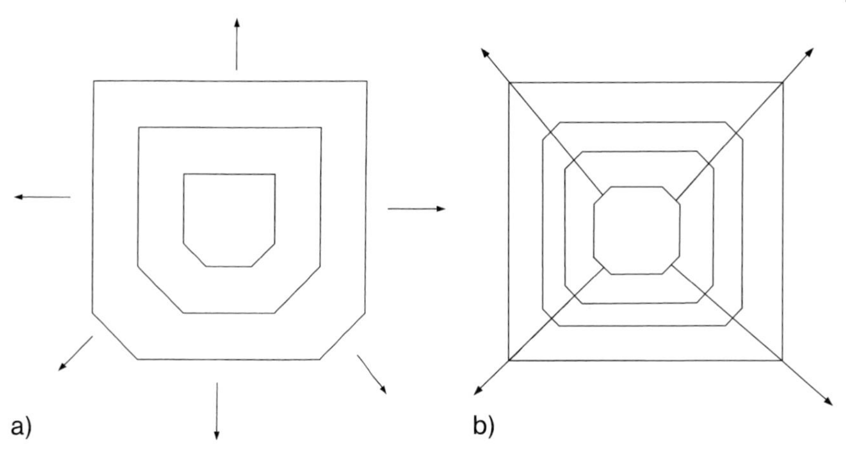

Fig. 14.23 Crystal growth: (a) invariant crystal growth, (b) overlapping crystal growth.

14.6.2
Equipment Used in Crystallisation Operations

Industrial crystallisers are classified according to the method of achieving supersaturation, i.e. by cooling, evaporation or mixed operations [50], as well as factors such as mode of operation (continuous or batch), desired crystal size distribution and purity. Detailed descriptions of design and operation are given by Mersmann and Rennie [51]. The crystallising suspension is often known as 'magma', while the liquid remaining after crystallisation is the 'mother liquor'. Fluidised beds are common with either cooling or evaporative crystallisation, in which the solution is desupersaturated as it flows through a bed of growing crystals (see Figs. 14.24, 14.25).

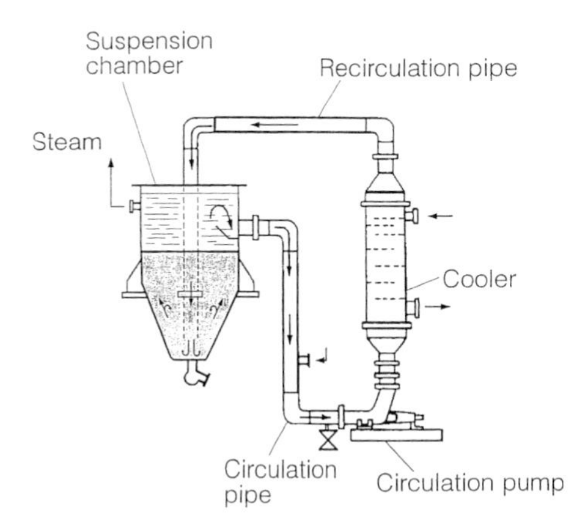

Fig. 14.24 Fluidised bed cooling crystalliser; from [51] with permission.

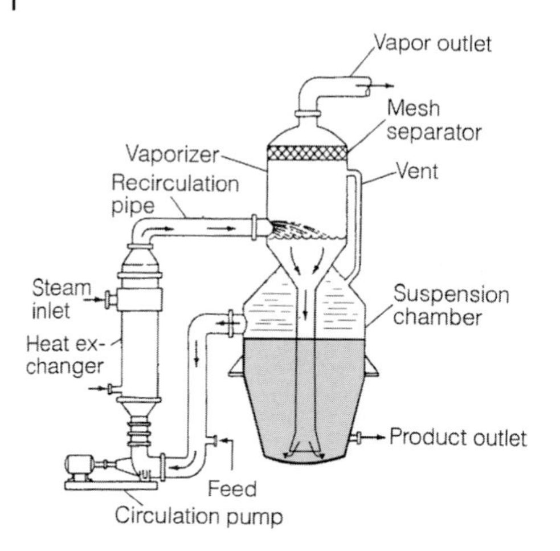

Fig. 14.25 Fluidised bed evaporative crystalliser; from [51] with permission.

Yield from a crystalliser (Y) can be calculated from:

$$Y = W[C_1 - C_2(1 - V)] \tag{14.9}$$

Where W is the initial mass of water, C_1 and C_2 are the concentrations of solute before and after crystallisation and V is the fraction of water evaporated.

Cooling crystallisers may incorporate continuous circulation through a cooling heat exchanger (e.g. Fig. 14.24) or for viscous materials, may incorporate a scraped surface heat exchanger (see Chapter 2).

Evaporative crystallisers are similar to forced circulation evaporators with a crystallisation vessel below the vapour/liquid separator (see Fig. 14.25). Simultaneous evaporation and cooling can be carried out without heat exchangers, the cooling effect being produced by vacuum evaporation of the saturated solution.

14.6.3 Food Industry Applications

14.6.3.1 Production of Sugar

Crystallisation is a major operation in sugar manufacture. Beet or cane sugar consist essentially of sucrose; and different grades of product require uniform crystals of different sizes. Supersaturation is effected by evaporation, but the temperature range is limited by the fact that sugar solutions are sensitive to caramelisation above 85 °C, while high viscosities limit the rate of crystallisation below 55 °C, so all operations are carried out within a fairly narrow temperature range. The supersaturated sugar solution is seeded with very fine sugar crystals and the 'massecuite', i.e. syrup/crystal mixture, is evaporated with further syrup addition. When the correct crystal size has been achieved, the crystals are removed and the syrup is passed to a second evaporator. Up to four evaporation stages may be used, with the syrup becoming less pure each time.

14.6.3.2 **Production of Salt**

Salt is much less of a problem then sugar, in that it is not temperature-sensitive and forms crystals more easily than sugar. Evaporation of seawater in lagoons is still carried out, but factory methods are more common. Continuous systems based on multiple effect evaporators (see Chapter 3) are widespread. Sodium chloride normally forms cubic crystals, but it is common to add potassium ferrocyanide as a habit-modifier, producing dendritic crystals which have better flow properties.

A number of other food materials are produced in pure form by crystallisation, including lactose, citric acid, monosodium glutamate and aspartame.

14.6.3.3 **Salad Dressings and Mayonnaise**

In some cases, crystallisation is carried out to remove unwanted components. Salad oils, such as cottonseed and soybean oil, are widely used in salad dressings and mayonnaise. They contain high melting point triglycerides which crystallise out on storage. This is not especially serious in pure oil, but would break an emulsion and hence lead to product deterioration. *Winterisation* is fractional crystallisation to remove higher melting point triglycerides. It is essential to remove this fraction while retaining the maximum yield of oil, which requires extremely slow cooling to produce a small number of large crystals, i.e. crystal growth with little nucleation. Crystallisation is carried out over 2–3 days with slow cooling to approximately 7 °C, followed by very gentle filtration. Agitation is avoided after the initial nucleation to prevent the formation of further nuclei. The winterised oil should then pass the standard test of remaining clear for 5.5 h at 0 °C.

14.6.3.4 **Margarine and Pastry Fats**

Fractional crystallisation of fats into high (stearin) and low (olein) melting point products is used to improve the quality of fats for specific purposes, e.g. margarine, pastry fats. Crystallisation may be accelerated by reducing the viscosity in the presence of solvents such as hexane. Adding detergents also improves the recovery of high melting point crystals from the olein phase.

14.6.3.5 **Freeze Concentration**

Ice crystallisation from liquid foods is a method of concentrating liquids such as fruit juices or vinegar without heating, or adjusting the alcohol content of beverages. Its use is limited as only a modest level of concentration is possible; and there is an inevitable loss of yield with the ice phase.

There are many applications of crystallisation in the manufacture of foods where a separation is not involved and will therefore not be dealt with here. These include ice cream, butter, margarine, chocolate and sugar confectionery.

14.7 Membrane Processes

Michael J. Lewis

14.7.1 Introduction

Over the last 50 years, a number of membrane processes have evolved which make use of a pressure driving force and a semipermeable membrane in order to effect a separation of components in a solution or colloidal dispersion. The separation is based mainly on molecular size, but to a lesser extent on shape and charge. The four main processes are reverse osmosis (hyperfiltration), nanofiltration, ultrafiltration and microfiltration. The dimensions of the components involved in these separations are typically in the range from <1 nm to >1000 nm. In fact, they can be considered to be a continuous spectrum of processes, with no clearcut boundaries between them. Most suppliers of membranes now offer a selection of membranes which cover this entire spectrum. Figure 14.26 illustrates these processes and how they also relate to traditional particle filtration.

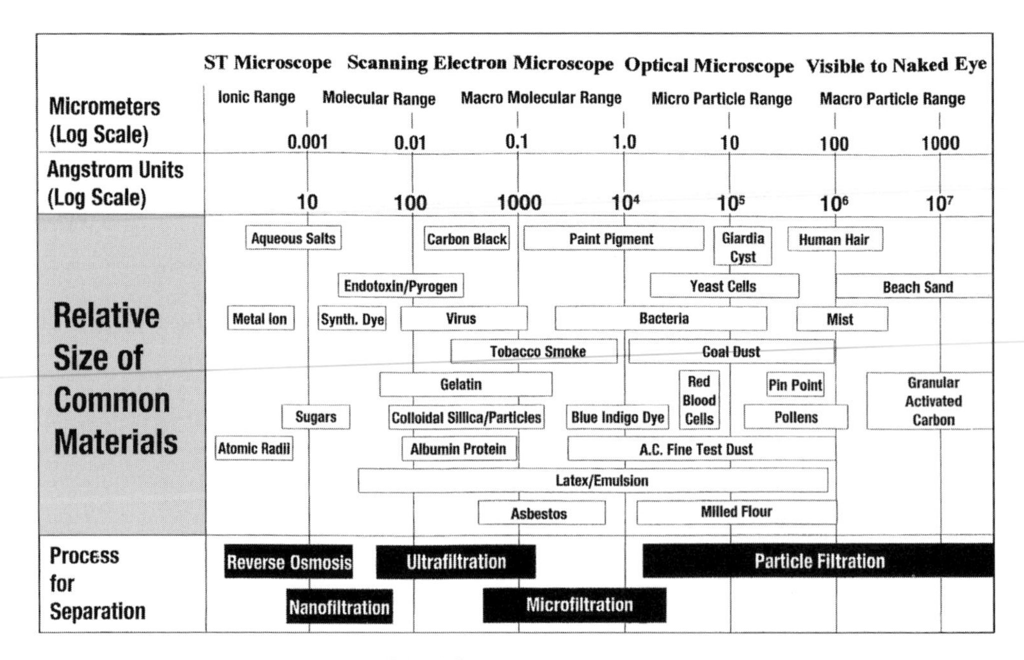

Fig. 14.26 Comparison of the different membrane processes with particle filtration; courtesy of GE Infrastructure – Water and Process Technologies.

14.7.2
Terminology

The feed material is applied to one side of a membrane and subjected to a pressure. In most cases, the feed flows in a direction parallel to the membrane surface and the term crossflow filtration is used to describe such applications. Dead-end systems are also used, but mainly for laboratory-scale separations. The stream which passes through the membrane under the influence of this pressure is termed the permeate (filtrate). The stream left after the required amount of permeate is removed is termed the concentrate or retentate.

If the membrane has a very small pore diameter (tight pores), the permeate will be predominantly water and the process is known as reverse osmosis (RO), similar in its effects to evaporation or freeze-concentration. In some cases, the permeate is the required material; for example in the production of 'potable' water from seawater or polluted waters or in the production of ultrapure water. In RO, the pressure applied needs to be in excess of the osmotic pressure of the solution. Osmotic pressures are highest for low molecular weight solutes, such as salt and sugar solutions; and large increases in their osmotic pressure occur during RO. The osmotic pressure, Π, of a dilute solution can be predicted from the Van t'Hoff equation:

$$\Pi = iRT\left(\frac{c}{M}\right) \tag{14.10}$$

where i is the degree of ionisation, R is the ideal gas constant, T is the absolute temperature, M is the molecular weight and c is the concentration (kg m^{-3}).

As the membrane pore size increases, the membrane becomes permeable to low molecular weight solutes in the feed. Lower pressure driving forces are required. However, larger molecular weight molecules, e.g. proteins, are still rejected by the membrane. It is this fractionation which makes ultrafiltration a more interesting process than RO. Membranes with even larger pore sizes allow smaller macromolecules to pass through, but retain particulate matter and fat globules; and this is termed microfiltration.

Concentration factor and rejection are two important processing parameters for all pressure activated processes. The concentration factor (f) is defined as follows:

$$f = \frac{V_F}{V_C} \tag{14.11}$$

where V_F is the feed volume and V_C is the final concentrate volume.

As soon as the concentration factor exceeds 2.0, the volume of permeate will exceed that of the concentrate. Concentration factors may be as low as 1.5 for some viscous materials and 5.0–50 for some dilute protein solutions. Generally higher concentration factors are used for ultrafiltration than for RO: over 50.0 can be achieved for UF treatment of cheese whey, compared to about to 5 for RO treatment of cheese whey.

The rejection or retention factor (R) of any component is defined as:

$$R = \frac{c_F - c_P}{c_F} \tag{14.12}$$

where c_F is the the concentration of component in the feed and c_p is the concentration in the permeate. It can easily be measured and is very important, as it influences the extent (quality) of the separation that can be achieved.

Rejection values normally range between 0 and 1.0; and sometimes they are expressed as percentages (0–100%). Occasionally negative rejections are found for some charged ions (Donnan effect).

1. When $c_p=0$, $R=1$, all the component is retained in the feed.
2. When $c_p=c_F$, $R=0$, the component is freely permeating.

If the concentration factor and rejection value are known, the yield of any component, which is defined as the fraction of that component present in the feed, which is recovered in the concentrate, can be estimated. Obviously for reverse osmosis, the yield for an ideal membrane is 1.0.

The yield (Y) can be calculated from:

$$Y = f^{R-1} \tag{14.13}$$

The derivation of this equation is provided in Lewis [52]. Thus for a component where $R=0.95$, at a concentration factor of 20, the yield is 0.86; i.e. 86% is retained in the concentrate and 14% is lost in the permeate.

14.7.3 Membrane Characteristics

The membrane itself is crucial to the process. The first commercial membranes were made of cellulose acetate and these are termed first generation membranes. However, temperatures had to be maintained below 30 °C and the pH range was 3–6. These constraints limited their use, as they could not be disinfected by heat or cleaned with acid or alkali detergents. These were followed in the mid-1970s by other polymeric membranes (second generation membranes), with polyamides (with a low tolerance to chlorine) and, in particular, polysulphones being widely used for foods. It is estimated that over 150 organic polymers have now been investigated for membrane applications. Inorganic membranes based on sintered and ceramic materials are also now available and these are much more resistant to heat and cleaning and disinfecting fluids.

The main terms used to describe membranes are microporous or asymmetric. Microporous membranes have a uniform porous structure throughout, although the pore size may not be uniform across the thickness of the membrane. They are usually characterised by a nominal pore size and no particle larger than this will pass through the membrane. In contrast to this, most membranes used for ultrafiltration are of the asymmetric type, having a dense active layer or skin of 0.5–1.0 micron in thickness, with a further support layer which is much more porous and of greater thickness. The membrane also has a chemical nature; and many materials have been evaluated. It may be hydrophilic or hydrophobic in nature. The hydrophobic nature can be characterised by measuring its contact angle, θ. The higher the contact angle, the more hydrophobic is the surface. Polysulphones are generally much more hydrophobic than cellulosic membranes. The surface may also be charged. All these factors give rise to interactions between the membrane and the components in the feed and influence the components passing through the membrane, as well as the fouling of the membrane.

14.7.4 Flux Rate

Permeate flux and power consumption are two important operating characteristics.

The permeate flux is usually expressed in terms of $\mathrm{l\,m^{-2}\,h^{-1}}$. This permits a ready comparison of different membrane configurations of different surface areas. Flux values may be from $<5\,\mathrm{l\,m^{-2}\,h^{-1}}$ to $>500\,\mathrm{l\,m^{-2}\,h^{-1}}$. Factors affecting the flux rate are the applied pressure, the volumetric flow rate of feed across the membrane surface, its temperature and its viscosity. The flux is also influenced by concentration polarization and fouling, which in turn are influenced by the flow conditions across the membrane. Inducing turbulence increases the wall shear stress and promotes higher flux rates [52].

The main energy consumption for membrane techniques is the power utilisation of the pumps. The power used, W, is related to the pressure (head) developed and the mass flow rate as follows:

$$W = m'hg \tag{14.14}$$

where m' is the mass flow rate ($\mathrm{kg\,s^{-1}}$), h is the head developed by the pump (m) and g is the acceleration due to gravity ($9.81\ \mathrm{m\,s^{-2}}$).

This energy is largely dissipated within the fluid as heat and results in a temperature rise. Cooling may be necessary if a constant processing temperature is required.

14.7.5 Transport Phenomena and Concentration Polarisation

A very important consideration for pressure driven membrane processes is that the separation takes place not in the bulk of solution, but in a very small region close to the membrane, known as the boundary layer, as well as over the membrane itself. This gives rise to the phenomenon of concentration polarisation over the boundary layer. It is manifested by a quick and significant reduction (2- to 10-fold) in flux when water is replaced by the feed solution, for example in a dynamic start.

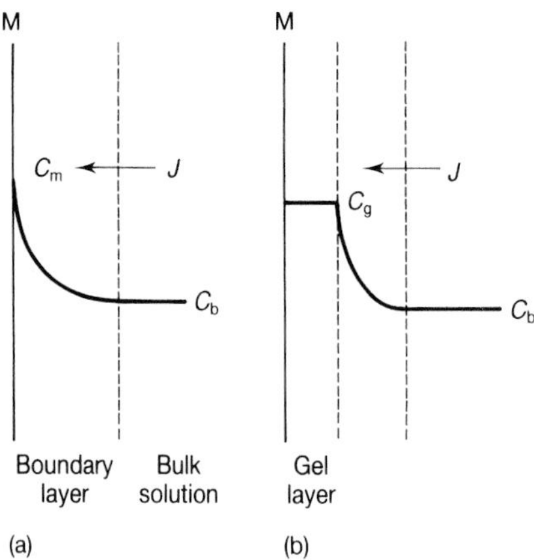

Fig. 14.27 (a) Concentration polarisation in the boundary layer, (b) concentration polarisation with a gel layer from [52] with permission.

Concentration polarisation occurs whenever a component is rejected by the membrane. As a result, there is an increase in the concentration of that component at the membrane surface, together with a concentration gradient over the boundary layer (see Fig. 14.27).

Eventually a dynamic equilibrium is established, where the convective flow of the component to the membrane surface equals the flow of material away from the surface, either in the permeate or back into the bulk of the solution by diffusion, due to the concentration gradient established. This increase in concentration, especially of large molecular weight components, offers a very significant additional resistance. It may also give rise to the formation of a gelled or fouling layer on the surface of the membrane. Whether this occurs will depend upon the initial concentration of the component and the physical properties of the solution. It could be very important, as it may affect the subsequent separation performance. Concentration polarisation itself is a reversible phenomenon, i.e. if the solution is then replaced by water, the original water flux should be restored. However, this rarely occurs in practice due to the occurrence of fouling, which is detected by a decline in flux rate at constant composition. Fouling is caused by the deposition of material on the surface of the membrane or within the pores of the membrane. Fouling is irreversible and the flux needs to be restored by cleaning. Therefore, during any membrane process, flux declines due to a combination of these two phenomena. More recently, it has been recognised that fouling can be minimised by operating at conditions at or below the critical flux. When operating below this value, fouling is minimised, but when operating above it, fouling deposits accumulate. This can be determined experimentally for different practical situations [53].

14.7.6
Membrane Equipment

Membrane suppliers now provide a range of membranes, each with different rejection characteristics. For ultrafiltration, different molecular weight cutoffs are available in the range 1000 Da to 500000 Da. Tight ultrafiltration membranes have a molecular weight cutoff value of around 1000–5000 Da, whereas the more 'open' or 'loose' membranes will have a value in excess of 100000 Da. However, because there are many other factors that affect rejection, molecular weight cutoff should only be regarded as giving a relative guide to its pore size and true rejection behaviour. Experimental determinations should always be made on the system to be validated, at the operating conditions to be used.
Other desirable features of membranes to ensure commercial viability are:

- reproducible pore size, offering uniformity both in terms of their permeate rate and their rejection characteristics
- high flux rates and sharp rejection characteristics
- compatible with processing, cleaning and sanitising fluids
- resistance to fouling
- an ability to withstand temperatures required for disinfecting and sterilising surfaces, which is an important part of the safety and hygiene considerations.

Extra demands placed upon membranes used for food processing include: the ability to withstand hot acid and alkali detergents (low and high pH), temperatures of 90 °C for disinfecting or 120 °C for sterilising and/or widely used chemical disinfectants, such as sodium hypochlorite, hydrogen peroxide or sodium metabisulphite. The membrane should be designed to allow cleaning both on the feed/concentrate side and the permeate side.

Membrane processing operations can range in their scale of operation, from laboratory benchtop units, with samples less than 10 ml through to large commercial-scale operations, processing at rates greater than 50 $m^3 h^{-1}$. Furthermore, the process can be performed at ambient temperatures, which allows concentration without any thermal damage to the feed components.

14.7.7
Membrane Configuration

The membranes themselves are thin and require a porous support against the high pressure. The membrane and its support, together are normally known as the module. This should provide a large surface area in a compact volume and must allow suitable conditions to be established, with respect to turbulence, high wall shear stresses, pressure losses, volumetric flow rates and energy requirements, thereby minimising concentration polarisation. Hygienic considerations are also important: there should be no dead spaces and the module should be capable of being cleaned in-place on both the concentrate and the permeate side. The membranes should be readily accessible, both for cleaning and replacement. It may also be an advantage to be able to collect permeate from individual membranes in the module to be able to assess the performance of each individually.

The three major designs are the tubular, flat plate and spiral wound configurations (see Figs. 14.28, 14.29).

Sintered or ceramic membranes can also be configured in the form of tubes. Tubular membranes come in range of diameters. In general tubes offer no dead spaces, do not block easily and are easy to clean. However, as the tube diameter increases, they occupy a larger space, have a higher hold-up volume and incur higher pumping costs. The two major types are the hollow fibre, with a fibre diameter of 0.001–1.2 mm and the wider tube with diameters up to 25 mm, although about 12 mm is a popular size.

For the hollow fibre system, the membrane wall thickness is about 250 μm and the tubes are self-supporting. The number of fibres in a module can be as few as 50 but sometimes >1000. The fibres are attached at each end to a tube sheet, to ensure that the feed is properly distributed to all the tubes. This may give rise to pore plugging at the tube entry point. Prefiltration is recommended to reduce this. They are widely used for desalination and in these RO applica-

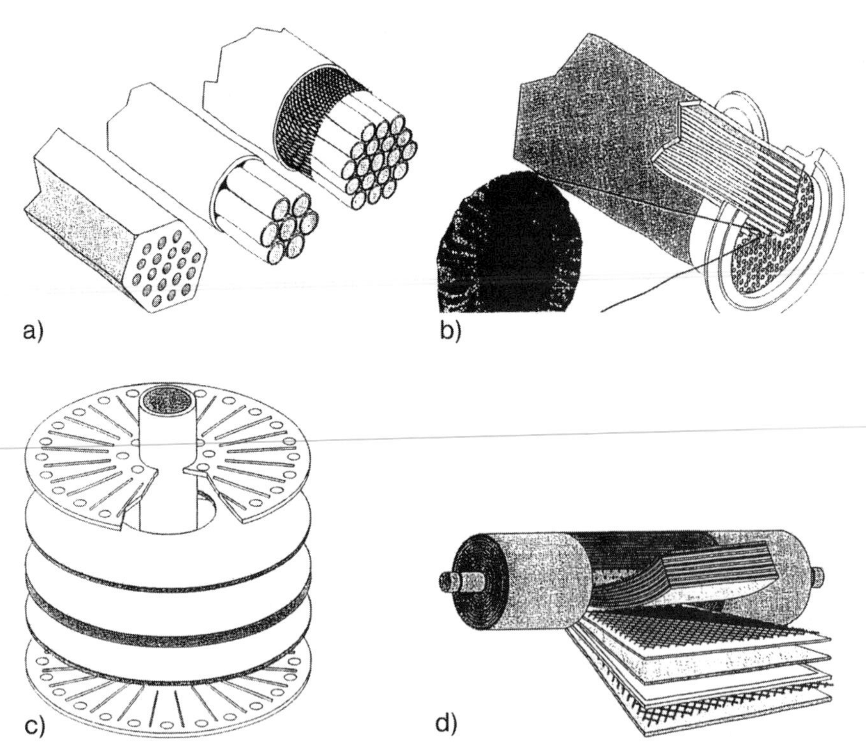

Fig. 14.28 This shows the tubular, hollow fibre, plate and frame and spiral wound configurations; courtesy of ITT Aquious.

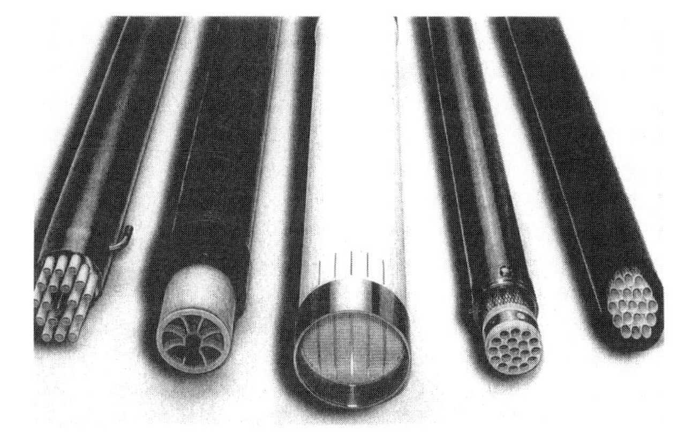

Fig. 14.29 Tubular, hollow fibre and spiral wound configurations; courtesy of ITT Aquious.

tions are capable of withstanding high pressures. It is the ratio of the external to internal diameter, rather than the membrane wall thickness which determines the pressure that can be tolerated. Hollow fibre systems usually operate in the streamline flow regime. However the wall shear rates are high. They tend to be expensive, because if one or several fibres burst, the whole cartridge needs to be replaced.

For wider tubes, the feed is normally pumped through the tube, which may be up to 25 mm in diameter, although a popular size is about 12 mm diameter. There may be up to 20 tubes in one module, tube lengths may be 1.2–3.8 m and tubes within the module may be connected in series or parallel. The membrane is cast or inserted into a porous tube which provides support against the applied pressure. Therefore they are capable of handling higher viscosity fluids and even materials with small suspended particles, up to one-tenth the tube diameter. They normally operate under turbulent flow conditions with flow velocities greater than 2 m s^{-1}. The corresponding flux rates are high, but pumping costs are also high, in order to generate the high volumetric flow rate required and the operating pressure.

The flat plate module can take the form of a plate-and-frame type geometry or a spirally wound geometry. The plate-and-frame system employs membranes stacked together, with appropriate spacers and collection plates for permeate removal, somewhat analogous to plate heat exchangers. The channel height can be 0.4–2.5 mm. Flow may be either streamline or turbulent and the feed may be directed over the plates in a parallel or series configuration. This design permits a large surface area to be incorporated into a compact unit. Membranes are easily replaced and it is easy to isolate any damaged membrane sandwich. Considerable attention has been devoted to the design of the plate to improve performance. This has been achieved by ensuring a more uniform distribution of fluid over the plate, by increasing the width of the longer channels and by reducing the ratio of the longest to the shortest channel length.

The spiral wound system is now widely used and costs for membranes are relatively low. In this case, a sandwich is made from two sheet membranes which enclose a permeate spacer mesh. This is attached at on end to a permeate removal tube and the other three sides of the sandwich are sealed. Next to this is placed a feed spacer mesh and the two together are rolled round the permeate collection tube in the form of a Swiss roll. The channel height is dictated by the thickness of the feed spacer. Wider channel heights will reduce the surface area to volume ratio, but reduce the pressure drop.

The typical dimensions of one spiral membrane unit would be about 12 cm in diameter and about 1 m in length. Up to three units may be placed in one housing, with appropriate spacers to prevent telescoping, which may occur in the direction of flow and could damage the sandwich. This configuration is becoming very popular and is relatively cheap. Again, the flow may be streamline or turbulent. Pressure drop/flow rate relationships suggest that flow conditions are usually turbulent.

Each system does and will continue to have its devotees. An alternative, much used, unit for simple laboratory separations is the stirred cell with agitation facilities. In contrast to the systems described earlier, this is a dead-end rather than a flow-through system.

As well as the membrane module, there will be pumps, pipeline, valves and fittings, gauges, tanks, heat exchangers, instrumentation and control and perhaps in-place cleaning facilities. For small installations, the cost of the membrane modules may only be a relatively small component of the total cost of the finished plant, once these other items have been accounted for. This may also apply to some large installations such as water treatment plants, where other separation processes are numerous and the civil engineering costs may also be high.

The simplest system is a batch process. The feed is usually recycled, as sufficient concentration is rarely achieved in one pass. Flux rates are initially high but decrease with time. Energy costs are high because the pressure is released after each pass. Residence times are long. Batch processes are usually restricted to small-scale operations. Batch processing with top-up is used in situations when the entire feed volume will not fit into the feed tank. Continuous processes may be single-stage (feed and bleed), or multistage processes, depending upon the processing capacity required. Figure 14.30 illustrates these different systems.

14.7.8
Safety and Hygiene Considerations

It is important that safety and hygiene are considered at any early stage when developing membrane processes. These revolve round cleaning and disinfecting procedures for the membranes and ancillary equipment, as well as the monitoring and controlling the microbial quality of the feed material. For many processes, thermisation or pasteurisation are recommended for feed pretreatment

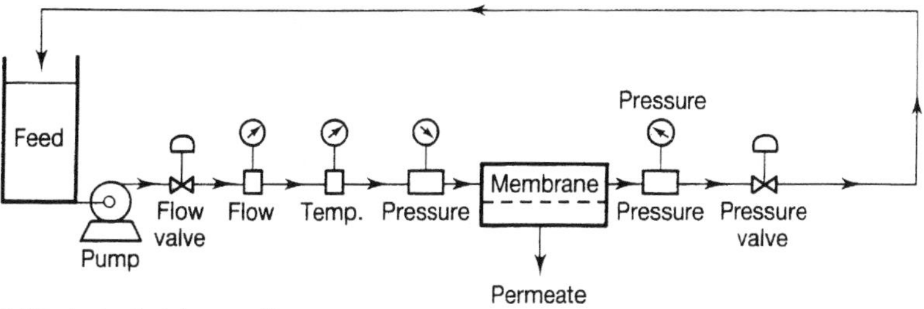

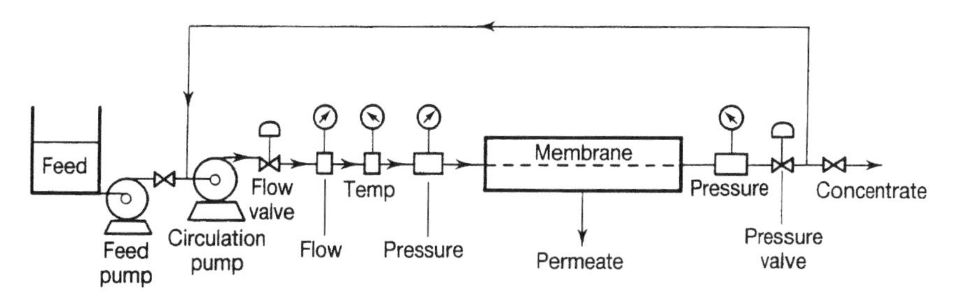

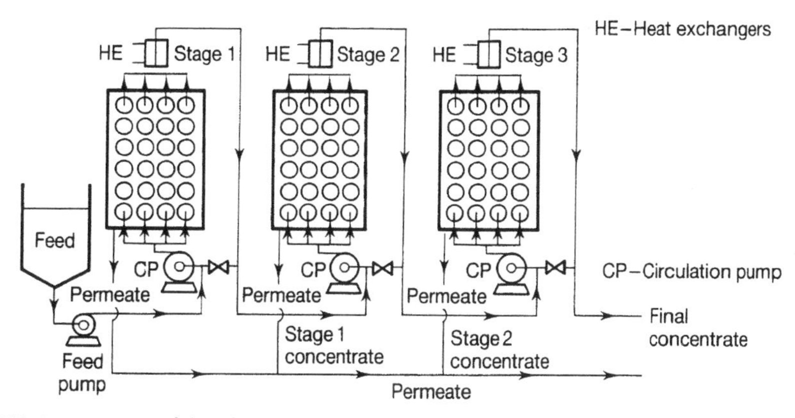

Fig. 14.30 Different plant layouts: (a) batch operation,
(b) continuous with internal recycle (feed and bleed),
(c) commercial scale with three stages; courtesy of ITT Aquious.

(see Chapter 2). Microfiltration may also be considered for heat-labile components. Relatively little has been reported about the microbiological hazards associated with membrane processing. All microorganisms will be rejected by the membrane and will therefore increase in number in the concentrate, by the concentration factor. There may also be some microbial growth during the process, so the residence time and residence time distribution should be known, as well as the operating temperature. If residence times are long, it may be advisable to operate either below 5 °C or above 50 °C, to prevent further microbial growth.

14.7.9
Applications for Reverse Osmosis

The main applications of RO are for concentrating fluids, by removal of water. RO permits the use of relatively low temperatures, even lower than vacuum evaporation. It reduces volatile loss caused by the phase change in evaporation and is very competitive from an energy viewpoint. RO uses much higher pressures than other membrane processes, in the range 2–8 MPa, and incurs greater energy costs and requires special pumps. Products of RO may be subtly different to those produced by evaporation, particularly with respect to low molecular weight solutes, which might not be completely rejected, and to volatile components, which are not completely lost.

Thin-film composite membranes, based on combinations of polymers have now largely replaced cellulose acetate, allowing higher temperatures, up to 80 °C and greater extremes of pH (3–11) to be used, thereby facilitating cleaning and disinfection. Therefore, the main applications of RO are for concentrating liquids, recovering solids and treatment of water.

14.7.9.1 Milk Processing

RO can be used for concentrating full cream milk up to a factor of 2–3 times. Flux rates for skimmed milk are only marginally higher than those for full cream milk. The product concentration attainable is nowhere near as high as that for evaporation, due to increasing osmotic pressure and fouling, due mainly to the increase in calcium phosphate, which precipitates out in the pores of the membrane. Therefore, most of the commercial applications have been for increasing the capacity of evaporation plant. Other possible applications that have been investigated and discussed include: (a) the concentration of milk on the farm for reducing transportation costs, (b) for yoghurt production at a concentration factor of about 1.5, to avoid the addition of skimmed milk powder, (c) for ice cream making, also to reduce the use of expensive skimmed milk powder, (d) for cheesemaking to increase the capacity of the cheese vats and (e) for recovering rinse water. Cheese whey can also be concentrated, to reduce transportation costs or prior to drying. Flux values for sweet whey are higher than for acid whey, which in turn are higher than for milk, for all systems tested [54]. Reviews of the use of RO and UF in dairying applications include El-Gazzar and Marth [55] and Renner [56].

14.7.9.2 **Other Foods**

Reverse osmosis has found application in the processing of fruit and vegetable juices, sometimes in combination with ultrafiltration and microfiltration. The osmotic pressure of juices is considerably higher than that of milk. It is advantageous to minimise thermal reactions, such as browning, and to reduce loss of volatiles. From a practical viewpoint, the flux rate and rejection of volatiles is important. RO modules can cope with single strength clear or cloudy juices and also fruit pulp. RO can be used to produce a final product, as in the case of tomato paste and fruit purees, or to partially concentrate, prior to evaporation. RO is a well established process for concentrating tomato juice from about 4.5° Brix, to 8–12° Brix. Other fruit juices which have been successfully concentrated are apple, pear, peach and apricot. Where juices have been clarified, osmotic pressure limits the extent of concentration and up to 25° Brix can be achieved. Unclarified juices may be susceptible to fouling. With purees and pulps, the viscosity may be the limiting factor and these can be concentrated to a maximum of 1.5 times. It is possible to concentrate coffee extract from about 13–36% total solids at 70 °C, with little loss of solids. Tea extracts can also be similarly concentrated.

Reverse osmosis is also used for waste recovery and more efficient use of processing water in corn wet-milling processes. Commercial plant is available for concentrating egg white to about 20% solids. In one particular application, egg white is concentrated and dried, after lysozyme has been extracted.

Dealcoholisation is an interesting application, using membranes which are permeable to alcohol and water. In a process akin to diafiltration, water is added back to the concentrated product, to replace the water and alcohol removed in the permeate. Such technology has been used for the production of low or reduced alcohol, beers, ciders and wine. Leeper [57] reported ethanol rejections for cellulose acetate ranged between 1.5% and 40.0%, for polyamides between 32.8% and 60.9% and, for other hybrid membranes, as high as 91.8%.

Reverse osmosis is used in many areas worldwide for water treatment, where there are shortages of fresh water, although it is still well exceeded by multistage fractional distillation. Potable water should contain less than 500 ppm of dissolved solids. Brackish water, e.g. bore-hole or river water, typically contains from 1000 ppm up to about 10 000 ppm of dissolved solids, whereas seawater contains upwards of 35 000 ppm of dissolved solids. If lower total solids are required, the permeate can be subjected to a second process, known as double reverse osmosis water.

14.7.10
Applications for Nanofiltration

Nanofiltration (NF) has been used for partially reducing calcium and other salts in milk and whey, with typical retention values of 95% for lactose and less than 50% for salts. Guu and Zall [58] have reported that permeate subject to NF gave improved lactose crystallisation. NF provides potential for improving the heat stability of the milk.

NF has also been investigated for removing pesticides and components responsible for the colour from ground water, as well as for purifying water for carbonation and soft drinks. For water production of high-grade purity, for analytical purposes, it may be double RO treated, as mentioned earlier.

NF is currently being investigated for fractionating oligosaccharides with prebiotic potential, produced by the enzymatic breakdown of different complex carbohydrates.

14.7.11 Applications for Ultrafiltration

14.7.11.1 Milk Products

Milk will be taken as an example to show the potentialities of ultrafiltration (UF). Milk is chemically complex, containing components of a wide range of molecular weights, such as protein, fat, lactose, minerals and vitamins. It also contains microorganisms, enzymes and perhaps antibiotics and other contaminants. Whole milk contains about 30–35% protein and about the same amount of fat (dry weight). Therefore, it is an ideal fluid for membrane separation processes, in order to manipulate its composition, thereby providing a variety of products or improving the stability of the colloidal system. The same principles apply to skimmed milk, standardised milk and some of its byproducts, such as cheese whey. Skimmed milk can be concentrated up to seven times and full cream milk up to about five times Kosikowski [59]. An IDF publication [60] gives a summary of the rejection values obtained during the ultrafiltration of sweet whey, acid whey, skimmed milk and whole milk, using a series of industrial membranes.

Bastian et al. [61] compared the rejection values during UF and diafiltration of whole milk. They found that the rejection of lactose, riboflavin, calcium, sodium and phosphorus was higher during diafiltration than UF. Diafiltration of acidified milk gave rise to lower rejections of calcium, phosphorus and sodium. Premaratne and Cousin [62] reported a detailed study on the rejection of vitamins and minerals during UF of skimmed milk. During a five-fold concentration, the following minerals were concentrated by the following factors: Zn (4.9), Fe (4.9), Cu (4.7), Ca (4.3), Mg (4.0) and Mn (3.0), indicating high rejection values. In contrast, most of the B vitamins examined were almost freely permeating.

Ultrafiltration has been used to concentrate cheese whey (~6.5% TS), which contains about 10–12% protein (dry weight), to produce concentrates which could then be dried to produce high protein powders (concentrates and isolates) which retain the functional properties of the proteins. Some typical concentration factors, *f*, used are as follows:

f=5: protein content (dry weight) about 35% (similar to skimmed milk)
f=20: protein content about 65%
f=20 plus diafiltration: protein content about 80%.

The product starts to become very viscous at a concentration factor of about 20, so diafiltration is required to further increase the protein in the final product.

The permeate from ultrafiltration of whey contains about 5% total solids, the predominant component being lactose. Since this is produced in substantial quantities, the economics of the process are dependent upon its utilisation. It can be concentrated by reverse osmosis and hydrolysed to glucose and galactose to produce sweeteners or fermented to produce alcohol or microbial protein. Skimmed milk has been investigated also but protein concentrates based on skimmed milk have not received the same amount of commercial interest as those based on whey proteins.

However, yoghurt and other fermented products have been made from skimmed milk and whole milk concentrated by ultrafiltration [55]. Whey protein concentrates have also been incorporated [63]. Production of labneh, which is a strained or concentrated yoghurt at about 21% total solids, has been described by Tamime et at. [64], by preconcentrating milk to 21% TS. Inorganic membranes have also been used for skimmed milk; and Daufin et al. [65] have investigated the cleanability of these membranes using different detergents and sequestering agents.

As well as exploiting the functional properties of whey proteins, full cream milk has been concentrated by UF prior to cheesemaking. The UF concentrate has been incorporated directly into the cheese vats. Some advantages of this process include: increased yield (particularly of whey protein), lower rennet and starter utilisation, smaller vats or even complete elimination of vats, little or no whey drainage and better control of cheese weights. Lawrence [66] suggests that concentration below a factor of 2.0 gives protein standardisation, reduced rennet and vat space, but no increased yield. Concentration factors greater than 2.0 result in an increased yield.

Some problems result from considerable differences in the way the cheese matures and hence its final texture and flavour. The types of cheese that can be made in this way include: Camembert type cheese, mozzarella, feta and many soft cheeses. Those which are difficult include the hard cheeses, such as Cheddar and also cottage cheese; and the problems are mainly concerned with poor texture. More discussion is given by Kosikowski [59]. Further reviews on the technological problems arising during the conversion of retentate into cheeses are discussed by Lelievre and Lawrence [67]. Quarg is also produced from ultrafiltered milk.

Ultrafiltration is an extremely valuable method of concentrating and recovering many the minor components, particularly enzymes from raw milk, many of which would be inactivated by pasteurisation. Such enzymes are discussed in more detail by Kosikowski [68]. Further reviews on membrane processing of milk are provided by Glover [53], El-Gazzar and Marth [54], Renner and El-Salam [55] and the International Dairy Federation [69].

14.7.11.2 **Oilseed and Vegetable Proteins**

There have been many laboratory investigations into the use of ultrafiltration for extracting proteins from oil seed residues, or for removing any toxic components. Lewis [70] and Cheryan [71] have reviewed the more important of these. The investigations include: for soya, the separation of low molecular weight peptides from soy hydrolysates (with the aim of improving quality), the dissociation of phytate from protein (followed by its removal by UF), the removal of oligosaccharides, the removal of trypsin inhibitor and performance of different membrane configurations. For cottonseed, the use of different extraction conditions has been evaluated, as have the functional properties of the isolates produced by UF. Investigations were performed with sunflower and alfalfa to remove the phenolic compounds responsible for the colour and bitter flavour and glucosinolates from rapeseed.

Many have been successful in terms of producing good quality concentrates and isolates, particularly with soyabean. However, few have come to commercial fruition, mainly because of the economics of the process, dictated by the relatively low value of products and the fact that acceptable food products can be obtained by more simple technology, such as isoelectric precipitation. A further problem arises from the fact that the starting residue is in the solid form, thereby imparting an additional extraction procedure. Extraction conditions may need to be optimised, with respect to time; temperature, pH and antinutritional factors. Another problem arises from the complexity of oil seed and vegetable proteins, compared to milk products, evidenced by electrophoretic measurement. It is likely that many of these proteins are near their solubility limits after extraction and further concentration will cause them to come out of solution and promote further fouling. Fouling and cleaning of membranes was found to be a serious problem during ultrafiltration of rapeseed meal extracts [72].

A further important area is the use of enzyme reactors. The earliest examples were to breakdown polysaccharides to simpler sugars in a continuous reactor and to use a membrane to continuously remove the breakdown products.

14.7.11.3 **Animal Products**

Slaughterhouse wastes contain substantial amounts of protein. Two important streams that could be concentrated by ultrafiltration are blood and waste water.

Blood contains about 17% protein. It can be easily separated by centrifugation into plasma (70%) and the heavier erythrocytes (red blood corpuscles or cells, ca. 30%). Plasma contains about 7% protein, whereas the blood corpuscles contain 28–38% protein. The proteins in plasma possess useful functional properties, particularly gelation, emulsification and foaming. They have been incorporated into meat products and have shown potential for bakery products, as replacers for egg white.

Whole blood, plasma and erythrocytes have all been subjected to UF processes [71, 73]. The process is concentration polarisation controlled and flux rates are low. High flow rate and low pressure regimes are best. Gel concentrations were approximately 45% for plasma protein and 35% for red blood cells. The fouling characteristics of different blood fractions have been investigated [74]. Whole blood was found to be the worst foulant, when compared to lysed blood and blood plasma.

Another important material is gelatin, which can be concentrated from very dilute solutions by UF. As well as concentration, ash is removed, which improves its gelling characteristics. This is one example where there have been some high negative rejections recorded for calcium, when ultrafiltered at low pH.

Eggs have also been processed by UF. Egg white contains 11–13% total solids (about 10% protein, 0.5% salts, 0.5% glucose). Large amounts of egg white are used in the baking industry. The glucose can cause problems during storage and causes excessive browning during baking. Whole egg contains about 25% solids and about 11% fat, whereas egg yolk contains about 50% solids. It is unusual to evaporate eggs prior to drying, because of the damage caused. Compared to RO, UF also results in the partial removal of glucose; and further removal can be achieved by diafiltration. Flux values during UF are much lower than for many other food materials, most probably due to the very high initial protein concentration; and rates are also highly velocity- and temperature-dependent [71].

Membrane-based bioreactors appear to be a very promising application for the production of ethanol, lactic acid, acetone, butanol, starch hydrolysates and protein hydrolysates.

14.7.12 Applications for Microfiltration

Microfiltration (MF) is generally used to separate particles suspended in liquid media and may frequently be considered as an alternative to conventional filtration or centrifugation. For industrial use, the aim is usually to obtain either a clear permeate or the concentrate. Therefore most applications are either clarification, or the recovery of suspended particles such as cells or colloids, or the concentration of slurries.

One application in the food industry has been in the treatment of juices and beverages. As MF is a purely physical process, it can have advantages over traditional methods, which may involve chemical additives, in terms of the quality of the product as well as the costs of processing.

Finnigan and Skudder [75] report that very good quality, clear permeate was found when MF processing cider and beer, with high flux rates and no rejection of essential components.

Clarification and biological stabilisation of wine musts and unprocessed wine have also been described for MF. This avoids the requirement for fining and, possibly, pasteurisation. Another section of the industry with several applications is dairy processing. Piot et al. [76] and Merin [77] have clarified sweet cheese whey using crossflow MF. This has the dual benefit of removing fat and reducing the bacterial population and could eliminate the need for fat separation and heat treatment in the production of whey protein powders prior to UF. The former authors reported that a 5-log reduction of microorganisms could be obtained in the microfiltrate compared to the whey, although some loss of whey protein was observed. Hanemaaijer [78] described a scheme for whey treatment incorporating MF and UF to produce 'tailor-made' whey products with specific properties for specific applications. The products include whey protein concentrates which are rich in whey lipids, as well as highly purified protein.

Bacterial removal from whole milk by MF is a problem because the size range of the bacteria overlaps with the fat globules and, to a lesser extent, with the casein micelles. However, some success has been achieved with skim milk. The 'Bactocatch' system can remove 99.6% of the bacteria from skimmed milk using ceramic membranes on a commercial scale [79]. The retentate (approximately 10% of the feed) can then be sterilised by a UHT process, mixed with the permeate and the mixture pasteurised, to give a product with 50% longer shelf life but no deterioration in organoleptic properties compared to milk that has only been pasteurised. One such product on the market in the UK is Cravendale milk [80]. The combination of MF and heat reduces the bacterial numbers by 99.99%.

Alternatively, the permeate could be used for cheesemaking, or the production of low-heat milk powder [81]. Piot et al. [82] described the use of membranes of pore diameter 1.8 μm to produce skimmed milk of low bacterial content. Recovery of fat from buttermilk has also been described [83].

Membranes have been used to concentrate milk prior to the manufacture of many cheese types. This results in improved yields and other associated benefits such as reduced requirement for rennet and starter and the ability to produce much more cheese per vat [84]. However, the use of MF is an attractive alternative. Rios et al. [83] have carried out extensive trials on this application and concluded that the use of 0.2 μm pore diameter membranes gave a product with better texture and yield than with centrifugation. The choice of ceramic membranes allowed the curd to be contacted directly with the membrane.

Other food applications have been reported with meat and vegetable products including the following. Devereux and Hoare [85] described the use of MF to recover precipitated soya protein. This could have advantages over recovery of the dissolved protein using UF. Gelatin is a proteinaceous material derived by hydrolysis of collagen. This is purified by filtration incorporating diatomaceous earth. The latter process can be replaced by crossflow microfiltration (CMF), which effectively removes dirt, coagulated proteins, fats and other particulate materials from the feed. Again, CMF gives higher yields of high quality product on a continuous basis. Short [86] calculated that incorporating plants for gelatin would have a payback time of 3 years for a capacity of 30 t h^{-1}.

Overall, MF has made significant advances in new applications in the food and biotechnology industries. However, it has not yet realised its full potential, largely due to the severe problems of flux decline due to fouling. It is believed that further developments in membrane design and a greater knowledge of fouling mechanisms will result in greater application in the future, especially in the field of downstream processing.

14.8 Ion Exchange

Alistair S. Grandison

14.8.1 General Principles

Ion exchange can be used for separations of many types of molecules, such as metal ions, proteins, amino acids or sugars. The technology is utilised in many sensitive analytical chromatography and laboratory separation procedures, frequently on a very small scale. However, industrial-scale production operations, such as demineralisation or protein recovery, are possible. More detailed information on the theory of ion exchange can be found elsewhere [87–89].

Ion exchange is the selective removal of a single, or group of, charged species from one liquid phase followed by transfer to a second liquid phase by means of a solid ion exchange material. This involves the process of adsorption – the transfer of specific solute(s) from a feed solution on to a solid ion exchanger. The mechanism of adsorption is electrostatic involving opposite charges on the solute(s) and the ion exchanger. After washing off the feed solution, the solute(s) is desorbed back into solution in a much purified form.

Ion exchange solids have fixed ions covalently attached to a solid matrix. There are two basic types of ion exchanger:

(a) *Cation exchangers* bear fixed negative charges (e.g. $-SO_3^-H^+$, $-PO_3^{2-}(H^+)_2$, -COOH) and therefore retain cations.
(b) *Anion exchangers* bear fixed positive charges (amines or imines, such as quaternary amine or diethylaminoethyl groups) and thus retain anions.

Ion exchangers can be used to retain simple ionised species such as metal ions, but may also be used in the separation of polyelectrolytes, such as proteins, which carry both positive and negative charges, as long as the overall charge on the polyelectrolyte is opposite to the fixed charges on the ion exchanger. This overall charge depends on the isoelectric point (IEP) of the polyelectrolyte and the pH of the solution. At pH values lower than the IEP, the net overall charge will be positive and vice versa. The main interaction is via electrostatic forces and, in the case of polyelectrolytes, the affinity is governed by the number of electrostatic bonds between the solute molecule and the ion exchanger. However, with large molecules such as proteins, size, shape and the degree of hydration of the ions may affect these interactions and hence the selectivity of the ion exchanger for different solutes. Figure 14.31 gives a generalised anion exchanger – i.e. bearing fixed positive charges.

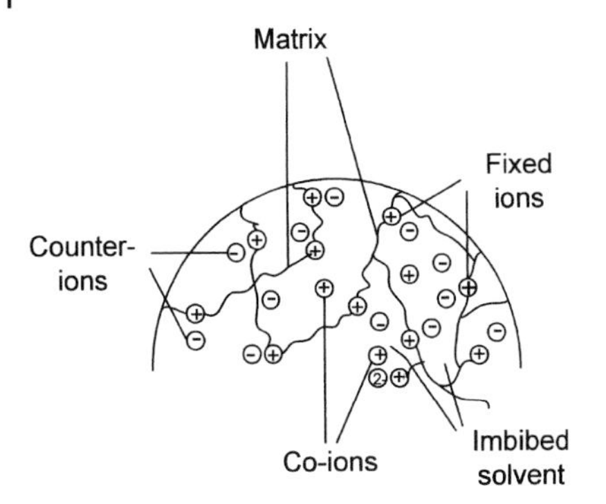

Fig. 14.31 Schematic diagram of a generalised anion exchanger.

To maintain electrical neutrality, these fixed ions must be balanced by an equal number of mobile ions of the opposite charge, i.e. anions, held by electrostatic forces. These mobile ions can move into and out of the porous molecular framework of the solid matrix and may be exchanged stoichiometrically with other dissolved ions of the same charge and are termed counterions. As the distribution of ions between the internal phase of the ion exchanger and the external phase is determined by the Donnan equilibrium, some co-ions (mobile ions having the same sign – positive in this example – as the fixed ions) will be present even in the internal phase. Therefore, if an anion exchanger, as in Fig. 14.31, is in equilibrium with a solution of NaCl, the internal phase contains some Na^+ ions, although the concentration is less than in the external phase because the internal concentration of Cl^- ions is much larger.

When an ion exchanger is contacted with an ionised solution, equilibration between the two phases rapidly occurs. Water moves into or out of the internal phase so that osmotic balance is achieved. Counter ions also move in and out between the phases on an equivalent basis. If two or more species of counter ion are present in the solution, they are distributed between the phases according to the proportions of the different ions present and the relative selectivity of the ion exchanger for the different ions. It is this differential distribution of different counter ions which forms the basis of separation by ion exchange. The relative selectivity for different ionised species results from a range of factors. The overall charge on the ion and the molecular or ionic mass are the primary determining factors, but selectivity is also related to degree of hydration, steric effects and environmental factors such as pH or salt content.

In the adsorption stage, a negatively charged solute molecule (e.g. a protein, P^-) is attracted to a charged site on the ion exchanger (R^+), displacing a counterion (X^-):

$$R^+X^- + P^- \rightarrow R^+P^- + X^-$$

This is shown schematically in Fig. 14.32 a.

In the desorption stage, the anion is displaced from the ion-exchanger by a competing salt ion (S^-) and hence is eluted:

$$R^+P^- + S^- \rightarrow R^+S^- + P^-$$

This is shown schematically in Fig. 14.32b.

Alternatively, desorption can be achieved by the addition of H^+ or OH^- ions. Ion exchangers are further classified, in terms of how their charges vary with changes in pH, into weak and strong exchangers. Strong ion exchangers are ionised over a wide range of pH and have a constant capacity within the range, whereas weak exchangers are only ionised over a limited pH range, e.g. weak cation exchangers may lose their charge below pH 6 and weak anion exchangers above pH 9. Thus, weak exchangers may be preferable to strong ones in some situations, for example where desorption may be achieved by a relatively small change in pH of the buffer in the region of the pKa of the exchange group. Regeneration of weak ion exchange groups is easier than with strong groups and therefore has a lower requirement of costly chemicals.

14.8.2 Ion Exchange Equipment

All ion exchangers consist of a solid insoluble matrix (termed the resin, adsorbent, medium, or just ion exchanger) to which the active, charged groups are attached covalently. The solid support must have an open molecular framework which allows the mobile ions to move freely in and out and must be completely insoluble throughout the process. Most commercial ion exchangers are based on an organic polymer network, e.g. polystyrene and dextran, although inorganic materials such as porous silica may be used. The latter are much more rigid and incompressible. The support material does not directly determine the ionic distribution between the two phases, but it does influence the capacity, the flow rate through a column, the diffusion rate of counterions into and out of the matrix, the degree of swelling and the durability of the material.

As the adsorption is a surface effect, the available surface area is a key parameter. For industrial processing, the maximum surface area to volume should be used to minimise plant size and product dilution. It is possible for a 1-ml bed of ion exchanger to have a total surface area $>100\ m^2$. The ion exchange material is normally deployed in packed beds and involves a compromise between large particles (to minimise pressure drop) and small particles to maximise mass transfer rates. Porous particles are employed to increase surface area/volume.

The capacity of an ion exchanger is defined as the number of equivalents (eq) of exchangeable ions per kilogram of exchanger, but is frequently expressed as $x\ meq\ g^{-1}$ (usually in the dry form). Most commercially available materials have capacities in the range 1–10 $eq\ kg^{-1}$ of dry material, but this may decline with age due to blinding or fouling, i.e. nonspecific adsorption of unwanted materials, such as lipids, onto the surface, or within the pores.

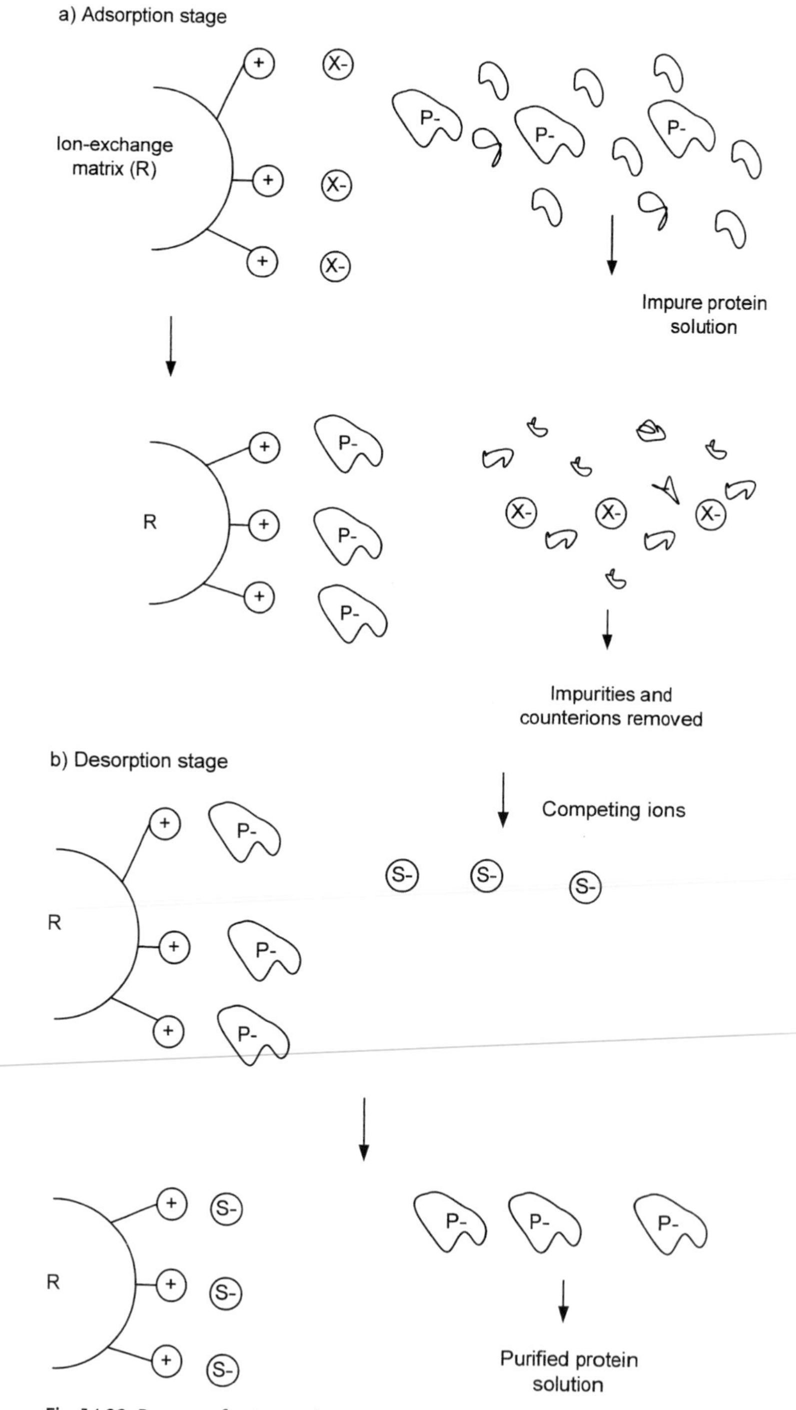

Fig. 14.32 Process of anion exchange: (a) adsorption stage, (b) desorption stage.

The choice of method of elution depends on the specific separation required. In some cases the process is used to remove impurities from a feedstock, while the required compound(s) remains unadsorbed. No specific elution method is required in such cases, although it is necessary to regenerate the ion exchanger with strong acid or alkali. In other cases the material of interest is adsorbed by the ion exchanger, while impurities are washed out of the bed. This is followed by elution and recovery of the desired solute(s). In the latter case, the method of elution is much more critical – for example, care must be taken to avoid denaturation of adsorbed proteins.

The adsorbed solute is eluted from the ion exchanger by changing the pH or the ionic strength of the buffer, followed by washing away the desorbed solute with a flow of buffer. Increasing the ionic strength of the buffer increases the competition for the charged sites on the ion exchanger. Small buffer ions with a high charge density displace polyelectrolytes, which can subsequently be eluted. Altering the buffer pH, so that the charge on an adsorbed polyelectrolyte is neutralised or made the same as the charges on the ion exchanger, results in desorption.

Fixed bed operations consisting of one or two columns connected in series (depending on the type of ions which are to be adsorbed) are used in most ion exchange separations. Liquids should penetrate the bed in plug flow, in either downward or upward direction. The major problems with columns arise from clogging of flow and the formation of channels within the bed. Problems may also arise from swelling of organic matrices when the pH changes. These problems may be minimised by the use of stirred tanks. However, these batch systems are less efficient and expose the ion exchangers to mechanical damage, as there is a need for mechanical agitation. The system involves mixing the feed solution with the ion exchanger and stirring until equilibration has been

Fig. 14.33 Commercial stirred tank reactors; by courtesy of Bio-Isolates plc.

achieved (typically 30–90 min in the case of proteins). After draining and washing the ion exchanger, the eluant solution is then contacted with the bed for a similar equilibration time before draining and further processing. Commercial stirred tank reactors for recovering protein from whey are shown in Fig. 14.33.

Mixed bed systems, containing both anion and cation exchangers, may be used to avoid prolonged exposure of the solutions to both high and low pH environments, as are frequently encountered when using anion and cation exchange columns in series, e.g. during demineralisation of sugar cane juice to prevent hydrolysis of sucrose, as described below.

14.8.3 Applications of Ion Exchange in the Food Industry

The main areas of the food industry where the process is currently used or is being developed are sugar, dairy and water purification. Ion exchange is also widely employed in the recovery, separation and purification of biochemicals and enzymes.

The main functions of ion exchange are:
(a) removal of minor components, e.g. de-ashing or decolourising
(b) enrichment of fractions, e.g. recovery of proteins from whey or blood
(c) isolating valuable compounds, e.g. production of purified enzymes.

Applications can be classified as follows.

14.8.3.1 Softening and Demineralisation

Softening of water and other liquids involves the exchange of calcium and magnesium ions for sodium ions attached to a cation exchange resin, e.g.:

$$R - (Na^+)_2 + Ca(HCO_3)_2 \rightarrow R - Ca^{++} + 2NaHCO_3$$

The sodium form of the cation exchanger is produced by regenerating with NaCl solution. Apart from the production of softened water for boiler feeds and cleaning of food and processing equipment, softening may be employed to remove calcium from sucrose solutions prior to evaporation (which reduces scaling of heat exchanger surfaces in sugar manufacture) and from wine (which improves stability).

Demineralisation using ion exchange is an established process for water treatment, but over the last 20 years it has been applied to other food streams. Typically the process employs a strong acid cation exchanger followed by a weak or strong base anion exchanger. The cations are exchanged with H^+ ions, e.g.:

$$2R^-H^+ + CaSO_4 \rightarrow (R^-)_2Ca^{++} + H_2SO_4$$

$$R^-H^+ + NaCl \rightarrow R^-Na^+ + HCl$$

and the acids thus produced are fixed with an anion exchanger, e.g.:

$$R^+OH^- + HCl \rightarrow R^+Cl^- + H_2O$$

Demineralised cheese whey is desirable for use mainly in infant formulations, but also in many other products such as ice cream, bakery products, confectionery, animal feeds etc. The major ions removed from whey are Na^+, K^+, Ca^{++}, Mg^{++}, Cl^-, HPO_4^-, citrate and lactate. Ion exchange demineralisation of cheese whey generally employs a strong cation exchanger followed by a weak anion exchanger. This can produce more than 90% reduction in salt content, which is necessary for infant formulae.

Demineralisation by ion exchange resins is used at various stages during the manufacture of sugar from either beet or cane, as well as for sugar solutions produced by hydrolysis of starch. In the production of sugar from beet, the beet juice is purified by liming and carbonatation and then may be demineralised by ion exchange. The carbonated juice is then evaporated to a thick juice prior to sugar crystallisation. Demineralisation may, alternatively, be carried out on the thick juice which has the advantage that the quantities handled are much smaller. To produce high quality sugar the juice should have a purity of about 95%. Ash removal or complete demineralisation of cane sugar liquors is carried out on liquors that have already been clarified and decolourised, so the ash load is at minimum. The use of a mixed bed of weak cation and strong anion exchangers in the hydrogen and hydroxide forms, respectively, reduces the prolonged exposure of the sugar to strongly acid or alkali conditions which would be necessary if two separate columns were used. Destruction of sucrose is thus minimised.

The cation and anion resins are sometimes used in their own right for dealkalisation or deacidification, respectively. Weak cation exchangers may be used to reduce the alkalinity of water used in the manufacture of soft drinks and beer, while anion exchangers can be used for deacidification of fruit and vegetable juices. In addition to deacidification, anion exchangers may also be used to remove bitter flavour compounds (such as naringin or limonin) from citrus juices. Anion or cation exchange resins are used in some countries to control the pH or titratable acidity of wine although this process is not permitted by other traditional wine-producing countries. Acidification of milk to pH 2.2, using ion exchange during casein manufacture by the Bridel process, has also been described.

Ion exchange processes can be used to remove specific metals or anions from drinking water and food fluids, which has potential application for detoxification or radioactive decontamination. Procedures have been described for the retention of lead, barium, radium, aluminium, uranium, arsenic and nitrates from drinking water. Removal of a variety of radionuclides from milk has been demonstrated. Radiostrontium and radiocaesium can be removed using a strongly acidic cation exchanger, while iodine 131 can be adsorbed on to a variety of anion exchangers. The production of low-sodium milk, with potential dietetic application, has been demonstrated.

14.8.3.2 **Decolourisation**

Sugar liquors from either cane or beet contain colourants such as caramels, melanoidins, melanins or polyphenols combined with iron. Many of these are formed during the earlier refining stages and it is necessary to remove them in the production of a marketable white sugar. The use of ion exchangers just before the crystallisation stage results in a significant improvement in product quality. It is necessary to use materials with an open, porous structure to allow the large colourant molecules access to the adsorption sites. A new approach to the use of ion exchange for decolourisation of sugar solutions is the application of powdered resin technology. Finely powdered resins (0.005–0.2 mm diameter) have a very high capacity for sugar colourants due to the ready availability of adsorption sites. The use of such materials on a disposable basis eliminates the need for chemical regenerants, but is quite expensive.

Colour reduction of fermentation products such as wine uses a strongly basic anion exchanger to remove colouring matter, followed by a strong cation exchanger to restore the pH. It is claimed that colour reduction can be achieved without substantially deleteriously affecting the other wine qualities.

14.8.3.3 **Protein Purification**

High purity protein isolates can be produced in a single step from dilute solutions containing other contaminating materials by ion exchange. The amphoteric nature of protein molecules permits the use of either anion or cation exchangers, depending on the pH of the environment. Elution takes place either by altering the pH or increasing the ionic strength. The eluate can be a single bulk, or a series of fractions produced by stepwise or linear gradients, although fractionation may be too complex for large-scale industrial production. Separation of a single protein may take place on the basis that it has a higher affinity to the charged sites on the ion exchanger compared to other contaminating species, including other proteins present in the feed. In such cases, if excess quantities of the feed are used, the protein of interest can be adsorbed exclusively, despite initial adsorption of all the proteins in the feed. Alternatively, it may be possible to purify a protein on the basis that it has a much lower affinity for the ion exchanger than other proteins present in the feed; and thus the other proteins are removed leaving the desired protein in solution. One limitation of the process for protein treatment is that extreme conditions of pH, ionic strength and temperature must be avoided to prevent denaturation of the protein.

An area of great potential is the recovery of proteins from whey, which is a byproduct of the manufacture of cheese and related products such as casein. Typically whey contains 0.6–0.8% protein which is both highly nutritious and displays excellent physical properties, yet the vast majority of this is wasted or under-utilised. Anion exchange materials can produce high purity functional protein from cheese whey, using a stirred tank reactor into which the whey is introduced at low pH. Following rinsing of nonadsorbed material, the protein fraction is eluted at high pH and further purified by ultrafiltration, so that the final protein content is approximately 97% (dry weight). It is further possible to fractionate the whey proteins into their separate components or groups of components. This approach has the potential of producing protein fractions with a range of functional properties which could be extremely valuable for use in the food industry. Lactoperoxidase and lactoferrin are valuable proteins with potential pharmaceutical applications which are present in small quantities in cheese whey. They may be purified from whey on the basis that these proteins are positively charged at neutral pH, whereas the major whey proteins are negatively charged. Another application of adsorption of whey protein by ion exchangers could be to improve the heat stability of milk. The use of ion exchange to recover or separate the caseins in milk is not carried out commercially, although it has been shown to be feasible.

This system has also been demonstrated for recovery of food proteins from waste streams resulting from the processing of soya, fish, vegetables and gelatine production, plus abattoir waste streams. Such protein fractions could be used as functional proteins in the food industry or for animal feeds. A variety of other food proteins have been purified or fractionated by ion-exchangers, including pea globulins, gliadin from wheat flour, egg, groundnut and soya protein.

Purification of proteins from fermentation broths usually involves a series of separation steps and frequently includes ion exchange. Large-scale purification of a variety of enzymes with applications in the food industry has been described, e.g. α-amylase, β-galactosidase.

14.8.3.4 **Other Separations**

Ion exchange has been used for various other separations in the food industry which do not fit into the above categories.

Fructose is sweeter than sucrose and glucose and can be used as a natural sweetener at reduced caloric intake. Although present in many natural sources, it is produced commercially from corn starch by hydrolysis to dextrose, which is then partially converted to fructose using the enzyme isomerase. The resulting high fructose corn syrup may be deionised by ion exchange and then a pure fructose fraction can be recovered with a sulphonic cation exchanger. Another application is the production of lactose-free milk. A process using sulphonated cation exchangers has been used to reduce the lactose level of skim milk to $<10\%$ of that in the feed, while retaining $>90\%$ of protein, minerals and citrate.

The purification of phenylalanine, which may be used in aspartame sweetener production, from fermentation broths using cationic zeolite material has been demonstrated. Ion exchange may also be used to purify enzymic reaction products such as flavour constituents from the enzymic degradation of fruit wastes.

14.8.4
Conclusion

There are many potential applications for ion exchange in the food industry, but few have been fully exploited in commercial practice. This is because of the complexity of the process and problems of scale-up. New applications are most likely to be developed in the food related aspects of biotechnology and in the production of high value protein fractions.

14.9
Electrodialysis

Alistair S. Grandison

14.9.1
General Principles and Equipment

Electrodialysis (ED) is a separations process in which membranes are used to separate ionic species from nonionic species. More detailed information on the theory and applications can be found elsewhere [88, 90, 91]. The process permits the separation of electrolytes from nonelectrolytes, concentration or depletion of electrolytes in solutions and the exchange of ions between solutions. Separation depends on ion-selective membranes, which are essentially ion exchange resins cast in sheet form, and electromigration of ions through ion-selective membranes depends on the electrical charge on the molecules, combined with their relative permeability through membranes. The membranes are composed of polymer chains which are crosslinked and intertwined into a network and bear either fixed positive or fixed negative charges. These may be heterogeneous membranes which consist of ion exchange resins dispersed in a polymer film or, more commonly, homogeneous membranes in which the ionic groups ($-NH_3^+$ or $-SO_3^-$) are attached directly to the polymer. Counter ions (see Section 14.8) are freely exchanged by the fixed charges on the membranes and thus carry the electric current through the membranes, while co-ions are repelled by the fixed charges and cannot pass through the membrane. Therefore cation membranes (with $-SO_3^-$ groups) allow the passage of positively charged ions, while anion membranes (with $-NH_3^+$ groups) allow the passage of negatively charged ions.

In practice, the cation and anion membranes are usually arranged alternately with plastic spacers to form thin solution compartments, as shown in Fig. 14.34.

In commercial practice 100–200 membranes may be assembled to form a membrane stack (Fig. 14.35) and an ED system may be composed of one or more stacks. Commercial ED membranes may be as large as 1–2 m^2.

The basic unit of a membrane stack is called a cell pair and comprises a pair of membranes and spacers, as illustrated in Figs. 14.34, 14.36.

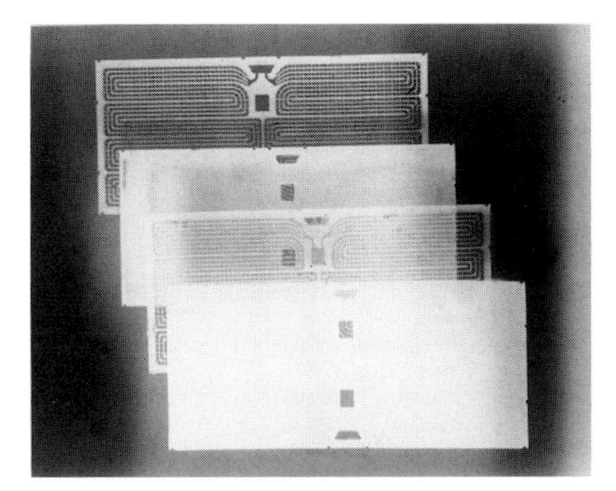

Fig. 14.34 Electrodialysis membranes and spacers; with permission of Ionics Inc.

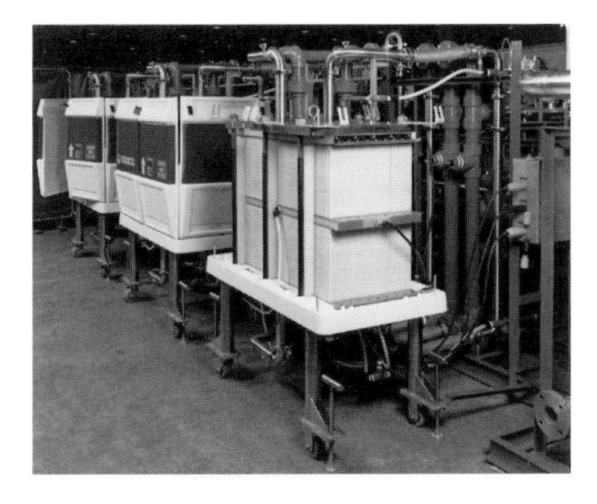

Fig. 14.35 Electrodialysis membrane stacks; with permission of Ionics Inc.

A positive electrode at one end and a negative electrode at the other permit the passage of a DC current. The electrical potential causes the anions to move towards the anode and the cations to move towards the cathode. However, the ion-selective membranes act as barriers to either anions or cations. Hence, anions migrating towards the anode will pass through anion membranes, but will be rejected by cation membranes, and vice versa. The membranes, therefore, form alternating compartments of ion-diluting (even-numbered compartments in Fig. 14.36) and ion-concentrating (odd-numbered) cells. If a feed stream containing dissolved salts, e.g. cheese whey, is circulated through the ion-diluting cells and a brine solution through the ion-concentrating cells, free mineral ions will leave the feed and be concentrated in the brine solution. Demineralisation of the feed is therefore achieved. Note that any charged macromolecules in the feed, such as proteins, will attempt to migrate in the electrical field, but will not pass through either anion or cation membranes, due to their molecular size. The efficiency of electrolyte transfer is determined by the current density and the residence time of the solutions within the membrane cells; and in practice efficiency this is limited to about 90% removal of minerals. The membranes are subject to concentration polarisation and fouling, as described in Section 14.7.

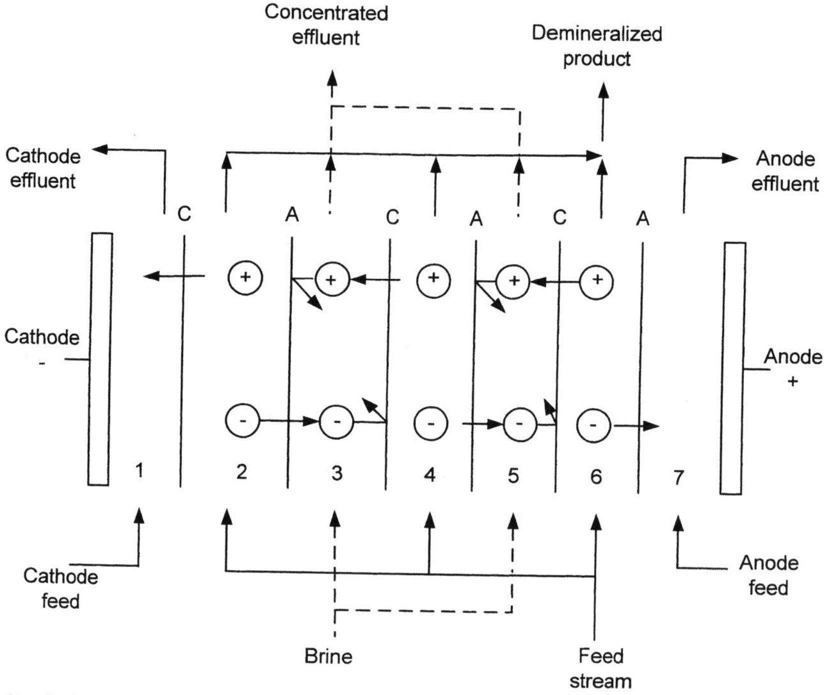

Fig. 14.36 Schematic diagram of electrodialysis process.

Alternative membrane configurations are possible, such as the use of cation membranes only for ion replacement. In Fig. 14.37, X^+ ions are replaced with Y^+ ions.

14.9.2
Applications for Electrodialysis

The largest application of ED has been in the desalination of brackish water to produce potable water. In Japan, all the table salt consumed is produced by ED of sea water. The major application of ED in the food industry is probably for desalting of cheese whey. Following ED, the demineralised whey is usually concentrated further and spray dried. ED could potentially be employed in the refining of sugar from either cane or beet but, in fact, commercial applications in these industries are limited by severe membrane fouling problems.

Other potential applications of ED in food processing include:
- demineralisation of ultrafiltration permeate to improve lactose crystallisation
- separation of lactic acid from whey or soybean stock
- removal of Ca from milk, either to improve protein stability during freezing, or to simulate human milk

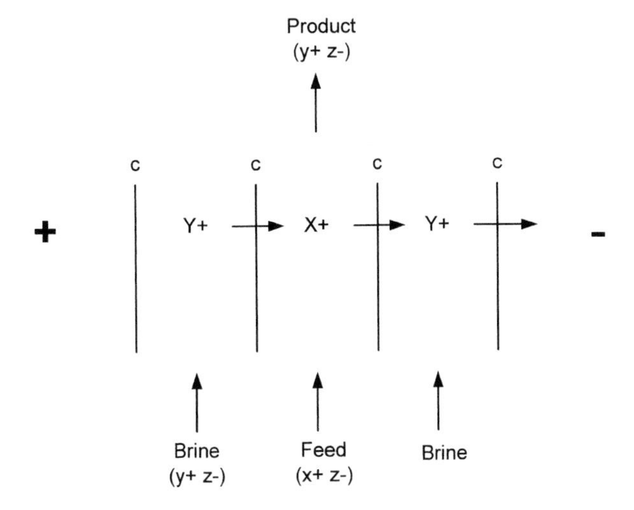

Fig. 14.37 Ion replacement using cation membranes.

- removal of radioactive metal ions from milk
- demineralisation of fermented milk products to improve flavour and textural quality
- extraction of salts from grape musts and wine to improve their stability
- controlling the sugar/acid ratio in wine either by deacidification of the grape musts by ion substitution ED using anionic membranes, or acidification using cationic membranes
- deacidification of fruit juices, either to reduce the sourness of the natural juices, or for the health food market
- desalination of spent pickling brine.

In addition, the process can be integrated into continuous fermentation or reactor designs.

References

1 Lewis, M. J. **1996**, Solids separation processes, in *Separation Processes in the Food and Biotechnology Industries*, ed. A. S. Grandison, M. J. Lewis, Woodhead, Cambridge, pp 245–286.

2 Brennan, J. G., Butters, J. R., Cowell, N. D., Lilly, A. E. V. **1990**, *Food Engineering Operations*, 3rd edn, Elsevier Applied Science, London.

3 Fellows, P. **2000**, *Food Processing Technology*, 2nd edn, Woodhead Publishing, Cambridge.

4 Toledo, R. T. B **1991**, *Fundamentals of Food Process Engineering*, 2nd edn, Van Nostrand Reinhold, New York.

5 McCabe, W. L., Smith, J. C., Harriott, P. **1985**, *Unit Operations of Chemical Engineering*, 4th edn, McGraw-Hill, New York.

6 Leniger, H.A., Beverloo, W.A. **1975**, *Food process Engineering*, Reidel Publishing, Dordrecht.
7 Purchas, D.B. **1971**, *Industrial Filtration of Liquids*, 2nd edn, Leonard Hill Books, London.
8 Akers, R.J., Ward, A.S. **1977**, Liquid Filtration Theory and Filtration Pre-Treatment, in *Filtration Principles and Practice, Part 1*, ed. C. Orr, Marcel Dekker, New York, pp 159–250.
9 Cheremisinoff, N.P., Azbel, D.S. **1983**, *Liquid Filtration*, Ann Arbor, Woburn.
10 Cheremisinoff, P.N. **1995**, *Solids/Liquids Separation*, Technomic Publishing, Lancaster, Penn.
11 Rushton, A., Griffiths, P.V.R. **1977**, Filter Media, in *Filtration Principles and Practice, Part 1*, ed. C. Orr, Marcel Dekker, New York, pp 251–308.
12 Lawson, H. **1994**, *Food Oils and Fats*, Chapman and Hall, New York.
13 De Greyt, W., Kellens, M. **2000**, Refining Practice, in *Edible Oil Processsing*, ed. W. Hamm, R.J. Hamilton, Sheffield Academic Press, Sheffield, pp 79–128.
14 Kellens, M. **2000**, Oil Modification Processes, in *Edible Oil Processing*, ed. W. Hamm, R.J. Hamilton, Sheffield Academic Press, Sheffield, pp 129–173.
15 O'Brien, R.D. **2004**, *Fats and Oils, Formulating and Processing for Applications*, 2nd edn, CRC Press, New York.
16 Hugot, E. **1986**, *Handbook of Cane Sugar Engineering*, 3rd edn, Elsevier, Amsterdam.
17 McGinnis, R.A. **1971**, Juice preparation III, in *Beet-Sugar Technology*, 2nd edn, ed. R. A. McGinnis, Beet Sugar Development Foundation, Fort Collins, Colo., pp 259–295.
18 Hough, J.S., Briggs, D.E., Stevens, R., Yound, T.W. **1982**, *Malting and Brewing Science*, vol. II, 2nd edn, Chapman and Hall, London.
19 Posada, J. **1987**, Filtration of Beer, in *Brewing Science*, vol. 3, ed. J. R. A Pollock, Academic Press, London, pp 379–439.
20 Farkas, J. **1988**, *Technology and Biochemistry of Wine*, vol. 2, Gordon and Breach Science Publishers, New York.
21 Amerine, M.A., Kunkee, R.E., Ough, C.S., Singleton, V.L., Webb, A.D. **1980**, *The Technology of Winemaking*, 4th edn, AVI Publishing, Westport.
22 Ribereau-Gayon, P., Glories, Y., Maujean, A., Dubourdieu, D. **2000**, *The Chemistry of Wine Stabilization and Treatments (Handbook of Eonology, vol. 2)*, John Wiley & Sons, Chichester.
23 Ambler, C.M. **1952**, The Evaluation of Centrifugal Performance, *Chem. Eng. Prog.*, 48, 150–158.
24 Walstra, P., Geirts, T.J., Noomen, A., Jellema, A., van Boekel, M.A.J.S. **1999**, *Dairy Technology, Principles of Milk Properties and Processes*, Marcel Dekker, New York.
25 Braddock, R.J. **1999**, *Handbook of Citrus By-Products and Processing Technology*, John Wiley & Sons, New York.
26 Nelson, P.E., Tressler, D.K. **1980**, *Fruit and Vegetable Juice Processing Technology*, 3rd edn, AVI Publishing, Westport.
27 Schwartzberg, H.G. **1987**, Leaching-Organic Materials, in *Handbook of Separation Process Technology*, ed. R.W. Rousseau, John Wiley & Sons, New York, pp 540–577.
28 Williams, M.A. **1997**, Extraction of lipids from natural sources, in *Lipid Technologies and Applications*, ed. F.D. Gunstone, F.B. Padley, Marcel Dekker, New York, pp 113–135.
29 Ebell, A., Storz, M. **1971**, Diffusion, in *Beet-Sugar Technology*, 2nd edn, ed. R.A. McGinnis, Beet Sugar Development Foundation, Fort Collins, pp 125–160.
30 Masters, K. **1991**, *Spray Drying Handbook*, 5th edn, Longman Scientific & Technical, Harlow.
31 Clarke, R.J. **1987**, Extraction, in *Coffee*, vol. 2, ed. R.J. Clarke, R. Macrae, Elsevier Applied Science, London, pp 109–145.
32 McPherson, A. **1987**, It was Squeeze or G, Now it's CCE, *Food Technol. Austral.* 39, 56–60.
33 Gardner, D.D. **1982**, Industrial Scale Hop Extraction with Liquid CO_2, *Chem. Ind.* 12, 402- 405.
34 Rizvi, S.S.H., Daniels, J.A., Benado, E.L., Zollweg, J.A. **1986**, Supercritical Fluid Extraction: Operating Principles

and Food Applications, *Food Technology*, 40, 56–64.

35 Brunner, G. **2005**, Supercritical Fluids: Technology and Application to Food Processing, *J. Food Eng.* 67, 21–33.

36 Temelli, F., Chen, C. S., Braddock, R. J. **1988**, Supercritical Fluid Extraction in Citrus Oil Processing, *Food Technol.* 46, 145–150.

37 Reverchon, E. **2003**, Supercritical Fluid Extraction, in *Encyclopedia of Food Science and Nutrition*, 2nd edn, ed. B. Caballero, L. C. Trugo, P. M. Finglas, Academic Press, London, pp 5680–5687.

38 Fair, J. R. **1987**, Distillation, in *Handbook of Separation Processes*, ed. R. W. Rousseau, John Wiley & Sons, New York, pp 229–339.

39 Foust, A. S., Wenzel, L. A., Clump, C. W., Maus, L., Andersen, L. B. **1980**, *Principles of Unit Operations*, 2nd edn, John Wiley & Sons, New York.

40 Anon. **2000**, Distillation: Technology and Engineering, in *Encyclopedia of Food Science and Technology*, 2nd edn, ed. F. J. Francis, John Wiley & Sons, New York, pp 509–518.

41 Schofield, T. **1995**, Natural Aroma Improvement by Means of the Spinning Cone, in *Food Technology International Europe*, ed. A. Turner, Sterling Publications, London, pp 137–139.

42 Owades, J. L. **2000**, Distilled Beverage Spirits, in *Encyclopedia of Food Science and Technology*, 2nd edn, ed. F. J. Francis, John Wiley & Sons, New York, pp 519–540.

43 Piggott, J. R., Connor, J. M. **2003**, Whisky, Whiskey and Bourbon, Products and Manufacture, in *Encyclopedia of Food Science and Nutrition*,2nd edn, ed. B. Caballero, L. C. Trugo, P. M. Finglas, Academic Press, London, pp 6171–6177.

44 Nicol, D. **1989**, Batch Distillation, in *The Science and Technologies of Whiskies*, ed. J. R. Piggott, R. Sharp, R. E. Duncan, Longman Scientific & Technical, Harlow, pp 118–149.

45 Panek, R. J., Boucher, A. R. **1989**, Continuous Distillation, in *The Science and Technologies of Whiskies*, ed. J. R. Piggott, R. Sharp, R. E. Duncan, Longman Scientific & Technical, Harlow, pp 150–181.

46 Mersmann, A. (ed.) **1994**, *Crystallisation Technology Handbook*, Marcel Dekker, New York.

47 Hartel, R. W. **2001**, *Crystallisation in Foods*, Aspen, Gaithersburg.

48 Singh, G. **1988**, Crystallisation from Solutions, in *Separation Techniques for Chemical Engineers*, 2nd edn, ed. P. A. Schweitzer, McGraw-Hill, London, pp 151–182.

49 Rajah, K. K. **1996**, Fractionation of Fat, in *Separation Processes in the Food and Biotechnology Industries*, ed. A. S. Grandison, M. J. Lewis, Woodhead Publishing, Cambridge, pp 207–241.

50 Saravacos, G. D., Kostaropoulos, A. E. **2002**, *Handbook of Food Processing Equipment*, Kluwer Academic, London.

51 Mersmann, A., Rennie, F. W. **1994**, Design of Crystallizers and Crystallization Processes, in *Crystallisation Technology Handbook*, ed. A. Mersmann, Marcel Dekker, New York, pp 215–325.

52 Lewis, M. J. **1996**, Ultrafiltration, in *Separation Processes in the Food and Biotechnology Industries*, ed. A. S. Grandison, M. J. Lewis, Woodhead Publishing, Cambridge, pp 97–154.

53 Youravong, W., Lewis M. J., Grandison, A. S. **2003**, Critical Flux in Ultrafiltration of Skim Milk, *Trans. Inst. Chem. Eng.* 81, 303–308.

54 Glover, F. **1985**, *Ultrafiltration and Reverse Osmosis for the Dairy Industry* (NIRD Technical Bulletin No. 5), National Institute for Research in Dairying, Reading.

55 El-Gazzar, F. E., Marth, E. H. **1991**, Ultrafiltration and Reverse Osmosis in Dairy Technology – a Review, *J. Food Prot.* 54, 801–809.

56 Renner, E., El-Salam, M. H. A. **1991**, *Application of Ultrafiltration in the Dairy Industry*, Elsevier Applied Science, London.

57 Leeper, S. A. **1987**, Membrane Separations in the Production of Alcohol Fuels by Fermentation, in *Membrane Separations in Biotechnology*, ed. W. C. McGregor, Marcel Dekker, New York.

58 Guu, Y. K., Zall, R. R. **1992**, Nanofiltration Concentration on the Efficiency of Lactose Crystallisation, *J. Food Sci.* 57, 735–739.

59 Kosikowski, F. V. **1986**, Membrane Separations in Food Processing, in *Membrane Separations in Biotechnology*, ed. W. C. MeGregor, Marcel Dekker, New York.

60 International Dairy Federation **1979**, Equipment available for membrane processing, *Int. Dairy Fed. Bull.* 115.

61 Bastian, E. D., Collinge, S. K., Ernstrom, C. A. **1991**, Ultrafiltraton: Partitioning of Milk Constituents into Permeate and Retentate, *J. Dairy Sci.* 74, 2423–2434.

62 Premaratne, R. J., Cousin, M. A. **1991**, Changes in the Chemical Composition During Ultrafiltration of Skim Milk, *J. Dairy Sci.* 74, 788–795.

63 de Boer, R., Koenraads, J. P. J. M. **1991**, Incorporation of Liquid Ultrafiltration – Whey Retentates in Dairy Desserts and Yoghurts, in *New Applications in Membrane Processes* (International Dairy Federation Special Issue 9201), International Dairy Federation, Brussels.

64 Tamime, A. Y., Davies, G., Chekade, A. S., Mahdi, H. A. **1991**, The effect of processing temperatures on the quality of labneh made by ultrafiltration, *J. Soc. Dairy Technol.* 44, 99–103.

65 Daufin, G., Merin, U., Kerherve, F. L., Labbe, J. P., Quemerais, A., Bousser, C. **1992**, Efficiency of Cleaning Agents for an Inorganic Membrane after Milk Ultrafiltration, *J. Dairy Res.* 59, 29–38.

66 Lawrence, R. C. **1989**, The Use of Ultrafiltration Technology in Cheese Making, *Int. Dairy Fed. Bull.* 240.

67 Lelievre, J., Lawrence, R. C. **1988**, Manufacture of Cheese from Milk Concentrated by Ultrafiltration, *J. Dairy Sci.* 55, 465–470.

68 Kosikowski, F. V. **1988**, Enzyme Behaviour and Utilisation in Dairy Technology, *J. Dairy Sci.* 71, 557–573.

69 International Dairy Federation **1991**, *New Applications of Membrane Processes* (Special Issue No. 9201), International Dairy Federation, Brussels.

70 Lewis, M. J. **1982**, Ultrafiltration of proteins, in *Developments in Food Proteins, vol. 1*, ed. B. J. F. Hudson, Applied Science Publishers, London, pp 91–130.

71 Cheryan M. **1986**, *Ultrafiltration Handbook*, Technomic Publishing, Lancaster.

72 Lewis, M. J., Finnigan, T. J. A. **1989**, Removal of Toxic Components Using Ultrafiltration, in *Process Engineering in the Food Industry*, ed. R. W. Field, J. A. Howell, Elsevier Applied Science, London, pp 291–306.

73 Ockerman, H. W., Hansen, C. L. **1988**, *Animal By-Product Processing*, Ellis Horwood, Chichester.

74 Wong, W., Jelen, P., Chang, R. **1984**, Ultrafiltration of Bovine Blood, in *Engineering and Food, vol. 1*, ed. B. McKenna, Elsevier Applied Science, London, pp 551–558.

75 Finnigan, T. J. A., Skudder, P. J. **1989**, The Application of Ceramic Microfiltration in the Brewing Industry, in *Processing Engineering in the Food Industry*, ed. R. W. Field, J. A. Howell, Elsevier Applied Science, London, pp 259–272.

76 Piot, M., Maubois, J. L., Schaegis, P., Veyre, R. **1984**, Microfiltration en Flux Tangential des Lactoserums de Fromagerie et al., *Le Lait* 64, 102–120.

77 Merin, U. **1986**, Bacteriological Aspects of Microfiltration of Cheese Whey, *J. Dairy Sci.* 69, 326–328.

78 Hanemaaijer, J. H. **1985**, Microfiltration in Whey Processing, *Desalination* 53, 143–155.

79 Malmbert, R., Holm, S. **1988**, Producing low-bacteria milk by ultrafiltration, *N. Eur. Food Dairy J.* 1, 1–4.

80 Cravendale Milk, available at: www.arlafoods.com.

81 Hansen R. **1988**, Better Market Milk, Better Cheese Milk and Better Low-Heat Milk Powder with Bactocatch Treated Milk, *N. Eur. Food Dairy J.* 1, 5–7

82 Piot, M., Vachot, J. C, Veaux, M., Maubois, J. L., Brinkman, G. E. **1987**, Ecremage et Epuration Bacterienne du Lait Entire Cru par Microfiltration sur Membrane en Flux Tangential, *Tech. Lait. Market.* 1016, 42–46.

83 Rios, G. M., Taraodo de la Fuente, B., Bennasar, M., Guidard, C. **1989**, Cross-Flowfiltration of Biological Fluids in Inorganic Membranes: A First State of the Art, in *Developments in Food Preservation 5*, ed. S. Thorne, Elsevier Applied Science, London, pp 131–175.

84 Grandison, A. S., Glover, F. A. **1994**, Membrane Processing of Milk, in *Modern Dairy Technology*, vol. 1, ed. R. K. Robinson, Elsevier Applied Science, London, pp 73–311.

85 Devereux, N., Hoare, M. **1986**, Membrane Separation of Proteins and Precipitates: Studies with Cross Flow in Hollow Fibres, *Biotechnol. Bioeng.* 28, 422–431.

86 Short, J. L. **1988**, Newer Applications for Crossflow Membrane Filtration, *Desalination* 70, 341–352.

87 Anderson, R. E. **1988**, Ion-Exchange Separations, in *Handbook of Separation Techniques for Chemical Engineers*, 2nd edn, ed. P. A. Schweitzer, McGraw-Hill, London, pp (1)387–(1)444.

88 Grandison, A. S. **1996**, Ion-Exchange and Electrodialysis, in *Separation Processes in the Food and Biotechnology Industries*, ed. A. S. Grandison, M. J. Lewis, Woodhead Publishing, Cambridge, pp 155–177.

89 Walton, H. F. **1983**, Ion-Exchange Chromatography, in *Chromatography, Fundamentals and Applications of Chromatographic Methods, Part A – Fundamentals and Techniques* (Journal of Chromatography Library, vol. 22A), ed. E. Heftmann, Elsevier Scientific, Amsterdam, pp A225–A255.

90 Lopez Leiva, M. H. **1988**, The Use of Electrodialysis in Food Processing Part 1: Some Theoretical Concepts, *Lebensm. Wiss. Technol.* 21, 119–125.

91 Lopez Leiva, M. H. **1988**, The Use of Electrodialysis in Food Processing Part 2: Review of Practical Applications, *Lebensm. Wiss. Technol.* 21, 177–182.

15
Mixing, Emulsification and Size Reduction

James G. Brennan

15.1
Mixing (Agitation, Blending)

15.1.1
Introduction

Mixing is a unit operation widely used in food processing. Many definitions of this term have been proposed. One of the simplest is: “an operation in which a uniform combination of two or more components is formed”. In addition to blending components together, mixing operations may bring about other desirable changes in the materials being mixed, such as mechanical working (as in dough mixing), promotion of heat transfer (as in freezing ice cream) facilitating chemical or biological reactions (as in fermentation). The components in a mixing operation may be liquids, pastes, dry solids or gases. The degree of uniformity attainable in a mixing operation varies, depending on the nature of the components. In the case of low viscosity miscible liquids or highly soluble solids in liquids a high degree of uniformity is attainable. Less intimate mixing is likely to occur in the case of viscous liquids, pastes and dry solids. Combining immiscible materials together usually requires specialised equipment, which is covered under emulsification in Section 15.2. Efficient utilisation of energy is another criterion of mixing. This is more easily attainable in the case of low viscosity liquids as compared with pastes and dry solids.

15.1.2
Mixing of Low and Moderate Viscosity Liquids

The impeller mixer is the most commonly used type of mixer for low viscosity liquids (viscosity less than 100 poise; 10 N s m^{-2}). Such a mixer consists of one or more impellers, fixed to a rotating shaft and immersed in the liquid. As the impellers rotate, they create currents within the liquid, which travel throughout the mixing vessel. If turbulent conditions are created within the moving

Food Processing Handbook. Edited by James G. Brennan

ISBN: 3-527-30719-2

streams of liquid, mixing will occur. Turbulence is usually most vigorous near the impeller and the liquid should pass through this region as often as possible. The fluid velocity in the moving streams has three components: (a) a radial component acting in a direction at right angles to the shaft, (b) a longitudinal component acting parallel to the shaft and (c) a rotational component acting in a direction tangential to the circle of rotation of the shaft. The radial and longitudinal components usually promote mixing but the rotational component may not.

If an impeller agitator is mounted on a vertical shaft located centrally in a mixing vessel, the liquid will flow in a circular path around the shaft. If laminar conditions prevail, then layers of liquid may form, the contents of the vessel rotate and mixing will be inefficient. Under these conditions a vortex may form at the surface of the liquid. As the speed of rotation of the impeller increases this vortex deepens. When the vortex gets close to the impeller, the power imparted to the liquid drops and air is sucked into the liquid. This will greatly impair the mixing capability of the mixer. Rotational flow may cause any suspended particles in the liquid to separate out under the influence of centrifugal force. Rotational flow, and hence vortexing, may be reduced by locating the mixer offcentre in the mixing vessel and/or by the use of baffles. Baffles usually consist of vertical strips fixed at right angles to the inner wall of the mixing vessel. These break up the rotational flow pattern and promote better mixing (Fig. 15.1). Usually four baffles are used, with widths corresponding to 1/18th (5.55%) to 1/12th (8.33%) of the vessel diameter.

Three main types of impeller mixers are used for liquid mixing.

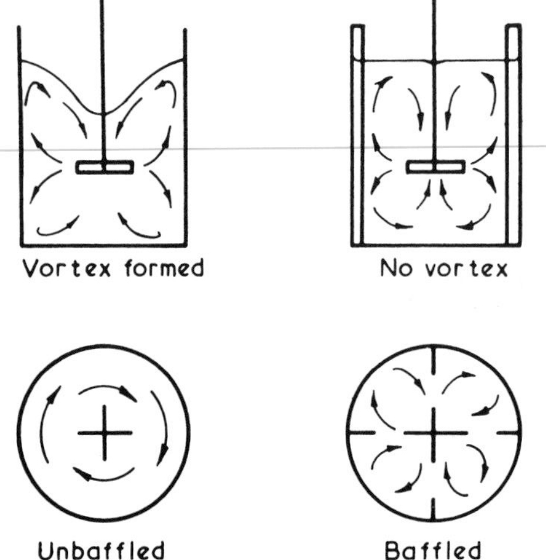

Fig. 15.1 Flow patterns in baffled and unbaffled vessels with paddle or turbine agitators; from [6] with permission of the authors.

15.1.2.1 **Paddle Mixer**

This type of mixer consists of a flat blade attached to a rotating shaft, which is usually located centrally in the mixing vessel (Fig. 15.2). The speed of rotation is relatively low, in the range 20–150 rpm. The blade promotes rotational and radial flow but very little vertical flow. It is usually necessary to fit baffles to the mixing vessel. Two or four blades may be fitted to the shaft.

Other forms of paddle mixer include: (a) the gate agitator (Fig. 15.2 b), which is used for more viscous liquids, (b) the anchor agitator (Fig. 15.2 c), which rotates close to the wall of the vessel and helps to promote heat transfer and prevent fouling in jacketed vessels and (c) counter-rotating agitators (Fig. 15.2 d), which develop relatively high shear rates near the impeller. Simple paddle agitators are used mainly to mix miscible liquids and to dissolve soluble solids in liquids.

15.1.2.2 **Turbine Mixer**

A turbine mixer has four or more blades attached to the same shaft, which is usually located centrally in the mixing vessel. The blades are smaller than paddles and rotate at higher speeds, in the range 30–500 rpm. Simple vertical blades (Fig. 15.3 a) promote rotational and radial flow. Some vertical flow develops when the radial currents are deflected from the vessel walls (Fig. 15.1). Swirling and vortexing are minimised with the use of baffles. Liquid circulation is generally more vigorous than that produced by paddles and shear and turbulence is high near the impeller itself.

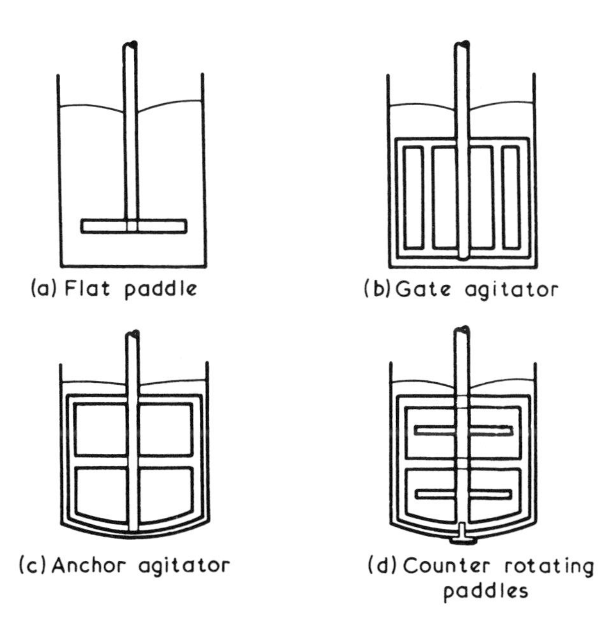

Fig. 15.2 Some typical paddle impellers; from [6] with permission of the authors.

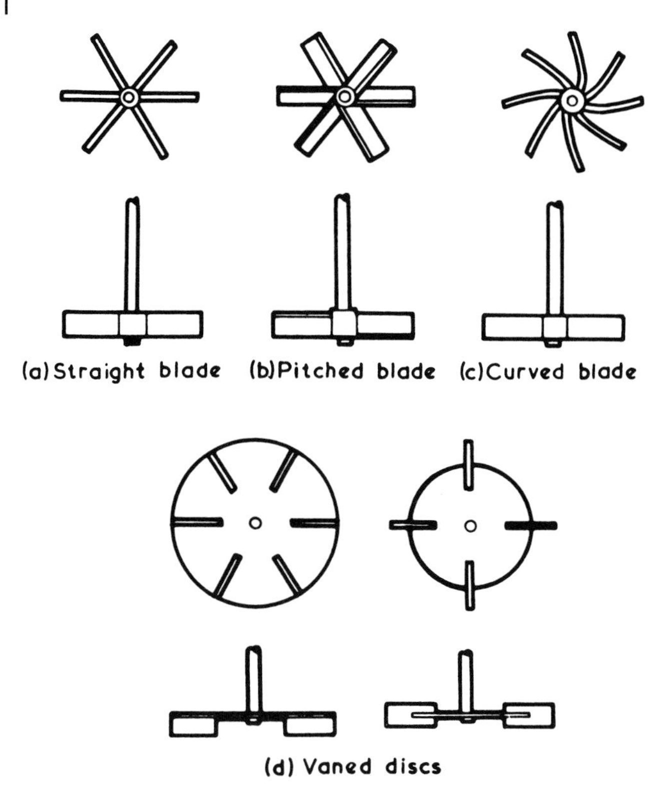

Fig. 15.3 Some typical turbine impellers; from [16] with permission of the authors.

Pitched blades (Fig. 15.3 b) increase vertical flow. Curved blades (Fig. 15.3 c) are used when less shear is desirable, e.g. when mixing friable solids. Vaned or shrouded discs (Fig. 15.3 d) control the suction and discharge pattern of the impeller and are often used when mixing gases into liquids. Turbine mixers are used for low and moderate viscosity liquids, up to 600 poise, for preparing solutions and incorporating gases into liquids.

15.1.2.3 **Propeller Mixer**

This type of mixer consists of a relatively small impeller, similar in design to a marine propeller, which rotates at high speed, up to several thousand rpm. It develops strong longitudinal and rotational flow patterns. If mounted on a vertical shaft and located centrally in the mixing vessel, baffling is essential (Fig. 15.4 a, b). Alternatively, the shaft may be located off centre in the vessel and/or at an angle to the vertical (Fig. 15.4 c). When mixing low viscosity liquids, up to 20 poise, the currents developed by propeller agitators can travel throughout large vessels. In such cases the shaft may enter through the side wall of the tank (Fig. 15.4 d). Special propeller designs are available which pro-

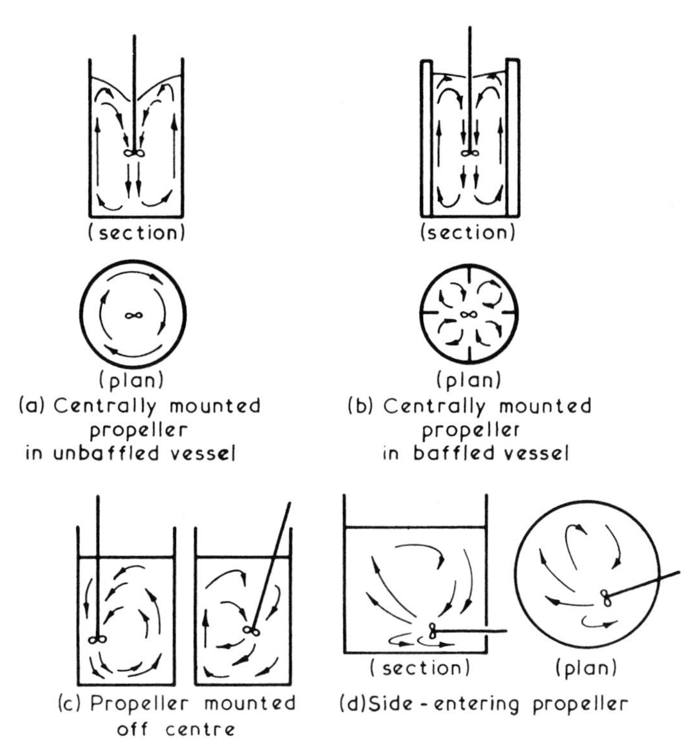

Fig. 15.4 Flow patterns in propeller agitated systems; from [16] with permission of the authors.

mote shear, for emulsion premixing. Others have serrated edges for cutting through fibrous solids.

Many other types of impeller mixers are used for low viscosity liquids, including discs and cones attached to shafts. These promote gentle mixing. More specialised mixing systems are available for emulsion premixing, dispersion of solids and similar duties. One such system is shown in (Fig. 15.5).

Other methods of mixing low viscosity liquids include: (a) pumping them through pipes containing bends, baffles and/or orifice plates, (b) injecting one liquid into a tank containing the other components and (c) recirculating liquids through a holding tank using a centrifugal pump [1–8].

15.1.3
Mixing of High Viscosity Liquids, Pastes and Plastic Solids

When mixing highly viscous and pastelike materials, it is not possible to create currents which will travel to all parts of the mixing vessel, as happens when mixing low viscosity liquids. Consequently, there must be direct contact between the mixing elements and the material being mixed. The mixing elements must

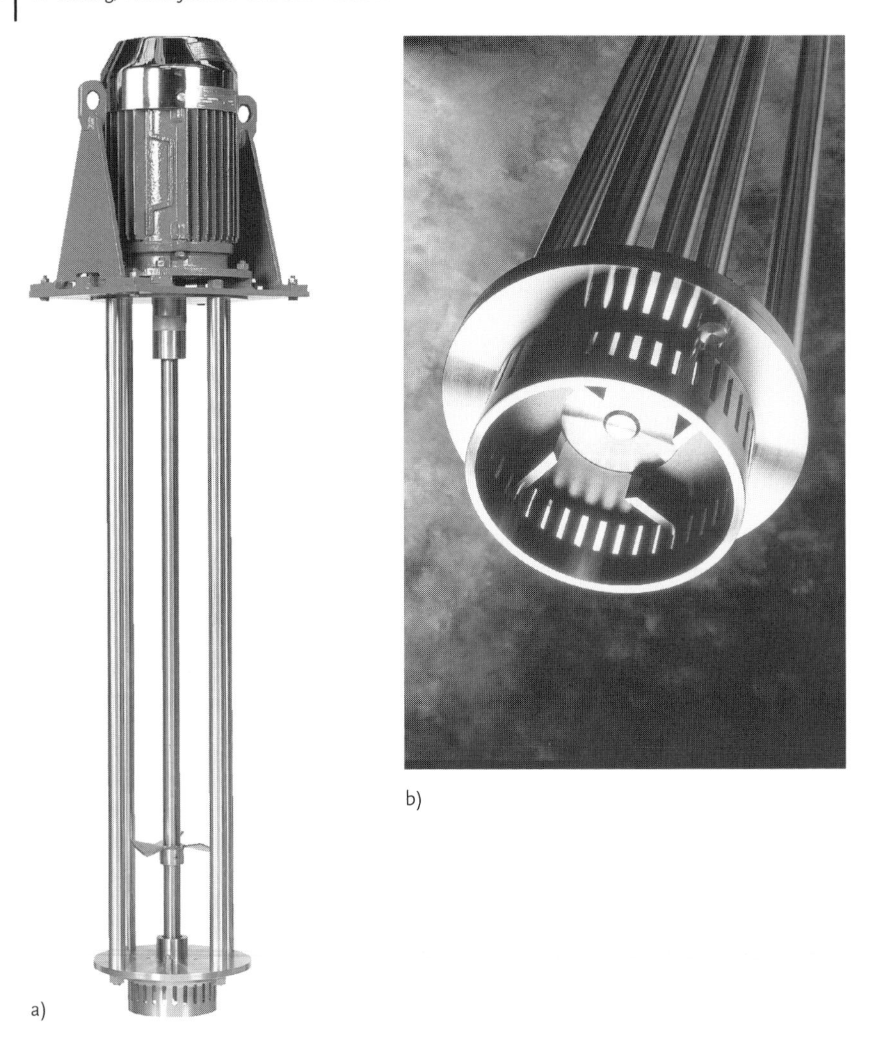

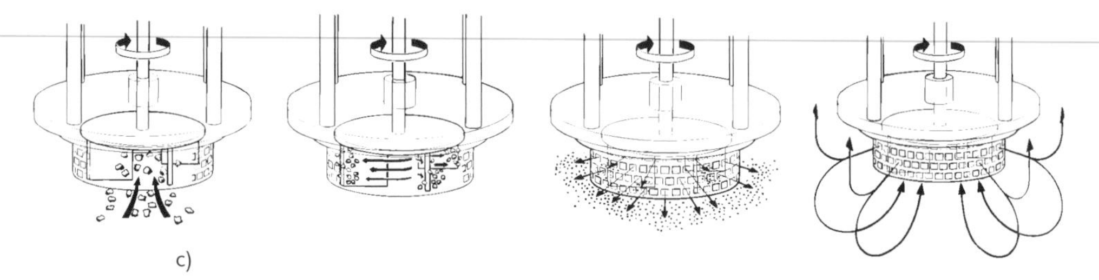

Fig. 15.5 A high shear mixer: (a) the complete mixer, (b) the mixing head, (c) high shear mixer in operation; by courtesy of Silverson Machines Ltd.

travel to all parts of the mixing vessel or the material must be brought to the mixing elements. Mixers for such viscous materials generally need to be more robust, have a smaller working capacity and have a higher power consumption than those used for liquid mixing. The speed of rotation of the mixing elements is relatively low and the mixing times long compared to those involved in mixing liquids.

15.1.3.1 Paddle Mixers
Some designs of paddle mixer, of heavy construction, may be used for mixing viscous materials. These include gate, anchor and counter-rotating paddles, as shown in Fig. 15.2 a, b and c, respectively.

15.1.3.2 Pan (Bowl, Can) Mixers
In one type of pan mixer the bowl rests on a turntable, which rotates. One or more mixing elements are held in a rotating head and located near the bowl wall. They rotate in the opposite direction to the pan. As the pan rotates it brings the material into contact with the mixing elements. In another design of pan mixer, the bowl is stationary. The mixing elements rotate and move in a planetary pattern, thus repeatedly visiting all parts of the bowl. The mixing elements are shaped to pass with a small clearance between the wall and bottom of the mixing vessel (Fig. 15.6). Various designs of mixing elements are used, including gates, forks, hooks and whisks, for different applications.

15.1.3.3 Kneaders (Dispersers, Masticators)
A common design of kneader consists of a horizontal trough with a saddle-shaped bottom. Two heavy blades mounted on parallel, horizontal shafts rotate towards each other at the top of their cycle. The blades draw the mass of material down over the point of the saddle and then shear it between the blades and the wall and bottom of the trough. The blades may move tangentially to each other and often at different speeds, with a ratio of 1.5 : 1.0. In some such mixers the blades may overlap and turn at the same or different speeds. Mixing times are generally in the range 2–20 min. One type of kneader, featuring Z- or Σ-blade mixing elements is shown in Fig. 15.7.

15.1.3.4 Continuous Mixers for Pastelike Materials
Many different devices are used to mix viscous materials on a continuous basis. Screw conveyors, rotating inside barrels, may force such materials through perforated plates or wire meshes. Passing them between rollers can effect mixing of pastelike materials. Colloid mills may be used for a similar purpose, see Section 15.2.3.7.

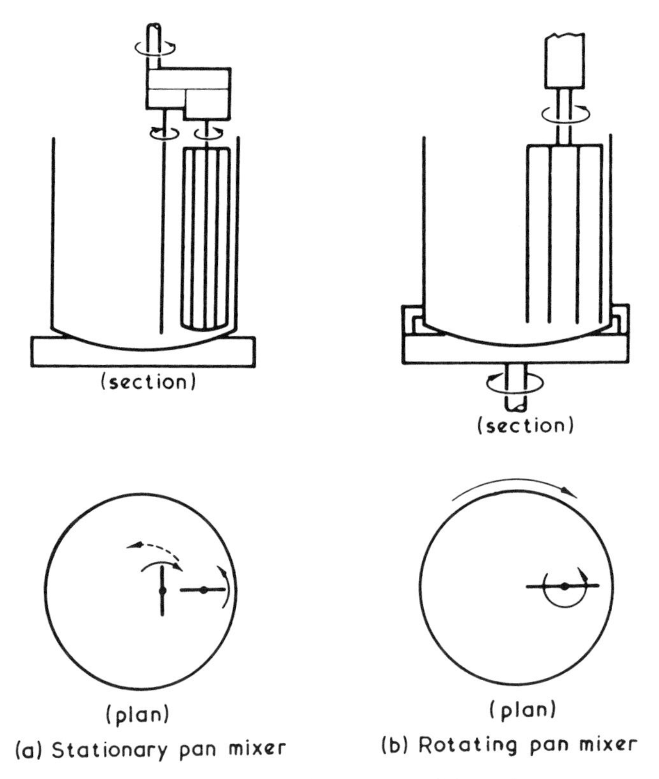

Fig. 15.6 Pan mixers; from [6] with permission of the authors.

15.1.3.5 **Static Inline Mixers**

When viscous liquids are pumped over specially shaped stationary mixing elements located in pipes, mixing may occur. The liquids are split and made to flow in various different directions, depending on the design of the mixing elements. Many different configurations are available. The energy required to pump the materials through these mixing elements is usually less than that required to drive the more conventional types of mixers discussed above [1, 5, 6, 8–12].

15.1.4 **Mixing Dry, Particulate Solids**

In most practical mixing operations involving dry particulate solids *unmixing* or *segregation* is likely to occur. Unmixing occurs when particles within a group are free to change their positions. This results in a change in the packing characteristics of the solid particles. Unmixing occurs mainly when particles of different sizes are being mixed. The smaller particles can move through the gaps between the larger particles, leading to segregation. Differences in particle shape

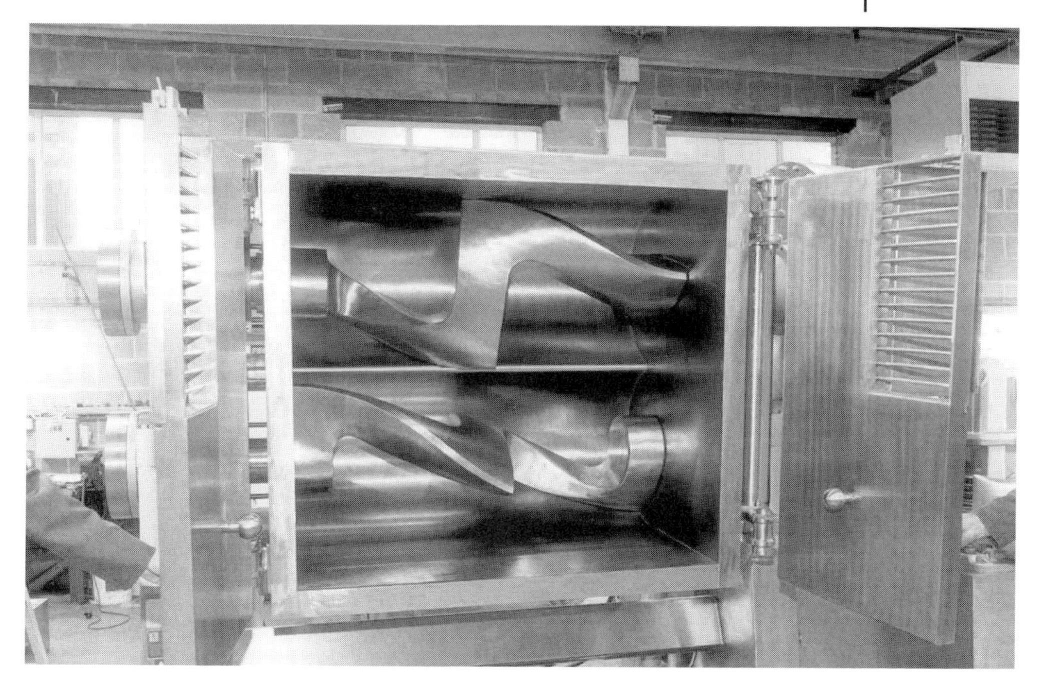

Fig. 15.7 Z-blade (Σ-blade) mixer; by courtesy of Winkworth Machinery Ltd.

and density may also contribute to segregation. Materials with particle sizes of 75 μm and above are more prone to segregation than those made up of smaller particles. Small cohesive particles, which bind together under the influence of surface forces, do not readily segregate. In mixing operations where segregation occurs, an equilibrium between mixing and unmixing will be established after a certain mixing time. Further mixing is not likely to improve the uniformity of the mix.

There are two basic mechanisms involved in mixing particulate solids, i.e. *convection*, which involves the transfer of masses or groups of particles from one location in the mixer to another, and *diffusion*, which involves the transfer of individual particles from one location to another arising from the distribution of particles over a freshly developed surface. Usually both mechanisms contribute to any mixing operation. However, one mechanism may predominate in a particular type of mixer. Segregation is more likely to occur in mixers in which diffusion predominates.

15.1.4.1 Horizontal Screw and Ribbon Mixers

These consist of horizontal troughs, usually semicylindrical in shape, containing one or two rotating mixing elements. The elements may take the form of single or twin screw conveyors. Alternatively, mixing ribbons may be employed. Two such ribbons may be mounted on a single rotating shaft. They may be continuous or interrupted. The design is such that one ribbon tends to move the solids in one direction while the other moves them in the opposite direction. If the rate of movement of the particles is the same in both directions, the mixer is operated on a batch principle. If there is a net flow in one direction, the mixer may be operated continuously. The mixing vessel may be jacketed for temperature control. If enclosed, it may be operated under pressure or vacuum. Convection is the predominant mechanism of mixing in this type of mixer. Some segregation may occur but not to a serious extent.

15.1.4.2 **Vertical Screw Mixers**

These consist of tall, cylindrical or cone-shaped vessels containing a single rotating screw, which elevates and circulates the particles. The screw may be located vertically at the centre of the vessel. Alternatively, it may be set at an angle to the vertical and made to rotate, passing close to the wall of the vessel (Fig. 15.8). The convective mechanism of mixing predominates in such mixers and segregation should not be a serious problem.

15.1.4.3 **Tumbling Mixers**

These consist of hollow vessels, which rotate about horizontal axes. They are partly filled, up to 50–60% of their volume, with the materials being mixed and then rotated, typically for 5–20 min, to bring about mixing of the contents. Because the main mechanism of mixing is diffusion, segregation can be a problem if the particles vary in size. Various designs are available, some of which are shown in Fig. 15.9. They may contain baffles or stays to enhance the mixing effect and break up agglomerates of particles.

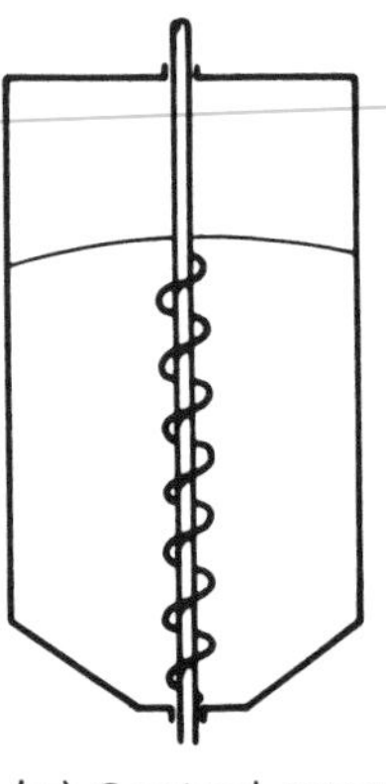

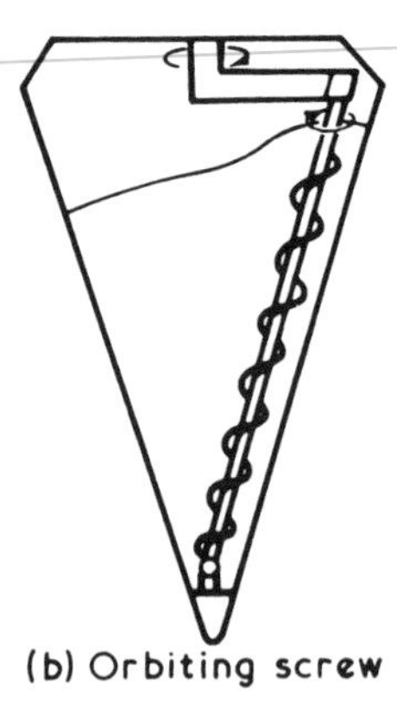

Fig. 15.8 Vertical screw mixers; from [6] with permission of the authors.

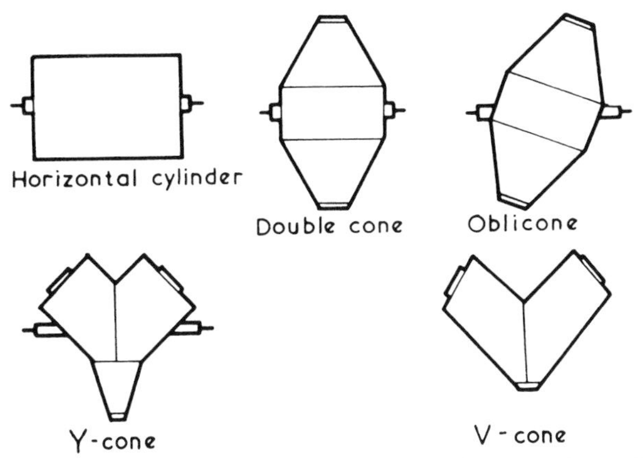

Fig. 15.9 Some typical tumbling mixer shapes; from [6] with permission of the authors.

15.1.4.4 **Fluidised Bed Mixers**

Fluidised beds, similar to those used in dehydration (see Section 3.2.3.5), may be used for mixing particulate solids with similar sizes, shapes and densities. The inclusion of spouting jets of air, in addition to the main supply of fluidising air, can enhance the mixing [1, 2, 5–8, 13–18].

15.1.5 Mixing of Gases and Liquids

Gases may be mixed with low viscosity liquids using impeller agitators in mixing vessels. Turbine agitators, as shown in Fig. 15.3, are generally used for this purpose. In unbaffled vessels, vortexing can draw gas into the liquid. However, the gas may not be well distributed throughout the liquid. It is more usual to use baffled vessels with relatively high-speed turbine impellers. Impellers with six, 12 and even 18 blades have been used. Pitched blades and vaned discs (Fig. 15.3 b and d, respectively) are particularly suited to this duty. Some special designs of vaned discs, featuring concave rather than flat blades, have been used for this purpose. Gas may be introduced into liquids using some designs of static inline mixers.

More heavy duty equipment, such as pan and Z-blade mixers, may be used to introduce gas into more viscous materials. For example whisk-like elements in pan mixers may be used to whip creams. Other types of elements, such as forks, hooks and gates, are used to introduce air into doughs and batters. Dynamic, inline systems, such a scraped surface heat exchangers (see Section 2.4.3.3) may be used to heat or cool viscous materials while at the same time introducing gas to aerate them, such as in the manufacture of ice cream, see Section 15.2.4.2 [19].

15.1.6
Applications for Mixing in Food Processing

15.1.6.1 Low Viscosity Liquids
Examples of applications for impeller mixers include: preparing brines and syrups, preparing liquid sugar mixtures for sweet manufacture, making up fruit squashes, blending oils in the manufacture of margarines and spreads, premixing emulsion ingredients.

15.1.6.2 Viscous Materials
Examples of applications for pan mixers and kneaders include: dough and batter mixing in bread, cake and biscuit making, blending of butters, margarines and cooking fats, preparation of processed cheeses and cheese spreads, manufacture of meat and fish pastes.

15.1.6.3 Particulate Solids
Examples of application for screw, ribbon and tumbling mixers include: preparing cake and soup mixes, blending of grains prior to milling, blending of flours and incorporation of additives into them.

15.1.6.4 Gases into Liquids
Examples of applications of this duty include: carbonation of alcoholic and soft drinks, whipping of dairy and artificial creams, aerating ice cream mix during freezing, supplying gas to fermenters.

15.2
Emulsification

15.2.1
Introduction

Emulsification may be regarded as a mixing operation whereby two or more normally immiscible materials are intimately mixed. Most food emulsions consist of two phases: (a) an aqueous phase consisting of water which may contain salts and sugars in solution and/or other organic and colloidal substances, known as hydrophilic materials and (b) an oil phase which may consist of oils, fats, hydrocarbons, waxes and resins, so called hydrophobic materials. Emulsification is achieved by dispersing one of the phases in the form of droplets or globules throughout the second phase. The material which broken up in this way is known as the dispersed, discontinuous or internal phase, while the other is referred to as the dispersing, continuous or external phase. In addition to the two main phases other substances, known as *emulsifying agents*, are usually included, in small quantities, to produce stable emulsions.

When the water and oil phases are combined by emulsification, two emulsion structures are possible. The oil phase may become dispersed throughout the aqueous phase to produce an oil in water (o/w) emulsion. Alternatively, the aqueous phase may become dispersed throughout the oil phase to produce a water in oil (w/o) emulsion. In general, an emulsion tends to exhibit the characteristics of the external phase. Two emulsions of similar composition can have different properties, depending on their structure. For example, an o/w emulsion can be diluted with water, coloured with water-soluble dye and has a relatively high electrical conductivity compared to a w/o system. The latter is best diluted with oil and coloured with oil-soluble dye. Many factors can influence the type of emulsion formed when two phases are mixed, including the type of emulsifying agent used, the relative proportions of the phases and the method of preparation employed.

Emulsions with more complex structures, known as *multiple emulsions*, may also be produced. These may have oil in water in oil (o/w/o) or water in oil in water (w/o/w) structures. The latter structure has the most applications. w/o/w emulsions may be produced in two stages. First a w/o emulsion is produced by homogenisation using a hydrophobic agent. This is then incorporated into an aqueous phase using a hydrophilic agent. In such a multiple emulsion, the oil layer is thin and water can permeate through it due to an osmotic pressure gradient between the two aqueous phases. This property is used in certain applications including prolonged drug delivery systems, drug overdose treatments and nutrient administration for special dietary purposes [20].

Emulsions with very small internal phase droplets, in the range 0.0015–0.15 μm in diameter, are termed *microemulsions*. They are clear in appearance, the droplets have a very large contact area and good penetration properties. They have found uses in the application of herbicides and pesticides and the application of drugs both orally and by intravenous injection [21].

In a two-phase system free energy exists at the interface between the two immiscible liquids, due to an imbalance in the cohesive forces of the two materials. Because of this *interfacial tension*, there is a tendency for the interface to contract. Thus, if a crude emulsion is formed by mixing two immiscible liquids, the internal phase will take the form of spherical droplets, representing the smallest surface area per unit volume. If the mixing is stopped, the droplets coalesce to form larger ones and eventually the two phases will completely separate. To form an emulsion, this interfacial tension has to be overcome. Thus, energy must be introduced into the system. This is normally achieved by agitating the liquids. The greater the interfacial tension between the two liquids the more energy is required to disperse the internal phase. Reducing this interfacial tension will facilitate the formation of an emulsion. This is one important role of emulsifying agents. Another function of these agents is to form a protective coating around the droplets of the internal phase and thus prevent them from coalescing and destabilising the emulsion [1, 6, 22, 23].

15.2.2
Emulsifying Agents

Substances with good emulsifying properties have molecular structures which contain both polar and nonpolar groups. Polar groups have an affinity for water, i.e. are hydrophilic, while nonpolar groups are hydrophobic. If there is a small imbalance between the polar and nonpolar groups in the molecules of a substance, it will be adsorbed at the interface between the phases of an emulsion. These molecules will become aligned at the interface so that the polar groups will point towards the water phase while the nonpolar groups will point towards the oil phase. By bridging the interface in this way, they reduce the interfacial tension between the phases and form a film around the droplets of the internal phase, thus stabilising them. A substance in which the polar group is stronger will be more soluble in water than one in which the nonpolar group predominates. Such a substance will tend to promote the formation of an o/w emulsion. However, if the nonpolar group is dominant the substance will tend to produce a w/o emulsion. If the polar and nonpolar groups are grossly out of balance the substance will be highly soluble in one or other of the phases and will not accumulate at the interface. Such materials do not make good emulsifying agents.

The emulsifying capability of an agent may be classified according to the hydrophile-lipophile balance (HLB) in its molecules. This is defined as the ratio of the weight percentage of hydrophilic groups to the weight percentage of hydrophobic groups in the molecule. HLB values for emulsifying agents range from 1 to 20. Agents with low values, 3–6, promote the formation of w/o emulsions,

Table 15.1 Examples of emulsifying agents and their HLB values

Ionic	HLB	Nonionic	HLB
Proteins (e.g. gelatin, egg albumin)		Glycerol esters	2.8
Phospholipids (e.g. lecithin)		Polyglycerol esters	
Potassium and sodium salts of oleic acid	18.0–20.0	Propylene glycol fatty acid esters	3.4
Sodium stearoyl-2-lactate		Sorbitol fatty acid esters	4.7
		Polyoxyethylene fatty acid esters	14.9–15.9
Hydrocolloids			
Agar		Carboxymethyl cellulose	
Pectin		Hydroxypropyl cellulose	
Gum tragacanth	11.9	Methyl cellulose	10.5
Alginates		Guar gum	
Carrageen		Locust bean gum	

while those with high values, 8–16, favour the formation of o/w types. Agents with even higher HLB values are used in detergents and solubilisers.

Protein, phospholipids and sterols, which occur naturally in foods, act as emulsifying agents. A wide range of manufactured agents is available, both ionic and nonionic. Ionic emulsifiers often react with other oppositely charged particles such as hydrogen and metallic ions to form complexes. This may result in a reduction in solubility and emulsifying capacity. Nonionic emulsifiers are more widely used in food emulsions. Hydrocolloids such as pectin and gums are often used to increase the viscosity of emulsions to reduce separation of the phases under gravity (creaming). Table 15.1 lists some commonly used emulsifying agents, together with their HLB values. The treatment of the principles of emulsification and the role of emulsifying agents given above is a very simplified one. More detailed accounts will be found in the references cited [6, 22–26].

15.2.3 Emulsifying Equipment

The general principle on which all emulsifying equipment is based is to introduce energy into the system by subjecting the phases to vigorous agitation. The type of agitation which is most effective in this context is that which subjects the large droplets of the internal phase to shear. In this way, these droplets are deformed from their stable spherical shapes and break up into smaller units. If the conditions are suitable and the right type and quantity of emulsifying agent(s) is present, a stable emulsion will be formed. One important condition which influences emulsion formation is temperature. Interfacial tension and viscosity are temperature-dependent, both decreasing with increase in temperature. Thus, raising the temperature of the liquids usually facilitates emulsion formation. However, for any system there will be an upper limit of temperature depending on the heat sensitivity of the components.

15.2.3.1 Mixers

In the case of low viscosity liquids, turbine and propeller mixers may be used to premix the phases prior to emulsification. In some simple systems, a stable emulsion may result from such mixing and no further treatment may be required. Special impellers designs are available which promote emulsification, such as the Silverson system shown in Fig. 15.5. In the case of viscous liquids and pastes, pan mixers, kneaders and some types of continuous mixers (see Section 15.1.3) may be used to disperse one phase throughout another. Tumbling mixers, such as those used for mixing powders (see Section 15.1.4.3) may also be used for this purpose.

15.2.3.2 Pressure Homogenisers

The principle of operation of all pressure homogenisers is that the premixed phases are pumped through a narrow opening at high velocity. The opening is usually provided between a valve and its seat. Therefore, a pressure homogeniser consists of one or two valves and a high pressure pump. As the liquids pass through the gap, 15–300 μm wide, between the valve and seat they are accelerated to speeds of 50–300 m s^{-1}. The droplets of the internal phase shear against each other, are distorted and break up into smaller units. As the liquids exit from the gap, there is a sudden drop in pressure. Some cavitation may occur. In many valve designs, the droplets impinge on a hard surface (breaker ring) set at 90° to the direction of flow of the liquids after they emerge from the gap. All of these mechanisms stress the droplets and contribute to their disruption. Droplets diameters of 0.1–0.2 μm are attainable in pressure homogenisers.

There are many different designs of valve available. Three examples are shown in Figure 15.10. As the liquids travels between the valve and its seat, the valve lifts against a heavy duty spring or torsion bar. By adjusting the tension on this spring or bar, the homogenising pressure may be set. This may range from 3.5–70.0 M N m^{-2}.

The literature suggests that there is an approximately inverse linear relationship between the logarithm of the homogenising pressure and the logarithm of the droplet diameter produced by a pressure homogeniser. Homogeniser valves are usually made of stainless steel or alloys such as stellite. More erosion-resistant materials such as tungsten carbide may be used, but not usually for food applications. It is important that a good fit is maintained between the valves and their seats. Even small amounts of damage to the surfaces can lead to poor

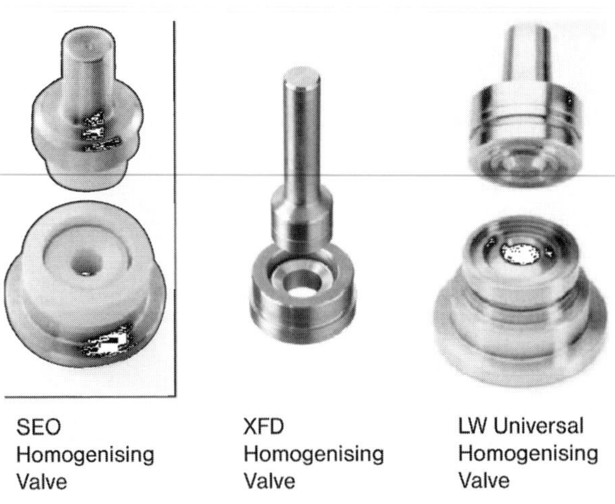

Fig. 15.10 Three different types of homogenising valve, by courtesy of APV-Rannie & Gaulin Homogenisers.

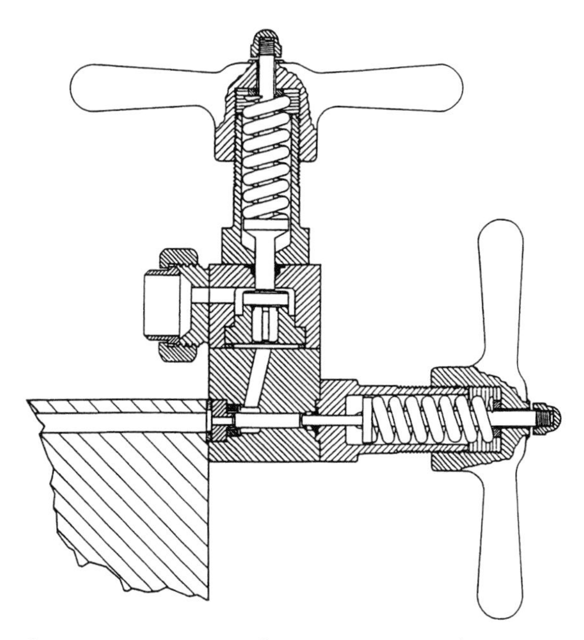

Fig. 15.11 Representation of a two-stage pressure homogeniser.

performance. Consequently, they should be examined regularly and reground or replaced when necessary. Valves made of compressed wire are also available. The liquids are pumped through the myriad of channels in the body of these valves. Such valves are difficult to clean and are discarded at the end of a day's run. They are known as single-service valves.

One passage through a homogenising valve may not produce a well dispersed emulsion. The small droplets of the internal phase may cluster together. These can be dispersed by passing them through a second valve. This is known as two-stage homogenisation. The first valve is set at a high pressure, 14–70 MPa, the second at a relatively low pressure, 2.5–7.0 MPa. A representation of a two-stage homogenising system is shown in Fig. 15.11.

The liquids are pumped through the homogenising valve(s) by means of a positive displacement pump, usually of the piston and cylinder type. In order to achieve a reasonably uniform flow rate, three, five or seven cylinders with pistons are employed, operating consecutively and driven via a crankshaft. The mixture is discharged from the chamber of each piston into a high-pressure manifold and exits from there via the homogenising valve. A pressure gauge is fitted to the manifold to monitor the homogenising pressure.

15.2.3.3 Hydroshear Homogenisers

In this type of homogeniser, the premixed liquids are pumped into a cylindrical chamber at relatively low pressure, up to 2000 kPa. They enter the chamber through a tangential port at its centre and exit via two cone-shaped discharge nozzles at the ends of the chamber. The liquids accelerate to a high velocity as they enter the chamber, spread out to cover the full width of the chamber wall and flow towards the centre, rotating in ever decreasing circles. High shear develops between the adjacent layers of liquid, destabilising the large droplets of the internal phase. In the centre of the cylinder a zone of low pressure develops and cavitation, ultrahigh frequency vibration and shock waves occur which all contribute to the break up of the droplets and the formation of an emulsion. Droplets sizes in the range 2–8 μm are produced by this equipment.

15.2.3.4 Microfluidisers

This type of homogeniser is capable of producing emulsions with very small droplet sizes directly from individual aqueous and oil phases. Separate streams of the aqueous and oil phases are pumped into a chamber under high pressure, up to 110 MPa. The liquids are accelerated to high velocity, impinge on a hard surface and interact with each other. Intense shear and turbulence develop which lead to a break up of the droplets of the internal phase and the formation of an emulsion. Very small emulsion droplets can be produced by recirculating the emulsion a number of times through the microfluidiser.

15.2.3.5 Membrane Homogenisers

If the internal phase liquid is forced to flow through pores in a glass membrane into the external phase liquid an emulsion can be formed. Glass membranes can be manufactured with pores of different diameters to produce emulsions with different droplet sizes, in the range 0.5–10 μm. Such membranes can produce o/w or w/o emulsions with very narrow droplet size distributions. In a batch version of this equipment the internal phase liquid is forced through a cylindrical membrane partly immersed in a vessel containing the external phase. In a continuous version, a cylindrical membrane through which the external phase flows is located within a tube, through which the internal phase flows. The internal phase is put under pressure forcing it through the membrane wall into the external phase. To date, membrane homogenisers are used mainly for the production of emulsions on a laboratory scale.

15.2.3.6 Ultrasonic Homogenisers

When a liquid is subjected to ultrasonic irradiation, alternate cycles of compression and tension develop. This can cause cavitation in any gas bubbles present in the liquid, resulting in the release of energy. This energy can be used to disperse one liquid phase in another to produce an emulsion. For laboratory-scale

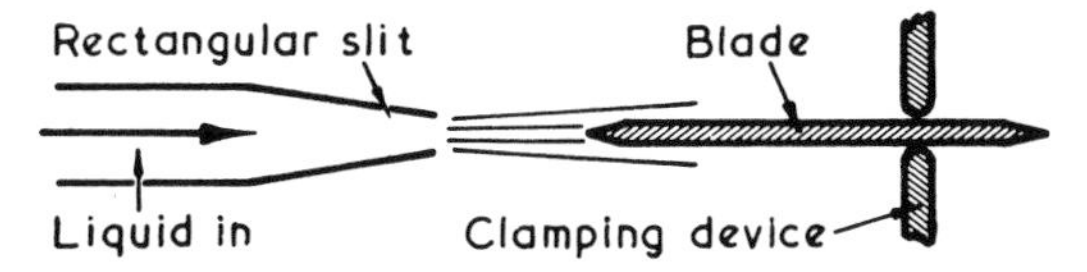

Fig. 15.12 Principle of the mechanical (wedge resonator) ultrasonic generator; from [6] with permission of the authors.

applications, piezoelectric crystal oscillators may be used. An ultrasonic transducer consists of a piezoelectric crystal encased in a metal tube. When a high-intensity electrical wave is applied to such a transducer, the crystal oscillates and generates ultrasonic waves. If a transducer of this type is partly immersed in a vessel containing two liquid phases, together with an appropriate emulsifying agent(s), one phase may be dispersed in the other to produce an emulsion.

For the continuous production of emulsions on an industrial-scale mechanical ultrasonic generators are used. The principle of a wedge resonator (liquid whistle) is shown in Fig. 15.12. A blade with wedge-shaped edges is clamped at one or more nodal points and positioned in front of a nozzle through which the premixed emulsion is pumped. The jet of liquid emerging from the nozzle impinges on the leading edge of the blade, causing it to vibrate at its natural frequency, usually in the range 18–30 kHz. This generates ultrasonic waves in the liquid which cause one phase to become dispersed in the other and the formation of an emulsion. The pumping pressure required is relatively low, usually in the range 350–1500 kPa, and droplet diameters of the order of 1–2 μm are produced.

15.2.3.7 **Colloid Mills**

In a colloid mill, the premixed emulsion ingredients pass through a narrow gap between a stationary surface (stator) and a rotating surface (rotor). In doing so the liquid is subjected to shear and turbulence which brings about further disruption of the droplets of the internal phase and disperses them throughout the external phase. The gap between the stator and rotor is adjustable within the range 50–150 μm. One type of colloid mill is depicted in Fig. 15.13. The rotor turns on a vertical axis in close proximity to the stator. The clearance between them is altered by raising or lowering the stator by means of the adjusting ring. Rotor speed ranges from 3000 rpm for a rotor 25 cm in diameter to 10 000 rpm for a smaller rotor 5 cm in diameter. Rotors and stators usually have smooth stainless steel surfaces. Carborundum surfaces are used when milling fibrous materials. Colloid mills are usually jacketed for temperature control. This type of mill, also known as a paste mill, is suitable for emulsifying viscous materials.

For lower viscosity materials the rotor is mounted on a horizontal axis and turns at higher speeds, up to 15 000 rpm. Mills fitted with rotors and stators with matching corrugated surfaces are also available. The clearance between the surfaces decreases outwardly from the centre. The product may be discharged

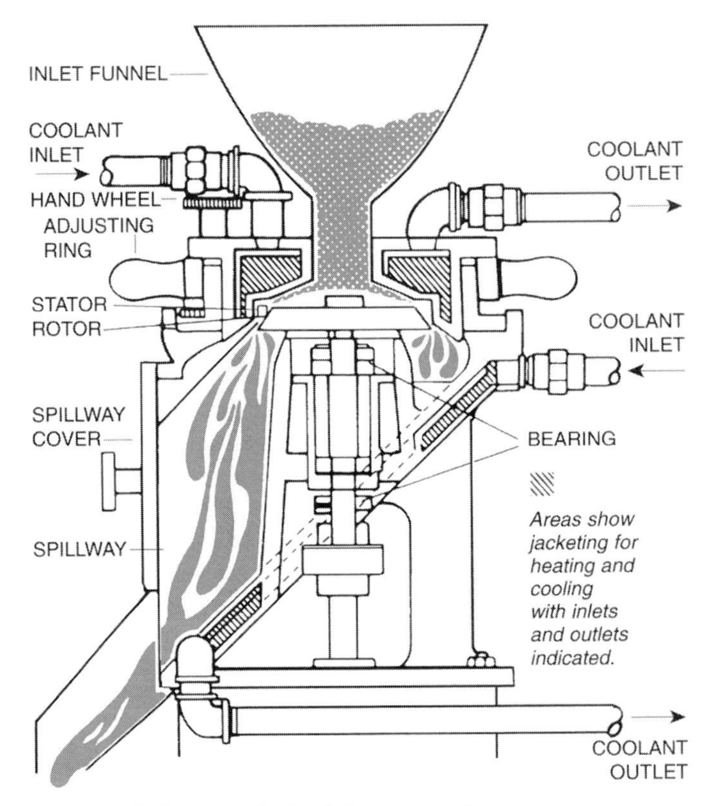

Fig. 15.13 Top-feed 'paste' colloid mill: by courtesy of Premier Mill Operation, SPX Process Equipment.

under pressure, up to 700 kPa. Incorporation of air into the product is limited and foaming problems reduced in this type of mill [1, 6, 22, 23, 27, 28].

15.2.4 Examples of Emulsification in Food Processing

Oil in water (o/w) emulsions of importance in food processing include the following.

15.2.4.1 Milk

This naturally occurring emulsion typically consists of 3.0–4.5% fat dispersed in the form of droplets throughout an aqueous phase which contains sugars and mineral salts in solution and proteins in colloidal suspension. The fat droplets, or globules, range in size from less than 1 μm to more than 20 μm in diameter. These are stabilised by a complex, multilayer coating made up of phospholipids, proteins, enzymes, vitamins and mineral salts, known as the *milk fat globule membrane.* Under the influence of gravity, these globules tend to rise to the surface to form a cream layer when milk is standing in vats or bottles. To prevent such separation milk may be subjected to two-stage homogenisation, reducing the fat globule size to not more than 2 μm in diameter. Pasteurised, homogenised milk is a widely used liquid milk product. Milk and cream which are to be UHT-treated are also homogenised to improve their stability. So also are evaporated milk and cream which are to be heat-sterilised in containers.

15.2.4.2 **Ice Cream Mix**

This is an o/w emulsion typically containing 10–12% fat dispersed in an aqueous phase containing sugars and organic salts in solution and proteins and some organic salts in colloidal suspension. The stability of the emulsion is important as it has to withstand the rigors of freezing and the incorporation of air to achieve an appropriate overrun. In addition to fat, which may be milk fat, vegetable oil or a combination of the two, ice cream mix contains about 10.5% milk solids/nonfat, usually in the form of skim milk powder, and 13% sucrose, dextrose or invert sugar. The milk solids/nonfat acts as an emulsifying agent but additional agents such as esters of mono- and diglycerides are usually included. Stabilisers such as alginates, carrageenan, gums and gelatin are also added. These increase the viscosity of the mix and also have an influence on the proportion of the aqueous phase which crystallises on freezing and the growth of the ice crystals. This in turn affects the texture and melting characteristics of the frozen product. The mix is pasteurised, usually subjected to two-stage pressure homogenisation and aged at 2–5 °C prior to freezing in scraped surface freezers. Some 50% or less of the aqueous phase freezes at this stage and air is incorporated to give an overrun of 60–100%. The product may then be packaged and hardened at temperatures of –20 °C to –40 °C.

15.2.4.3 **Cream Liqueurs**

These are further examples of o/w dairy emulsions. They need to have longterm stability in and alcoholic environment. Soluble sodium caseinate can be used to stabilise the finely dispersed emulsion.

15.2.4.4 **Coffee/Tea Whiteners**

These substitutes for cream or milk are also o/w emulsions typically containing vegetable oil, sodium caseinate, corn syrup, high HLB emulsifying agents and potassium phosphate. They are usually prepared by pressure homogenisation and are available in UHT-treated liquid form or in spray dried powder form.

15.2.4.5 **Salad Dressings**

Many *'French' dressings* consist of mixtures of vinegar, oil and various dry ingredients. They are not emulsified as such and the liquid phase separates after mixing. They need to be mixed or shaken thoroughly before use. Other dressings are true o/w emulsions. *Salad cream,* for example, contains typically 30–40% oil, sugar, salt, egg (either yolk or whole egg in liquid or dried form), mustard, herbs, spices, colouring and stabiliser(s). The cream is acidified with vinegar, and/or lemon juice. The lecithin present in the egg usually is the main emulsifying agent, but some additional o/w agents may be added. Gum tragacanth is the stabiliser most commonly used. This increases the viscosity of the emulsion. Being thixotropic, it thins on shaking and facilitates dispensing of the cream. The gum is dispersed in part of the vinegar and water, allowed to stand for up to 4 days until it is fully hydrated and then beaten and sieved. The rest of the aqueous ingredients are premixed, heated to about 80 °C, cooled to about 40 °C and sieved. An emulsion premix is prepared by adding the oil gradually to the aqueous phase with agitation. This premix is then further emulsified by means of a pressure homogeniser, colloid mill or ultrasonic device. It is then vacuum-filled into jars or tubes. *Mayonnaise* is also an o/w emulsion with similar ingredients to salad cream but containing 70–85% oil. The high oil content imparts a high viscosity to the product and stabiliser(s) are usually not required. The premixing is usually carried out at relatively low temperature, 15–20 °C, and a colloid mill used to refine the emulsion.

15.2.4.6 **Meat Products**

Emulsification of the fat is important in the production of many meat products such as sausages, pastes and pates. Efficient emulsification can prevent fat separation, influence the texture of the product and its behaviour on cooking. Meat emulsions are relatively complex systems. They are usually classed as o/w emulsions but differ in many ways from those discussed above. They are two phase systems consisting of fairly coarse dispersions of solid fat in an aqueous phase containing gelatin, other proteins, minerals and vitamins in solution or colloidal suspension and insoluble matter, including meat fibre, filling materials and seasonings. The emulsifying agents are soluble proteins. Emulsification is brought about simultaneously with size reduction of the insoluble matter in a variety of equipment typified by the mincer and bowl chopper. In the mincer the material is forced by means of a worm through a perforated plate with knives rotating in contact with its surface. It is assumed that some shearing occurs which contributes to the emulsification of the fat. The bowl chopper consists of a hemispherical bowl which rotates slowly about a vertical axis. Curved knives rotate rapidly on a horizontal axis within the bowl. As the bowl rotates is brings the contents into contact with the rotating knives which simultaneously reduce the size of the solid particles and mix the ingredients. Soluble protein is released and emulsification of the fat takes place.

15.2.4.7 **Cake Products**

Cake has been defined as a protein foam stabilised by gelatinised starch and containing fat, sugar, salt, emulsifiers and flavouring materials. It is aerated mainly by gases evolved by chemical reaction involving raising agents. The fat comes from milk, eggs, chocolate and/or added shortenings. The protein comes in the flour, eggs and milk. Emulsifying agents are available in the milk, eggs and flour. Additional agents may be added separately and/or included in the shortenings. Cake batters have an o/w structure. Efficient emulsification of the fat in such batters is essential. Free liquid fat adversely affects the stability of the foam formed on beating and aeration and, consequently, is detrimental to the crumb structure, volume and shape of the baked product. Simultaneous mixing and emulsification is attained in various types of mixers such as pan mixers, operating at relatively high speed, and some continuous mixers. Colloid mills and even pressure homogenisers have been used to ensure good emulsification in low viscosity batters.

Water in oil (w/o) emulsions of importance in food processing include:

15.2.4.8 **Butter**

This is usually described as a w/o emulsion but, in fact, it has quite a complex structure. The continuous phase of free fat in liquid form contains fat crystals, globular fat, curd granules, crystals of salt, water droplets and gas bubbles. The water droplets remain dispersed due to the semi-solid nature of the continuous phase rather than being stabilised by a layer of emulsifying agent.

Pasteurised milk is separated by centrifugation to give cream containing 30–40% fat. This cream is aged by holding at a low temperature for several hours. The purpose of ageing is to achieve the optimum liquid:solid fat ratio in the butter. Butter may also be made from cultured cream which has been inoculated with lactic acid producing microorganisms and ripened, typically for 20 h at 14 °C, to develop the flavour. In batch churning, the aged cream is tumbled inside a hollow vessel known as a churn. Air is incorporated into the cream and the fat globules concentrate in the surface of the air bubbles. As these break and reform, some of the fat globules break open releasing free fat. The remaining globules form clumps, known as butter grains, held together by some of the free fat. As churning proceeds these grains grow in size. When they reach an optimum size, about 1 cm in diameter, churning is stopped, the aqueous phase, known as buttermilk, is drained off and the grains are washed free of curd with chilled water. The moisture content is adjusted, salt is then added if required and the mass of butter grains is tumbled in the churn for a further period. During this stage of the process, known as working, more fat globules break open, releasing more free fat, water droplets and salt crystals are dispersed throughout the bulk of the fat phase and the texture of the product develops. Working may be carried out under a partial vacuum to reduce the air content of the butter. When working is complete, about 40% of the fat remains in globular form. The butter is discharged from the churn, packaged and transferred to chilled or frozen storage.

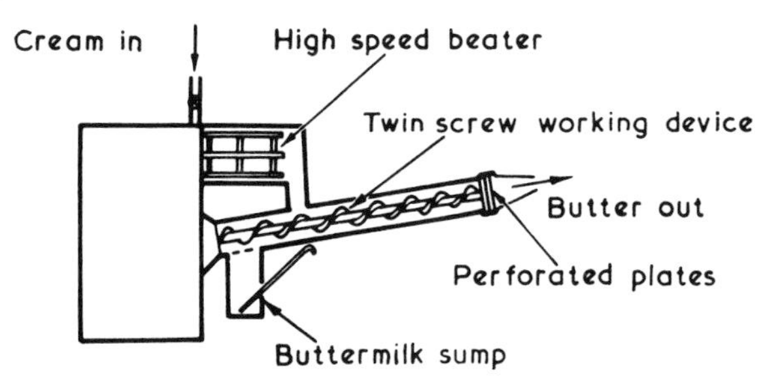

Fig. 15.14 General principle of accelerated a churning and working device; from [6] with permission of the authors.

Most continuous methods of buttermaking work on the principles represented in Fig. 15.14. Aged cream containing 30–40% fat is metered into the churning section of the equipment where it is acted on by high-speed beaters. These bring about the rupture of some of the fat globules and the formation of small butter grains. As the grains and buttermilk exit the churning section, the buttermilk drains into a sump and the grains are carried by twin screws up the sloping barrel and extruded through a series of perforated plates. The action of the screws combined with the perforated plates brings about further disruption of fat globules and disperses the remaining water as droplets throughout the fatty phase. The butter is extruded in the form of a continuous ribbon from the working section. Salt may be introduced into the working section, if required.

Other continuous buttermaking methods involve concentrating the fat in the cream to about 80% by a second centrifugation step. This unstable, concentrated, o/w emulsion is then converted into a semi-solid, w/o system by simultaneous agitation and cooling in a variety of equipment.

15.2.4.9 **Margarine and Spreads**

These w/o products are made from a blend of fats and oils together with cultured milk, emulsifying agents, salt, flavouring compounds and other additives. The blend of fats and oils is selected according to the texture required in the final product. The emulsifying agents used have relatively low HLB values. A typical combination is a mixture of mono- and diglycerides and lecithin. The flavour, originally derived from the cultured milk, is now usually supplemented by the addition of flavouring materials such as acetyl methyl carbinols or aliphatic lactones.

In a typical manufacturing process, the fats and oils are measured into balance tanks. Other ingredients are added and an emulsion premix formed by high-speed agitation. This premix is then pumped through a series of refrigerated, scraped surface heat exchangers where it is simultaneously emulsified and cooled. A three-dimensional network of long, thin, fat crystals is formed. It finally passes through a working/holding device, similar to the working section in Fig. 15.14, where the final texture develops.

Margarine contains 15% water, similar to butter. Low calorie spreads are also made from a blend of oils and contain up to 50% water. They may be based on milk fat or combinations of milk fat and vegetable oils [6, 29–33, 35–38].

15.3 Size Reduction (Crushing, Comminution, Grinding, Milling) of Solids

15.3.1 Introduction

Size reduction of solids involves creating smaller mass units from larger mass units of the same material. To bring this about, the larger mass units need to be subjected to stress by the application of force. Three types of force may be applied, i.e. *compression, impact* and *shear.* Compressive forces are generally used for the coarse crushing of hard materials. Careful application of compressive forces enables control to be exercised over the breakdown of the material, e.g. to crack open grains of wheat to facilitate separation of the endosperm from the bran (see Section 15.3.3.1). Impact forces are used to mill a wide variety of materials, including fibrous foods. Shear forces are best applied to relatively soft materials, again including fibrous foods. All three types of force are generated in most types of mill, but generally one predominates. For example, in most roller mills compression is the dominant force, impact forces feature strongly in hammer mills and shear forces are dominant in disc attrition mills (see Section 15.3.2).

The extent of the breakdown of a material may be expressed by the *reduction ratio,* which is the average size of the feed particles divided by the average size of the products particles. In this context, the term average size depends on the method of measurement. In the food industry, screening or sieving is widely used to determine particle size distribution in granular materials and powders. In this case, the average diameter of the particles is related to the aperture sizes of the screens used. Size reduction ratios vary from below 8:1 in coarse crushing to more than 100:1 in fine grinding.

The objective in many size reduction operations is to produce particles within a specified size range. Consequently, it is common practice to classify the particles coming from a mill into different size ranges. Again, screening is the technique most widely used for this purpose. To achieve a specified reduction ratio, it may be necessary to carry out the size reduction in a number of stages. A different type of mill may be used in each stage and screens employed between stages. An example of a multistage operation is depicted in Fig. 15.15.

When a solid material is subjected to a force, its behaviour may be represented by a plot of stress versus strain, as shown in Fig. 15.16. Some materials exhibit elastic deformation when the force is first applied. The strain is linearly related to stress (see curve 2 in Fig. 15.16). If the force is removed the solid object returns to its original shape. Elastic deformations are valueless in size re-

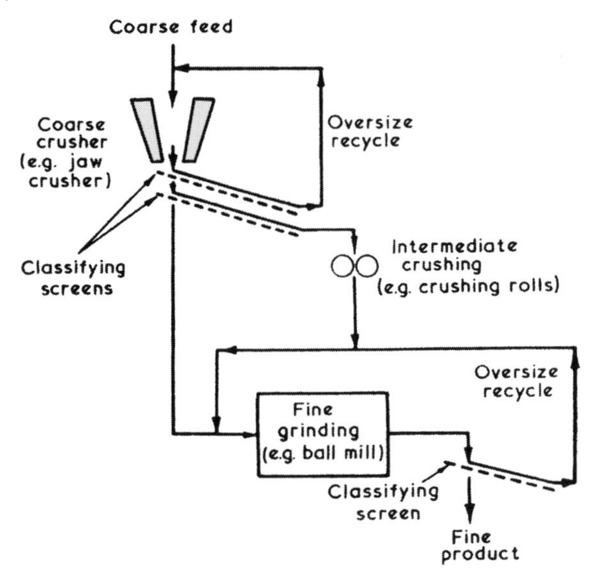

Fig. 15.15 A typical size reduction flow sheet; from [6] with permission of the authors.

duction. Energy is used up but no breakdown occurs. Point E is known as the elastic limit. Beyond this point, the material undergoes permanent deformation until it reaches its yield point Y. Brittle materials will rupture at this point. Ductile materials will continue to deform, or flow, beyond point Y until they reach the break point B, when they rupture. The behaviour of different types of material is depicted by the five curves in Fig. 15.16 and explained in the caption to that figure.

The breakdown of friable materials may occur in two stages. Initial fracture may occur along existing fissures or cleavage planes in the body of the material. In the second stage new fissures or crack tips are formed and fracture occurs along these fissures. Larger particles will contain more fissures than smaller ones and hence will fracture more easily. In the case of small particles, new crack tips may need to be created during the milling operation. Thus, the breaking strength of smaller particles is higher than the larger ones. The energy required for particle breakdown increases with decrease in the size of the particles. In the limit of very fine particles, only intermolecular forces must be overcome and further size reduction is very difficult to achieve. However, such very fine grinding is seldom required in food applications.

Only a very small proportion of the energy supplied to a size reduction plant is used in creating new surfaces. Literature values range from 2.0% down to less than 0.1%. Most of the energy is used up by elastic and inelastic deformation of the particles, elastic distortion of the equipment, friction between particles and between particles and the equipment, friction losses in the equipment and the heat, noise and vibration generated by the equipment.

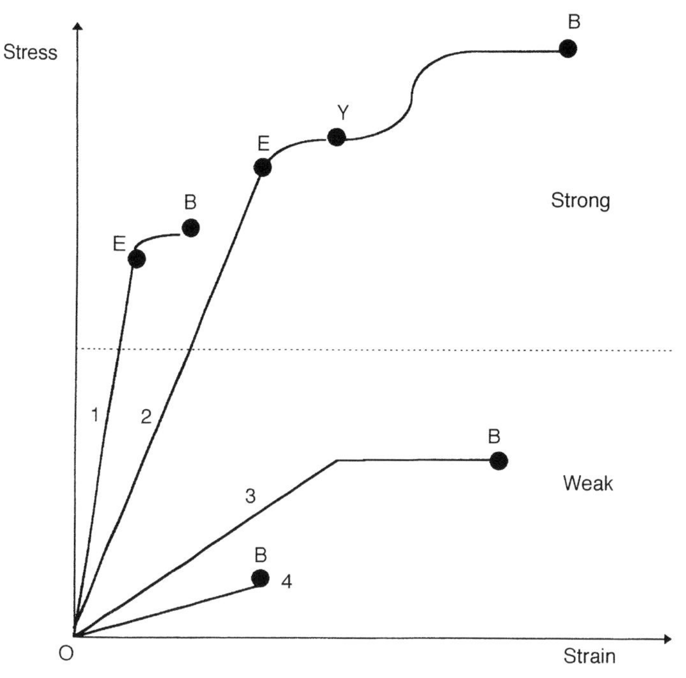

Fig. 15.16 Stress-strain diagram for various foods: E, elastic limit; Y, yield point; B, breaking point; O-E, elastic region; E-Y, inelastic deformation; Y-B, region of ductility. 1. Hard, strong, brittle material. 2. Hard strong ductile material. 3. Soft, weak, ductile material. 4. Soft, weak, brittle material. Adapted from [39] with permission.

Mathematical models are available to estimate the energy required to bring about a specified reduction in particle size. These are based on the assumption that the energy dE required to produce a small change dx in the size of a unit mass of material can be expressed as a power function of the size of the material. Thus:

$$\frac{dE}{dx} = -\frac{K}{x^n} \tag{15.1}$$

Rittinger's Law is based on the assumption that the energy required should be proportional to the new surface area produced, i.e. $n=2$. So:

$$\frac{dE}{dx} = -\frac{K}{x^2} \tag{15.2}$$

or, integrating:

$$E = K\left[\frac{1}{x_2} - \frac{1}{x_1}\right] \tag{15.3}$$

Where x_1 is the average initial size of the feed particles, x_2 is the average size of the product particles, E is the energy per unit mass required to produce this increase in surface area and K is a constant, known as Rittinger's constant. Rittinger's law has been found to hold better for fine grinding.

Kick's Law is based on the assumption that the energy required should be proportional to the size reduction ratio, i.e. $n = 1$. So:

$$\frac{dE}{dx} = -\frac{K}{x} \tag{15.4}$$

or, integrating:

$$E = -\ln\frac{x_1}{x_2} \tag{15.5}$$

Where is the size reduction ratio (see above).

Kick's law has been found to apply best to coarse crushing.

In *Bond's Law*, n is given the value 3/2. So:

$$\frac{dE}{dx} = -\frac{K}{x^{3/2}} \tag{15.6}$$

or, integrating:

$$E = 2K\left[\frac{1}{(x_2)^{1/2}} - \frac{1}{(x_1)^{1/2}}\right] \tag{15.7}$$

Bond's law has been found to apply well to a variety of materials undergoing coarse, intermediate and fine grinding [1, 6–8, 26, 39, 40].

15.3.2 Size Reduction Equipment

15.3.2.1 Some Factors to Consider When Selecting Size Reduction Equipment

Mechanical Properties of the Feed Friable and crystalline materials may fracture easily along cleavage planes. Larger particles will break down more readily than smaller ones. Roller mills are usually employed for such materials. Hard materials, with high moduli of elasticity, may be brittle and fracture rapidly above the elastic limit. Alternatively, they may be ductile and deform extensively before breakdown. Generally, the harder the material, the more difficult it is to break down and the more energy is required. For very hard materials, the dwell time in the action zone must be extended, which may mean a lower throughput or the use of a relatively large mill. Hard materials are usually abrasive and so the working surfaces should be made of hard wearing material, such as manganese steel, and should be easy to remove and replace. Such mills are relatively slow moving and need to be of robust construction. Tough materials have the ability to resist the propagation of cracks and are difficult to breakdown. Fibres tend to increase toughness by relieving stress concentrations at the ends of the cracks. Disc mills, pin-disc mills or cutting devices are used to break down fibrous materials.

Moisture Content of the Feed The moisture content of the feed can be of importance in milling. If it is too high, the efficiency and throughput of a mill and the free flowing characteristics of the product may be adversely affected. In some cases, if the feed material is too dry, it may not breakdown in an appropriate way. For example, if the moisture content of wheat grains are too high, they may deform rather than crack open to release the endosperm. Or, if they are too dry, the bran may break up into fine particles which may not be separated by the screens and may contaminate the white flour. Each type of grain will have an optimum moisture content for milling. Wheat is usually 'conditioned' to the optimum moisture content before milling (see Section 15.3.3.1). Another problem in milling very dry materials is the formation of dust, which can cause respiratory problems in operatives and is a fire and explosion hazard.

In *wet milling*, the feed materials is carried through the action zone of the mill in a stream of water.

Temperature Sensitivity of the Feed A considerable amount of heat may be generated in a mill, particularly if it operates at high speed. This arises from friction and particles being stressed within their elastic limits. This heat can cause the temperature of the feed to rise significantly and a loss in quality could result. If the softening or melting temperatures of the materials are exceeded the performance of the mill may be impaired. Some mills are equipped with cooling jackets to reduce these effects.

Cryogenic milling involves mixing solid carbon dioxide or liquid nitrogen with the feed. This reduces undesirable heating effects. It can also facilitate the milling of fibrous materials, such as meats, into fine particles.

15.3.2.2 **Roller Mills (Crushing Rolls)**

A common type of roller mill consists of two cylindrical steel rolls, mounted on horizontal axes and rotating towards each other. The particles of feed are directed between the rollers from above. They are nipped and pulled through the rolls where they are subjected to compressive forces, which bring about their breakdown. If the rolls turn at different speeds shear forces may be generated which will also contribute to the breakdown of the feed particles. The roll surfaces may be smooth, corrugated, grooved, fluted or they may have intermeshing teeth or lugs. In food applications smooth, grooved or fluted rolls are most often used. Large rolls, with diameters greater than 500 mm, rotate at speeds in the range 50–300 rpm. Smaller rolls may turn at higher speeds. Usually the clearance between the rolls, the *nip*, is adjustable. An overload compression spring is usually fitted to protect the roll surfaces from damage should a hard object try to pass between them (see Fig. 15.17).

The surface of the rolls may be cooled or heated by circulating water or some other thermal fluid within their interior. Large, smooth surfaced rolls are used for relatively coarse crushing, usually achieving a reduction ratio of 4 or lower. Smaller rolls, with different surface configurations, operating at higher speeds can achieve higher ratios. Smooth or fluted rolls, operating at the same or slightly different speeds, are used to crack open grains and seeds. For finer milling, shallow grooves and larger differential speeds are employed. For very fine milling, smooth surfaced rolls, operating at high differential speeds are used. To achieve high reduction ratios, the material being milled may be made to pass between two or more pairs of rolls in sequence, with the clearance decreasing from one pair to the next. Some separation of the particles into different size ranges may take place between each pair of rolls. This principle is employed in the milling of wheat grains (see Section 15.3.3.1). Machines consisting of two, three or more smooth-surfaced rolls, arranged in either a horizontal or vertical sequence, are used to mill liquid products. The product passes between the rolls in a zigzag flow pattern. The clearance between the rolls decreases in the direction of flow of the product and the speeds of consecutive rolls differ, generating shear forces.

Instead of two rolls operating against each other, one roll may operate against a flat or curved hard surface. An example is the *edge mill* (see Fig. 15.18). The heavy rolls roll and slip over a table. Usually the rolls are driven, but in some machines the table is driven instead. The rolls and table surfaces may be smooth or grooved.

In the case of a mill consisting of two rolls rotating towards each other, the *angle of nip*, α, is the term used to describe the angle formed by the tangents to

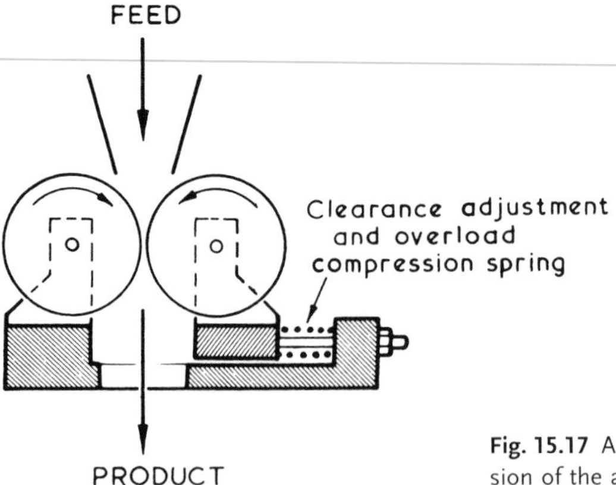

Fig. 15.17 A roller mill; from [6] with permission of the authors.

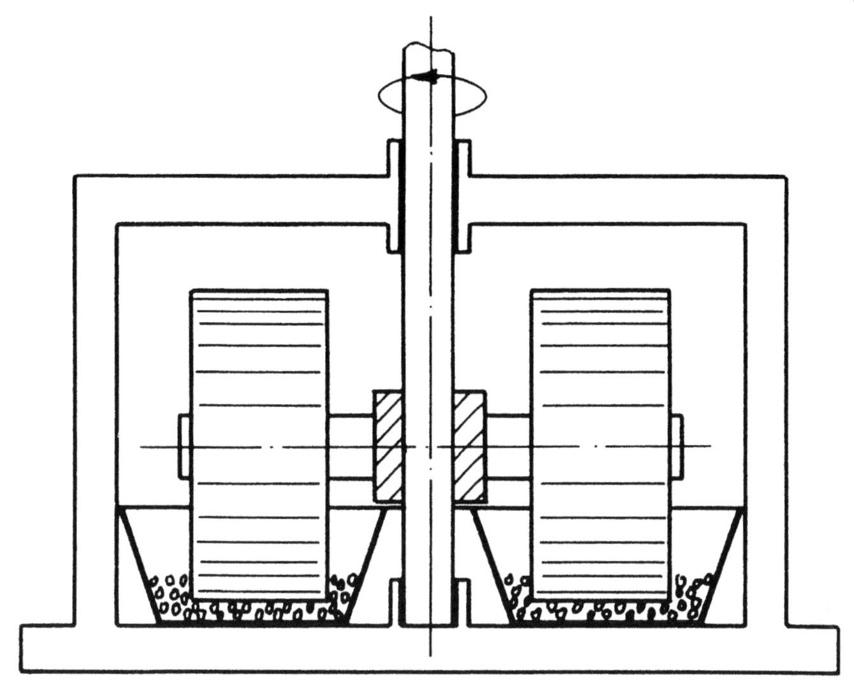

Fig. 15.18 Edge mill; from [8] with kind permission of Springer Science and Business Media.

the roll faces at the point of contact between the particle and the rolls. It can be shown that:

$$\cos\frac{a}{2} = \frac{D_r + D_p}{D_r + D_f} \tag{15.8}$$

Where D_f is the average diameter of the feed particles, D_p is the average diameter of the product particles, corresponding to the clearance between the rolls, and D_r is the diameter of the rolls.

For the particle to be 'nipped' down through the rolls by friction:

$$\tan\frac{a}{2} = \mu \tag{15.9}$$

Where μ is the coefficient of friction between the particle and the rolls.

Equations (15.8) and (15.9) can be used to estimate the largest size of feed particle that a pair of rolls will accommodate.

The theoretical mass flow rate M (kg s^{-1}) of product from a pair of rolls of diameter D_r (m) and length l (m) when the clearance between the rolls is D_p (m), the roll speed is N (rpm) and the bulk density of the product is ρ (kg m^{-3}) is given by:

$$M = \frac{\pi D_r N D_p l \rho}{60} \qquad (15.10)$$

The literature suggests that the actual mass flow rate is usually between 0.1 and 0.3 of the theoretical value [6–8, 39–41].

15.3.2.3 **Impact (Percussion) Mills**

When two bodies collide, i.e. impact, they compress each other until they have the same velocity and remain in this state until restitution of the compression begins. Then the bodies push each other apart and separate. If one of the bodies is held in position, the other body has to conform with this position for a short interval of time. During the very short time it takes for restitution of compression to occur, a body possesses strain energy which can lead to fracture. The faster the bodies move away from each other the more energy is available to bring about fracture. The faster the rate at which the force is applied the more quickly fracture is likely to occur [40].

Hammer Mill In this type of mill, a rotor mounted on an horizontal shaft turns at high speed inside a casing. The rotor carries hammers which pass within a small clearance of the casing. The hammers may be hinged to the shaft so that heads swing out as it rotates. In some designs, the hammers are attached to the shaft by rigid connections. A toughened plate, known as the breaker plate, may be fitted inside the casing, see Fig. 15.19. The hammers and breaker plate are made of hard wearing materials such as manganese steel. The hammers drive the feed particles against the breaker plate. Fracture of the particles is brought about mainly by impact forces, but shear forces also play a part. The casing may be fitted with a screen through which the product is discharged. The size of the screen aperture determines the upper limit particle size in the product. This way of operating a mill is known as *choke feeding*. When milling friable materials, choke feeding may result in a high proportion of very small particles in the product. When milling fibrous or sticky materials, the screen may become blocked. In some such cases the screen may be removed. The mill casing may be equipped with a cooling jacket. The hammer mill is a general purpose mill used for hard, friable, fibrous and sticky materials.

Beater Bar Mill In this type of mill, the hammers are replaced by bars in the form of a cross. The tips of the bars pass within a small clearance of the casing. Beater bars are mainly used in small machines.

Comminuting Mill Knives replace the hammers or bars in this type of mill. They may be hinged to the shaft so that the swing out as it rotates. Alternatively, they may be rigidly fixed to the shaft. Such mills are used for comminuting relatively soft materials, such as fruit and vegetable matter. In some designs, the knives are sharp on one edge and blunt on the other. When the shaft rotates

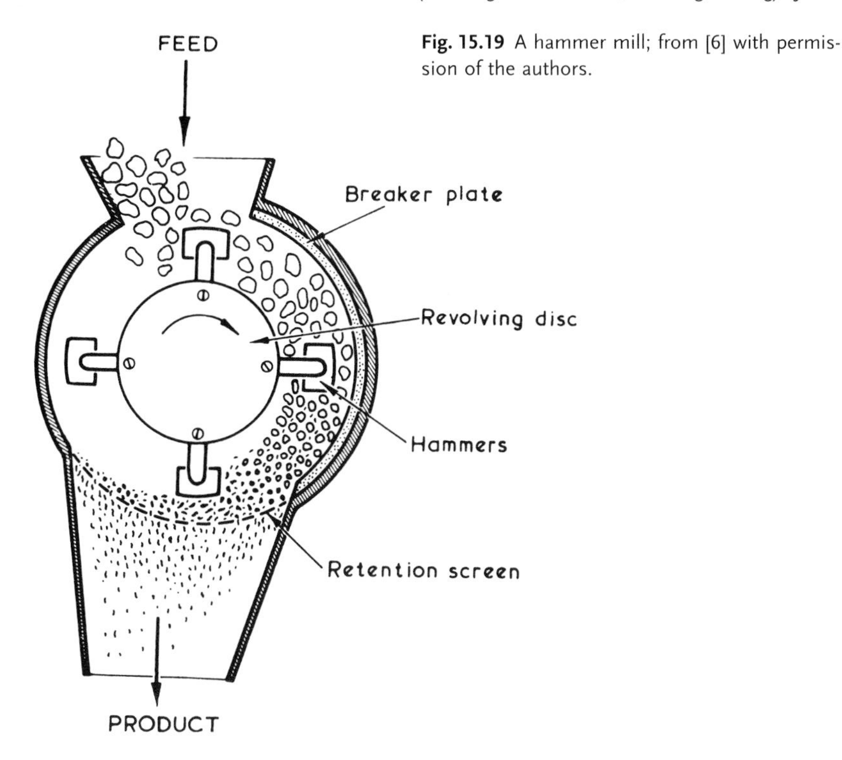

Fig. 15.19 A hammer mill; from [6] with permission of the authors.

in one direction the machine has a cutting action. When the direction of rotation of the shaft is reversed, the blunt edges of the knives act as beater bars.

Pin (Pin-Disc) Mill In one type of pin mill a stationary disc and rotating disc are located facing each other, separated by a small clearance. Both discs have concentric rows of pins, pegs or teeth. The rows of one disc fit alternately into the rows of the other disc. The pins may be of different shapes; round, square or in the form of blades. The feed in introduced through the centre of the stationary disc and passes radially outwards through the mill where it is subjected to impact and shear forces between the stationary and rotating pins. The mill may be operated in a choke feed mode by having a screen fitted over the whole or part of the periphery (see Fig. 15.20). Alternatively, it may not have a screen and the material is carried through the mill by an air stream (see Fig. 15.21). In another type of mill, both discs rotate either in the same direction at different speeds or in opposite directions. In some mills the clearance between the discs is adjustable. Disc speeds may be up to 10000 rpm. Pin mills may be fitted with jackets for temperature control. Such mills are suitable for fine grinding friable materials and for breaking down fibrous substances.

Fluid Energy (Jet) Mill In this type of mill, the solid particles to be comminuted are suspended in a gas stream travelling at high velocity into a grinding cham-

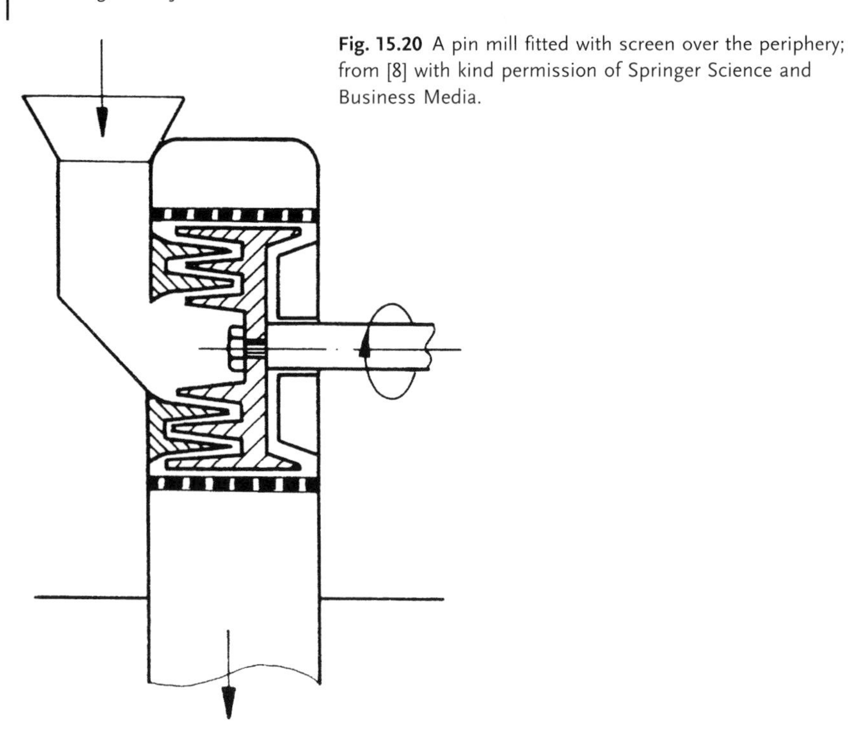

Fig. 15.20 A pin mill fitted with screen over the periphery; from [8] with kind permission of Springer Science and Business Media.

ber. Breakdown occurs through the impact between individual particles and with the wall of the chamber. The gases used are compressed air or superheated steam, which are admitted to the chamber at a pressure of the order of 700 kPa. An air-solids separation system, usually a cyclone, is used to recover the product. Particles up to 10 mm can be handled in these mills but usually the feed consists of particles less than 150 μm. The product has a relatively narrow size range. Since there are no moving parts or grinding media involved, product contamination and maintenance costs are relatively low. However, the energy efficiency of such mills is relatively low [6–8, 39–41].

15.3.2.4 Attrition Mills

The principle of attrition mills is that the material is rubbed between two surfaces. Both pressure and frictional forces are generated. The extent to which either of these forces predominates depends on the pressure with which both surface are held together and the difference in the speed of rotation of the surfaces.

Buhrstone Mill This is the oldest form of attrition mill. It consists of two stones, one located above the other. The upper stone is usually stationary while the bottom one rotates. Matching grooves are cut in the stones and these make

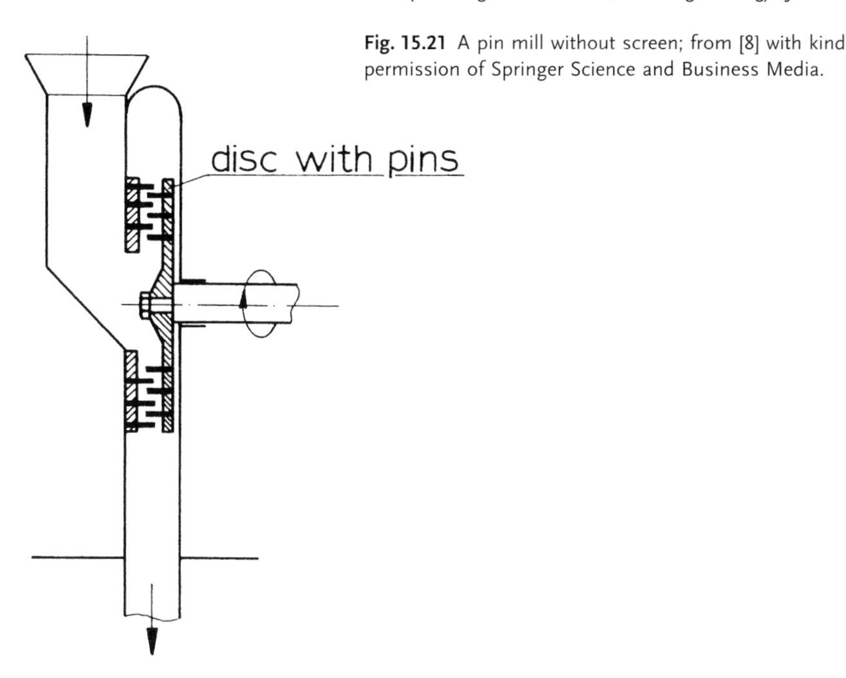

Fig. 15.21 A pin mill without screen; from [8] with kind permission of Springer Science and Business Media.

a scissor action as the lower one rotates. The feed is introduced through a hole in the centre of the upper stone and gradually moves outwards between the stones and is discharged over the edge of the lower stone. In some such mills both stones rotate, in opposite directions. Siliceous stone is used. In more recent times toughened steel 'stones' have replaced the natural stone for some applications. Such stone mills were used for milling of wheat for centuries and are still used to produce wholegrain flour today. They are often driven by windmills. They are also used for wet milling of corn.

Single-Disc Attrition Mill In this type of mill a grooved disc rotates in close proximity to a stationary disc with matching grooves. The feed is introduced through the centre of the stationary disc and makes its way outwards between the discs and is discharged from the mill via a screen (see Fig. 15.22a). Shear forces are mainly responsible for the breakdown of the material, but pressure may also play a part. The clearance between the two discs is adjustable.

Double-Disc Attrition Mill This attrition mill consists of two counter-rotating discs, with matching grooves, located close to each other in a casing. The feed is introduced from the top and passes between the discs before being discharged, usually through a screen at the bottom of the casing (see Fig. 15.22b). Shear forces, again, predominantly cause the breakdown of the material.

Both types of disc attrition mills are used for milling fibrous materials such as corn and rice.

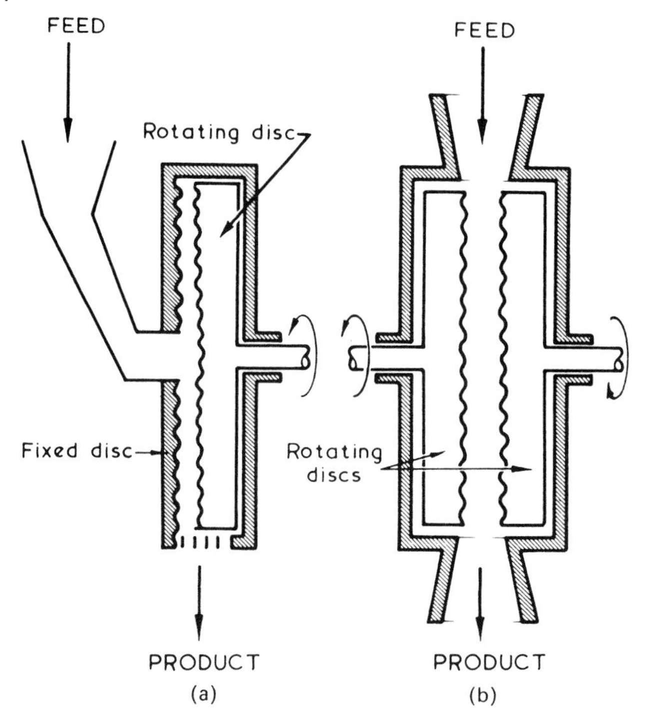

Fig. 15.22 Disc mills. (a) A single disc mill. (b) A double disc mill. From [6] with permission of the authors.

The Foos Mill This design of mill has studs fitted to the discs instead of grooves. It could be regarded also as a modified pin mill (see Section 15.3.2.3). It is used for similar applications as the other disc mills.

Colloid Mill This is another example of an attrition mill which is used for emulsification and in the preparation of pastes and purees (see Section 15.2.3.7) [6–8, 39–41].

15.3.2.5 **Tumbling Mills**

A typical tumbling mill consists of a cylindrical shell, sometimes with conical ends, which rotates slowly about a horizontal axis and is filled to about 50% of its volume with a solid grinding medium. As the shell rotates the loose units of the grinding medium are lifted up on the rising side of the shell to a certain height. They then cascade and cataract down the surface of the other units. The material being comminuted fills the void spaces between the units of the grinding medium. Size reduction takes place between these units in the jostling as they are lifted up and in the rolling action and impact as they fall down. The most commonly used grinding media are balls and rods. Contamination of the feed material due to wear of the grinding medium is a problem with tumbling mills and needs to be monitored.

Ball Mills In ball mills the grinding medium consists of spherical balls, 25–150 mm in diameter, and usually made of steel. Alternatively, flint pebbles or porcelain or zircon spheres may be used. Mills employing such grinding materials are known as *pebble mills.* At low rotation speeds the balls are not lifted very far up the wall of the cylindrical and tumble over each other as they roll down. In such circumstances, shear forces predominate. At higher speeds, the balls are lifted up higher and some fall back down generating impact forces which contribute to the size reduction of the feed material. Above a certain *critical speed,* the balls will be carried round against the cylinder wall under the influence of centrifugal force and comminution ceases. This critical speed can be calculated from the expression:

$$N = \frac{42.3}{(D)^{1/2}} \tag{15.11}$$

Where N is the critical speed (rpm) and D is the diameter of the cylinder (m). Ball mills are run at 65–80% of their critical speeds.

Ball mills may be batch or continuous. In the latter case, the feed material flows steadily through the revolving shell. It enters at one end through a hollow trunnion and leaves at the other end through the trunnion or through peripheral openings in the shell. Ball mills are used for fine grinding and may be operated under dry or wet conditions.

In the *vibration ball mill,* the shell containing the balls is made to vibrate by means of out of balance weights attached to each end of the shaft of a double-ended electric motor. In such mills, impact forces predominate and very fine grinding is attainable.

Another variation of the ball mill, known as an *attritor,* consists of a stationary cylinder filled with balls. A stirrer keeps the balls and feed material in slow motion generating shear and some impact forces. It is best suited to wet milling and may be operated batchwise or continuously. This type of mill is used in chocolate manufacture (see Section 15.3.3.2).

Rod Mills Grinding rods, usually made of high carbon steel, are used instead of balls in rod mills. They are 25–125 mm in diameter and may be circular, square or hexagonal in cross-section. They extend to almost the full length of the shell and occupy about 35% of the shell volume. In such mills, attrition forces predominate but impacts also play a part in size reduction. They are classed as intermediate grinders and are more useful than ball mills for milling sticky materials [6–8, 39–41].

15.3.3 Examples of Size Reduction of Solids in Food Processing

15.3.3.1 Cereals

Wheat The structure of the wheat grain is complex but, in the context of this section, it can be assumed to consist of three parts, i.e. the outer layer or bran, the white starchy endosperm and the embryo or germ. According to Meuser [41]: "The objectives of milling white flour are: 1. To separate the endosperm, which is required for the flour, from the bran and embryo, which are rejected, so that the flour shall be free from bran specks, and of good colour, and so that the palatability and digestibility of the product shall be improved and its storage life lengthened. 2. To reduce the maximum amount of endosperm to flour fineness, thereby obtaining the maximum extraction of white flour from the wheat." The number of parts of flour by weight produced per 100 parts of wheat is known as the *percentage extraction rate*. The wheat grain contains about 82% of endosperm, but it is not possible to separate all of it from the bran and embryo. In practice, the extraction rate is in the range 70–80%.

Prior to milling, the wheat grains are cleaned to remove metal fragments, stones, animal matter and unwanted vegetable matter and conditioned to the optimum moisture content for milling. A mill consists of two sections, a *break section* and a *reduction section*. The clean grain is fed to the break section, which usually consists of four or five pairs of fluted rolls. The rolls rotate towards each other at different speeds. As the grains pass through the first pair of break rolls they are split open. Some large fragments of endosperm (semolina) are released together with a small amount of small particles, less than 150 μm in size, which is collected as flour. The fragments of bran coming from these rolls will have endosperm attached to them. The fractions from the first break rolls are separated by sieving (see Fig. 15.23). The bran passes through a second set of fluted rolls where more semolina and flour is released. This is repeated two or three more times, until relatively clean bran is collected as a byproduct from the last set of sieves. The clearance between the pairs of break rolls and the depth of the fluting decrease in the direction of flow of the bran. The semolina, which contains the germ and some particles, goes to the reduction section. This section consists of up to 16 pairs of smooth surfaced rolls, rotating towards each other at different speeds. The speed differential is less than that between the break rolls. As the semolina passes through the first set of reduction rolls, some of the large fragments are broken down into flour; and the germ, which is relatively soft, is pressed into flakes. These fractions are separated by sieving. The germ is usually discharged from this set of sieves as another byproduct. The large particles of semolina pass through another set of reduction rolls where more flour is produced. This is repeated up to fourteen more times. In addition to some flour, another byproduct, consisting mainly of fine bran particles and known as wheat feed, is discharged from the last set of sieves. The clearance between the pairs of reduction rolls decreases in the direction of flow of the semolina. The greatest proportion of flour is collected from the early reduc-

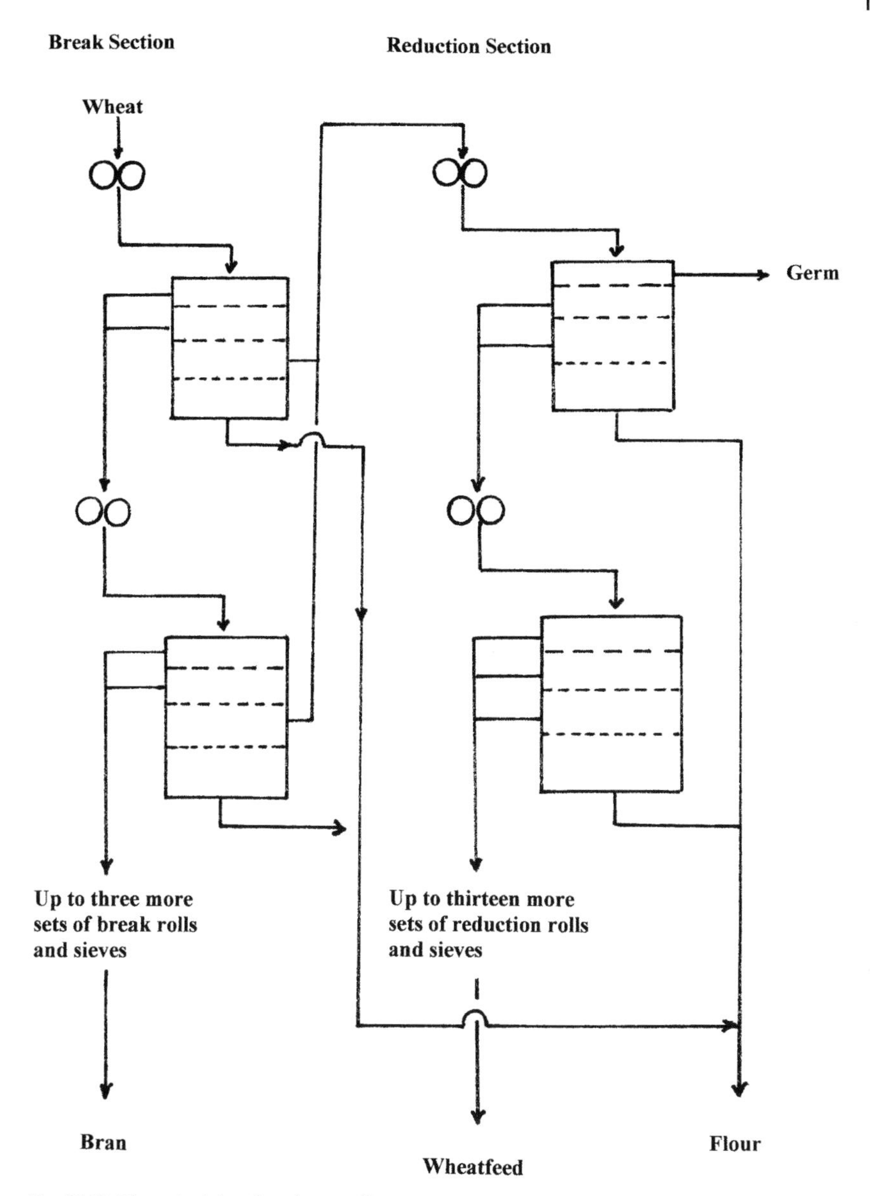

Fig. 15.23 The principle of a wheat milling process.

tion rolls. The flour coming from the later reduction rolls contains more fine bran particles and is darker in colour than that collected from the early rolls.

This is a rather simplified description of white flour milling. More detailed accounts are available in the literature [41, 42]. Recent developments in flour milling are discussed in [43]. Wholemeal flour contains all the products of milling cleaned wheat. It is produced using millstones (see Section 15.3.2.4), by roller mills using a shortened system or by attrition mills. Brown flours, which have extraction rates of 85–98%, are produced by a modified white flour milling system. The wheat may be conditioned to a lower moisture content than that used for white flour so that the bran breaks into smaller particles. Durum wheat is milled to produce coarsely ground endosperm, known as semolina, which is used in the manufacture of pasta and as couscous.

Sorghum, millets and maize and rye may be dry milled using modified flour milling systems.

Rice Unlike most milled cereal products, in the production of milled rice the endosperm is kept as intact as possible. Brown rice is dehulled using rubber covered rolls or by means of abrasive discs. The bran and embryo are then removed, a process known as whitening. This may be achieved by means of abrasive cones or rolls. A further step, known as polishing, removes the aleurone layer, which is rich in oil. This extends the storage life of the rice by reducing the tendency for oxidative rancidity to occur. Polishing is achieved by means of a rotating cone covered with leather strips.

Maize This cereal is wet milled to produce a range of products including starch, oil and various types of cattle feed. Cleaned maize is steeped in water, containing 0.1–0.2% sulphur dioxide, at 50 °C for 28–48 h. The steeping softens the kernel and facilitates separation of the hull, germ and fibre from each other. The SO_2 may disrupt the -SS- bonds in the protein, enabling starch/protein separation. After steeping, the steep water is concentrated by vacuum evaporation and the protein it contains is recovered by settling. The steeped maize is coarsely milled in a Foos Mill (see Section 15.3.2.4). The grain is cracked open and the germ released. The germ is recovered by settling or by means of hydrocyclones. Oil may be extracted from the germ by pressing. The degermed material is strained off from the liquid and milled in an impact or attrition mill. The hulls and fibre are separated from the protein and starch by screening. The suspension of starch and protein coming from the screen is fed to high-speed centrifuges where they are separated from each other. The starch is purified in hydrocyclones, filtered and dried. The protein is also filtered and dried. The products from the wet milling of maize include about 66% starch, 4% oil and 30% animal feed. Most of the starch is converted into modified starches, sweeteners, alcohol and other useful products [42–46].

15.3.3.2 Chocolate

Size reduction equipment is employed at several stages in the manufacture of chocolate. The first stage is breaking the shell of the cocoa bean to release the nib. The beans are predried in heated air, or steam puffed, or exposed to infrared radiation to weaken the shell and loosen the nib. The bean shells are generally broken by a type of impact mill. The beans are fed onto plate or disc, rotating at high speed. They are flung against breaker plates by centrifugal force. Sometimes they fall onto a second rotating disc, which flings them back against the plate again. The size distribution of the product from the mill should be as narrow as possible, ideally with the size range 7.0–1.5 mm. The nibs are separated from the fragments of shell by a combination of sieving and air classification (winnowing). It is important that this separation is as complete as possible, as the presence of shell can cause excessive wear on the equipment used subsequently to grind the nibs and refine the chocolate. The next application for size reduction equipment is the grinding of cocoa nibs. These are ground to cocoa liquor for the removal of cocoa butter from within the cellular structure. Cocoa nib contains about 55% butter, contained within cells which are 20–30 μm in size. These cells must be broken to release the butter. The cell wall material and any shell and germ present are fibrous and tough. The particle size range after grinding should be in the range 15–50 μm. The grinding of the nibs is usually carried out in two steps. They are first preground in a hammer, disc or pin mill. The second step is carried out in an agitated ball mill or attrition mill. The first of these mills consists of a cylinder filled to as much as 90% of its volume with balls made of steel or ceramic material. For large throughputs, a series of ball mills may be used, with the size of the balls decreasing in the direction of flow of the liquor, from 15 mm down to 2 mm. Some contamination of the liquor is to be expected due to wear of the equipment. Three single-disc attrition mills, working in series, may replace the ball mill. The three pairs of discs may be housed in the one machine. These can be fed directly with dry nibs, but pregrinding is more usual.

A further stage in chocolate manufacture is refining. The purpose of refining is to ensure that the particle size of the dispersed phase is sufficiently small so as not to impart a gritty texture to the chocolate when it is eaten. Usually this means that the largest particles should be in the range 15–30 μm. If the cocoa liquor has been properly ground, the purpose of refining is to reduce the size of the added sugar crystals and, in the case of milk chocolate, the particles of milk powder. Refining is usually carried out using a five-roll machine. The gap between the rolls decreases from bottom to top while the speed of rotation increases. The cocoa liquor is introduced between the bottom two rolls. The film of chocolate leaving each gap is transferred to the faster roll and so moves upwards. The product is scraped off the top roll.

Uncompressed rolls are barrel-shaped. Hydraulic pressure is applied to the roll stack to compress the camber on the rolls and obtain an even coating across the length of each roll. The reduction ratio attainable is 5–10, which should result in the particles in the product being in the correct size range. The temperature of the rolls is controlled by circulation of water through them. It is important that the feed material to the rolls is well mixed. In some systems, a two-roll refiner known as a prerefiner is employed to prepare the feed for the main roll stack.

Cocoa powder is produced by partially defatting ground cocoa nibs in a press. The broken cake from the press is blended prior to milling to standardise the colour. The blend is fed to a pin mill were it is broken down to a powder. The powder is cooled to about 18 °C by being conveyed through jacketed tubes by compressed air. At the end of the cooling line, the powder is recovered from the air by means of a cyclone. It is transported to a stabiliser, where the latent heat is removed to solidify the remaining fat [47–49].

15.3.3.3 **Coffee Beans**

After roasting, coffee beans are cooled and then ground. The extent of the grinding depends on the how the coffee is to be brewed. Generally, coffee which is to be percolated will have a larger particles size than filter coffee. The actual particle sizes vary from country to country. In Europe, coarse ground coffee for percolators has an average particle size of about 850 μm, while that of a medium grind, intended to be filtered, is about 450 μm. Finely ground coffee, about 50 μm, is required for some types of Turkish coffee and for expresso machines. In the latter case, the beans are ground in small amounts just prior to use. Ground beans for large-scale percolation (extraction) for the production of instant coffee may be somewhat coarser than the normal coarse ground product.

The original method of grinding coffee beans was by means of a buhrstone-type mill (see Section 15.3.2.4). These are still in use in some developing countries, particularly in markets where the customer grinds his roast beans at the time of purchase. The stones have been replaced by serrated, cast iron discs. For large-scale grinding of coffee beans, multiroll mills are mostly used. Three or four pairs of rolls are located one above the other. In each pair, the rolls rotate at different speeds. The fast roll has U-shaped corrugations running lengthwise while the other has peripheral corrugations. As the beans pass down between successive pairs of rolls, the clearance between them decreases and the roll speed increases. A three-stand mill is used to produce coarse or medium ground coffee. A fourth pair of rolls is used for fine grinding. Chaff remaining after roasting is released during grinding. This may be removed or incorporated with the ground coffee in a type of ribbon mixer. The latter procedure is known as *normalising*. Various types of attrition and hammer mills are used to grind roasted beans on a smaller scale.

A considerable amount of carbon dioxide is release during grinding of roasted beans. Quite high concentrations of the gas can develop in the vicinity of the milling equipment. If ventilation is poor, operatives may be physiologically affected. The main effect is that the person becomes silly and acts oddly, without realising why. Water vapour and volatile flavour compounds are also released during grinding. Cryogenic grinding, using solid carbon dioxide or liquid nitrogen, may be employed to maximise the retention of the flavour volatiles [50, 51].

15.3.3.4 **Oil Seeds and Nuts**

It is common practice to flake or grind oilseeds prior to extraction, either by pressing or solvent extraction, in order to rupture the oil cells. Hammer mills and attrition mills are sometimes used for the preliminary breakdown of large oil seeds such as copra or palm. However, flaking rolls are usually used to prepare seeds for extraction. A common arrangement used for cottonseed, flaxseed and peanuts consists of five rolls located one above the other. The seed is fed in between the top two rolls and then passes back and forth between the rolls down to the bottom of the stack. Each roll supports the weight of the rolls above it, so the seeds are subjected to increasing pressure as they move down the stack. The top roll is grooved, but the others have smooth surfaces [52].

15.3.3.5 **Sugar Cane**

Large, heavy duty, two- or three-roll crushers are used to break and tear up the cane to prepare it for subsequent milling for the extraction of the sugar. The surfaces of the rolls are grooved to grip the cane and very high pressures are applied to the cane, 8–12 t dm^{-2}. Following this initial crushing, the cane may be further broken down in shredding devices, which tear open the cane cells and release some of the juice. These devices are essentially large-capacity hammer mills, with rotors turning at 1000–1200 rpm, carrying hammers pivoted to discs or plates. The hammers pass very close to an anvil plate made up of rectangular bars. The material leaving the shredders is made up of cell material and long thread-like fibres which hold it together when it is subjected to pressure during extraction of the juice by milling.

To extract the juice, the shredded cane is passed through a series, usually five, of triple-roll mills. These are heavy-duty rolls, with grooved surfaces which exert high pressure, 15–30 t dm^{-2} on the cane, expressing the juice. This is not a size reduction operation as such, but is an example of another use of roller mills. Water is added to the material after each set of rolls. Alternatively, the dilute juice from the last mill is sent back before the preceding mill. This procedure, known as *imbibition*, increases the amount of sugar extracted, compared with dry crushing [53].

Numerous other food materials are size-reduced in mills. Mustard seeds are milled in a similar fashion to wheat grains, passing through grooved break rolls and then through smooth-surfaced reduction rolls. Spices are milled using a variety of impact and attrition mills. Considerable heat may be generated and result in a loss of volatile oils and development of undesirable aromas and flavours. Mills may be jacketed and cooled. Cryogenic milling, using liquid nitrogen, can result in higher quality products. Sugar crystals may be ground in impact mills to produce icing sugar. Particles of milk powder, lactose and dry whey may be reduced in size in impact mills.

References

1 Fellows, F.P. **2000**, *Food Processing Technology*, 2nd edn, Woodhead Publishing, Cambridge.

2 Nienow, A.W., Edwards, M.F., Harnby, N. **1992**, Introduction to mixing problems, in *Mixing in the Process Industries*, 2nd edn, ed. N. Harnby, M.F. Edwards, A.W. Nienow, Butterworth-Heinemann, Oxford, pp 1–24.

3 Edwards, M.F., Baker, M.R. **1992**, A Review of Liquid Mixing Equipment, in *Mixing in the Process Industries*, 2nd edn, ed. N. Harnby, M.F. Edwards, A.W. Nienow, Butterworth- Heinemann, Oxford, pp 118–136.

4 Edwards, M.F., Baker, M.R. **1992**, Mixing of Liquids in Stirred Tanks, in *Mixing in the Process Industries*, 2nd edn, ed. N. Harnby, M.F. Edwards, A.W. Nienow, Butterworth- Heinemann, Oxford, pp 137–158.

5 Lindley, J.A. **1991 a**, Mixing Processes for Agricultural and Food Materials 1: Fundamentals of Mixing, *J. Agric. Eng. Res.* 48, 153–170.

6 Brennan, J.G., Butters, J.R., Cowell, N.D., Lilly, A.E.V. **1990**, *Food Engineering Operations*, 3rd edn, Elsevier Applied Science, London.

7 McCabe, W.L., Smith, J.C., Harriott, P. **1985**, *Unit Operations of Chemical Engineering*, 4th edn, McGraw-Hill, New York.

8 Leniger, H.A., Beverloo, W.A. **1975**, *Food Process Engineering*, Reidel Publishing, Dordrecht.

9 Edwards, M.F. **1992**, Laminar Flow and Distributive Mixing, in *Mixing in the Process Industries*, 2nd edn, ed. N. Harnby, M.F. Edwards, A.W. Nienow, Butterworth-Heinemann, Oxford, pp 200–224.

10 Lindley, J.A. **1991 b**, Mixing Processes for Agricultural and Food Materials: Part 2, Highly Viscous Liquids and Cohesive Materials, *J. Agric. Eng. Res.* 48, 229–247.

11 Richards, G. **1997**, Motionless Mixing-Efficiency with Economy, *Food Process.* 1997, 29–30.

12 Godfrey, J.C. **1992**, Static Mixers, in *Mixing in the Process Industries*, 2nd edn, ed. N. Harnby, M.F. Edwards, A.W. Nienow, Butterworth-Heinemann, Oxford, pp 225–249.

13 Harnby, N. **2003**, Mixing of Powders, in *Encyclopedia of Food Science and Nutrition*, 2nd edn, ed. B. Caballero, L.C. Trugo, P.M. Finglas, Academic Press, London, pp 4028–4033.

14 Harnby, N. **1992**, Characterization of Powder Mixtures, in *Mixing in the Process Industries*, 2nd edn, ed. N. Harnby, M.F. Edwards, A.W. Nienow, Butterworth-Heinemann, Oxford, pp 25–41.

15 Harnby, N. **1992**, The Selection of Powder Mixers, in *Mixing in the Process Industries*, 2nd edn, ed. N. Harnby, M.F. Edwards, A.W. Nienow, Butterworth-Heinemann, Oxford, pp 42–61.

16 Lindley, J.A. **1991c**, Mixing Processes for Agricultural and Food Materials: 3 Powders and Particulates, *J. Agric. Eng. Res.* 49, 1–19.

17 Geldart, D. **1992**, Mixing in Fluidized Beds, in *Mixing in the Process Industries*, 2nd edn, ed. N. Harnby, M.F. Edwards, A. W. Nienow, Butterworth-Heinemann, Oxford, pp 62–78.

18 Harnby, N. **1992**, The Mixing of Cohesive Powders, in *Mixing in the Process Industries*, 2nd edn, ed. N. Harnby, M.F. Edwards, A.W. Nienow, Butterworth-Heinemann, Oxford, pp 79–98.

19 Middleton, J.C. **1992**, Gas-Liquid Dispersion and Mixing, in *Mixing in the Process Industries*, 2nd edn, ed. N. Harnby, M.F. Edwards, A. W. Nienow, Butterworth-Heinemann, Oxford, pp 322–363

20 Matsumo, S. **1986**, Review Paper – W/O/W Type Emulsions with a View to Possible Food Applications, *J. Texture Stud.* 17, 141–159.

21 Friberg, S.E., Kayali, I. **1991**, Surfactant Association Structures, Microemulsions, and Emulsions in Foods: an Overview, in *Microemulsions and Emulsions in Foods*, ed. M. El-Nokaly, D. Cornell, American Chemical Society, Washington, D.C., pp 7–24.

22 Robins, M.M., Wilde, P. J. **2003**, Colloids and Emulsions, in *Encyclopedia of Food Sciences and Nutrition*, 2nd edn, B.

Caballero, L.C. Trugo, P.M. Finglas, Academic Press, London, pp 1517–1524.
23 McClements, D.J. **1999**, *Food Emulsions*, CRC Press, New York.
24 Kinyanjul, T., Artz, W.E., Mahungu, S. **2003**, Emulsifiers, in *Encyclopedia of Food Sciences and Nutrition*, 2nd edn, ed. B. Caballero, L.C. Trugo, P.M. Finglas, Academic Press, London, pp 2070–2077.
25 Grindsted Products **1988**, *Emulsifiers and Stabilisers for the Food Industry*, Grindsted Products, Braband.
26 Lewis, M.J. **1987**, *Physical Properties of Foods and Food Processing Systems*, Ellis Horwood, Chichester.
27 Wilbey, R.A. **2003**, Homogenisation, in *Encyclopedia of Food Sciences and Nutrition*, 2nd edn, ed. B. Caballero, L.C. Trugo, P.M. Finglas, Academic Press, London, pp 3119–3125.
28 Walstra, P. **1983**, Formation of emulsions, in *Encyclopedia of Emulsion Technology*, vol. 1, ed. P. Becher, Marcel Dekker, New York, pp 57–127.
29 Yang, S.C., Lai, L.S. **2003**, Dressing and Mayonnaise, in *Encyclopedia of Food Sciences and Nutrition*, 2nd edn, ed. B. Cabellero, L.C. Trugo, P.M. Finglas, Academic Press, London, pp 1892–1896.
30 Walstra, P., Geurts, T.J., Noomen, A., Jellema, A., van Boekel, M.A.J.S. **1999**, *Dairy Technology*, Marcel Dekker, New York.
31 Lane, R. **1998**, Butter and Mixed Fat Spreads, in *The Technology of Dairy Products*, 2nd edn, ed. R. Early, Blackie Academic and Professional, London, pp. 158–107.
32 Bucheim, W., Dejmek, P. **1997**, Milk and Dairy-Type Emulsions, in *Food Emulsions*, 3rd edn, ed. S. E. Friberg, K. Larsson, Marcel Dekker, New York, pp 235–278.
33 Ford, L.D., Borwanker, R., Martin, R.W. Jr, Holcomb, D.N. **1997**, Dressings and Sauces, in *Food Emulsions*, 3rd edn, ed. S.E. Friberg, K. Larsson, Marcel Dekker, New York, pp 361–412.
34 Berger, K.G. **1997**, Ice Cream, in *Food Emulsions*, 3rd edn, ed. S.E. Friberg, K. Larsson, Marcel Dekker, New York, p. 490
35 Eliasson, A-C., Silvero, J. **1997**, Fat in baking, in *Food Emulsions*, 3rd edn, ed. S.E. Friberg, K. Larsson, Marcel Dekker, New York, pp 525–548.
36 Mizukoshi, M. **1997**, Baking Mechanism in Cake Production, in *Food Emulsions*, 3rd edn, ed. S.E. Friberg, K. Larsson, Marcel Dekker, New York, pp 549–575.
37 Krog, N.J., Riison, T.H., Larsson, K. **1985**, Applications in the Food Industry: I, in *Encyclopedia of Emulsion Technology*, vol. 2, ed. E. Becher, Marcel Dekker, New York, pp 321–365.
38 Jaynes, E. **1985**, Applications in the Food Industry: II, in *Encyclopedia of Emulsion Technology*, vol. 2, ed. E. Becher, Marcel Dekker, New York, pp 367–384.
39 Loncin, M., Merson, R.L. **1979**, *Food Engineering – Principles and Selected Applications*, Academic Press, London.
40 Lowrison, G.C. **1974**, *Crushing and Grinding – the Size Reduction of Solid Materials*, Butterworths, London.
41 Meuser, F. **2003**, Types of Mill and Their Uses, in *Encyclopedia of Food Science and Nutrition*, 2nd edn, ed. B. Caballero, L.C. Trugo, P.M. Finglas, Academic Press, London, pp 3987–3997.
42 Kent, N.L., Evers, A.D. **1993**, *Technology of Cereals*, 4th edn, Elsevier Science, Oxford.
43 Barnes, P.J. **1989**, Wheat Milling and Baking, in *Cereal Science and Technology*, ed. G.H. Palmer, Aberdeen University Press, Aberdeen, pp 367–412.
44 Owens, W. G. **2001**, Wheat, Corn and Coarse Grains Milling, in *Cereals Processing Technology*, ed. G. Owens, Woodhead Publishing, Cambridge, pp 27–52.
45 Posner, E.S. **2003**, Principles of Milling, in *Encyclopedia of Food Science and Nutrition*, 2nd edn, ed. B. Caballero, L.C. Trugo, P.M. Finglas, Academic Press, London, pp 3980–3986.
46 Morrison, W.R., Barnes, P.J. **1983**, Distribution of Wheat Acyl Lipids and Tocols into Flour Millstreams, in *Lipids in Cereal Technology*, ed. P.J. Barnes, Academic Press, London, pp 149–163.
47 Heemskerk, R.F.M. **1999**, Cleaning, Roasting and Winnowing, in *Industrial Chocolate Manufacture and Use*, 3rd edn,

ed. S.T. Beckett, Blackwell Science Ltd., Oxford, pp 79–100.

48 Meursing, E.H., Zijderveld, J.A. **1999**, Cocoa Mass, Cocoa Butter and Cocoa Powder, in *Industrial Chocolate Manufacture and Use*, 3rd edn, ed. S.T. Beckett, Blackwell Science, Oxford, pp 101–114.

49 Ziegler, G., Hogg, R. **1999**, Particle Size Reduction, in *Industrial Chocolate Manufacture and Use*, 3rd edn, ed. S.T. Beckett, Blackwell Science, Oxford, pp 115–136.

50 Sivetz, M., Desrosier, N.W. **1979**, *Coffee Technology*, AVI Publishing, Westport.

51 Clarke, R.J. **1987**, Roasting and Grinding, in *Coffee Technology*, vol. 2, ed. R.J. Clarke, R. Macrae, Elsevier Applied Science, London, pp 73–107.

52 Williams, M.A., Hron, R.J. Sr. **1996**, Obtaining Oils and Fats from Source Materials, in *Bailey's Industrial Oil and Fat Products*, 5th edn, vol. 4, ed. Y.H. Hui, John Wiley & Sons, New York, pp 61–155.

53 Hugot, E. **1986**, *Handbook of Cane Sugar Engineering*, 3rd edn, Elsevier Science, Amsterdam.

Subject Index

Food Processing Handbook. Edited by James G. Brennan

ISBN: 3-527-30719-2

f

i

r